Non-Traditional and Advanced Machining Technologies

Non-Traditional and Advanced Machining Technologies

Second Edition

Helmi Youssef and Hassan El-Hofy

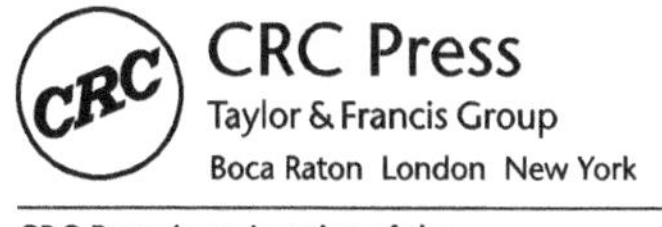

CRC Press is an imprint of the
Taylor & Francis Group, an **informa** business

by CRC Press
6000 Broken Sound Parkway NW, Suite 300, Boca Raton, FL 33487-2742

and by CRC Press
2 Park Square, Milton Park, Abingdon, Oxon, OX14 4RN

CRC Press is an imprint of Taylor & Francis Group, LLC

Library of Congress Cataloging-in-Publication Data

Names: Youseff, Helmi, author. | El-Hofy, Hassan, author.
Title: Non-traditional and advanced machining technologies : machine tools and operations / Helmi Youseff and Hassan El-Hofy.
Description: Second edition. | Boca Raton, FL : CRC Press, [2020] | Includes bibliographical references and index.
Identifiers: LCCN 2020011453 (print) | LCCN 2020011454 (ebook) | ISBN 9780367431341 (hardback) | ISBN 9781003055310 (ebook)
Subjects: LCSH: Machining. | Machining--Equipment and supplies.
Classification: LCC TJ1185 .Y683 2020 (print) | LCC TJ1185 (ebook) | DDC 671.3/5--dc23
LC record available at https://lccn.loc.gov/2020011453
LC ebook record available at https://lccn.loc.gov/2020011454

ISBN: 978-0-367-43134-1 (hbk)
ISBN: 978-0-367-51812-7 (pbk)
ISBN: 978-1-003-05531-0 (ebk)

Typeset in Times
by Deanta Global Publishing Services, Chennai, India

Dedication

To our grandsons and granddaughters,

Helmi Youssef: Youssef, Nour, Anourine, Fayrouz, and Yousra

Hassan El-Hofy: Omar, Zainah, Youssef,
Hassan, Hana, Ali, and Hala

Contents

UNIT II Advanced Machining Technology

Preface

Non-Traditional and Advanced Machining Technologies consists of 11 chapters. Every chapter has been updated emphasizing new information on the relevant topics. Today's interests such as machining of DTC materials, assisted machining technologies, and hybrid processes are also featured in brand-new chapters. Accordingly, this book provides a comprehensive description of non-traditional and advanced machining technologies, from the basic to the most advanced, in today's industrial applications. It is a fundamental textbook for undergraduate students enrolled in production, materials and manufacturing, industrial, and mechanical engineering programs. Students from other disciplines can also use this book while taking courses in the area of manufacturing and materials engineering. It should also be useful to graduates enrolled in high-level machining technology courses and professional engineers working in the field of the manufacturing industry. The book covers the technologies, machine tools, and operations of several non-traditional machining processes. The treatment of the different subjects has been developed from the basic principles of machining processes, machine tool elements, and control systems, and extends to ecological machining and the most recent machining technologies, including non-traditional methods, and machining DTC materials. The book presents environment-friendly machine tools and operations; as well as design for machining, accuracy, and surface integrity realized by both traditional and non-traditional machining operations. Solved examples, problems, and review questions are provided.

Design for accurate and economic machining, ecological machining, levels of accuracy, and surface finish attained by machining methods are also presented. The topics covered throughout the book chapters reflect the rapid and significant advances that have occurred in various areas in machining technologies, and they are organized and described in such a manner as to draw the interest of students. The chapters of the book are aimed at motivating and challenging students to explore technically and economically viable solutions to a variety of important questions regarding product design and optimum selection of machining operation for a given task.

Outline of *Non-Traditional and Advanced Machining Technologies*

Unit I: Non-Traditional Machining Operations and Non-Traditional Machine Tools

Chapter 1 presents an introduction and classification of the non-traditional machine tools.

Chapter 2 presents the mechanical machining processes such as ultrasonic, jet machining, and abrasive flow machining.

Chemical milling, electrochemical machining and electrochemical grinding machine tools are described in Chapter 3.

Machine tools for thermal processes such as electric discharge, laser beam, electron beam, and plasma arc machining are presented in Chapter 4. Machine tools, basic elements, accessories, operations, removal rate, accuracy, and surface integrity are covered for each case.

Unit II: Advanced Machining Technology

Chapter 5 presents machining of difficult-to-cut materials such as stainless steels and super alloys, while in Chapter 6 the machinability of composites and ceramics, are treated by traditional and non-traditional means.

Chapter 7 covers assisted machining technologies. These technologies include US, thermal, EC, and magnetic effects to enhance the machinability of DTC materials.

An introduction to design recommendations for economic machining and sources of dimensional variations by traditional and non-traditional processes is covered in Chapter 8.

Dimensional accuracy and surface integrity by traditional and non-traditional machining processes are dealt with in Chapter 9. Sources of surface alterations, their effects on the functional properties of machined parts, and recommendations for minimizing surface effects are also given.

The environment-friendly machine tools and operations are described in Chapter 10 which tend to detect the source of hazards and minimize their effect on the operator, machine tool, and environment.

Hexapod mechanisms, design features, constructional elements, characteristics, control, and their applications in traditional and non-traditional machining, manufacturing, and robotics are dealt with in Chapter 11.

Acknowledgements

Many individuals have contributed to the development of the second edition of this book. It is a pleasure to express our deep gratitude to Professor Dr. Ing. A. Visser, Bremen University, Germany, for supplying valuable materials during the preparation of this new edition. We would like to appreciate the efforts of Dr. Khaled Youssef for his continual assistance in tackling software problems during the preparation of the manuscript. Special thanks are offered to Saied Teileb of Lord Alexandria Razor Company for his fine Auto-CAD drawings.

Heartfelt thanks are due to our families for their great patience, support, encouragement, enthusiasm, and interest during the preparation of the manuscript. We extend our heartfelt gratitude to the editorial and production staff at Taylor & Francis Group for their efforts to ensure that this book is as accurate and well-designed as possible.

We very much appreciate the permissions from all publishers to reproduce many illustrations from a number of authors as well as the courtesy of many industrial companies that provided photographs and drawings of their products to be included in this new edition of the book. Their generous cooperation is a mark of sincere interest in enhancing the level of engineering education. The credits for all this great help are given in the captions under the corresponding illustrations, photographs, and tables. It is with great pleasure that we, therefore, acknowledge the help of the following organizations:

Alexandria Engineering Journal, Alexandria, Egypt.
American Society of Mechanical Engineers.
ASM International, Materials Park, OH.
Cassell and Co. Ltd., London, UK.
Chapman and Hall, London, UK.
CIRP, Paris, France.
Dar Al-Maaref Publishing Co., Alexandria, Egypt.
El-Fath Press, Alexandria, Egypt.
Elsevier Ltd, Oxford, UK.
Hodder and Stoughton Educational, London, UK.
Industrial Press Inc., New York, NY.
Industrie-Anzeiger, Aachen, Germany.
John Wiley & Sons, Inc., New York, NY.
Khanna Publisher, Delhi, India.
Leuze Verlag, Bad Saulgau, Germany.
Machinability Data Center, Cincinnati, OH.
Marcel Dekker Inc., New York, NY.
McGraw Hill Co., New York, NY.
Mir Publishers, Moscow, Russia.
Oxford University Press, UK.

Pearson Education Inc., NJ.
Peter Peregrines Ltd., Stevenage, UK.
Prentice Hall Publishing Co., New York, NY.
SME, Dearborn, MI.
Springer Verlag, Berlin, Germany.
Tata McGraw Hill Co., New Delhi, India.
TH-Aachen, Germany.
TH-Braunschweig, Germany.
VDI Verlag, Düsseldorf, Germany.
VEB-Verlag Technik, Berlin, Germany.
Alfred Herbert Ltd., Coventry, UK.
All Metals & Forge Group, NJ.
American Gear Manufacturing Association (AGMA), VA.
British Stainless Steel Association, Sheffield, UK.
Carpenter Technology Corporation, PA.
Charmilles Technologies, Geneva, Switzerland.
Cincinnati Machines, OH.
DeVlieg Machine Co., MI.
Encyclopedia Britannica Inc., UK.
Falcon Metals Group, Waldwick, NJ.
Geodetic Inc., Melbourne, FL.
Hardinge Incorporation, Berwyn, PA.
Heald Machine Company, Worcester, MA.
Heinemann Machine Tool Works-Schwarzwald, Germany.
Herbert Machine Tools Ltd., UK.
High Performance Alloys Inc., USA.
Hoffman Co., Carlisle, PA.
Hottinger-Baldwin Meβ-technik, Darmstadt, Germany.
Index-Werke AG, Esslingen/Neckar, Germany.
Indian Institute of Technology, Kanpur, India.
Ingersoll Waldrich Siegen Werkzeugmachinen GmbH, Germany.
Kennametal Incorporation, Pittsburgh, PA.
Kistler Instrumente AG, Switzerland.
Krupp, Widia, GmbH, Essen, Germany.
Lehfeld Works, Heppenheim, Germany.
Liebherr Verzahntechnik, Kempten, Germany.
MAZAK Corporation Florence, KY.
MG Industries/Steigerwald, Berlin, Germany.
Mitsubishi EDSCAN, Toyohashi, Japan.
Nassovia-Krupp, Werkzeugmaschinenfabrik, Langen/Frankfurt, Germany.
ONSRUD tool company, USA.
PittlerMachinenfabrik AG, Langen/Frankfurt, Germany.
Sandvik Coromant, Sweden.
Seco Tools, UAE.
Standard Tool Co., Athol, MA.
Thermal-Dynamic Corp., Chesterfield, MO.
VEB-Drehmaschinenwerk/Leipzig, Germany.
WMW Machinery Co, New York, NY.

Author Biographies

Helmi Youssef, born in August, 1938 in Alexandria, Egypt, acquired his BSc degree with honors in Production Engineering from Alexandria University in 1960. He then consolidated his scientific experience in the Carolo-Welhelmina, TH Braunschweig, in Germany during the period 1961–1967. In June 1964, he acquired his Dipl.-Ing. degree, and in December 1967, he completed his Dr.-Ing. degree in the domain of Nontraditional Machining. In 1968, he returned to Alexandria University, Production Engineering Department, as an assistant professor. In 1973, he was promoted to associate, and in 1978, to full professor. In the period 1995–1998, Professor Youssef was the chairman of the Production Engineering Department, Alexandria University. Since 1989, he has been a member of the scientific committee for the promotion of professors in Egyptian universities.

Based on several research and educational laboratories, which he had built, Professor Youssef founded his own scientific school in both Traditional and Nontraditional Machining Technologies. In the early 1970s, he established the first Nontraditional Machining Technologies research laboratory in Alexandria University, and maybe in the whole region. Since that time, he has carried out intensive research in his fields of specialization and has supervised many PhD and MSc theses.

Between 1975 and 1998, Professor Youssef was a visiting professor in Arabic universities, such as El-Fateh University in Tripoli, the Technical University in Baghdad, King Saud University in Riyadh, and Beirut Arab University in Beirut. Beside his teaching activities in these universities, he established laboratories and supervised many MSc theses. Moreover, he was a visiting professor in different academic institutions in Egypt and abroad. In 1982, he was a visiting professor in the University of Rostock, Germany, and during the years 1997–1998, he was a visiting professor in the University of Bremen, Germany.

Professor Youssef has organized and participated in many international conferences. He has published many scientific papers in specialized journals. He has authored many books in his fields of specialization, two of which are single authored. The first is in Arabic, titled *Nontraditional Machining Processes, Theory and Practice*, published in 2005, and the other is titled *Machining of Stainless Steels and Superalloys, Traditional and Nontraditional Techniques*, published by Wiley in 2016. Another two coauthored books were published by CRC in 2008 and 2011, respectively. The first is on *Machining Technology*, while the second deals with *Manufacturing Technology.*

Currently, Professor Youssef is an emeritus professor in the Production Engineering Department, Alexandria University. His work currently involves developing courses and conducting research in the areas of metal cutting and nontraditional machining.

Hassan El-Hofy received his BSc in Production Engineering from Alexandria University, Egypt in 1976 and his MSc in 1979. Following his MSc, he worked as an assistant lecturer in the same department. In October 1980, he left for Aberdeen University in Scotland, UK and began his PhD work with Professor J. McGeough in electrochemical discharge machining. He won the Overseas Research Student award during the course of his doctoral degree, which he duly completed in 1985. He then returned to Alexandria University and resumed his work as an assistant professor. In 1990, he was promoted to the rank of associate professor. He was on sabbatical as a visiting professor at Al-Fateh University in Tripoli between 1989 and 1994.

In July 1994, he returned to Alexandria University and was promoted to the rank of full professor in November 1997. From September 2000, he worked as a professor for Qatar University. He chaired the accreditation committee for the mechanical engineering program toward ABET Substantial Equivalency Recognition, which was granted to the College of Engineering programs, Qatar University in 2005. He received the Qatar University Award and a certificate of appreciation for his role in that event.

Professor El-Hofy's first book, entitled *Advanced Machining Processes: Nontraditional and Hybrid Processes*, was published by McGraw Hill Co. in 2005. The third edition of his second book, entitled *Fundamentals of Machining Processes—Conventional and Nonconventional Processes*, was published in November 2018 by Taylor & Francis Group, CRC Press. He also coauthored the book entitled *Machining Technology—Machine Tools and Operations*, which was published by Taylor & Francis Group, CRC Press in 2008. In 2011, he released his fourth book, entitled *Manufacturing Technology—Materials, Processes, and Equipment*, which again was published by Taylor & Francis Group, CRC Press. Professor El-Hofy has published over 80 scientific and technical papers and has supervised many graduate students in the area of advanced machining. He serves as a consulting editor to many international journals and is a regular participant in many international conferences.

Between August 2007, and August 2010, he became the chairman of the Department of Production Engineering, Alexandria University. In October 2011, he was nominated as the vice dean for Education and Student's affairs, College of Engineering, Alexandria University. Between December 2012 and February 2018, he was the dean of the Innovative Design Engineering School at the Egypt-Japan University of Science and Technology in Alexandria, Egypt. He worked as acting Vice President of Research from December 2014 to April 2017 at the Egypt-Japan University of Science and Technology. Currently, he is the Professor of Machining Technology at the Department of Industrial and Manufacturing Engineering at Egypt-Japan University of Science and Technology.

List of Symbols

Symbol	Definition	Unit
$A(x)$	Area of acoustic horn at position x	mm^2
A_0	Area of acoustic horn at position 0	mm^2
A_t	Horizontal vibration amplitude	μm
B_t	Vertical vibration amplitude	μm
c	Acoustic speed in horn material	m/s
C	Capacitance	μF
C'	Modified acoustic speed in horn material	km/s
$c1$	Specific heat of workpiece material	N m/kg °C
C_d	Coefficient of thermal diffusivity	m^2/s
c_i	Constraints	
D	Drilled hole diameter	mm
D_{max}	Maximum delamination diameter	mm
$D(x)$	Diameter of acoustic horn at position x	mm
d_c	Fixation hole diameter of horn	mm
d_f	Electron beam focusing diameter	mm
D_0	Diameter of acoustic horn at position 0	mm
E	Young's modulus	MPa
E_d	Energy of individual discharge	J
F	Force	N
F_a	Two-dimensional delamination factor	%
F_d	Delamination factor	%
f, f_r	Frequency	s^{-1}
F_x	Radial magnetic force	N
F_y	Tangential magnetic force	N
H	Magnetic field strength	Tesla
h	Ascent factor of exponential horn	m/s
h_g	Frontal gap thickness in EDM	mm
i_b	Electron beam current	A
i_c	Charging current	A
i_d	Discharging current	A
I_p	Premagnetizing current	A
kr	Coefficient of magneto-mechanical coupling	
k_t	Thermal conductivity	N/s °C
l	Length	mm
M	Mobility	–
m	Mass	kg
n_e	Number of elements in the hexapod system	
P	Laser power	W
R	Resistance	Ω
R_0	Initial level position	mm
R_a	Average surface roughness	μm
R_m	Magnification factor	μm

R_t, R_{max}	Peak-to-valley surface roughness	μm
t	Workpiece thickness in laser cutting	mm
T	Depth of cut (time)	mm (s)
$t(x)$	Thickness function	m
T_1	Input torque	N mm
t_1	Plate thickness	mm
t_c	Charging time	μs
t_d	Discharging time	μs
T_e	Chemical etch depth	mm
t_i	Pulse duration	°C
v	Laser cutting rate	m/s
V	Volume of conglomerate	mm^3
V_A	Anodic dissolution rate	mm/min
V_b	Electron beam accelerating voltage	V
V_c	Capacitor voltage	V
v_f	Feed rate in ECM	mm/min
V_o	Open circuit voltage	V
V_s	Breakdown voltage	V
W_{ave}	Average power	W
w_0	Width	mm
$x(t)$	Horizontal position at time t	μm
$\dot{x}(t)$	Horizontal velocity at time t	m/s
x_n	Nodal point location	m
$z(t)$	Vertical position at time t	μm
$z(t)$	Vertical velocity at time t	m/s
χ	Susceptibility of conglomerates	
β_m	Abrasive/air weight mixing ratio	%
ξ	Oscillation amplitude	μm
ω	Angular speed	radian/s
ε_{ms}	Coefficient of magnetostrictive elongation	
η	Current efficiency	%
ρ	Density of the magnetostriction material	kg/m^3
θ_m	Melting point of workpiece material	°C
σ	Stress	kg/mm^2
λ	Wavelength	μm

List of Acronyms

Abbreviation	Description
ac	Alternating current
AFM	Abrasive flow machining
AGMA	American Gear Manufacturing Association
AISI	American Iron and Steel Institute
AJECM	Abrasive jet electrochemical machining
AJM	Abrasive jet machining
AMZ	Altered material zone
ANSI	American National Standards Institute
ASME	American Society of Mechanical Engineers
ASTM	American Society for Testing and Materials
ATM	Atmosphere
AWJ	Abrasive water jet
AWJD	Abrasive water jet deburring
AWJM	Abrasive water jet machining
BHN	Brinell hardness number
BUE	Built-up edge
CBN	Cubic boron nitride
CD	Conventional drilling
CFG	Creep feed grinding
CFRP	Carbon fiber reinforced polymer
CG	Conventional grinding
CHM	Chemical machining
CH milling	Chemical milling
CI	Cast iron
CMC	Ceramic matrix composite
CNC	Computer numerical control
CW	Continuous wave
CY	Cyaniding
dC	Direct current
DFM	Design for manufacturing
DIN	Deutsches Institut für Normung
DOF	Degrees of freedom
DOT	Department of Transportation
DTC	Difficult-to-cut
EB	Electron Beam
EBM	Electron beam machining
ECA	Electrochemical abrasion
ECAM	Electrochemical arc machining
ECD	Electrochemical dissolution
ECDB	Electrochemical deburring
ECDG	Electrochemical discharge grinding

ECDM	Electrochemical discharge machining
ECG	Electrochemical grinding
ECH	Electrochemical honing
ECM	Electrochemical machining
ECS	Electrochemical sharpening
ECUSM	Electrochemical ultrasonic machining
ED milling	Electrical discharge milling
EDG	Electrodischarge grinding
EDM	Electrodischarge machining
EDS	Electrodischarge sawing
EDT	Electrodischarge texturing
EDWC	Electrodischarge wire cutting
EEDM	Electroerosion dissolution machining
EF	Etch factor
EHS	Environmental health and safety
ELID	Electrolytic in-process grinding
ELP	Electropolishing
EMF	Electromagnetic field
EMS	Environmental Management System
EOB	End of block
EP	Extreme pressure
EPA	Environmental Protection Agency
FEA	Finite element analysis
FRP	Fiber reinforced polymer
GFRP	Glass fiber reinforced polymer
H1, H2	Hardness values
HAZ	Heat-affected zone
HB	Hardness Brinell
HF	High frequency
HMIS	Hazardous Material Identification System
HMP	Hybrid machining processes
HP	Hybrid process
HRC	Hardness Rockwell
HSS	High-speed steel
HT	High temperature
IBM	Ion beam machining
IEG	Inter-electrode gap
IGA	Intergranular attack
IMPS	Integrated manufacturing production system
ipr	Inches per revolution
IR	Infrared
ISO	International Organization for Standardization
L and T	Laps and tears
Laser	Light amplification by stimulated emission of radiation
LAM	Laser-assisted machining/milling
LAT	Laser-assisted turning
LBM	Laser beam machining

LBT	Laser beam torch
LECM	Laser-assisted electrochemical machining
LSG	Low-stress grinding
LVDT	Linear variable displacement transducer
MA	Mechanical abrasion
MAF	Magnetic assisted finishing
MAP	Magnetic assisted polishing
MCK	Microcracks
MEMS	Micro-electro-mechanical systems
MFG	Magnetic fluid grinding
MFP	Magnetic float polishing
MMC	Metal matrix composites
MPE	Maximum permissible exposure
MQL	Minimum quantity lubrication
MR	Machinability rating
MRAFF	Magnetorheological abrasive flow finishing
MRF	Magnetorheological finishing
MRR	Material removal rate
MS	Manufacturing system
MSDS	Material safety data sheets
NASA	National Aeronautics and Space Administration
NC	Numerical control
Nd	Neodymium
NdY:	AG Neodymium-doped yttrium aluminum garnet
NFPA	National Fire Protection Association
NHZ	Nominal hazard zone
NTD	Nozzle-tip distance
NTM	Non-traditional machining
OA	Overaging
OCV	Open circuit voltage
OSHA	Occupational Safety and Health Administration
OTM	Overtempered martensite
PAC	Plasma arc cutting
PAH	Polycyclic aromatic hydrocarbons
PAM	Plasma arc machining/assisted milling
PAT	Plasma-assisted turning
PCBN	Poly-cubic boron nitride
PCD	Polycrystalline diamond
PMC	Polymer matrix composites
PM-HSS	Powder metallurgy HSS
PSZ	Partially stabilized zirconia
PBM	Plasma beam machining
PCB	Printed circuit board
PCD	Polycrystalline diamond
PCM	Photochemical machining
PD	Plastic deformation
PEO	Polyethylene oxide

PH Precipitation hardened
PIV Positive infinitely variable
PKM Parallel kinematic mechanism
PKS Parallel kinematic system
PVD Physical vapor deposition
PM Pulsed mode
RBSN Reaction bonded silicon nitride
RC Recast
RUM Rotary ultrasonic machining
SAE Society of Automotive Engineers
SA Super alloys
SB Sand blasting
SE Selective etching
SI Surface integrity
SMAF Semi-magnetic abrasive finishing
SOD Stand-off distance
SRR Stock removal rate
SS Stainless steel
STEM Shaped tube electrolytic machining
TAM Thermally assisted machining
TEM Transverse excitation mode
TIR Total indicator reading
TTZ Transformation-toughened-zirconia
UAW United Auto Workers
UNS Unified National Standard
US Ultrasonic
USM Ultrasonic machining
UTM Untempered martensite
UTS Ultimate tensile strength
UV Ultraviolet
VA Vibration assisted
VAD Vibration-assisted drilling
VAG Vibration-assisted grinding
VAM Vibration-assisted machining
VAM Vibration-assisted machining/milling
VAT Vibration-assisted turning
VECP Vibration-assisted electrochemical polishing
VESP Vibratory-enhanced shear processing
VRR Volumetric removal rate
WC Tungsten carbide
WECM Wire electrochemical machining
WHO World Health Organization
WJM Water jet machining
WP Workpiece
YAG Yttrium aluminum garnet
YTS Yield tensile strength

Unit I

Non-Traditional Machining Operations and Non-Traditional Machine Tools

1 Non-Traditional Machining Processes

1.1 INTRODUCTION

Over the previous decades, engineering materials have been greatly developed. The cutting speed and the material removal rate when machining such materials using traditional methods like turning, milling, grinding, etc. tend to fall. In many cases, it is impossible to machine hard materials to certain shapes using these traditional methods. Sometimes, it is required to machine alloy steel components of high strength in the hardened condition. It is no longer possible to find tool materials which are sufficiently hard to cut at economical speeds, materials such as hardened steels, austenitic steels, super alloys, carbides, ceramics, and fiber-reinforced composite materials. The traditional methods are unsuitable to machine such materials economically and there is no possibility that they can be further developed to do so, because most of these materials are harder than the materials available for use as cutting tools.

By adopting a unified program, and utilizing the results of basic and applied research, it has now become possible to process many of the engineering materials which were formerly considered to be unmachinable using traditional methods. The new machining processes, so developed, are often called modern machining processes or non-traditional machining processes (NTMP). These are non-traditional in the sense that traditional tools are not employed; instead energy in its direct form is utilized.

The NTM processes specifically have the following characteristics as compared to traditional processes:

- They are capable of machining a wide spectrum of metallic and non-metallic materials irrespective of their hardness or strength.
- The hardness of cutting tools is of no relevance, especially as in much of NTMP there is no physical contact between the work and the tool.
- Complex and intricate shapes in hard and extra-hard materials can be readily produced with high accuracy and surface quality and free of burrs.
- Simple kinematic movements are needed in NTM equipment.
- Microholes and miniature holes and cavities can be readily produced by NTM processes.

It should be, however, concluded that:

1. NTM methods cannot replace the traditional machining (TM) methods. They can only be used when they are economically justified, or it is impossible to use TM processes.

2. A particular NTM process found suitable under given conditions may not be equally efficient under other conditions. A careful selection of the NTM process for a given machining job is therefore essential (Pandey and Shan 1980). The following aspects must be considered in that selection:
   - Properties of the work material and the form geometry to be machined
   - Process parameters
   - Process capabilities
   - Economic and environmental considerations

1.2 CLASSIFICATION OF NON-TRADITIONAL MACHINING PROCESSES

NTM processes are generally classified according to the type of energy utilized in material removal. They are classified into three main groups, Table 1.1:

1. Mechanical processes, in which the material removal depends on mechanical abrasion or shearing.
2. Chemical and electrochemical processes. In chemical processes the material is removed in layers due to ablative reaction where acids or alkalis are used as etchants. The electrochemical machining is characterized by high removal rate. The machining action is due to anodic dissolution caused by the passage of high-density dc current in the machining cell.
3. Thermoelectric processes where the metal removal rate depends upon the thermal energy acting in the form of controlled and localized power pulses leading to melting and evaporation of the work material.

An important and latest development has been realized by adopting what is called the hybrid machining processes (HMP). These are new processes produced by integrating two or more NTM processes to improve the performance and promoting the removal rate and accuracy of the hybrid process. Examples of these processes are electrochemical grinding (ECG), electrochemical honing (ECH), laser beam texturing (LBT), electrochemical deburring (ECD), electrochemical ultrasonic machining (ECUSM), abrasive water jet machining (AWJM), etc.

TABLE 1.1
Classification of NTMP According to the Type of Fundamental Energy

Fundamental Energy	Removal Mechanism	NTMP
Mechanical	Erosion	AJM, WJM, USM, MFM, AFM
Chemical	Ablative reaction (etching)	CHM, PCM
Electrochemical	Anodic dissolution	ECM, ECT, ECG, ECH
Thermoelectric	Fusion and vaporization	EDM, LBM, EBM, IBM, PBM

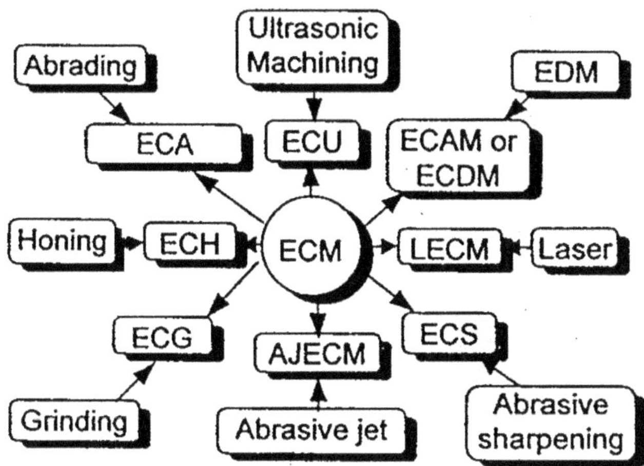

FIGURE 1.1 Hybrid NTM processes integrated with ECM (Rajurkar, Zhu, McGeough, Kozak, and De Silva, 1999).

Scientific research is still carried out in this field to check the capabilities of HMP. Some of them realized remarkable success especially the hybrid processes that are integrated with ECM as shown in Figure 1.1. A sample of some important NTM processes and HM processes and their relevant machines are dealt with in Chapters 2, 3, and 4.

1.3 REVIEW QUESTIONS

1.3.1 Briefly explain the non-transitional machining process. Give examples of applications.
1.3.2 Discuss the dimensional tolerance that can be achieved by non-traditional machining.
1.3.3 What is the importance of non-traditional machining?
1.3.4 Non-traditional machining processes yield low rates of material removal compared to traditional machining processes even though they have gained wide popularity. Discuss why. ECM, EDM, USM, etc. are commonly referred to as non-traditional machining processes. What is non-traditional in these processes? Explain.
1.3.5 Justify the need for non-traditional machining process in today's industry.
1.3.6 What are the basic limitations of non-traditional machining processes? Explain.
1.3.7 What are the basic factors upon which the non-traditional machining processes are classified? Explain. List five of them.
1.3.8 Explain this statement: NTM should not be considered as a replacement for TM.
1.3.9 Which of the NTMP causes thermal damage? What is the consequence of such damage to the WP?

REFERENCES

Pandey, PC & Shan, HS 1980, *Modern machining processes*, Tata McGraw Hill Co, New Delhi.

Rajurkar, KP, Zhu, D, McGeough, JA, Kozak, J & De Silva, AK 1999, 'New developments in ECM', *Annals of CIRP*, vol. 48, no. 2, pp. 569–579.

2 Mechanical Non-Traditional Machining Operations and Machine Tools

2.1 JET MACHINES AND OPERATIONS

In jet machining, high-velocity stream of water (WJM) or water mixed with abrasive materials (AWJM) is directed to the workpiece to cut the material. If a mixture of gas and abrasive particles is used, the process is referred to as abrasive jet machining (AJM) and is used in machining and finishing operations such as deburring, cleaning, and polishing.

2.1.1 Abrasive Jet Machining

2.1.1.1 Process Characteristics and Applications

In abrasive jet machining (AJM), a fine stream of abrasives is propelled through a special nozzle by a gas carrier (CO_2, Ni, or air) of a pressure ranging from 1 to 9 bar. Thus, the abrasives attain a high speed ranging from 150 to 350 m/min, exerting impact force causing mechanical abrasion of the workpiece (target material). The workpiece is positioned from the nozzle at a distance called the stand-off distance (SOD), or the nozzle tip distance (NTD) as shown in Figure 2.1.

In AJM, Al_2O_3 or SiC abrasives, of grain size ranging from 10 to 80 μm, are used. The nozzles are generally made of sintered carbides (WC) or synthetic sapphire of diameters 0.2 to 2 mm. To limit the jet flaring, nozzles may have rectangular orifices ranging from 0.1×0.5 mm to 0.18×3 mm. The optimum jet angle is determined according to the ductility or brittleness of the workpiece material to be machined, Figure 2.2.

AJM is not considered as a gross material removal process. Its removal rate when machining the most brittle materials such as glass, quartz, and ceramic is about 30 mg/min, whereas only a fraction of that value is realized when machining soft and ductile materials (Youssef, 2005). Due to the limited removal rate, and also the significant taper, AJM is not suitable for machining deep holes and cavities. However, the process is capable of producing holes and profiles in sheets of thicknesses comparable to the nozzle diameter. AJM is applicable for cutting, slitting, surface cleaning, frosting, and polishing. The process advantages and limitations are listed below.

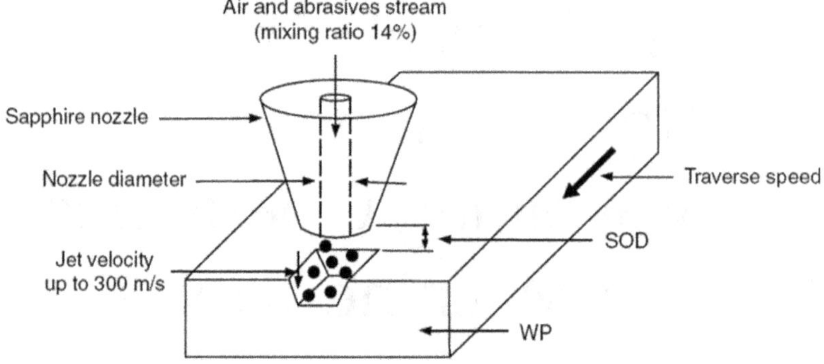

FIGURE 2.1 AJM terminology (El-Hofy, 2005).

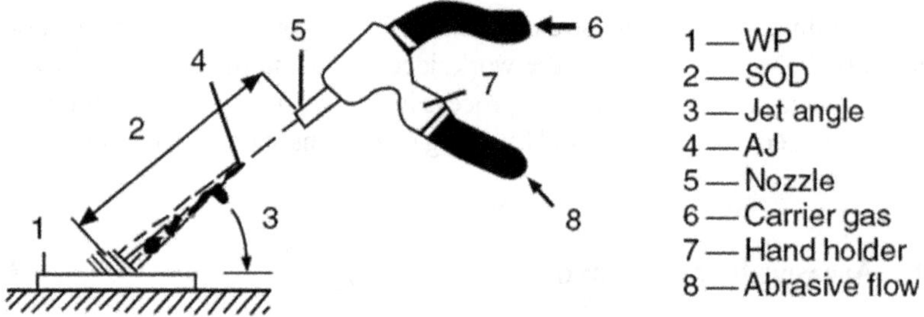

FIGURE 2.2 AJM inclination angle (Düniβ, Neumann, and Schwartz, 1979).

Advantages

- Capable to produce holes and intricate shapes in hard and brittle materials
- Used to cut fragile materials of thin walls
- Heat-sensitive materials such as glass and ceramics can be machined without affecting their physical properties and crystalline structure, since no or little heat is generated during machining
- Safe in operation
- Characterized by low capital investment and low power consumption
- Can be used to clean surfaces, especially in areas which are inaccessible by ordinary methods
- Produced surfaces after cleaning by AJM are characterized by their high wear resistance

Limitations

- The application of AJM is restricted to brittle materials. It is not recommended for machining soft and malleable materials.
- Abrasives cannot be reused because they lose their sharpness and, hence, cutting ability.
- Nozzle clogging occurs if fine grains with a diameter dg <10 μm are used.
- The process accuracy is poor due to the flaring effect of the abrasive jet.
- Deep holes are produced by significant taper.

- Sometimes, machined parts have to undergo an additional operation of cleaning to get rid of grains sticking to the surface.
- Excessive nozzle wear causes additional machining cost.
- The process tends to pollute the environment.

Fields of Applications

AJM has been successfully applied in the following domains:

- Deflashing and trimming of parting lines of injection molded parts and forgings
- Cleaning metallic molds and cavities
- Cutting thin sectioned fragile components made of glass, refractoriness, mica, and so on
- Cleaning surfaces from corrosion, paints, glue, and contaminants, especially those which are inaccessible
- Marking on glass
- Frosting interior or exterior surfaces of glass tubes
- Engraving on glass using metallic or rubber masks

Some typical applications are:

- Beveling of electronic wafer disk composed of silicon disk (0.4 mm thick) welded to a tungsten disk (0.7 mm thick), Figure 2.3a. A trimming rotating fixture is shown in Figure 2.3b. The disk rotates slowly (n = 5–10 rpm), while the nozzle is directed at an angle of 45°. One minute is required to bevel a disk.
- Engraving registration numbers on glass windows of cars
- Deburring fine internal intersecting holes in plastic components needed for medical applications
- Deburring of surgical needles and hydraulic valves
- Deburring parts of nylon, Teflon, and derlin

2.1.1.2 Work Station of Abrasive Jet Machining

Figure 2.4 shows a typical work station of AJM which is connected to a gas supply (gas bottles or compressed air). The carrier gas must not flare excessively when discharged from nozzle to atmosphere. Furthermore, it should be non-toxic, cheap, available, and capable to be dried and filtered. Air is widely used owing to its availability. In small stations, CO_2 and N_2 gas bottles are commonly used. After filtering, the pressure of the compressed gas of 7 to 9 bar is regulated, to suit the working conditions. The gas is then introduced to the mixing chamber containing the abrasives. The chamber is equipped with a vibrator providing amplitude ξ of 1 to 2 mm at a frequency f_r from 5 to 50 Hz. The abrasive flow rate is controlled through the adjustment of ξ and f_r. From the mixing chamber the gas/abrasive mixture is directed to the nozzle that directs the jet onto the target or workpiece. The jet velocity of 150 to 350 m/min depends upon the gas pressure at the nozzle, the orifice diameter of the nozzle, and the mixing ratio. The flow rate of a typical working station is about 0.6 m^3/hr which is controlled through a foot control valve.

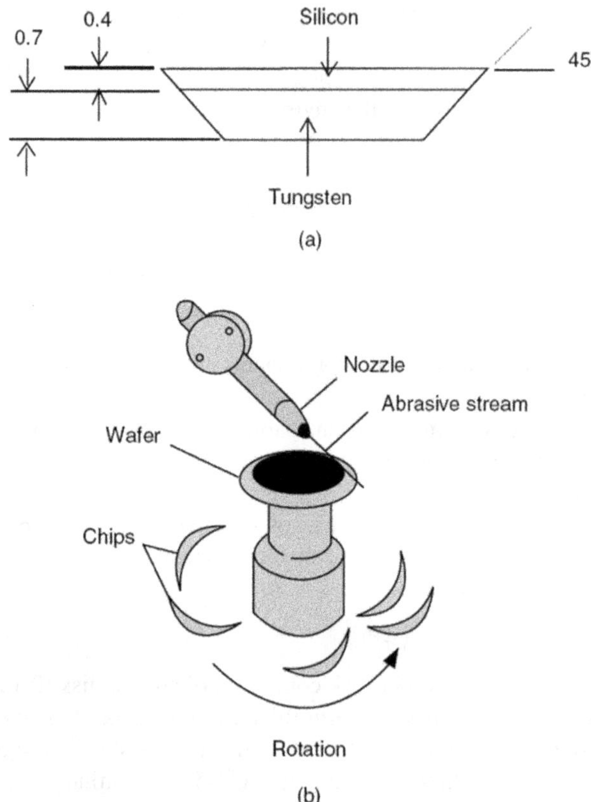

FIGURE 2.3 Edge trimming by AJM: (a) Wafer disk (silicon/tungsten) and (b) trimming fixture (Benedict, 1987).

The nozzle is mounted in a special fixture, and sometimes held in hand, depending on the type of operation required (cutting, trimming, engraving, frosting, and cleaning). When machining thin-walled fragile materials, it may be necessary to control the relative motion between the nozzle and the work by a cam and pantograph depending upon the required size and shape of cut. The AJM station must be equipped by a vacuum dust collector to limit the pollution. Strict measures and precautions should be undertaken in case of machining toxic materials such as beryllium to collect produced dust and debris.

2.1.1.3 Process Capabilities

The performance of AJM in terms of material removal rate (MRR) and accuracy is affected by the selected machining conditions. The MRR for a certain material is mainly affected by the kinetic energy of the abrasives, i.e. the speed with which the abrasive bombards the work material. This speed depends upon:

- Gas pressure at nozzle
- Nozzle diameter
- Abrasive grain size

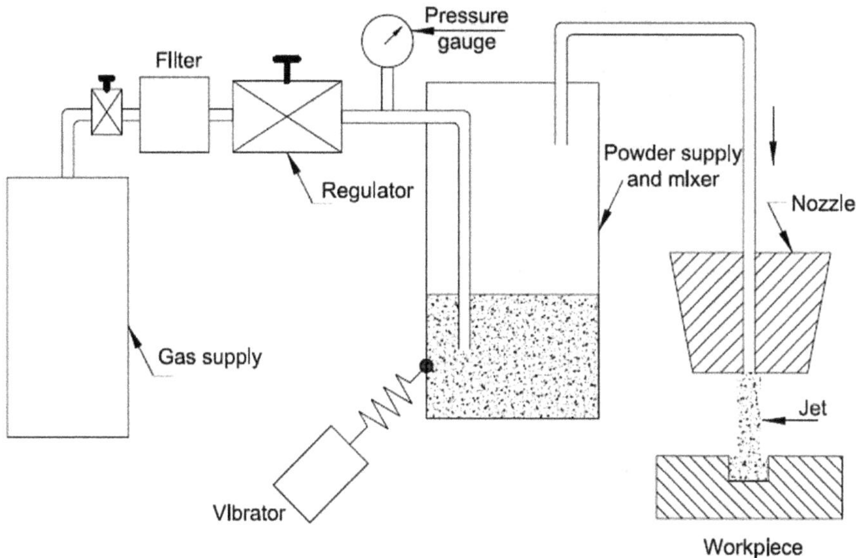

FIGURE 2.4 Typical AJM workstation (El-Hofy, 2005).

- Weight mixing ratio $\beta_m \left(\dfrac{\text{abrasive flow rate}}{\text{air flow rate}}\right)$
- Stand-off distance (SOD)

MRR attains a maximum value at a mixing ratio $\beta_m = 0.15$ (Düniβ, 1979), and a stand-off distance from 15 to 17 mm (Verma and Lal, 1984). It increases with increasing gas pressure at the nozzle. The type of material to be machined, and the abrasive grain size, have an influence on the MRR. The latter increases with increasing grain size (Machinability Data Handbook, 1980). Sharp-edged abrasives of irregular shape, dry, and well classified (non-commercial) are best suited to perform the job. The limiting size of abrasive grains which permits the grain to be suspended in the carrier gas is about 80 μm. SiC and Al_2O_3 abrasives are used for cutting and slitting operations, whereas sodium bicarbonate, dolomite, and glass beads are used for cleaning, frosting, and polishing.

When selecting the best working conditions ($\beta_m = 0.15$, abrasives Al_2O_3 of grain size 50 μm, SOD = 14 mm, nozzle pressure = 7 bar), the MRR achieved in case of machining is in the order of 30 mg/min (*Machining Data Handbook*, 1980).The accuracy improves by selecting smaller SOD, Figure 2.5, which reduces the material removal rate. The grain size is the decisive factor for determining the surface finish.

2.1.2 Water Jet Machining (Hydrodynamic Machining)

2.1.2.1 Process Characteristics and Applications

Over the last five decades, a number of studies using high-pressure water jets (pulsed or continuous) in mining applications have been made (Farmer and Attewell, 1965;

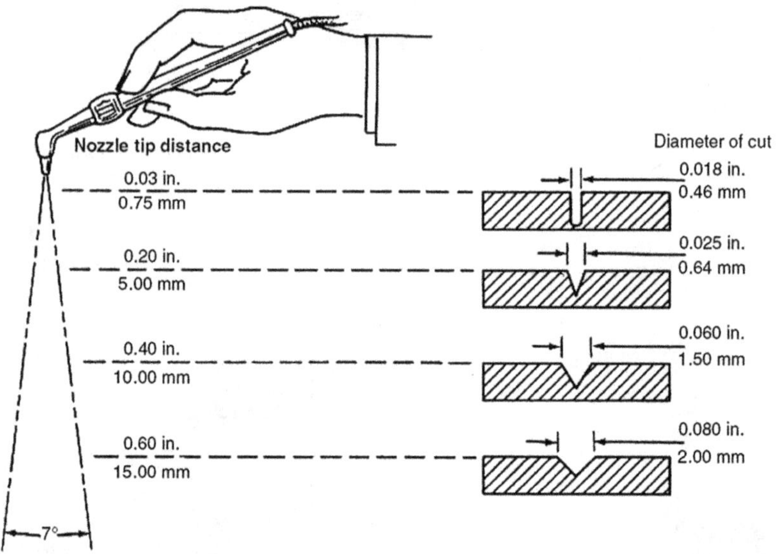

FIGURE 2.5 Effect of SOD on kerf width and accuracy (From Machinability Data Center *Machining Data Handbook*, Cincinnati, OH, 1980).

Brook and Summers 1969). Franz (1972) of the University of Michigan succeeded in cutting wood using high-velocity water jets; he reported the importance of improved coherence of the water jet with the addition of polymers. Since then, the cutting capability of liquid jets has been reported for a wide spectrum of target materials, including lead, Al, Cu, Ti, steels, and granite. It is hard to believe that a jet of water may successfully cut steel and granite. However, in scientific terms, this is acceptable, as illustrated in Figure 2.6, where a stream of water is propelled at high pressure (2000–8000 bar) through a converging nozzle to give a coherent jet of water of high speed of 600 to 1400 m/s. At the target, the kinetic energy of the jet is converted spontaneously to high-pressure energy, inducing high stresses exceeding the flow strength of target material, causing mechanical abrasion. WJM has the following advantages and limitations.

Advantages

- Water is cheap, non-toxic, and can be easily disposed and recirculated.
- The process requires limited volume of water (100–200 l/hr).
- The tool (nozzle) does not wear and, therefore, does not need sharpening.
- No thermal degrading of the work material, since the process does not generate heat. For this reason, WJM is best suited to explosive environments.
- It is ideal for cutting asbestos, glass fiber insulation, beryllium, and fiber-reinforced plastics (FRP), because the process provides a dustless atmosphere. For this reason, the process is not hazardous and environmentally safe.
- The process provides clean and sharp cuts, free from burrs.
- It is applicable for laser reflective materials such as, glass, copper, and aluminum.

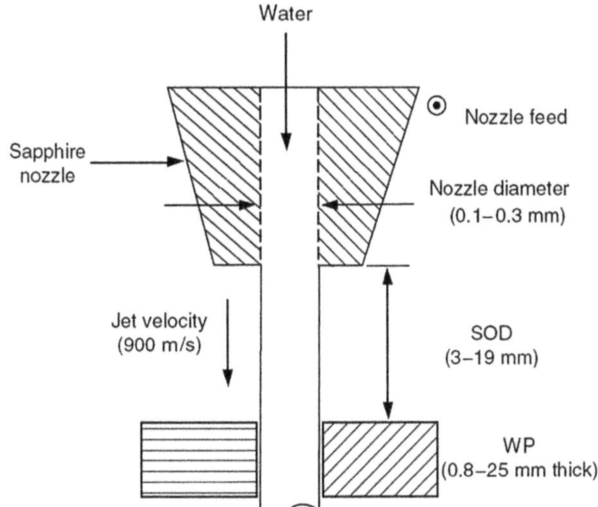

FIGURE 2.6 WJM terminology (El-Hofy, 2005).

- Starting holes are not needed to perform the cut.
- Wetting of the workpiece material is minimal.
- Noise is minimized as the power unit and intensifier can be kept away from the cutting station.
- While AJM is commonly used to cut only brittle materials, it is applicable to machine both brittle and ductile materials.
- The workpiece is subjected to a limited mechanical stress, as the force exerted by the jet does not generally exceed 50N. Therefore, cutting is performed without the need of using elaborate fixturing of the workpiece.
- WJ approaches the ideal single point tool.

Limitations of WJM

- WJM is not safe in operation if safety precautions are not strictly followed.
- The process is characterized by a high production cost due to:
    - High capital cost of the machine
    - The need of highly qualified operators
- WJM is not adapted to mass production because of the high maintenance requirement.

Applications of WJM: It is used in many industrial applications comprising the following:

- Cutting of metals and composites applied in aerospace industries
- Underwater cutting and shipbuilding industries
- Cutting of rocks, granite, and marble
- It is ideal in cutting soft materials such as wood, paper, cloth, leather, rubber, and plastics

- Slicing and processing of frozen foods, baked foods, and meat. In such cases, alcohol, glycerin, and cooking oils are used as alternative cutting fluids
- WJM is also used in:
  - Cleaning, polishing, and degreasing of surfaces
  - Removal of nuclear contaminations
  - Cleaning of tubes and castings
  - Surface preparation for inspection purposes
  - Surface strengthening
  - Deburring

2.1.2.2 Equipment of WJM

Figure 2.7 visualizes a simplified layout of WJM equipment. It consists of the following stations:

1. ***Multi-stage filtering station***: The function of which is to filter the solid particles down to 0.5 μm. In this stage, it is also recommended to perform deionization and demineralization of water to allow for better performance of machine elements and extended nozzle life. After filtering, water is mixed by polymers to obtain a coherent jet.
2. ***Oil pump and water high pressure–intensifier station***: It consists of a hydraulic pump powered by an electric motor that provides oil at about 120 bar. Such a pressure is needed to drive a double acting plunger pump (intensifier) that pumps water from 4 bar to about 4000 bar or more. Figure 2.8 illustrates the operation of the high-pressure intensifier that consists of two terminal small cylinders for water and a large central cylinder for oil. A limit switch, located at each end of the terminal cylinders, signals the electronic controls to shift the directional control valve and reverses the

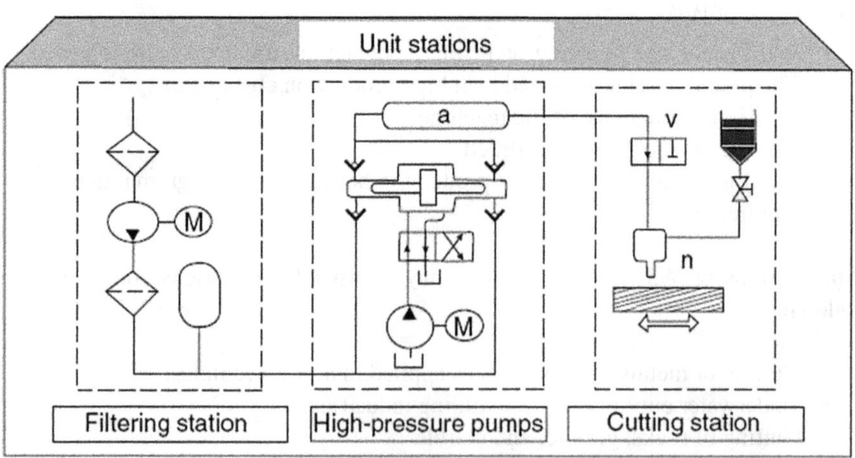

FIGURE 2.7 Simplified layout of WJM equipment (König, 1990).

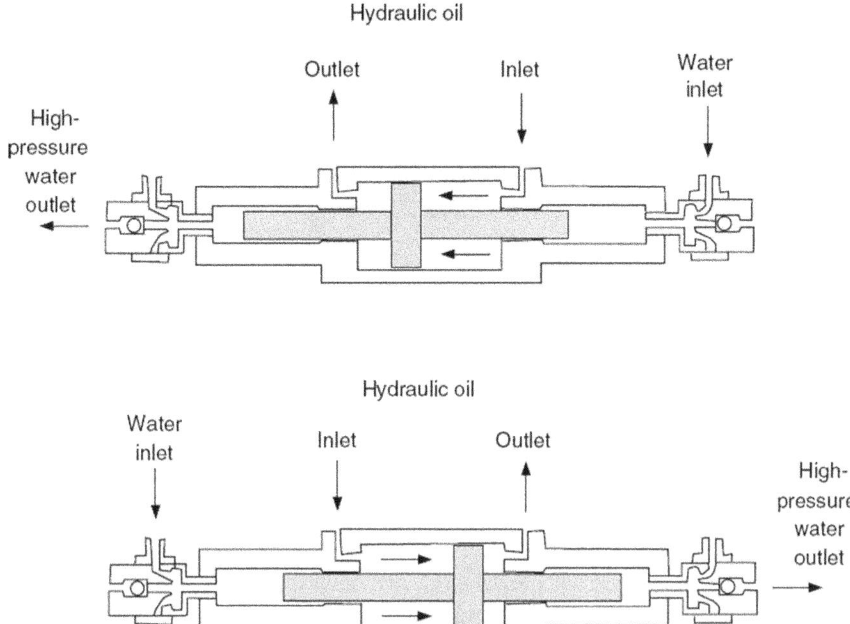

FIGURE 2.8 HP intensifier (Nordwood and Johnston, 1984).

piston direction. As one side of the intensifier is in the inlet position, the opposite side generates an ultrahigh pressure output, and vice versa. The ultrahigh-pressure water is delivered to an accumulator tank, Figure 2.7, to provide the water pressure free of fluctuation and hydraulic spikes to the cutting station. During idle times, the water is stored in the accumulator under pressure to be ready at any time to perform the cutting. The intensifier offers complete flexibility for both cutting and cleaning applications. It also supports single or multiple cutting nozzles for increased productivity.

3. ***Cutting station***: The conversion cutting nozzle, Figure 2.9, converts the ultrahigh pressure (about 4000 bar) into a high speed of 400 to 1400 m/s. The nozzle provides a coherent water jet stream for optimum cutting. The jet coherency can be enhanced by adding long chain polymers such as polyethylene oxide (PEO) with a molecular weight of four million. Such addition provides the water with higher viscosity and hence increases the coherent length up to 600 d_n, where d_n is the nozzle orifice diameter that falls between 0.1 and 0.35 mm. For optimum cutting, the stand-off distance (SOD) is selected within this range. Even if SOD is selected beyond this range, the stream is still capable of performing non-cutting operations such as cleaning, polishing, degreasing, etc., due to the existence of the concentrated liquid cone in the growing spray envelop, (Youssef, 2005).

Nozzles are generally made from very hard materials such as WC, synthetic sapphire, or diamond. Diamond provides the largest nozzle life, whereas WC gives the lowest

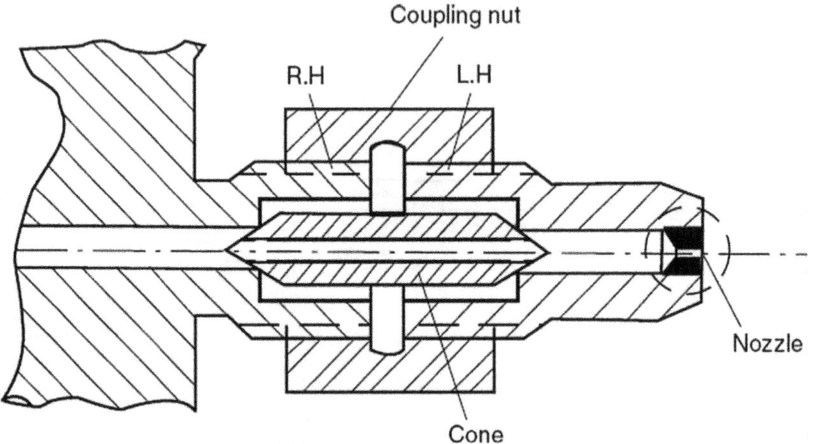

FIGURE 2.9 Nozzle assembly of WJM equipment (Youssef, 2005).

one. About 200 hours of operation is expected from a nozzle of synthetic sapphire, which becomes damaged by particles of dirt and the accumulation of mineral deposits if the water is not filtered and treated. High-pressure tubing, Figure 2.10, transports pressured water to the cutting nozzle. Thick tubes of diameters ranging from 6 to 14 mm and of diameter ratio 1/5 to 1/10 are used, Figure 2.10a. For severe pressure which may exceed the yielding stress of the tube material, shrink-fit tubes should be used, Figure 2.10b. To achieve the best sealing conditions, metal-to-metal, line (not surface) contact should be used in high-pressure tube fittings. It is preferable that the on-off valves for such machines operating at high pressures are of the needle type. The compact design of the nozzle head promotes integration with a motion-control system ranging from two-axis x–y tables to sophisticated multi-axis CNC installations.

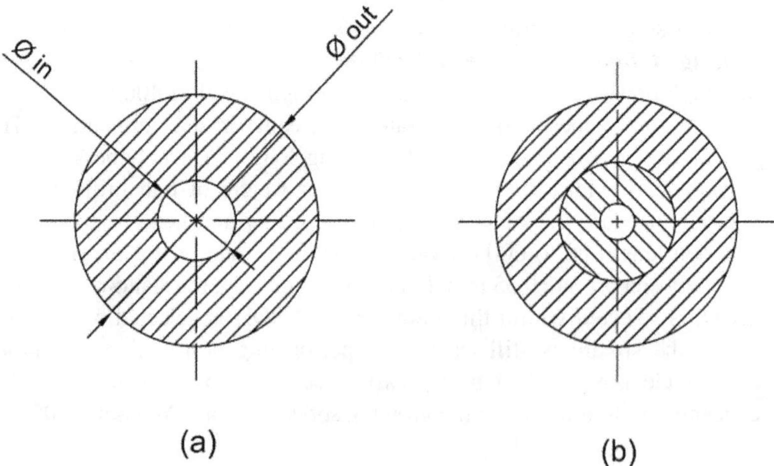

FIGURE 2.10 High-pressure tubing: (a) Thick-wall tubing and (b) shrink-fit tubing.

The cutting station must be equipped with a catcher which acts as a reservoir for collecting the machining debris entrained in the jet. Moreover, it absorbs the rest energy after cutting which is estimated to be 90% of the total jet energy. It reduces the noise levels (105 dB) associated with the reduction of the water jet from Mach 3 to subsonic levels. Figure 2.11 shows a schematic illustration of the WJM equipment.

2.1.2.3 Process Capabilities

The material removal rate, accuracy, and surface quality are influenced by the workpiece material and the machining parameters. Brittle materials fracture, while ductile ones are cut well. Material thickness ranges from 0.8 mm to 25 mm or more. Table 2.1 illustrates the cutting rates for different material thicknesses. For a given nozzle diameter, the increase of pressure allows more power to be used, which in turn increases the penetration depth or the traverse speed. The quality of cutting improves at higher pressures and lower traverse speeds. Under such conditions, greater thicknesses can be cut.

2.1.3 Abrasive Water Jet Machining

2.1.3.1 Process Characteristics and Applications

Abrasive water jet machining (AWJM) is a hybrid process (HP) since it is an integration of AJM and WJM processes. The addition of abrasives to the water jet drastically increases the range of materials which can be cut with a water jet

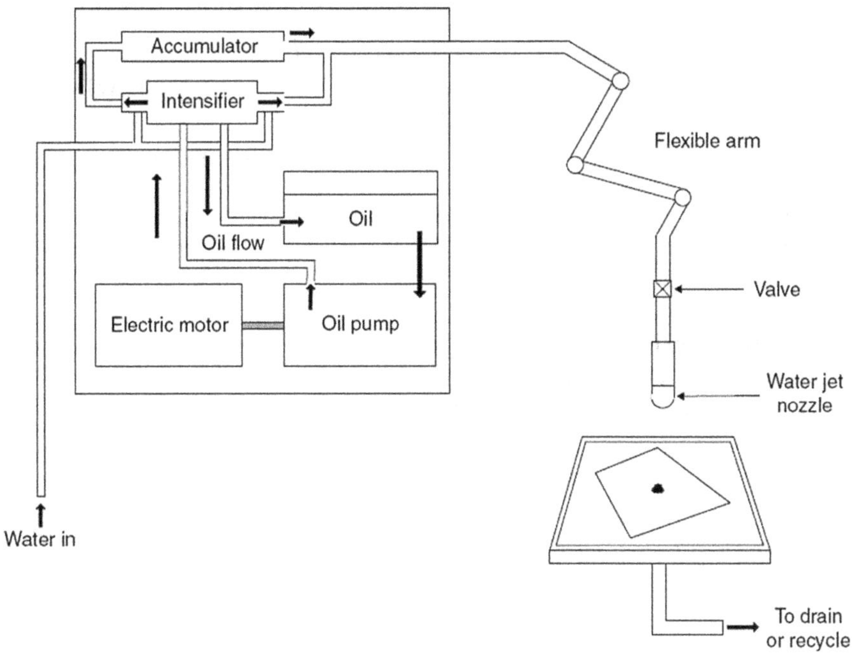

FIGURE 2.11 Schematic illustration of WJM equipment (El-Hofy, 2005).

TABLE 2.1
Traverse Speeds and Thicknesses of Various Materials Cut by Water Jet

Material	Thickness (mm)	Traverse Speed (mm/min)
Leather	2.2	20
Vinyl chloride	3.0	0.5
Polyester	2.0	150
Kevlar	3.0	3.0
Graphite	2.3	5.0
Gypsum board	10.0	0.6
Corrugated board	7.0	200
Pulp sheet	2.0	120
Plywood	6.0	1.0

Source: (Tlusty, 2000).

and maximizes the material removal rate of this hybrid process. The MRR is based, therefore, on using the kinetic energies of the abrasives and water in the jet. Intensive research works have been carried out during the last three decades to explore the capabilities of this new process. It has been reported that the AWJM process is capable of machining both soft and hard materials at very high speeds compared with those realized by WJM. It cuts ten to 50 times faster than the WJM process. Moreover, the cuts performed by AWJM have better edge and surface qualities.

AWJM uses a comparatively lower water pressure than that used by WJM (about 80%) to accelerate the AWJ. The typical mixing ratio of abrasive to water in the jet is about 3/7 by volume, Figure 2.12. Abrasives (garnet, sand, Al_2O_3, and so on) of a grain size 10–180 μm are often used.

As previously mentioned, apart from its capability to machine soft and hard materials at very high speeds, the AWJM process has the same advantages of WJM. However, it has the following two limitations:

- Due to the existence of the abrasives in the jet, there is an excessive wear in the machine and its elements.
- The process is not environmentally safe as compared to WJM.

The AWJM process has many fields of application such as:

- Cutting of metallic materials: Cu, Al, Pb, Mo, Ti, W
- Cutting carbides and ceramics
- Cutting concrete, marble, and granite, Figure 2.13a
- Cutting plastics and asbestos, Figure 2.13b
- Cutting large casting with very narrow kerf which reduces material and energy wastages, Figure 2.13c

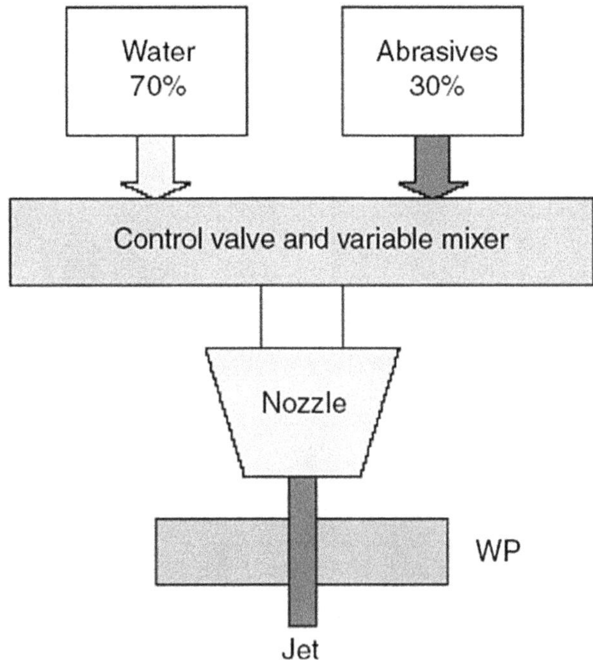

FIGURE 2.12 AWJM elements terminology (El-Hofy, 2005).

- Cutting composites such as FRP, and sandwiched Ti-honeycomb without burr formation. The latter is used in the aerospace industry
- Cutting of acrylic and glass

In the field of machining technology, AWJM has two promising applications as illustrated in Figure 2.14. These are milling of flat surfaces and turning of cylindrical surfaces.

Abrasive water jet cutting is extensively used in the aerospace, automotive, and electronics industries. In the aerospace industry, parts such as titanium bodies for military aircrafts, engine components (aluminum, titanium, heat-resistant alloys), aluminum body parts, and interior cabin parts are made using abrasive water jet cutting. In the automotive industry, parts such as interior trim (head liners, trunk liners, door panels) and fiberglass body components and bumpers are made by this process. Similarly, in the electronics industry, circuit boards and cable stripping are made by abrasive water jet cutting. The process is also applicable in deburring (AWJD), sharpening of grinding wheels, and surface strengthening to increase the fatigue strength.

Advantages of AWJM

- In most of the cases, no secondary finishing required
- Can cut both hard and soft materials
- No cutter-induced distortion

AUTOLINE™ abrasive cutting – marble (hard materials)

(a)

Variety of waterjet cut parts

(b)

(c)

FIGURE 2.13 Cutting by AWJM: (a) Marble and (b) plastics and asbestos (Ingersoll-Rand, 1996); (c) abrasive water jet cutting of cylinder casting (Dhakal, 2007).

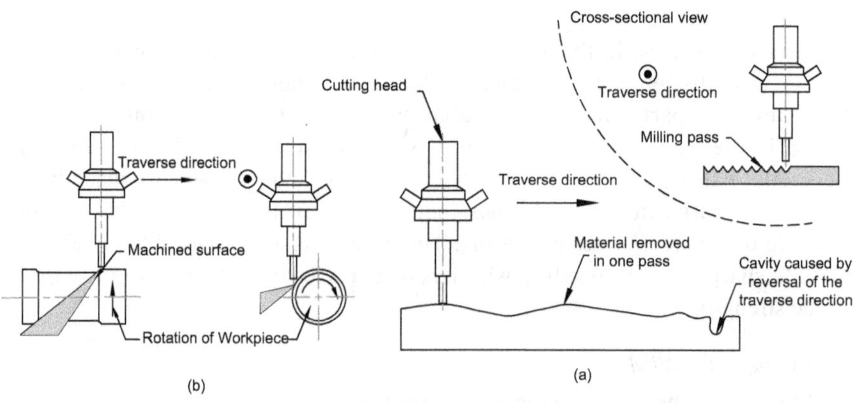

FIGURE 2.14 Two promising applications of AWJM: (a) Milling and (b) turning (Hoogstrate and van Luttervelt, 1997).

- Low cutting forces on workpieces
- Limited tooling requirements
- Little to no cutting burr
- Smaller kerf size reduces material and energy wastages
- No heat-affected zone
- Eliminates thermal distortion
- No slag or cutting dross
- Precise, multiplane cutting of contours, and bevels of any angle

Limitations of abrasive water jet cutting

- Cannot drill flat bottom
- Cannot cut materials that degrade quickly with moisture
- Surface finish degrades at higher cut speeds which are frequently used for rough cutting
- The major disadvantages of abrasive water jet cutting are high capital cost and high noise levels during operation

2.1.3.2 AWJM Equipment

The equipment of AWJM does not differ greatly from the basic WJM equipment. So, it is composed of:

1. **Water Filtering Station**
   It is the same as that of the WJM, but in AWJM cutting polymers are not commonly used because the general opinion is that the increased coherence of the jet prevents the abrasive particles from being mixed with the water jet and, therefore, the accelerating process of the abrasive is less efficient, although Swiss glass manufacturers and French crystal manufacturers are of a different opinion and use a polymer additive in combination with abrasives, thereby increasing the cutting speed considerably and reducing the abrasive consumption drastically. Research in this area is still needed.
2. **Pressure Generation Station**
   A double acting intensifier is designed to deliver less pressure than that used in WJM. The usual range of pressure in AWJM is from 250 to 350 MPa at a discharge rate of 5 l/min, accordingly the pressure loss is decreased, and the system piping is less stressed.
3. **Cutting Station**
   The cutting nozzle in the machining station of WJM equipment is replaced by what is called the jet former in AWJM equipment.

Jet former: In the jet former, the pressure energy of the water is first converted into kinetic energy of the water which in turn is partially converted into kinetic energy of the abrasive particles. Figure 2.15a illustrates a jet former of Ingersoll-Rand, while Figure 2.15b illustrates a sectional view of the same jet former. At the end of the high-pressure tubing, an orifice is installed, which consists commonly of a hexagonal-rhombohedral sapphire Al_2O_3, a ruby, or diamond with a hole of 0.08 to 0.8 mm inner diameter. Diameters under 0.25 mm are mainly used in high-pressure pure

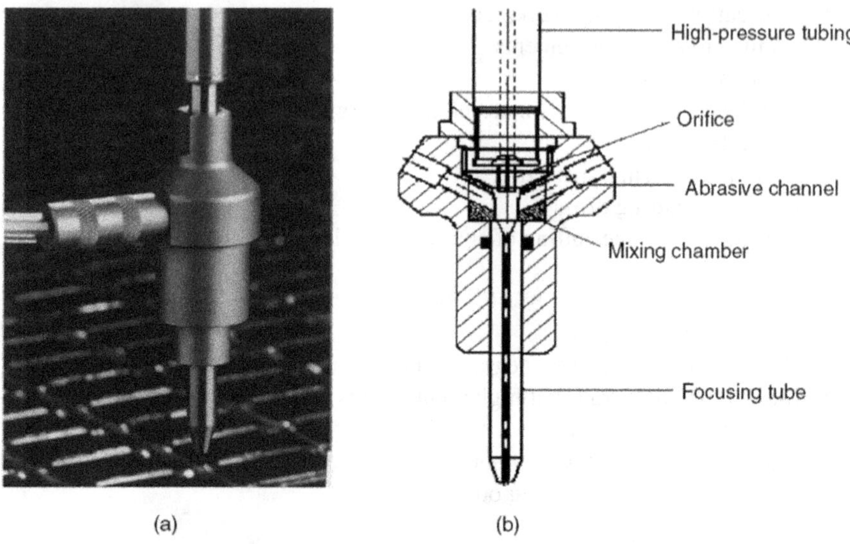

FIGURE 2.15 Jet former: (a) Ingersoll-Rand (1996) and (b) cross-sectional view.

water jet cutting applications. Orifice diameters between 0.25 and 0.40 mm are used in AWJM applications. Diameters over 0.40 mm are mainly used in low-pressure cleaning applications of AWJ cleaning (Hoogstrate and van Luttervelt, 1997).

Through the orifice, Figure 2.15b, the high-pressure water is expelled, and pure water jet is formed and directed into the mixing chamber. Through the interaction of the pure water jet and the surrounding air a vacuum is created in the mixing chamber causing airflow from outside through the abrasive channels to the mixing chamber. In the mixing chamber, the jet loses its coherency; therefore, a focusing tube, Figure 2.15b, is installed below the mixing chamber to restore the coherency of the AWJ. The resulting diameter of the AWJ is nearly equal to the focusing tube diameter. Figure 2.16 illustrates an assembly chart of the jet former of Ingersoll-Rand. The design of the jet former is specified by the following parameters:

- Water orifice diameter
- Distance along jet axis from orifice to entrance point
- Entrance direction (angle) of abrasives
- Cross-section of the abrasive feed channel
- Mixing chamber length/diameter ratio
- Diameter of focusing tube
- Length of focusing tube

Focusing tube (also called abrasive tube or accelerator tube): The performance of the focusing tube depends upon:

- Geometry of the inlet zone
- Inner diameter of the tube: The smaller this diameter, the more concentrated the total energy is. For reliable functioning, the focusing tube diameter should be at least five times the particle diameter.

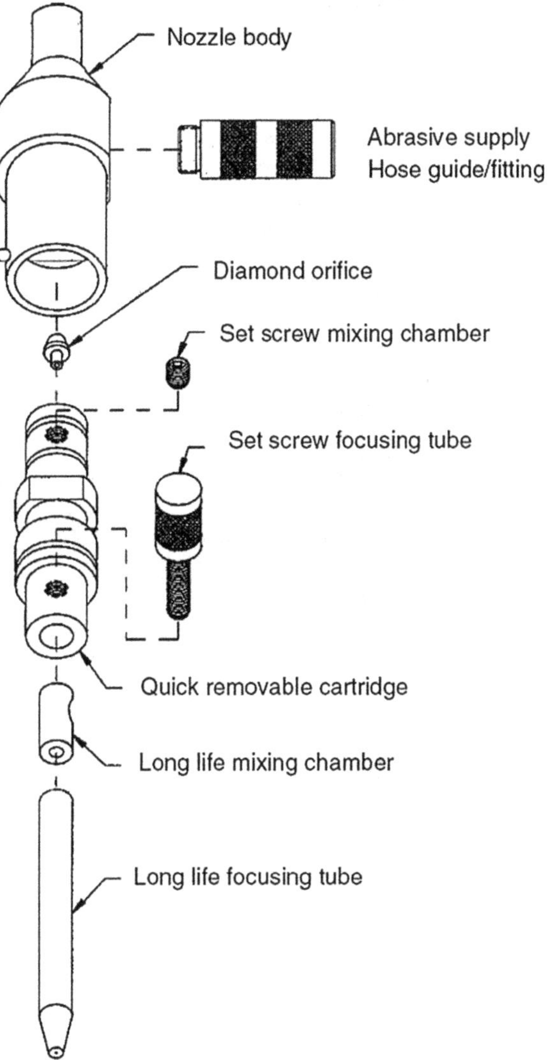

FIGURE 2.16 Assembly chart of jet former AWJM (Ingersoll-Rand, 1996).

- Length of the tube: Longer tubes produce a more coherent jet but cause more friction between the jet and the tube wall, resulting in lower abrasive jet velocities. Longer tubes are also more difficult to align.

Mixing abrasives with water: Due to the complex turbulent nature of the mixing process, not much modeling has been carried out. However, an acceptable theory for the mixing of the particles and the jet in the focusing tube was developed. It is assumed that each particle enters the water jet with a negligible velocity. It is accelerated and pushed out of the water jet, hits the inner wall of the focusing tube, rebounces and enters the water jet again. This happens until the velocity direction of the particle is nearly parallel to the direction of the water jet, Figure 2.17. As a consequence of this acceleration process, two effects are encountered:

1. The abrasive particles are fragmented due to collisions with the focusing tube and other abrasive particles. This causes a significant diameter reduction of the abrasive particles after the focusing process, Figure 2.18. Recycling of abrasives does not seem interesting due to this particle fragmentation. Nevertheless, abrasive recycling units have been recently introduced into the market.
2. The focusing tube is exposed to extremely abrasive conditions. Therefore, it should be made of advanced wear-resistant materials which provide a

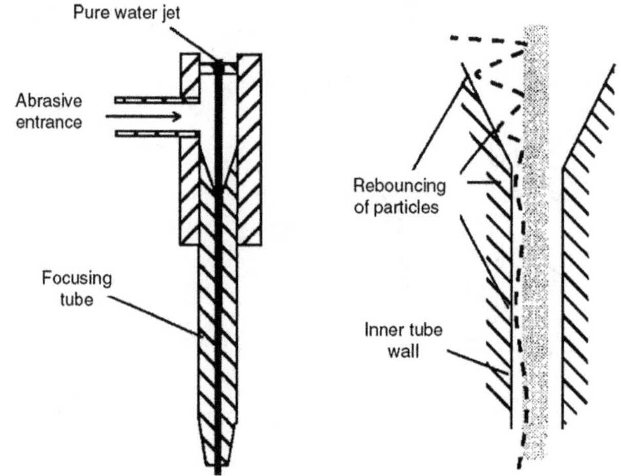

FIGURE 2.17 Abrasive acceleration in the focusing tube (Hoogstrate and van Luttervelt, 1997).

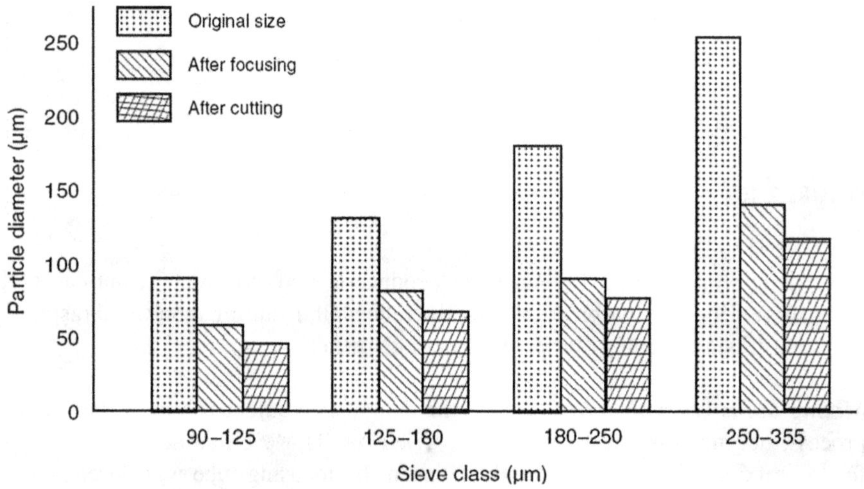

FIGURE 2.18 Wear of garnet grit in AWJM (Schmelzer, 1994).

reliable and stable cutting over a longer service life as shown in Figure 2.19. An intelligent nozzle system is proposed, in which a wear monitoring system is implemented using a grid pattern of electrical conductive wires, Figure 2.20, which is mounted on a focusing tube-tip to record the wear of the inner diameter of the tube (Kovacetic, 1994). When the tube diameter wears out, the wires will successively be interrupted which can be easily signaled. This diameter information can be used to trigger the end of the focusing tube life.

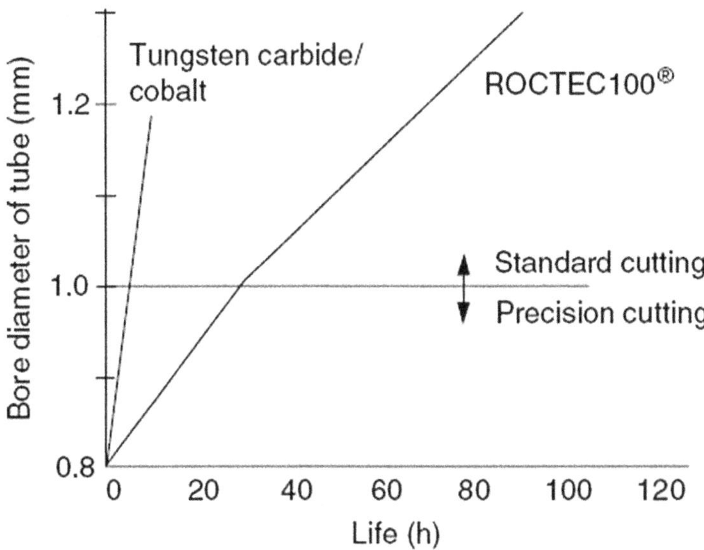

FIGURE 2.19 Wear of focusing tubes made of different materials (Product Information of Allfi AG, 1997).

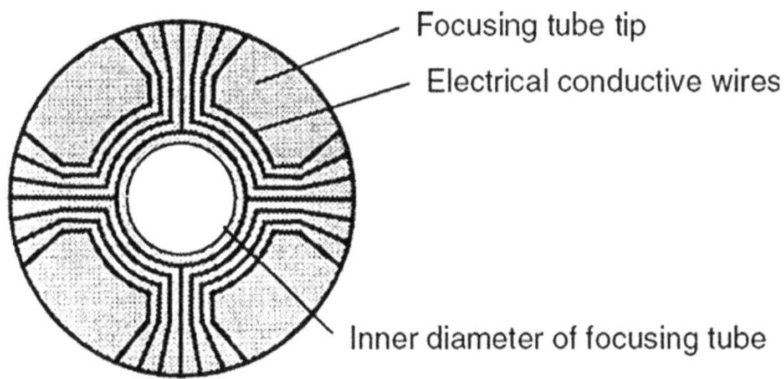

FIGURE 2.20 Wear sensor for focusing tube exit diameter (Kovacevic, 1994).

Important characteristics of the AWJ: Five important AWJ characteristics must be realized to perform effectively:

- Jet velocity determines the cutting capability
- Jet coherence determines the kerf shape and the quality of cut
- Abrasive/water mass ratio ensures optimum cutting efficiency
- Rotational jet symmetry determines cutting capability in different directions
- Establishing a time-independent jet structure gives a uniform quality along a cut in a workpiece

Computerized WJ and AWJ machines are now available on the market. They are capable of loading a CAD drawing from another system, and to determine starting and end points and the sequence of operations. Other CNC machines operate with a modem and CAD/CAM capabilities that permit transfer from AUTOCAD, DXF formats.

2.1.3.3 Process Capabilities

The typical machining variables of the AWJM process include:

- Water pressure
- Water nozzle diameter
- Geometry of focusing tube (length and diameter)
- Stand-off distance (SOD)
- Size and type of abrasive grits
- Abrasive/water ratio
- Hardness and strength of the workpiece material
- Type of workpiece material (metallic, non-metallic, or composite)

When machining glass by AWJ, a cutting rate of about 16–20 mm^3/min is achieved. An AWJ cuts through 360-mm-thick slabs of concrete or 76-mm-thick tool steel plates at a traverse speed of 38 mm/min in a single pass. When cutting steel plates (or metallic materials), the surface roughness R_t ranges from 3.8 to 6.4 μm while tolerances of ±130 μm are obtainable. Repeatability of ±40 μm, squariness of 43 μm/m, and straightness of 50 μm per axis are expected. Sand and garnet are frequently used as abrasive materials. However, garnet is preferred because it is 30% more effective then sand. A carrier liquid consisting of water with anticorrosive additives contributes to higher acceleration of abrasives with a consequent higher abrasive speed and increased material removal rate (El-Hofy, 2005). The penetration depth increases with increasing water pressure and decreasing traverse velocity, provided other working conditions are being constant.

The stand-off distance (SOD) has an important effect on MRR and the accuracy. It attains values between 0.5 and 5 mm. The smallest value (0.5 mm) realizes higher accuracy and smallest kerf width, whereas the largest value (5 mm) realizes the maximum MRR. Beyond 5 mm, the jet gradually loses gradually its cutting capability till it reaches 50 to 80 mm, at which the jet is used efficiently in surface cleaning and peening.

Table 2.2 illustrates the traverse velocities when cutting different materials of different thicknesses using AWJ. Accordingly, it can be depicted that:

- Pure metals (Ti, Al) have the same machinability.
- Glass is cut at eight to ten times faster than metals and alloys.

Surface roughness depends on the workpiece material, grit size, and type of abrasives. A material with a high removal rate produces large surface roughness. For this reason, fine grains are used for machining soft metals to obtain the same roughness as hard ones. Additionally, the larger the abrasive/water ratio, the higher the MRR will be (El-Hofy, 2005).

In the domain of composites, the abrasive water jet machining process is particularly good as the cutting rates are considerably higher, and it does not delaminate the layered material. A comparison study was carried out by König and Schmezler (1990) to investigate the performance of WJM and AWJM when cutting a 5-mm-thick plate of fiber-reinforced plastic (FRP) under optimum working conditions:

WJ and AWJ	
Water pressure	= 300 MPa
Nozzle diameter	= 0.225 mm
SOD	= 5 mm
AWJ	
Abrasives: Garnet # 80 mesh	
Abrasive flow rate = 300 g/min	

The outcome of this study shows that AWJM has realized a traverse velocity of 2000 mm/min, which is 40 times that realized by WJM. The surface roughness as obtained by AWJM (R_a = 4.4 μm is about 30% less than that obtained by WJM (R_a = 6.4 μm).

TABLE 2.2
Traverse Velocity (mm/min) when Machining Different Materials by AWJM

Material Thickness	6 mm	15 mm	19 mm	25 mm	50 mm
Titanium	250	150	100	50	16
Aluminum	250	150	100	50	16
FRP	500	280	130	75	25
Stainless steel	200	90	60	40	15
Glass	2000	1000	700	500	150

Source: (Youssef, 2005).

2.2 ULTRASONIC MACHINING

2.2.1 Definition, Characteristics, and Applications

Ultrasonic machining (USM) is an economically viable operation in which a hole or cavity can be pierced in hard and brittle materials, whether electric conductive or not, using an axially oscillating tool. The tool oscillates with small amplitude of 10 to 50 μm at high frequencies of 18 to 40 kHz to avoid unnecessary noise (the audio threshold of the human ear is 16 kHz). During tool oscillation, abrasive slurry (B_4C and SiC) is continuously fed into the working gap between the oscillating tool and the stationary workpiece. The abrasive particles are, therefore, hammered by the tool into the workpiece surface, and, consequently, they abrade the workpiece into a conjugate image of the tool form. Moreover, the tool imposes a static pressure ranging from 1N to some kilograms depending on the size of the tool tip, Figure 2.21. The static pressure is necessary to sustain the tool feed during machining.

The process productivity is realized by the large number of impacts per unit time (frequency), whereas the accuracy is achieved by the small oscillation amplitude employed.

The tool tip, usually made of a relatively soft material, is also subjected to an abrasion action caused by the abrasives; thus, it suffers from wear which may affect the accuracy of the machined holes and cavities. Due to the fact that the tool oscillates and moves axially, USM is not limited to the production of circular holes. The tool can be made to the shape required, and hence extremely complicated shapes can be produced in hard materials. The process is characterized by the absence of any deleterious or thermal effects on the metallic structure of the workpiece. Outside the machining domain, ultrasonic (US) techniques are applied in non-destructive testing (NDT), welding, surface cleaning, as well as diagnostic and multi-medical applications. The USM process is characterized by the following advantages and disadvantages:

Advantages

- Intricate and complex shapes and cavities in electric or non-electric conductive materials can be readily machined ultrasonically.
- Since the tool exhibits no rotational movement, the process is not limited to produce circular holes.
- High dimensional accuracy and surface quality are the main features of USM.
- In particular, in the sector of electric non-conductive materials, the USM process is not in competition with other NTM processes regarding accuracy and removal rates.
- Since there is no temperature rise of the workpiece, no changes in physical properties or micro-structure, whatsoever, can be expected.

Disadvantages

- When machining electric conductive materials (except carbon), a limited material removal rate, as compared with ECM and EDM, is realized.
- USM is not capable of machining holes and cavities with a lateral extension of more than 25 to 30 mm with a limited depth of cut.

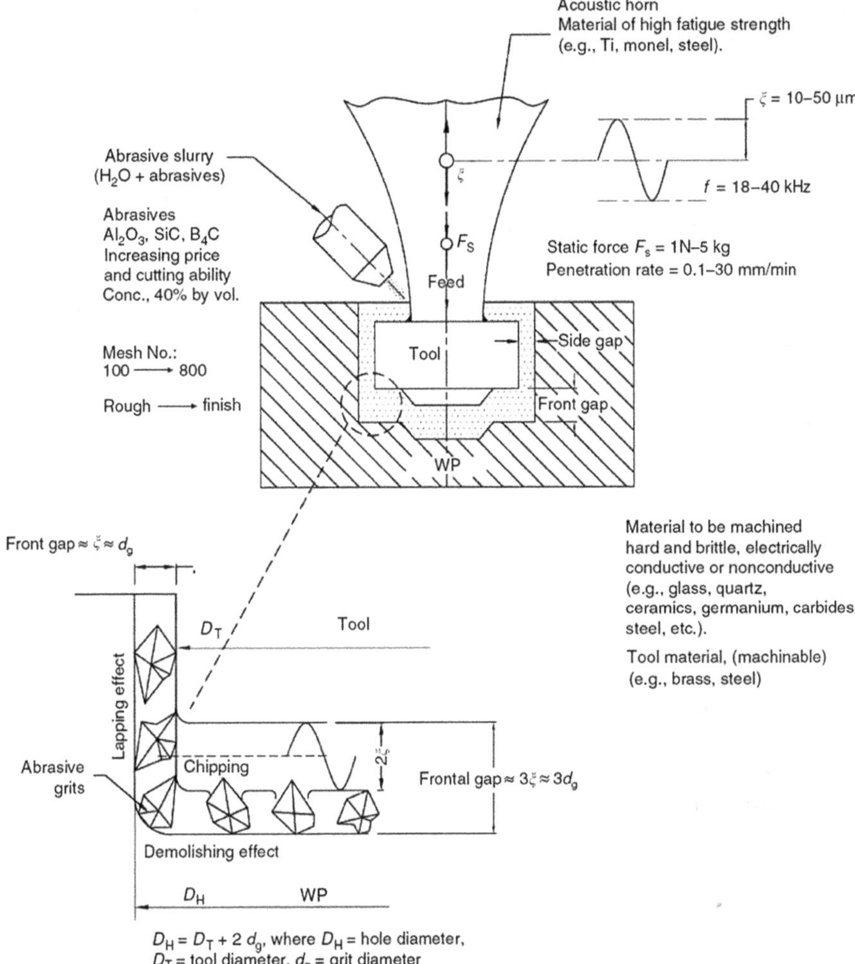

FIGURE 2.21 Characteristics of the USM process.

- The tool suffers excessive frontal and side wear in the case of machining conductive materials such as steels and carbides. The side wear destroys the accuracy of holes and cavities and leads to a considerable conicity.
- Every job needs a special tool of high cost which is added to the machining cost.
- USM is characterized by the high rate of power consumption.
- When machining through holes, the workpiece should be supported by a pad of machinable material to prevent breaking out.
- In case of blind holes, the designer should not allow sharp corners, because these cannot be produced by USM.
- The abrasive slurry should be regularly changed to get rid of worn abrasives, which means additional cost.

Applications: It should be understood that USM is generally applied to machine shallow cavities and forms in hard and brittle materials having a surface area less than 10 cm^2.

Some typical applications of USM are listed below.

- Manufacturing of forming dies in hardened steel and sintered carbides
- Manufacturing of wire drawing dies, cutting nozzles for jet machining applications in sapphire, and sintered carbides
- Slicing hard brittle materials such as glass, ceramics, and carbides
- Coining and engraving applications
- Boring, sinking, blanking, and trepanning
- Thread cutting in ceramics by rotating the tool or the workpiece

Figure 2.22 illustrates some products produced ultrasonically.

- Engraving a medal made of agate (König, 1990)
- Producing a fragile graphite electrode for EDM (König, 1990)
- Piercing and blanking of glass (König, 1990)
- Sinking a shearing die in hardened steel or WC (Lehfeld, 1967)
- Piercing slots and central cross in glass or glass/graphite composite (Kalpakjian, 1984)
- Drilling fine holes $\phi = 0.4$ mm in glass (Kalpakjian, 1984)
- Production outside contour and holes of master cutters made of zirconium oxide ZrO_2 of a textile machine (König, 1990)
- Drilling of thermocouple holes for measuring the temperature distribution in a cutting tip made of silicon nitride (SiN) for research purposes (König, 1990)

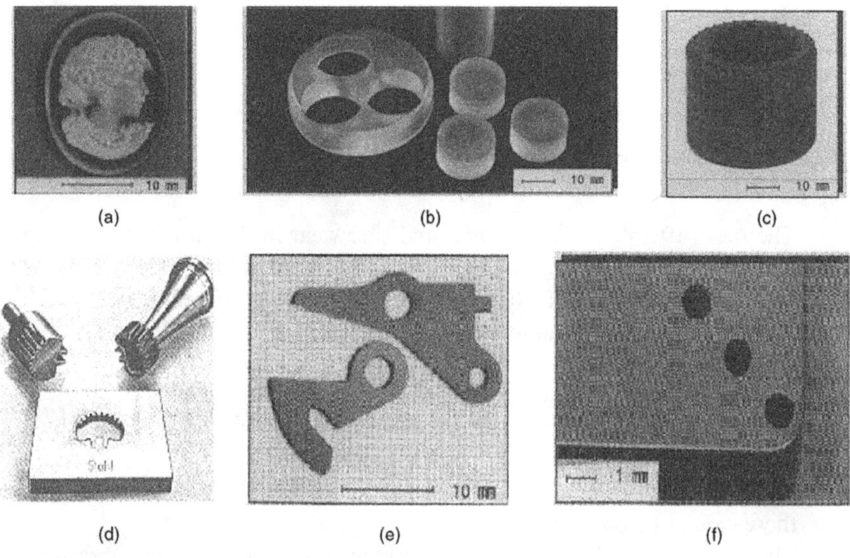

FIGURE 2.22 Typical products by USM.

2.2.2 USM Equipment

The USM equipment shown in Figure 2.23 has a table capable of orthogonal displacement in the X and Y directions, and a tool spindle carrying the oscillating system is moving in the direction of Z perpendicular to the plane X–Y. The machine is equipped with a high frequency generator 2 of a rating power of 600 watt, and a two-channel recording facility 3 to monitor important machining variables (tool displacement Z and oscillation amplitude ξ). A centrifugal pump is used to supplement the abrasive slurry into the working zone. Figure 2.24 shows schematically the main elements of the equipment which consist of the oscillating system, the tool feeding mechanism, and the slurry system.

Oscillating system and magnetostriction effect: The oscillating system comprises the transducer contained in the acoustic head 1, the primary acoustic horn 2, and the secondary acoustic horn 3, Figure 2.25.

1. **Acoustic Transducer**
   It transforms the electrical energy to mechanical energy in the form of mechanical oscillations. Magnetostrictive transducers are generally employed in USM, but piezo-electric ones may also be used.

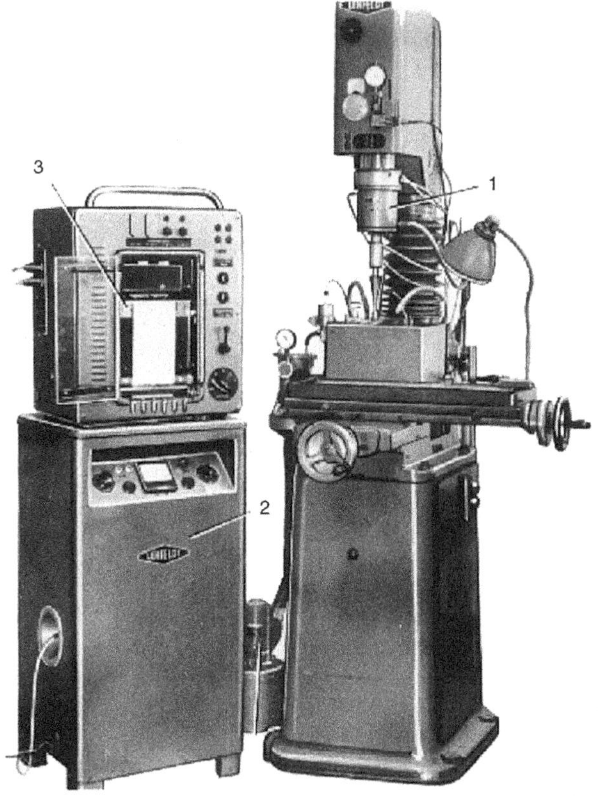

FIGURE 2.23 USM equipment (Lehfeld Works, 1967).

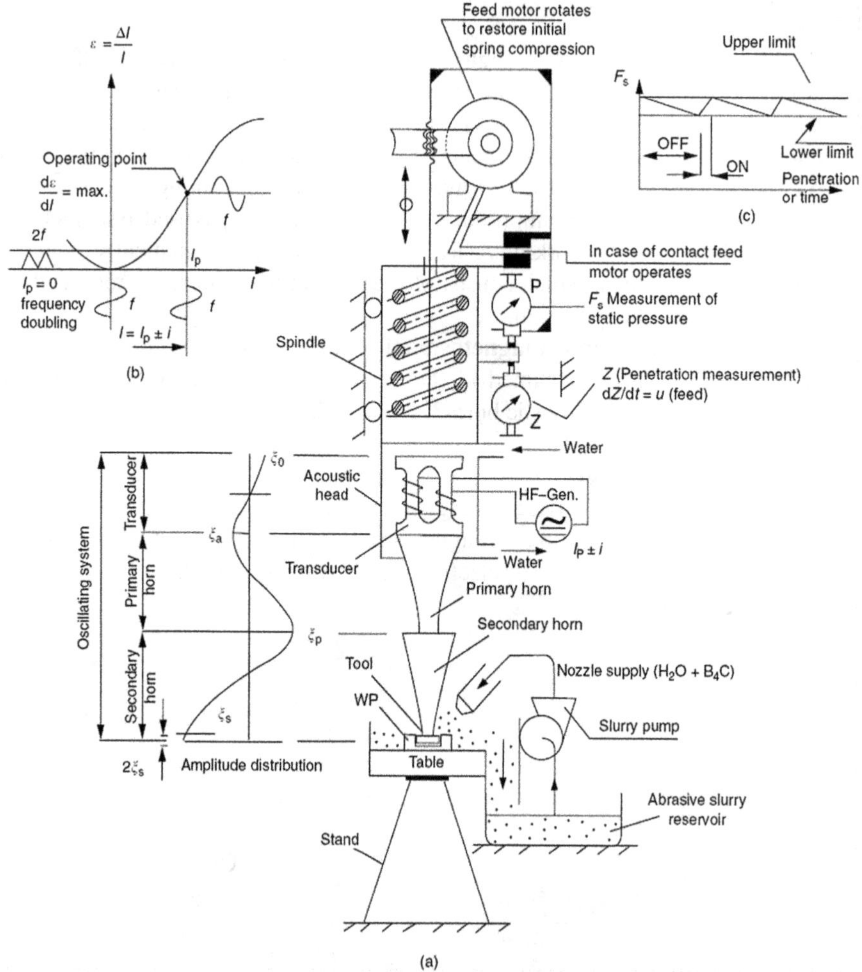

FIGURE 2.24 Schematic of complete vertical USM equipment.

a. ***Magnetostrictive transducer***: The magnetostriction effect was first discovered by Joule in 1874. According to this effect, in the presence of an applied magnetic field, ferromagnetic metals and alloys change in length. The deformation can be positive or negative depending on the ferromagnetic material. An electric signal of US frequency f_r is fed into a coil which is wrapped around a stack made of magnetostriction character (iron-nickel alloy). This stack is made of laminates to minimize eddy current and hysteresis losses; moreover, it must be cooled to dissipate the generated heat, Figure 2.24a. The alternating magnetic field produced by the high-frequency (HF) AC generator causes the stack to expand and contract at same frequency.

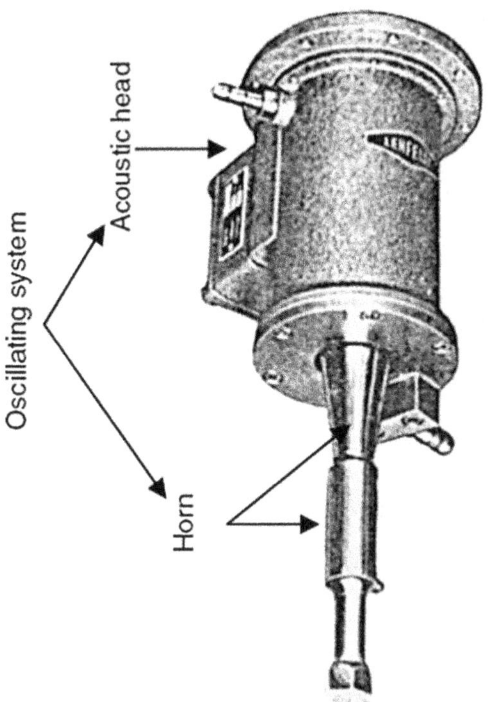

FIGURE 2.25 Oscillating system of USM equipment (Firma Dr. Lehfeldt & Co., 1967).

To achieve the maximum magnetostriction effect, the HF-AC current i must be superimposed on an appropriate dc premagnetizing current I_p that must be exactly adjusted to attain an optimum or working point. This point corresponds to the inflection point $\left(\frac{d\varepsilon^2}{dI^2}=0\right)$ of the magnetostriction curve, Figure 2.24b. Without the application of the premagnetizing direct current I_p, it is evident that the magnetostriction effect occurs in the same direction for a given ferromagnetic material irrespective of the field polarity, and hence the deformation will vary at twice the frequency $2f_r$ of the oscillating current providing the magnetic field, Figure 2.24b. Therefore, the premagnetizing direct current I_p has the following functions:

- When precisely adjusted, it provides the maximum magnetostriction effect (maximum oscillating amplitudes).
- It prevents the frequency doubling phenomenon.

If the frequency of the AC signal and hence that of the magnetic field is tuned to be the same as the natural frequency of the transducer (and the whole oscillating system), so that it will be at mechanical resonance, then the resulting oscillation amplitude becomes quite large and the exciting power attains its minimum value. For tuning purposes, the

HF ultrasonic generator is provided with a tuning system to provide a maximum amplitude for a given resonant frequency.

Transducer length: The resonance condition is realized if the transducer length l is equal to half of the wavelength λ (or positive integer number n of it).

Therefore,

$$\ell = \frac{n}{2}\lambda = \frac{\lambda}{2}, \quad \text{if } n = 1 \tag{2.1}$$

and

$$\lambda = \frac{c}{f_r} = \frac{1}{f_r}\sqrt{\frac{E}{\rho}} \tag{2.2}$$

where

c = acoustic speed in magnetostrictive material (m/s)

f_r = resonant frequency (1/s)

E, ρ = Young's modulus (MPa) and density (kg/m^3) of the magnetostrictive material

Hence,

$$\ell = \frac{c}{2f_r} = \frac{1}{2f_r} = \sqrt{\frac{E}{\rho}} \tag{2.3}$$

Characteristics of some magnetostrictive alloys: Since the magnetostrictive materials convert the magnetic energy to mechanical energy, a high coefficient of magneto-mechanical coupling k_r and coefficient of magnetostrictive elongation ε_{ms} are essential.

$$k_r = \left[\frac{\text{magnitude of mechanical energy}}{\text{magnitude of magnetic energy}}\right]^{1/2} \tag{2.4}$$

and

$$\varepsilon_{ms} = \frac{\Delta\ell}{\ell} \tag{2.5}$$

Alfer (13% Al, 87% Fe) is characterized by high coefficients, k_r and ε_{ms} as shown in Table 2.3.

b. ***Piezo-electric transducers***:

A main disadvantage of magnetostriction transducers is the high power loss (η = 55%). The power loss is converted into heat which necessitates the cooling of the transducer. In contrast, piezo-electric transducers are more efficient (η = 90%), even at higher frequencies (f = 25–40 kHz). Piezo-electric transducers utilize crystal like quartz which undergoes dimensional changes proportional to the voltage applied. Similar to

TABLE 2.3
Coefficients ε_{ms} and k_r of a Magnetostrictive Alloy

Magnetostrictive Alloy	$\varepsilon_{ms} \times 10^6$	k_r
Alfer (13% Al, 87% Fe)	40	0.28
Hypernik (50% Ni, 50% Fe)	25	0.20
Permalloy (40% Ni, 60% Fe)	25	0.17
Permendur (49% Co, 2% V, 49% Fe)	9	0.20

Source: (Kaczmarek, 1976).

magnetostrictors, the length of the crystal should be equal to half the wavelength of the sound in the crystal to produce resonant condition. At a frequency of 40 kHz, the resonant length l of the quartz crystal ($E = 5.2 \times 10^4$ MPa, $\rho = 2.6 \times 10^3$ kg/m^3) equal to 57 mm. Sometimes a polycrystalline ceramic transducer like barium titanate is used.

2. **Acoustic Horns (Mechanical Amplifiers or Concentrators)**
   The oscillation amplitude ξ_0 as obtained from the magnetostrictive transducer does not exceed 5 μm which is too small for effective removal rates. The amplitude at the tool should, therefore, be increased to practical limits of 40 to 50 μm, fitting one or more amplifiers into the output end of the transducer, Figure 2.25a.

The acoustic horn (concentrator) should have the following functions:

- Transmit the mechanical energy to the tool
- Amplify the amplitudes to practical limits
- Concentrate the power on a small machining area

In order to attain resonance, the acoustic horns, like transducers, should be half-wavelength resonators, whose terminals oscillate axially in an opposite direction relative to each other. The nodal points (points of zero amplitude $\xi_n = 0$) are little displaced toward the upper end in case of tapered concentrators. Figure 2.24a illustrates the amplitude distribution of the cascaded oscillating system along its longitudinal axis. Table 2.4 shows the amplitude and magnification factors of each oscillating element.

Accordingly, the overall magnification factor R_m of the system is given by:

$$R_m = R_{tr} \times R_P \times R_S = 1 \times \frac{\xi_p}{\xi_0} \times \frac{\xi_s}{\xi_p} = \frac{\xi_s}{\xi_0} \tag{2.6}$$

The acoustic head (the transducer and the primary acoustic horn) is delivered by the manufacturer as an integral part with the machine, Figure 2.25. The tool is attached to the free end of the secondary acoustic horn by threading, brazing, or press-fitting.

TABLE 2.4
Amplitudes and Magnification Factors of Individual Elements of the Cascaded Oscillating System Shown in Figure 2.24

Oscillating Element	Amplitude	Magnification Factor
Transducer	ξ_0	$R_{tr} = \xi_0/\xi_0 = 1$
Primary horn	ξ_p	$R_p = \xi_p/\xi_0$
Secondary horn	ξ_s	$R_s = \xi_s/\xi_p$

The oscillation amplitude of the primary horns is small enough, such that they are durable and not easily discarded.

Tool feeding mechanism: The tool feeding mechanism should perform the following functions:

- Bring the tool slowly to the workpiece
- Provide adequate static pressure and sustain it during cutting
- Decrease the pressure before the end of cut to eliminate sudden fracture
- Overrun a small distance to ensure the required hole size at the exit
- Retract the tool upwards rapidly after machining

Figure 2.24c illustrates an automatic tool feed mechanism, which operates precisely, through the application of roller frictionless guides. When the oscillating system is freely suspended (no contact between the tool and the workpiece), the static pressure on the workpiece equals zero. When machining starts, the tool comes into contact with the workpiece; the spring in the machine spindle expands giving a measure for the static pressure. The static force is indicated by the dial gauge *P*. As machining proceeds, the spring is compressed and the static force decreases, Figure 2.24c, until the contact switch is actuated allowing the feed motor to rotate, and rapidly recovers the set-value of static force. The dial gauge *Z* indicates the tool displacement.

2.2.3 Design of Acoustic Horns

The general differential equation for longitudinal oscillation of acoustic horns can be derived by considering the equilibrium of an infinitesimal element *dx* under the action of elastic and inertia forces, Figure 2.26.

$$\text{Elastic force} = F + \frac{\partial F}{\partial F}\mathrm{d}x - F \tag{2.7}$$

$$\text{Inertial force} = A(x)\cdot \mathrm{d}x \cdot \rho \cdot \frac{\partial^2 F}{\partial t^2} \tag{2.8}$$

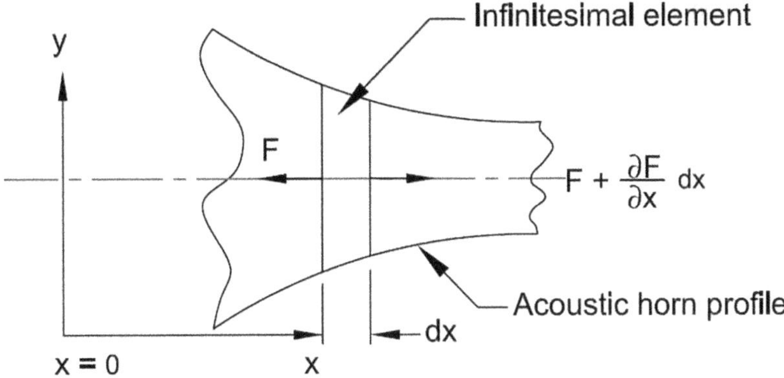

FIGURE 2.26 Equilibrium of infinitesimal element under the action of elastic and inertial forces.

where

$$F = \text{elastic force} = EA(x) \cdot \frac{\partial y}{\partial x}$$

$A(x)$ = shape function = cross-sectional area function of horn at axial position x
y = displacement, depending on x and $t = \xi \sin \omega t$
t = time
x = axial position as measured from the horn fixed end
ξ = oscillation amplitude = $f(x)$
ω = angular speed = $2\pi f$
f_r = ultrasonic frequency
E = Young's modulus of horn material
ρ = density of horn material

$$\frac{\partial y}{\partial x} = \text{strain} = f(x,t)$$

$$\frac{\partial^2 y}{\partial t^2} = \text{accleration} = f(x,t)$$

Equating the elastic force (Equation 2.7) and the inertia force (Equation 2.8), then

$$\frac{\partial F}{\partial x} \cdot \mathrm{d}x = A(x) \cdot \mathrm{d}x \cdot \rho \cdot \frac{\partial^2 y}{\partial t^2} \tag{2.9}$$

Substituting the values of F and y in Equation 2.9, the general differential equation becomes

$$\frac{d^2\xi}{dx^2}+\frac{d\ln A(x)}{dx}\cdot\frac{d\xi}{dx}+\left[\frac{\omega}{c}\right]^2\xi=0 \quad (2.10)$$

where$c=\sqrt{E/\rho}$ = acoustic speed in horn material.

The general, differential equation 2.10 can be solved after substituting the shape function $A(x)$. Four shape functions are available for acoustic horns, Figure 2.27. These are:

- Cylindrical stepped horn
- Exponential horn
- Conical horn
- Hyperbolic horn

The choice of the shape function $A(x)$ controls the magnification factor R_m, Figure 2.28. However, exponential and stepped types are frequently used; the conical and hyperbolic horns are difficult to design.

1. **Design of the Cylindrical Stepped Acoustic Horns, ($A(x) = c$)**
   Stepped horns are mainly employed in machining brittle materials such as glass, germanium, and ceramics, where there is no need to use high amplitudes. Accordingly, fatigue at nodal points due to stress concentration can be avoided. Moreover, stepped horns are easily designed and produced. Substituting $A(x) = c$ in the general differential Equation 2.10, then

$$\frac{d^2\xi}{dx^2}+\left[\frac{\omega}{c}\right]^2\xi=0 \quad (2.11)$$

   Figure 2.29 shows the amplitude, strain, and stress distributions of the stepped acoustic horn. Assuming f_r = 20 kHz, D_o / D_ℓ,= 5, the table in Figure 2.29 determines the resonant lengths for different horn materials. The magnification factor R_m can be calculated according to Equation 2.12.

$$R_m=\frac{\xi_\ell}{\xi_0}=\left(D_0/D_l\right)^2=25 \quad (2.12)$$

2. **Design of Exponential Acoustic Horns, [$A(x) = A_0\, e^{-2hx}$]**
   Exponential horns are mainly employed to machine hard and tough materials such as carbides and hardened steels, using large oscillation amplitudes without the risk of fatigue failure. They can be easily designed, and their contours can be easily produced on CNC lathes.

The area of an exponential horn varies according to the function,

$$A(x)=A_0e^{-2hx} \quad (2.13)$$

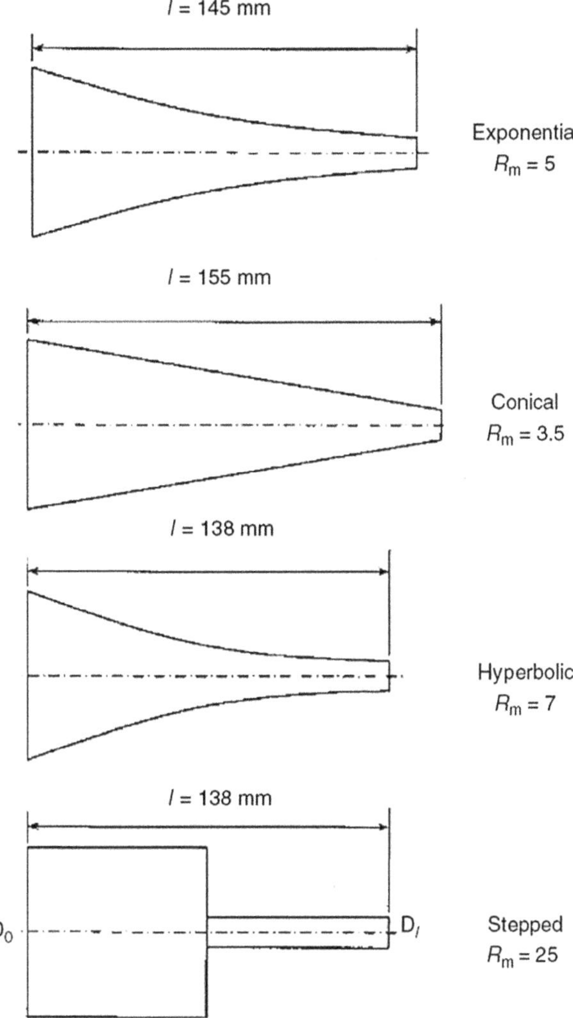

FIGURE 2.27 $\lambda/2$ – Resonators of different shape functions.

If the exponential horn has a circular cross-section, then

$$D(x) = D_0 e^{-hx} \tag{2.14}$$

where

$A(x), D(x)$ = cross-sectional area, and horn diameter at location x

A_0, D_0 = cross-sectional area, and horn diameter at $x = 0$

h = exponential ascent factor

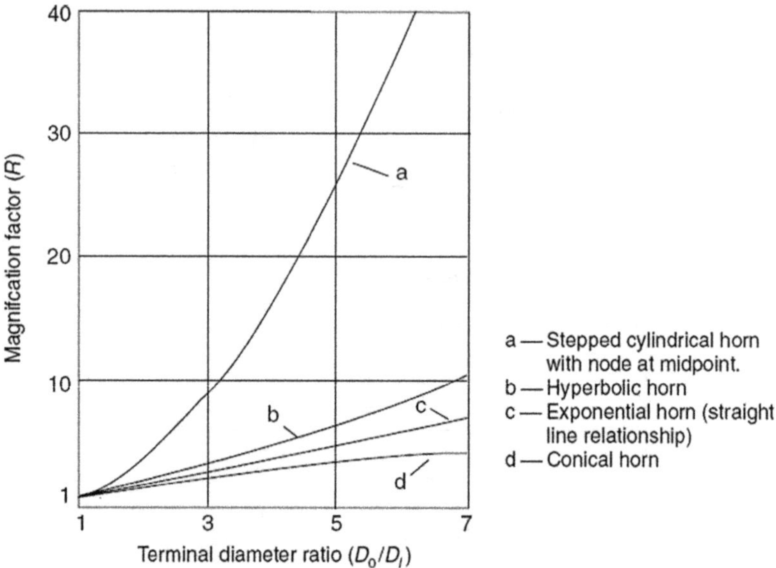

FIGURE 2.28 Effect of terminal diameter ratio and shape function on the magnification factor (Youssef, 2005).

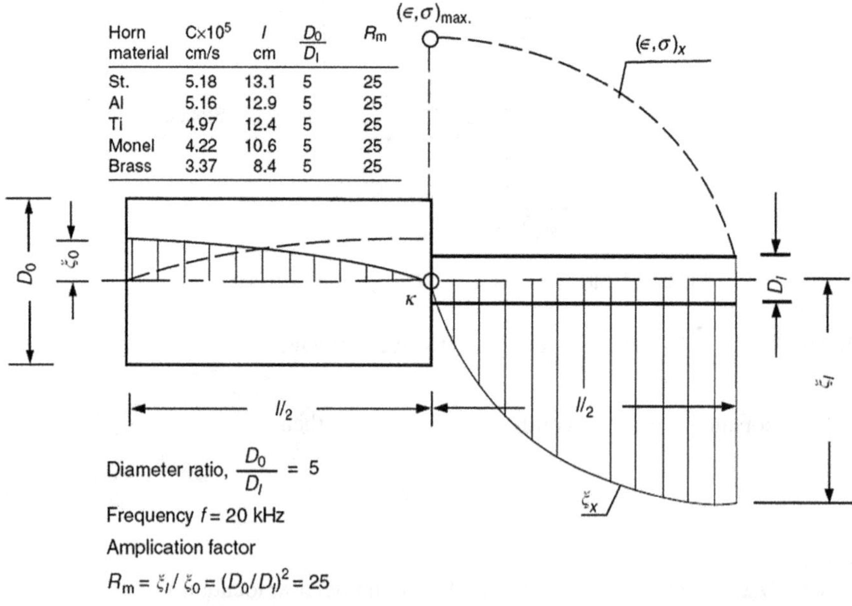

Horn material	$C \times 10^5$ cm/s	l cm	$\frac{D_0}{D_l}$	R_m
St.	5.18	13.1	5	25
Al	5.16	12.9	5	25
Ti	4.97	12.4	5	25
Monel	4.22	10.6	5	25
Brass	3.37	8.4	5	25

FIGURE 2.29 Amplitude, strain, and stress distribution of stepped acoustic horn (Youssef, 2005).

Substituting the value of $A(x)$ according to Equation 2.13 in the general differential Equation 2.10, the differential equation of exponential horn is obtained by

$$\frac{d^2\xi}{dx^2} - 2h \cdot \frac{d\xi}{dx^2} + \left[\frac{\omega}{c}\right]^2 \xi = 0 \tag{2.15}$$

from which the amplitude distribution ξ *is* given by

$$\xi = -\xi_0 \sqrt{1 + \left(\frac{h\ell}{\pi}\right)^2} \cdot e^{hx} \cdot \sin\left(\sqrt{\left(\frac{\omega}{c}\right)^2 - h^2} \cdot x - \hat{\varphi}\right) \tag{2.16}$$

where

ξ_0 = amplitude at $x = 0$
ℓ = horn length
$\hat{\varphi}$ = arc tan $(\pi/h\ell)$

Resonance condition:

$$\sqrt{\left[\frac{\omega}{c}\right]^2 - h^2} = \frac{\pi}{\ell} \tag{2.17}$$

Horn length ℓ:

$$\ell = \frac{\pi}{\sqrt{\left[\frac{\omega}{c}\right]^2 - h^2}} \tag{2.18}$$

Ascent factor h:

$$h = \frac{\omega}{c} / \sqrt{1 + \left[\frac{\pi}{h\ell}\right]^2} \tag{2.19}$$

The horn length ℓ can also be expressed by Equations 2.20 and 2.21:

$$\ell = \frac{c'}{2f_r} = \frac{c}{2f_r} \cdot \frac{1}{\sqrt{1 - (hc/\omega)^2}} \tag{2.20}$$

$$\ell = \frac{c}{2f_r} \sqrt{1 + \left[\frac{\ln D_0 / D_t}{\pi}\right]^2} \quad \text{for circular exponential horns} \tag{2.21}$$

where c' is the modified acoustic speed in horn material.

Figure 2.30 shows the distribution of the oscillation amplitude along the axis of an exponential acoustic horn, where ξ_0 is the amplitude at $x = 0$, and ξ_ℓ is the amplitude

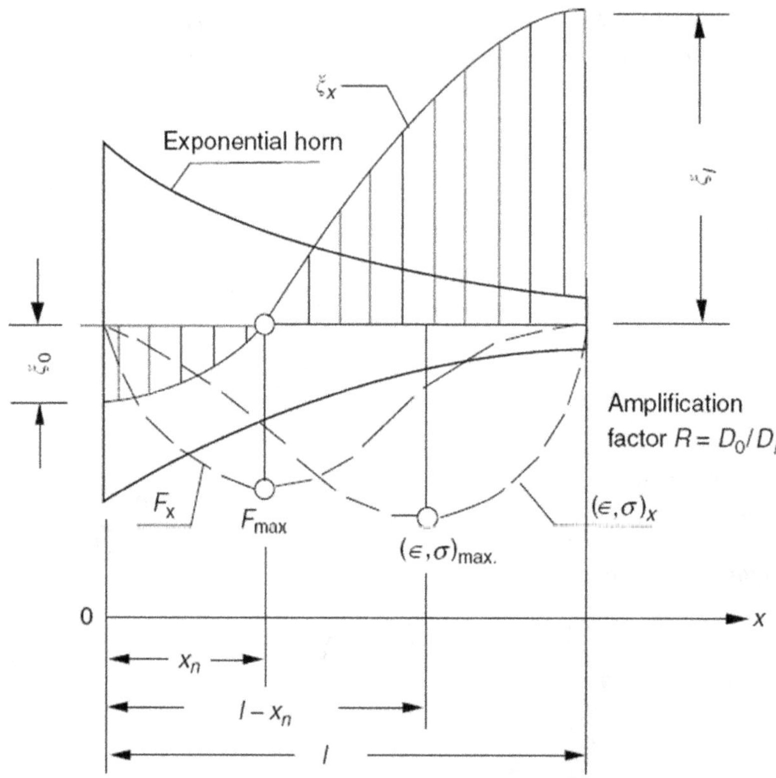

FIGURE 2.30 Amplitude, strain, and stress distribution along the axis of an exponential acoustic horn (Youssef, 2005).

at $x = \ell$ (tool amplitude). Figure 2.31a illustrates a monogram that determines the resonant length of a circular exponential horn in terms of the acoustic speed c, frequency f_r, and diametric ratio of its terminals D_o / D_ℓ.

Magnification Factor R_m

Referring to Equation 2.16, then

$$R_m = \left|\frac{\xi_\ell}{\xi_0}\right| = \frac{e^{hl}\sin\pi - \hat{\varphi}}{\sin\hat{\varphi}} = e^{hl} \tag{2.22}$$

From Equation 2.13

$$A_\ell = A_0 e^{-2hl}$$

$$\sqrt{\frac{A_0}{A_\ell}} = e^{hl} \tag{2.23}$$

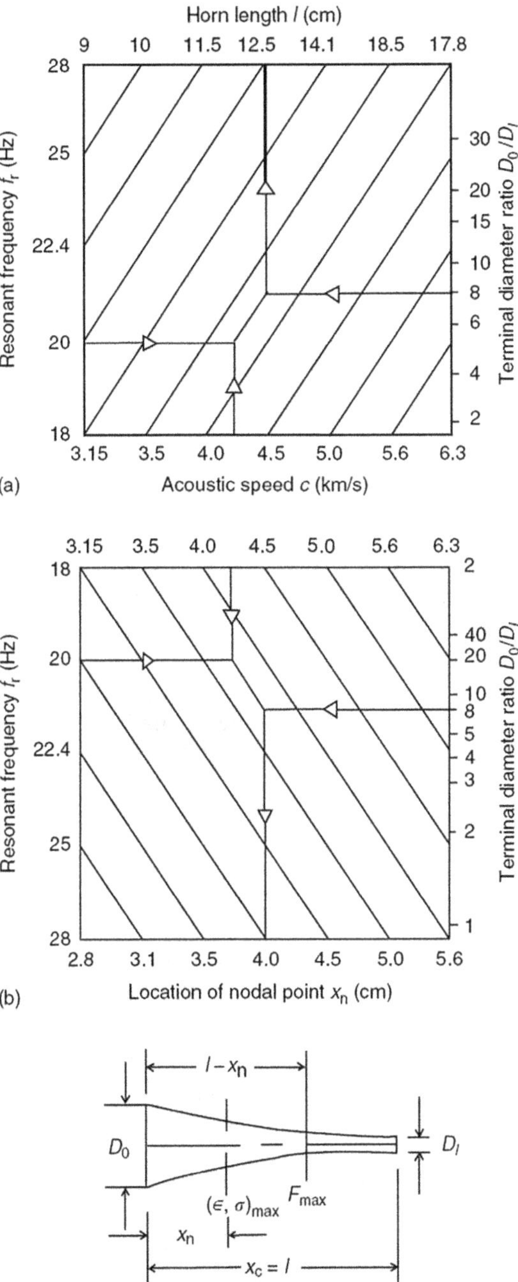

FIGURE 2.31 Nomogram to determine the length and nodal point of exponential acoustic horn (Blanck, 1961).

From Equations 2.22 and 2.23, the magnification factor R_m of non-circular acoustic horns is given by

$$R_m = e^{hl} = \sqrt{\frac{A_0}{A_\ell}} \tag{2.24}$$

and for circular acoustic horns

$$R_m = e^{hl} = \frac{D_0}{D_\ell} \tag{2.25}$$

From the foregoing discussion, it can be concluded that:

- The magnification factor R is independent of horn material.
- In case of exponential circular acoustic horn, the magnification factor R depends upon diameter ratio of its terminals D_o / D_ℓ
- For a given diameter ratio, D_o / D_ℓ, the stepped horn possesses the highest magnification factor $R_m = (D_o / D_\ell)^2$, followed by the hyperbolic, the exponential, and finally the conical type, Figure 2.28.

Determination of the Nodal Point x_n

It is the point of zero amplitude ($\xi_{xM} = 0$). It is important to determine this point exactly, in order to eliminate damping of the oscillating system when the primary horn is suspended in the acoustic head, Figure 2.24a. Substituting $\xi_{xn} = 0$ in Equation 2.16, it follows that,

$$x_n = \frac{\ell}{\pi} \cdot \hat{\varphi}$$

$$= \frac{\ell}{\pi} \cdot \tan^{-1} \frac{\pi}{\ln(D_0 / D_\ell)} = \frac{c}{2\pi f} \cdot \sqrt{1 + \left[\frac{\ln(D_0 / D_\ell)}{\pi}\right]^2} \cdot \tan^{-1} \frac{\pi}{\ln(D_0 / D_\ell)} \tag{2.26}$$

The nodal point x_n of the exponential horns can be determined from the monogram in Figure 2.31b in terms of the same parameters c, f_r, and D_o / D_ℓ.

Distributions of strain $\mathcal{E}_s$, stress σ, and force F along the axis of exponential horn

$$\varepsilon_x = \frac{d\xi}{dx} = \xi_0 h \sqrt{\left[1 + \left(\frac{hl}{\pi}\right)^2\right]\left[1 + \left(\frac{\pi}{hl}\right)^2\right]} \cdot e^{hx} \cdot \sin \frac{\pi}{\ell} x$$

$$\sigma = E \Delta \varepsilon_x \tag{2.27}$$

$$= E \xi_0 h \sqrt{\left[1 + \left(\frac{hl}{\pi}\right)^2\right]\left[1 + \left(\frac{\pi}{hl}\right)^2\right]} \cdot e^{hx} \cdot \sin \frac{\pi}{\ell} x \tag{2.28}$$

$$F_x = A_{(x)} \cdot \sigma_x$$

$$= E\xi_0 h A_0 \sqrt{\left[1+\left(\frac{hl}{\pi}\right)^2\right]\left[1+\left(\frac{\pi}{hl}\right)^2\right]} \cdot e^{hx} \cdot \sin\frac{\pi}{\ell}x \qquad (2.29)$$

Figure 2.30 shows the distribution of ε_s, σ, and F along the axis of the exponential acoustic horn.

Selection of Horn Material

Finally, the maximum induced stress as expressed by Equation 2.30 occurs at the nodal point, Figure 2.30. It should be less than the allowable fatigue strength of the horn material.

$$\sigma_{max} = E\xi_0 h \sqrt{\left[1+\left(\frac{hl}{\pi}\right)^2\right]\left[1+\left(\frac{\pi}{hl}\right)^2\right]} \cdot e^{hl\left(1-\frac{\hat{\varphi}}{\pi}\right)\sin(\pi-\hat{\varphi})} \qquad (2.30)$$

Other Shapes of Exponential Acoustic Horns

The same equations are valid for acoustic horns of internal exponential form shown in Figure 2.32a. These are used for machining large holes. Another horn is shown in Figure 2.32b with external exponential form and adapted with internal conical form to allow machining of the largest holes.

For machining of rectangular form or slitting, a horn of rectangular cross-section, of constant width w_0, while its thickness $t(x)$ varies according to an exponential function is recommended, Figure 2.32c.

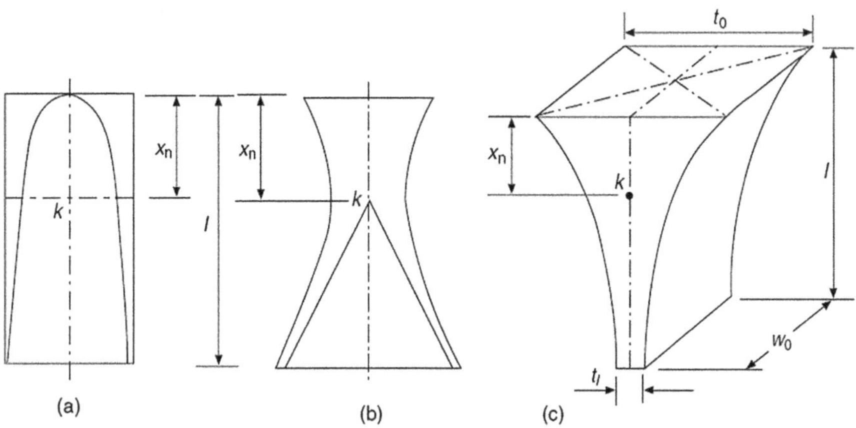

FIGURE 2.32 Different shapes of exponential acoustic horns (Youssef, 2005).

Therefore,

$$A(x) = A_0 e^{-2hx}$$

$$w_0 t(x) = w_0 \cdot t_0 e^{-2hx}$$

$$t(x) = t_0 e^{-2hx} \tag{2.31}$$

Illustrative Example 1

Use the chart in Figure 2.31 to design an exponential acoustic horn made of monel (acoustic speed in horn material $c = 4.22 \times 10^5$ cm/sec). Its natural frequency f_r = 20 kHz, and its terminal diameters are D_0 = 40 mm and $D_\ell = 5$ mm.

SOLUTION

Horn length:

Referring to Figure 2.31a,

$$D_0 / D_\ell = 8, \; c = 4.22 \,\text{km/s}, \; f_r = 20 \,\text{kHz},$$

$$\text{then, } \ell = 12.5 \text{ cm}$$

Location of nodal point x_n:

Referring to Figure 2.31b,

$$D_0 / D_\ell = 8, \; c = 4.22 \,\text{km/s}, \; f_r = 20 \text{kHz},$$

$$\text{then, } x_n = 4 \text{ cm}$$

Magnification factor R_m:

$$R_m = D_0 / D_\ell = 40 / 5 = 8$$

Exponential ascent factor h:

$$R_m = e^{hl}$$

$$8 = e^{12.5h}$$

from which

$$h = 0.166 \frac{1}{cm}$$

Determination of horn contour:

$$D_x = D_0 e^{-hx}$$

$$= 40 e^{-0.166x}$$

x (cm)	0	2	$x_n=4$	7	10	12.5
$D_{(x)}$(mm)	40	28.7	20.6	12.5	7.6	5.0

Illustrative Example 2

The same acoustic horn as in Example 2.1 is required to be designed (ℓ = 12.5 cm, R_m = 8), but it should be provided by a conical cavity as illustrated in Figure 2.33, to accommodate a tool of 20 mm diameter.

SOLUTION

Zone I (x = 0–4 cm):

The horn contour is exactly the same as in Example 1.

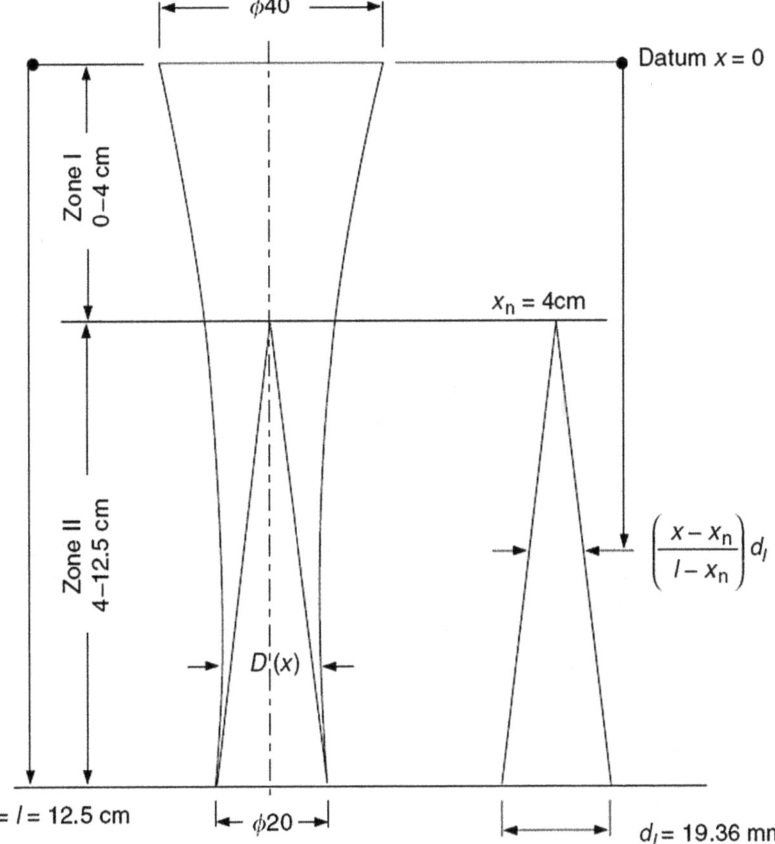

FIGURE 2.33 Exponential acoustic horn with conical cavity.

Zone II (x = 4–12.5 cm)

$$A_{(x)} = A_0 e^{-2hx}$$

$$D_{(x)}^2 - \left(\frac{x - x_n}{\ell - x_n} d_\ell\right)^2 = D_0^2 e^{-2hx}$$

Therefore,

$$D_{(x)} = \sqrt{D_0^2 e^{-2hx} + \left(\frac{x - x_n}{\ell - x_n} d_\ell\right)^2}$$

d_l can be calculated from the equation

$$d_l = \sqrt{(D_l)_{hollow} - (D_l)_{solid}}$$

$$= \sqrt{(20)^2 - (5)^2} = 19.36 \text{ mm}$$

If x = 7 cm, then

$$D_7 = \sqrt{(40)^2 e^{-2\times0.166\times7} + \left(\frac{7-4}{12.5-4} \times 1936\right)^2} = 14.24 \text{ cm}$$

The following table illustrates the horn diameters at different lengths.

	Zone I			Zone II		
x (cm)	0	2	x_n=4	7	10	12.5
$D_{(x)}$(mm)	40	28.7	20.6	14.25	15.63	20

Illustrative Example 3

For a given USM equipment, operating at a resonant frequency f_r = 20 kHz, it is required to design an exponential acoustic horn, of ball bearing steel 100 Cr 6 ($c = 5.05 \times 10^5$ m/s), provided with a 3 mm diameter hole for the suction of the abrasive slurry from the nodal point and another one for the fixation in the primary horn. Assume D_0 = 39 mm, and fixation hole diameter d_c = 16 mm and depth 18 mm, Figure 2.34.

SOLUTION

Magnification factor:

$$R_m = e^{hl} = \sqrt{\frac{A_0}{A_\ell}} = \sqrt{\frac{39^2 - 16^2}{8^2 - 3^2}}$$

$$= 4.8$$

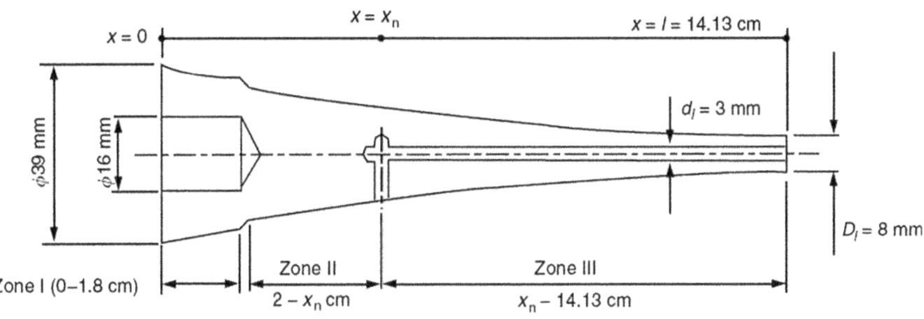

FIGURE 2.34 Exponential acoustic horn with suction and fixation holes.

Ascent factor h and horn length:

$$h = \frac{\omega}{c} \cdot \frac{1}{\sqrt{1 + (\pi / hl)^2}}$$

$$hl = \ln R = \ln 4.8 = 1.57$$

$$h = \frac{2\pi \times 20000}{5.05 \times 10^5} \cdot \frac{1}{\sqrt{1 + (\pi / 1.57)^2}}$$

$$= 0.111 \frac{1}{\text{cm}}$$

$$\ell = \frac{1.57}{0.111} = 14.13 \text{ cm}$$

Nodal point $X_{\hat{n}}$:

$$x_{\hat{n}} = \frac{\ell}{\pi} \tan^{-1} \pi / \ln\left(D_0 / D_l\right) = 4.98 \text{ cm}$$

Determination of horn contour:

Zone I ($x = 0$–1.8 cm)

$$D_{(x)}^2 - d_c^2 = \left(D_0^2 - d_c^2\right) e^{-2hx}$$

$$D_{(x)} = \sqrt{d_c^2 - \left(D_0^2 - d_c^2\right) e^{-2hx}}$$

$$d_c = 16 \text{ mm}, \; D_0 = 39 \text{ mm}, \; h = 0.111 \frac{1}{\text{cm}}$$

Zone II ($x = 2 - x_n$)

$$D_{(x)}^2 = \left(D_0^2 - d_c^2\right)e^{-2hx}$$

$$D_{(x)} = \sqrt{\left(D_0^2 - d_c^2\right)e^{-hx}}$$

Zone III ($x = x_n - \ell$)

$$D_{(x)}^2 - d_\ell^2 = \left(D_0^2 - d_c^2\right)e^{-2hx}$$

$$D_{(x)} = \sqrt{d_\ell^2 + \left(D_0^2 - d_c^2\right)e^{-2hx}}$$

$$d_\ell = 3 \text{ mm}$$

The following table illustrates the horn diameters at different zones and lengths.

	Zone I			Zone II			Zone III				
x (cm)	0	1	1.8	2	4	4.98	6	8	10	12	14.13
$D_{(x)}$(mm)	39	35.6	33.2	28.5	22.8	20.5	18.5	14.9	12.1	9.9	8.0

2.2.4 Process Capabilities

Stock removal rate: It seems that the dominant factor involved in USM is the direct hammering of the abrasive grains, caused by the oscillating tool. Therefore, the stock removal rate (SRR) depends mainly upon:

- The work material
- Amplitude and frequency of tool oscillation
- The abrasive size and type
- The static pressure
- Abrasive concentration (mixing ratio) in the slurry

Last but not least, the efficiency of the slurry supplement in the working gap affects the stock removal rate considerably. The conventional method of supplying the abrasive slurry is the nozzle supply system a, Figure 2.35, in which the slurry is directly supplied at the oscillating tool. Pumping in or suction from the working gap through a central hole in the horn are found to be more effective in regime b; Figure 2.35c shows schematically a comparison between regimes a and b, regarding the penetration rate u. In the nozzle supply regime a, the penetration rate is much less than that in regime b. Moreover, in regime a the penetration rate decreases continuously with the hole depth, whereas in regime b it remains unaffected by the hole depth.

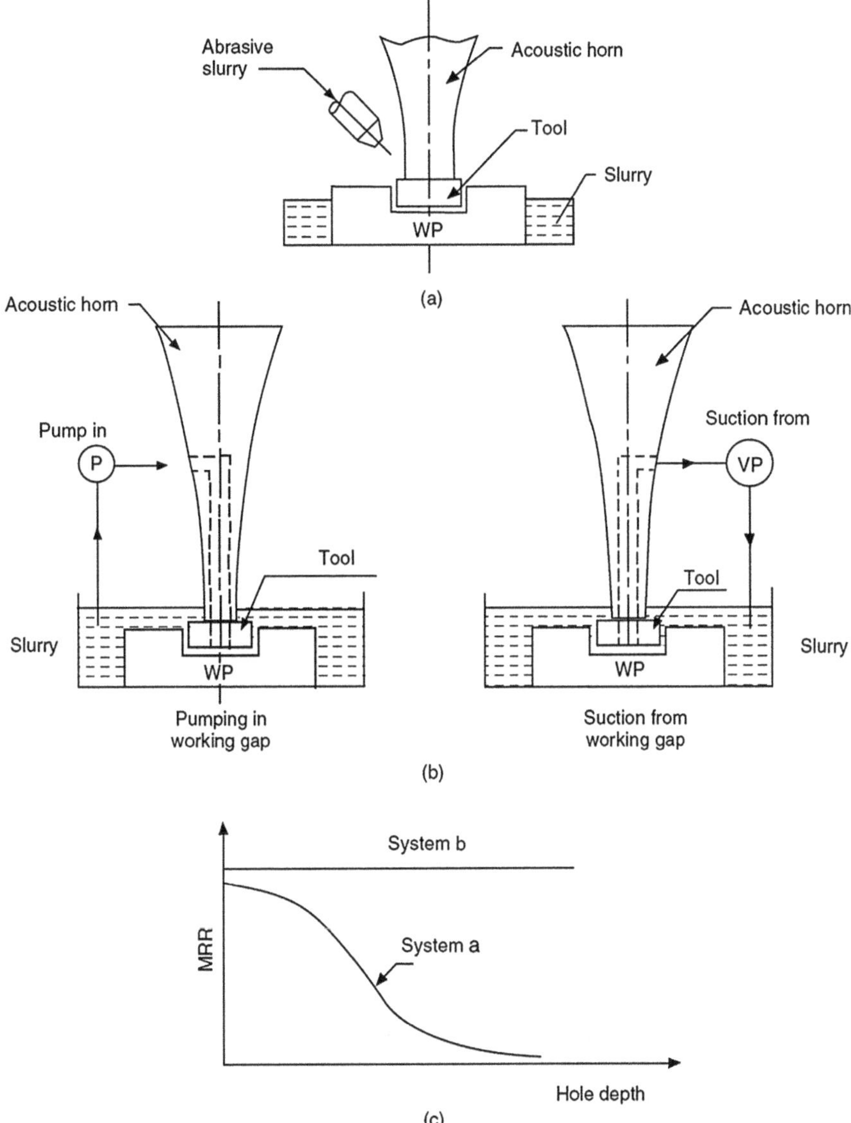

FIGURE 2.35 Slurry supplying system: (a) Nozzle supply system, (b) pumping in or suction from working gap, and (c) comparison of MRR of different supply systems.

Regarding the work material, the specific removal rate is affected by the ratio of the tool hardness to the workpiece hardness. The higher the ratio, the lower the material removal rate will be. That explains why soft and tough materials are recommended for USM tools (El-Hofy, 2005). The highest machining rates are realized when machining brittle materials such as glass, quartz, ceramics, and germanium, whereas the lowest machining rates are expected when machining hard and tempered steels and carbides.

The USM process is not applicable for soft and ductile materials, such as copper, lead, ductile steels, and plastics, which absorb energy by deformation. Moreover, some of the particles become embedded in soft faces of the work material and the cutting action is further retarded. In practice, oscillation amplitude is mainly selected with reference to the size of abrasive grits used. It should be selected to be approximately the same as the grit size. SRR increases with increasing oscillation amplitude (or abrasive grit size). The maximum amplitude value is governed by the maximum allowable strength of the material from which the acoustic horn is designed.

The removal rate increases with the frequency of the oscillating system. However, the frequency is constant and exactly equal to the natural frequency of the system, and hence the frequency is not considered a factor. The specific removal rate increases with the applied static pressure. It attains a maximum value after which it decreases with a further increase of the static pressure. As previously mentioned, two types of abrasives are commonly used in USM; these are B_4C and SiC. B_4C is more expensive, however it is economically recommended for USM due to its increased cutting ability and resistance to abrasion. Moreover, grits of B_4C have less specific gravity and, hence, more capability to suspend in the slurry as compared with SiC. It is found that the maximum SRR is achieved if a slurry mixing ratio (abrasives/water) of 40% by volume is used.

Accuracy and surface quality: Factors affecting the accuracy and surface quality of holes and cavities produced ultrasonically are:

- Work material
- Tool material and tool design
- Oscillation amplitude and grain size of abrasives
- Hole depth and machining time
- Cavitations effect

A main feature of the USM operation is that the abrasives start to cut for themselves a side way between the tool and the workpiece (side gap) in order to move through it downwards to the frontal gap, in which the material removal takes place, Figure 2.21.

From the foregoing discussion, it is understood that the ultrasonically produced holes are somewhat larger than the tool used by a certain oversize (overcut) which approximately equals the size of the abrasive grains used. This oversize is affected more or less by the machining time, which in turn depends upon the depth of hole, material of workpiece, tool land (tool design), as well as the other machining parameters. Furthermore, it should be emphasized that the hole accuracy does not mean the hole oversize. It means the repeatability of the oversize. In USM, it is good practice to perform a pilot test at the desired machining conditions, from which the actual oversize is precisely determined, and accordingly the tool diameter is calculated. Tolerances of ± 25 μm can be easily obtained by USM. However, it is possible to obtain tolerances as close as ± 5 μm in the case of some provisions being taken.

Deep holes, especially those produced in difficult-to-machine materials, suffer from considerable conicity depending on the depth of hole. This conicity is due to the side tool wear and the prolonged machining time.

The wall roughness of ultrasonically machined holes is mainly governed by:

- The material to be machined
- The roughness is larger when machining brittle materials such as glass and germanium
- The oscillation amplitude and the grain size
- The roughness increases with increasing oscillation amplitude and grain size of abrasives used

The surface quality is deteriorated if cavitation conditions are prevailing. From this point of view, the use of pumping regime in the working gap is preferred to using the suction arrangement. Moreover, the rotation of the workpiece in case of circular holes may improve the surface quality and the hole roundness.

2.2.5 Recent Developments

1. **Contouring USM**
   When sinking a three-dimensional cavity by conventional USM, a form tool is used, which is generally complex and costly, Figure 2.36a. The same cavity, however, can be produced by what is called contouring USM, Figure 2.36b, in which the machining is implemented by a simple tool in accordance with a tool path which is determined by CNC facility.
2. **Rotary Ultrasonic Machining**
   A modified version of USM is shown in Figure 2.37 where a tool bit is rotated at 5000 rpm against the workpiece. The process is, therefore, called rotary ultrasonic machining (RUM). This process is used for machining non-metallic materials such as glass, ceramic, carbides, ferrite, quartz, zirconium oxide, ruby, sapphire, beryllium oxide, and some composites. RUM ensures high removal rate, lower tool pressures for delicate and fragile parts, less breakout of through holes, and improved aspect ratio. When machining small holes, RUM allows non-interrupted drilling, while conventional drilling necessitates a tool retraction, which increases machining time.

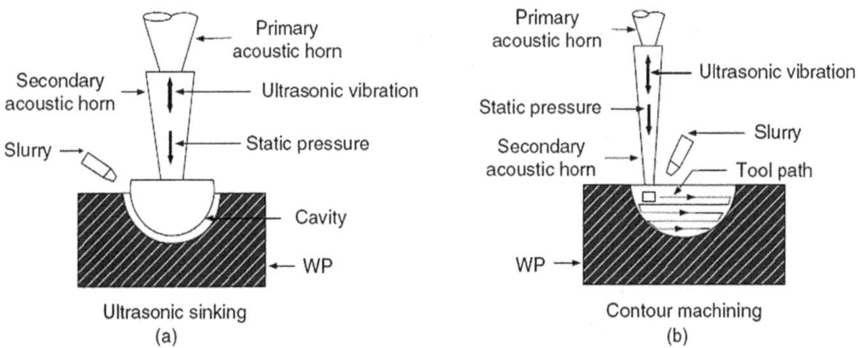

FIGURE 2.36 Sinking (conventional) and contouring USM: (a) Ultrasonic sinking and (b) contour machining (El-Hofy, 2005).

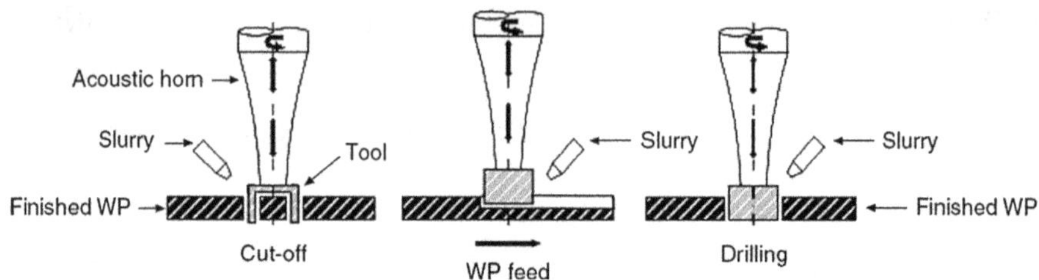

FIGURE 2.37 Rotary ultrasonic machining (RUM) (El-Hofy, 2005).

2.3 ABRASIVE FLOW MACHINING

2.3.1 PRINCIPLES

Abrasive flow machining (AFM) is a purely mechanical NTM process that finishes surfaces and edges by extruding a viscous abrasive media flowing, under pressure, through or across a workpiece. This process provides a high level of surface finish and close tolerances with an economically acceptable rate of surface generation for a wide range of industrial components. A common setup is to position the work part between two opposing cylinders, one containing media (visco-elastic polymer) and the other empty. The polymer has the consistency of putty, which is forced under pressures ranging between 0.7 and 20 MPa to flow through the part from the first cylinder to the other, and then back again, as many times as necessary to achieve the desired material removal and finish, Figure 2.38. Many types of abrasives are used in AFM. Mostly B_4C and SiC, however, A12O3 or diamond abrasives are sometimes used. Most abrasive action occurs during the process if the hole changes size or direction.

Media viscosity in AFM is directly related to the size of the restricting passage of the workpiece to be abraded. Fundamentally, viscosity must be high enough to maintain extrusion-type flow, and high enough to hold the abrasive grains at the outermost surface of the slug with sufficient force to allow abrasives to cut edges and surfaces of the restricting passage. At the same time, the media must be soft enough to flow at a reasonable rate to perform the operation in acceptable cycle time. Machining action using AFM is gentle and continuous, and the debris or chips are retained in the media, which can tolerate as much as 10% of its volume in chips and impurities. The media life is limited at the point at which a substantial number of the grains become dull. Reconditioning of the media can be accomplished by additions of abrasive grit and lubricant.

AFM is not a mass material removal process, but it is particularly useful for polishing or de-burring inaccessible internal passages and hard-to-reach locations. Materials from soft aluminum to tough Ni-base super alloys are processed with AFM. Holes of diameter smaller than 0.4 mm are sometime difficult to process with AFM. Polishing of blind holes is impractical because AFM requires flowing media. AFM is capable of removing EDM, EBM, and LBM heat-damaged zones and significantly improving surface roughness. Moreover, it is capable of inducing high compressive residual stresses to the machined surface, in a very thin sub-layer of about 10 μm, and so proves that AFM offers an alternative finishing process, benefiting from the surface integrity and productivity point of view (Kenda, Pusavec, Kermouche, and Kopac, 2011).

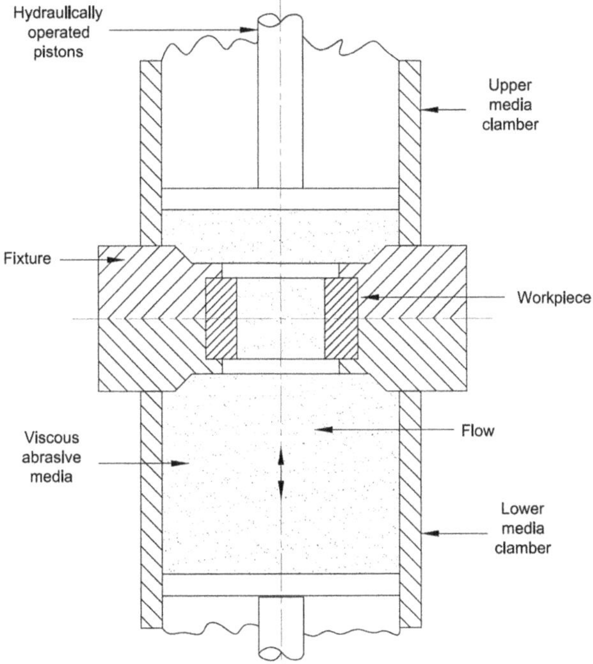

FIGURE 2.38 AFM schematic.

Table 2.5 illustrates the designations of the abrading media for AFM regarding their viscosity grades, and the types and grit size of abrasives. The same table includes three grit sizes of Al_2O_3. Table 2.6 shows the grit (Mesh #) and the corresponding grain size in μm. In AFM, it is recommended to use different sizes of abrasives. Table 2.7 is a guide for selecting the viscosity of the abrading media in AFM, depending on the passage widths (for a length to width ratio = 2:1).

AFM is characterized by the following advantages:

- AFM can be applied to any difficult-to-cut material.
- Material can be removed from targeted and hard-to-reach locations.
- Media can be engineered to match the application requirements.
- AFM improves air, gas, or liquid flow behavior, reduces cavitations tendencies, and generating desirable laminar flow.
- More uniform de-burring than that which could be accomplished with hand tools.

2.3.2 Process Parameters of Abrasive Flow Machining

The efficiency of the AFM process depends on a number of process parameters, which can be classified in three groups, namely,

1. The polishing media parameters (viscosity, abrasive material, abrasive mesh, abrasive concentration and temperature)

TABLE 2.5

Designations of the Abrading Media for AFM Regarding their Viscosity Grades, and the Types and Grit Size of Abrasives (*Machining Data Handbook*, 1980)

Viscosity Grade or Base:

LV	or 500 – low
LMV	or 300 – low-medium
MV	or 200 – medium
HMV	or 100 – high-medium
HV	or 50 – high

Grit Type:

A – Aluminum oxide
B – Boron carbide
C – Silicon carbide (or S in some manuf. nomenclatures)
D – Diamond (in certain cases)

Media Nomenclature:

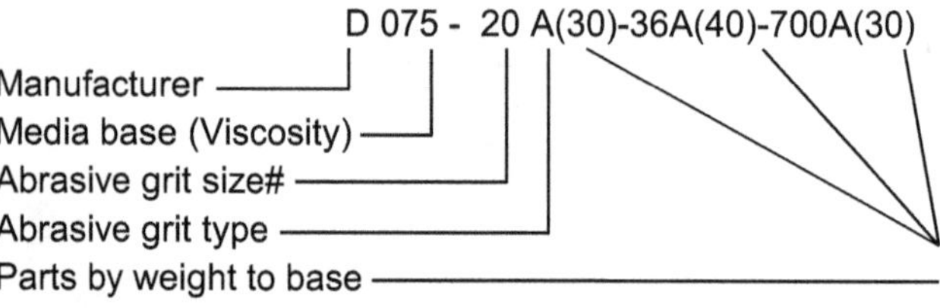

TABLE 2.6

Grit (Mesh #) and the Corresponding Grain Size of Abrasives Used in AFM

Fine		Medium		Coarse	
Mesh #	**Aver. Size µm**	**Mesh #**	**Aver. Size µm**	**Mesh #**	**Aver. Size µm**
1000	8	270	44	90	216
800	11	200	66	70	328
500	19	170	86	50	432
400	23	120	142	35	710
325	33	100	173	25	1035
				20	1340

TABLE 2.7
A Guide for Selecting the Viscosity of the Abrading Media in AFM

Passage Width		Extrusion Pressure				
in	**mm**		**Low**		**to**	**High**
1/64	0.4	——————	——————	——————	——————	LV
1/32	0.8	——————	——————	——————	LV——	LMV
1/16	1.6	——————	——	LV——	LMV——	MV
1/8	3.2	——	LV——	LMV——	MV——	HMV
1/4	6.4	——	LMV——	MV——	HMV——	HV
1/2	12.8	——	MV——	HMV——	MV	
1	25.4	——	HMV——	HV		
2	50.8	——	HV			

Remark: Viscosity grades for passage ways with 2:1 length to width ratio.

2. The AFM process parameters (pressure, volume flow, number of cycles, and machining time)
3. The workpiece parameters (material, hardness, roughness, pre-machining process, texture orientation, and workpiece shape)

The performance of the process as reflected in its material removal rate (MRR), accuracy, and surface finish of the AFM machined part depends upon the following main parameters, which can be accurately controlled to suit any machining requirements:

- Extrusion pressure and viscous media flow rate
- Media viscosity and its viscosity
- Size of abrasive particles
- Abrasive concentration
- Particle hardness
- Work material hardness
- Machining time or number of strokes
- Geometry and dimensions of the machined part

Table 2.8 provides typical parameters of the AFM process.

One of the attractive features of AFM is the wide variety of its easily adjustable and controllable process parameters. Laboratory investigations have been carried out to study the performance of AFM when machining cold-rolled-steel samples. The related data should only be taken as *guidelines* for the actual production situations.

TABLE 2.8
Typical Parameters of AFM

Parameters	Values
Media:	
Viscosity	See Tables 2–1, 2–3
Grit size	# 20 to # 1000
Grit type	Al_2O_3, SiC, B_4C, diamond
Starting temp.	30–50°C
Flow:	
Extrusion Pressure	7–200 bar
Vol. rate	7–225 lit/min
Number of strokes	1–100

Effect of process parameters of AFM on:

1. **Material Removal**
   Figure 2.39 illustrates the effect of the machining time, the extrusion pressure, and the media temperature on the material removal, expressed as a diametric increase of the machined hole. The material removal increases as the machining time and extrusion pressure increase, Figure 2.39a. It also increases with increasing media temperature, due to the increase of the media flow rate, Figure 2.39b. The related operating conditions are given in the figure. The material removal depends upon the restriction passage. It decreases with increasing length of the passage, since the media flow rate decreases.

   The type of abrasive grains influences the material removal. From the comparison shown in Table 2.9, it is depicted that the highest material removed as expressed in diametric increase is achieved when B4C abrasive

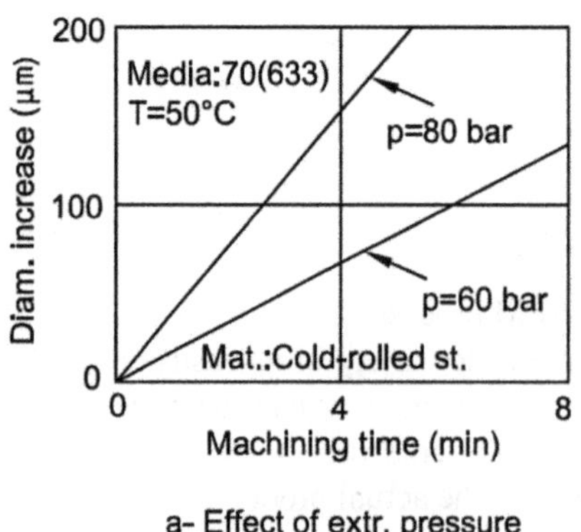

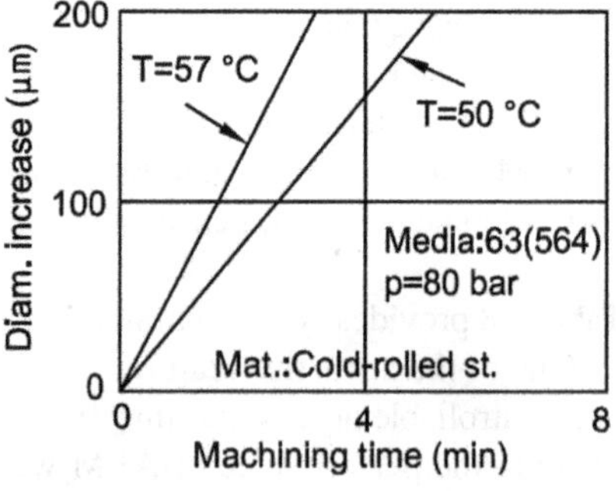

FIGURE 2.39 Effect of process parameters of AFM (Rhoades, 1977).

TABLE 2.9
Comparison Between Three Types of Abrasive Grits Regarding Material Removal Capability

Main Type of Abrasives in Media (37.8%)	Extr. Pressure (bar)	Media Temp. (°C)	Diam. Increase (mm)
Al_2O_3			0.45
SiC	27	30	0.50
B_4C			0.98

Working conditions:
Work material: Cold-rolled steel
Hole diam.: 3 mm
Hole length: 6 mm
No. of reverse strokes: 10
Machining time: 4 min

Abrading media:
Base polymer 100 (HMV) 56.5%
36-grit size Al_2O_3, or SiC, or B_4C 37.8%
700-grit size SiC 5.7%

is used, while Al_2O_3 provides the lowest material removal, provided the same working conditions.

2. **Surface Quality**
AFM is very efficient process, suitable for finishing of complex surfaces. An important application, that will be treated afterwards, is AFM of pre-machined EDMed surfaces. It has been shown that the EDM induces undesired high tensile stresses in and beneath the machined surface.

Table 2.10 illustrates the improvement of the surface quality when AFM is used after the EDM process. Moreover, an increase in the media pressure leads to a decrease of the surface roughness R_a. From a quantitative point of view, it is found that EDM generates surface with average value of R_a parameter of approximately 1.68 μm. On the other hand, AFM reduces it to the value between 0.94 and 0.23 μm. Roughness along the workpiece is lower than the transverse one that corresponds to the flow direction of polishing media. Along the media, R_a reaches values between 0.47 and 0.23 μm, while in the transverse direction, values are higher (R_a is between 0.94 and 0.55 μm) (Kenda, Pusavec, Kermouche, and Kopac, 2011).

TABLE 2.10
Surface Roughness by EDM and AFM at Different Extrusion Pressures

Operation – Pressure	Measuring Direction	Roughness R_a (μm)
EDM	along	1.7
	transverse	1.7
EDM/AFM – 3.5 MPa	along	0.47
	transverse	0.94
EDM/AFM – 6.0 MPa	along	0.23
	transverse	0.55

2.3.3 Applications of AFM

As already mentioned, AFM is used to burr, polish, radius edges, remove recast layers, reduce compressive residual stresses, and provide smooth surfaces. The process embraces a wide range of applications – from critical aerospace and medical components to the high-production volume of parts. Other applications include finishing impellers, integrally bladed rotors (IBRs), compressor wheels, turbine disks, and gears. Milled surfaces of an IBR, as finished by AFM, need only 15 to 30 min of operation time. Their surface finishes are considerably improved while eliminating hours of hand finishing. The process, which was initially developed for the critical de-burring of aircraft valve bodies and components, can yield production rates of up to hundreds, or even thousands, of parts per hour.

2.4 REVIEW QUESTIONS AND PROBLEMS

2.4.1 What are parameters affecting the stock removal rate (SRR) in AJM? Give three examples of typical materials that can be effectively machined by AJM.
2.4.2 In AWJM, at what stage are the abrasives introduced in the water jet? Draw a schematic outline of equipment involved.
2.4.3 Discuss the effect of the following parameters on production accuracy and the removal rate in AJM: Grain size, jet velocity, SOD.
2.4.4 Describe at least three typical applications of AJM.
2.4.5 Discuss why the AJM technique, when applied to ductile materials, leads to a low metal removal rate.
2.4.6 Describe at least three typical applications of AJM.
2.4.7 Explain the mechanism of material removal in WJM.
2.4.8 Using a block diagram or a line sketch, show the main components of a WJM plant.
2.4.9 Explain how material is removed in USM.
2.4.10 Show diagrammatically the main elements of an USM machine.
2.4.11 Explain the advantages and disadvantages of USM.
2.4.12 What are the main applications of USM?
2.4.13 Explain the effect of USM parameters on the removal rate.
2.4.14 Define the magnetostriction effect as applied in USM. What are the aims of using a premagnetizing dc current in magnetostrictive transducers?
2.4.15 Give three examples of materials that can be machined economically by USM.
2.4.16 List three types of abrasive materials that are frequently used in USM and arrange them according to their cutting ability.
2.4.17 What are the applications of ultrasonic machining? Why can very hard material be cut better by the ultrasonic process than soft material?
2.4.18 Describe the process of USM and discuss the effects of the various parameters, showing the advantages and limitations of the process.
2.4.19 State the suitability of USM to cut: Rubber, glass, HSS, ceramics.
2.4.20 Use a neat sketch to show the principle of USM and show on your sketch how oscillation amplitude can be amplified using acoustic horns. Give the reason why oscillation amplitude should be limited in USM.

2.4.21 Explain, with the help of a sketch, the magnetostriction effect as applied in USM. Draw the US machine showing its main parts and describe the function of each.

2.4.22 What are the constituents of slurry used in USM? Name the characteristics of a good suspension medium. Which fluid satisfies most of these requirements?

2.4.23 Derive the general differential equation to express the longitudinal vibration of an exponential type acoustic horn.

2.4.24 Why is abrasive slurry used in ultrasonic machining?

2.4.25 Sketch and explain ultrasonic machining.

2.4.26 Explain the principle of ultrasonic machining. What are the limitations of USM?

2.4.27 Why is frequency tuning a must in ultrasonic machining?

2.4.28 What is the function of a concentrator in ultrasonic machining?

2.4.29 Design and construct the exponential horns for an ultrasonic drilling machine operating at a resonant frequency of 20 kHz for the following cases:

a. Horn material: Steel 100 Cr6, c = 5.05 × 105 cm/s. Resonant length l = 14 cm. Initial diameter $D0$ = 40 mm.
b. Same as (a), but the horn has a rectangular cross-section of constant width b = 50 mm and $A0$ = 40 × 50 mm^2.
c. Horn material: Brass, c = 3.3 × 105 cm/s. Resonant length l = 10 cm, initial diameter $D0$ = 40 mm.
d. Same as (c), but the horn is hollow and provided with a conical hole starting from the nodal point to secure an outside horn diameter. $D\ell$ = 50 mm to accommodate a large trepanning tool.

2.4.30 An exponential horn is made of monel of acoustic speed c = 4.2 × 105 cm/s; the horn has the following features: Maximum diameter D_0 = 36 mm, resonant length ℓ = 12 cm, operating frequency = 20 kHz. The horn should be provided with a central hole for slurry suction of 2.5 mm diameter from nodal point to x = 12 cm. Calculate the magnification factor, location of the node, horn diameter at nodal point, horn outside diameter at x = 2, 7, 12 cm.

2.4.31 What is AFM? Draw a neat sketch to show its principle.

2.4.32 What are the main applications of AFM?

2.4.33 Discuss the effect of process parameters of the AFM process on the abrading rate.

2.4.34 What are the main advantages of the AFM process?

REFERENCES

Blanck, D (1961) *Getzmäßigkeiten beim Stoßläppen mit Ultraschallfrequenz*, Dissertation, Braunschweig, Germany.

Brook, N & Summers, DA 1969, 'The penetration of rock by high speed water jets', *International Journal of Rock Mechanics and Mineral Science*, vol. 6, no. 3, pp. 249.

Corporation Dr. Lehfeldt 1967, GmbH, Heppenheim, Germany.

Dhakal, HN 2007, *Lecture notes on NTMPs*, Mechanical and Marine Engineering, School of Engineering, University of Plymouth, UK.

Dünipβ, W, Neumann, M & Schwartz, H 1979, *Trennen-Spanen and abtragen*, VEB-Verlag Technik, Berlin.

El-Hofy, H 2005, *Advanced machining processes, nontraditional and hybrid processes*, McGraw-Hill Co., New York.

Farmer, IW & Attewell, PB 1965, 'Rock penetration by high speed water jets', *International Journal of Rock Mechanics and Mineral Science*, vol. 2, no. 2, pp. 165–169.

Franz, NC 1972, 'Fluid additives for improving high velocity jet cutting', *First International Symposium of Jet Cutting Technology*, England.

Hoogstrate, AM & Van Luttervelt, CA 1997, 'Opportunities in AWJM', *Annals of CIRP*, vol. 46, no. 2, pp. 679–714.

Ingersoll-Rand 1996, *Technical data*, Hannover Exhibition.

Kaczmarek, J 1976, *Principles of machining by cutting, abrasion, and erosion*, Peter Peregrines Ltd, Stevenage, Hertfordshire.

Kalpakjian, S 1984, *Manufacturing processes for engineering materials*, Addison Wesley Publishing Co, Reading, MA.

Kenda, J, Pusavec, F, Kermouche, G & Kopac, J 2011, 'Surface integrity in abrasive flow machining of hardened tool steel AISI D2', *Procedia Engineering*, vol. 19, pp. 172–177. doi:10.1016/j.proeng.2011.11.097.

König, W 1990, *Fertigungsverfahren-Band 3 Abtragen*, VDI Verlag, Düsseldorf.

König, W & Schmelzer, M 1990, 'Schneiden mit feststoffbeladenen Wasserstrahlen als leistungsfähiges Bearbeitungsverfahren für faserverstärkte Kunsstoffe', *Industrie-Anzeiger*, vol. 109 H91, pp. 70–71.

Kovacevic, R 1994, 'Sensing the AWJ nozzle wear', *International Journal of WJ Technology*, vol. 2, no. 1.

Machinability Data Center, *Machining data handbook* 1980, vol. 2, 3rd edn, Machinability Data Center, Cincinnati, OH.

Rhoaden LJ 1977, 'Extrude hone-edge and surface finishing-capabilities and costs', Technical paper MR77-366, Society of Mechanical Engineers, Dearborn, MI, p. 14.

Verma, AP & Lal, KG 1984, 'An experimental study of AJM', *International Journal of Machine Tool Design and Research*, vol. 24, no. 1, pp. 19–29.

Youssef, HA 2005, *Non-traditional machining processes-theory and practice*, 1st edn, El-Fath Press, Alexandria.

3 Chemical and Electrochemical Non-Traditional Machining Operations and Machine Tools

3.1 CHEMICAL MACHINING

Chemical machining (CHM), also called chemical etching, is the oldest non-traditional process which was previously used in zincograph preparation. CHM depends on controlled chemical dissolution of the work material by contact with a strong reagent (etchant). At the present time, the process is mainly used to produce shallow cavities of intricate shapes in materials independent on their hardness or strength. CHM embraces two main applications. These are chemical milling (CH milling), Figure 3.1a, and photochemical machining (PCM), also called spray etching, Figure 3.1b.

3.1.1 Chemical Milling

This process has a special significance in the aircraft and aerospace industries, where it is used to reduce the thickness of plates enveloping walls of rockets and airplanes, with the aim of striving to improve the stiffness to weight ratio, Figure 3.2. CH milling is also used in metal industries to thin out walls, webs, and ribs of parts that have been produced by forging, sheet metal forming, or casting, Figure 3.3. Furthermore, the process has many applications related to improving surface characteristics including:

- Elimination of Ti-oxide (α-case) from Ti-forgings and superplastic formed parts.
- Elimination of the decarburized layer from low alloy steel forgings.
- Removal of the recast layer from parts machined by EDM.
- Removal of burrs from conventionally machined parts of complex shapes.

Figure 3.4 shows the production of a tapered disk by gradual immersion of the disk in the etchant while it is rotating. The process is also capable of producing burr-free printed circuit boards. In CH milling, a special coating called maskant or resist

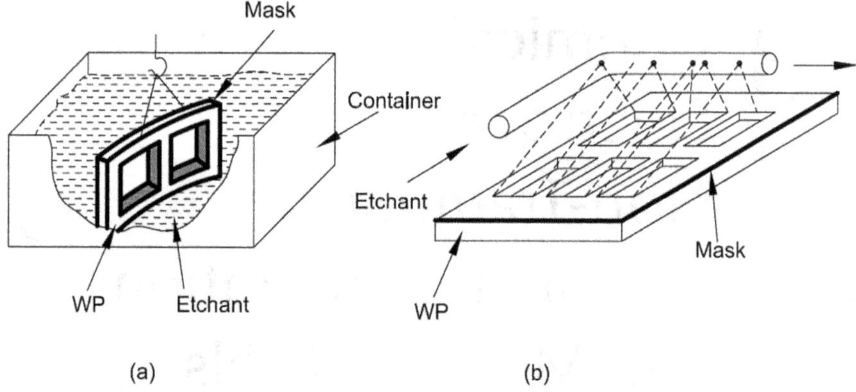

FIGURE 3.1 CHM processes: (a) CH milling and (b) PCM (spray etching).

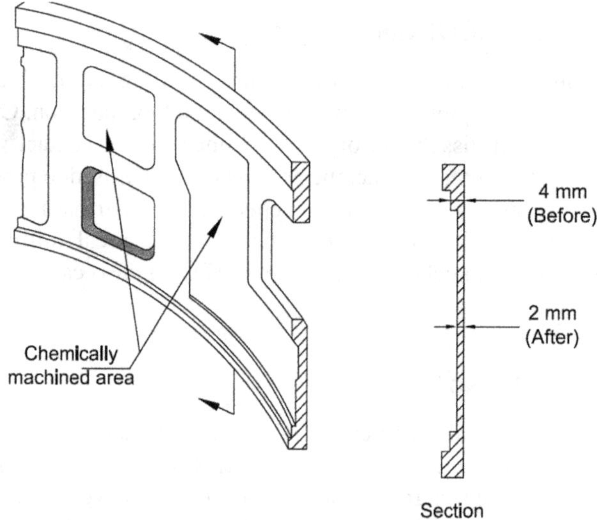

FIGURE 3.2 CH milling striving to improve the stiffness-to-weight ratio of Al alloy plates for space vehicles (ASM International, 1989).

protects areas from which the metal is not to be removed. The process is used to produce pockets and contours. CH milling consists of the following steps:

1. Preparing the workpiece surface by cleaning, mechanically or chemically, to provide good adhesion of the masking material.
2. Masking using strippable mask that adheres to the surface and withstands chemical abrasion during etching.
3. Scribing of the mask using special templates to expose areas to be etched. The type of selected mask depends on the work size, number of parts, and the desired resolution of details. Silk screens are recommended for shallow cuts of close dimensional tolerances.

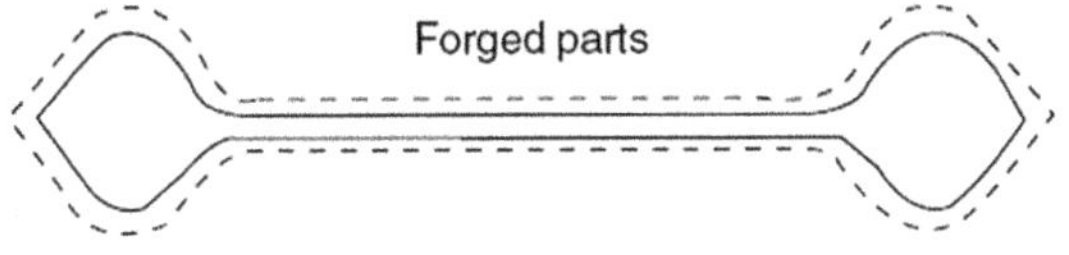

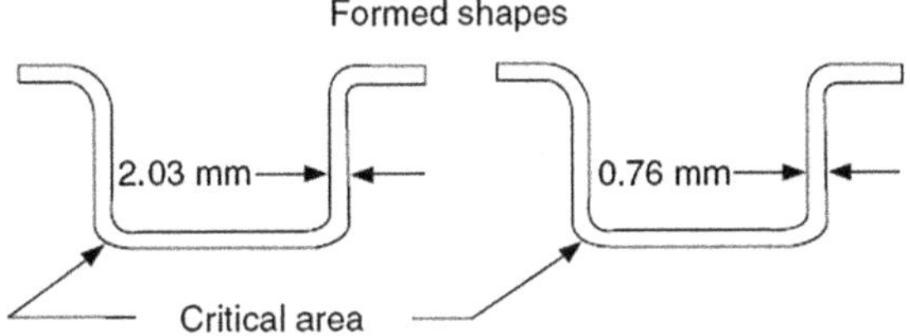

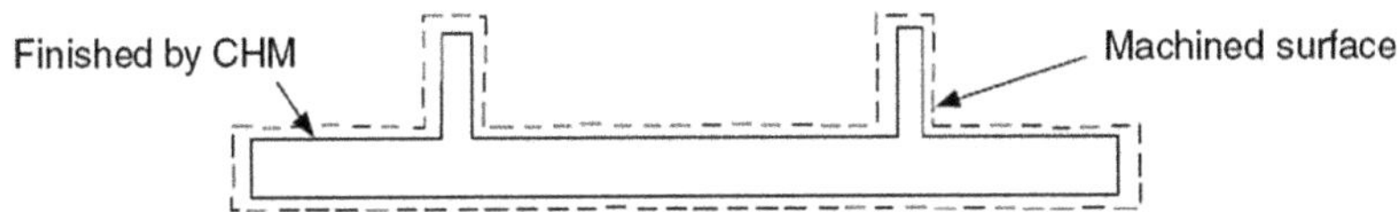

FIGURE 3.3 Thinning of parts by CH milling.

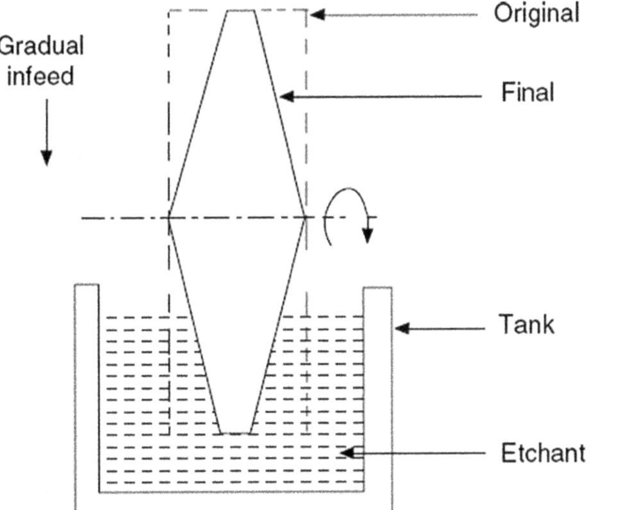

FIGURE 3.4 CH milling of tapered disk by gradual immersion in etchant while rotating.

4. After etching, the work is rinsed, and the mask is stripped manually, mechanically, or chemically.
5. The work is washed by deionized water then dried by nitrogen.

During CH milling, Figure 3.5, the etching depth is controlled by the time of immersion. The etchants used are very corrosive and, therefore, must be handled

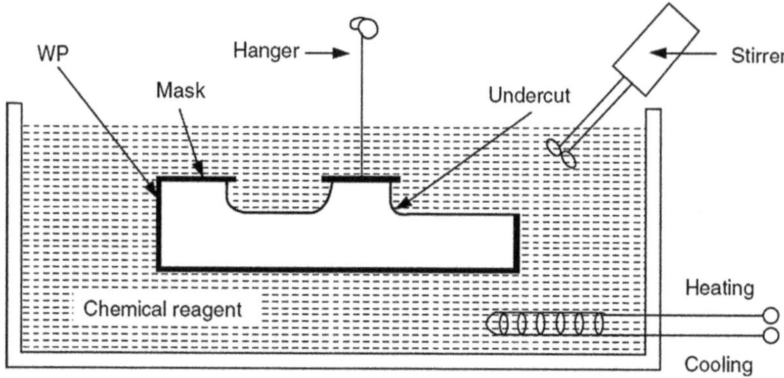

FIGURE 3.5 CHM setup.

with adequate safety precautions. Vapors and gases produced from the chemical reaction must be controlled for environmental protection. A stirrer is used for agitation of fluid. Typical reagent temperatures range from 37 to 85°C which should be controlled within ±5°C in order to attain a uniform machining. Faster etching rates occur at higher etchant temperatures and concentrations.

When the mask is used, the machining action proceeds both inwardly from the mask opening and laterally beneath the mask thus creating the etch factor (EF), which is the ratio of the undercut d_u to the depth of etch T_e (EF = d_u/T_e), Figure 3.6; this ratio must be considered when scribing the mask using templates. A typical etch of 1:1 occurs at a cut depth of 1.27 mm. Deeper cuts can reduce this ratio to 1:3.

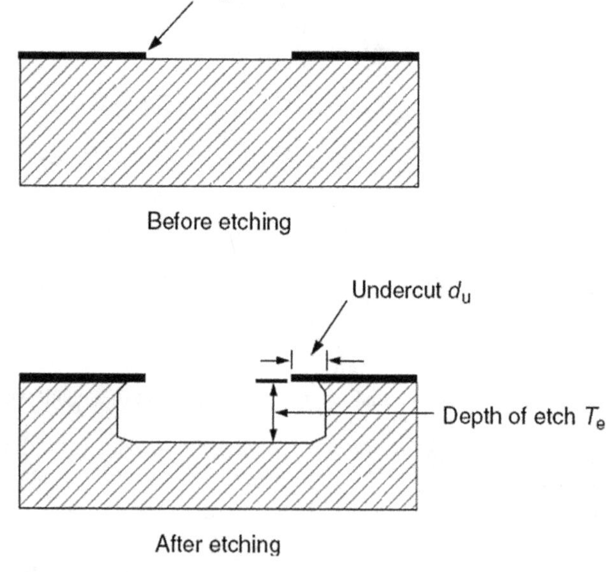

FIGURE 3.6 EF after CHM.

Tooling for CH Milling

Tooling for CH milling is relatively inexpensive and simple to modify. Four types of tools are required: Maskants, etchants, scribing templates, and accessories.

Maskants: Synthetic or rubber base materials are frequently used. They should possess the following properties:

- Tough to withstand handling
- Inert to the chemical reagent used
- Able to withstand heat generated by etching
- Adhere well to the work surface
- Scribe easily
- Easy to remove after etching

Table 3.1 shows the recommended maskants used for different types of etchants as well as the characteristics of these maskants.

Multiple coats of the maskant are frequently used to increase the etchant resistance and to avoid the formation of pinholes on the machined surfaces. Deeper cuts that require longer exposure time to the etchant can also be achieved. Dip, brush, roller, and electro coating as well as adhesive tapes can be used to apply masks.

Spraying the mask on the workpieces through a silk screen, on which the desired design is imposed, combines the maskant application with the scribing operation since no peeling is required. The product quality is, therefore, improved due the ability to generate finer details.

TABLE 3.1
Recommended Maskants for Different Types of Etchants

	Maskant Material Exposed to				
	Oxidizing Acids		All Types	Acids and Alkalines	
Property	**Polyvinyl Chloride**	**Polyethylene**	**Butyl Rubber**	**Acrylonitrile Rubber**	**Neoprene Rubber**
Ease to manufacture	Good	Good	Fair	Fair	Good
Shelf life (months)	6–12	6–12	4–6	3–6	6–8
Ease of application					
Dipping	Good/fair	Poor/fair	Good	Good	Good
Flow coating	Good/fair	Poor/fair	Good	Good	Good
Air spraying	Good	Good	Poor	Poor	Fair
Type of cure	Air/heat	Heat	Heat	Air/heat	Air/heat
Resist to etchant	Very good	Very good	Very good	Very good	Ver good
Up to temperature limit (°C)	70	60	145	120	80

Source: Reproduced from *Machining data handbook*, vol. 2, 3rd edn, Machinability Data Center, Cincinnati, OH, 1980.

However, the thin coating applied when using silk screens will not resist etching for a long time as compared to the cut-and-peel method. Photoresist masks, which are used in photochemical machining (spray etching), also combine both the coating and scribing operations. The relatively thin coats applied as dip or spray coats will not withstand rough handling or long exposure times. However, photoresist masks ensure high accuracy and ease of modification. Typical tolerances for different masks are given in Table 3.2.

Moreover, the tolerance depends also on the etching depth and the material of workpiece machined. Cu and Cu alloys provide the closest tolerances, whereas Al alloys are machined with large tolerances (Machinability Data Handbook, 1980).

Etchants: Etchants are high concentrated acidic or alkaline solutions maintained within a controlled range of chemical composition and temperature. They are capable of reacting with the workpiece material to produce a metallic salt, which dissolves in the solution. Table 3.3 shows the machined material, the recommended etchant, its concentration and temperature, etch factor, and etch rate. When machining glass or germanium (Ge), acidic solution HF or HF+ HNO_3 are used as etchants. When machining tungsten (W), it is recommended to use either:

- Alkaline solution: K_3 $Fe(CN)_6$: NaOH = 20:3 (by volume) or
- Acidic solution: HF: HNO_3 = 30:70 (by volume) Kalpakjian, 1984).

A suitable etchant should provide the following requirements:

- Good surface finish of the work
- Uniformity of metal removal
- Control of selective and intergranular attack
- Low cost and availability
- Ability to regenerate and/or readily neutralize and dispose of its waste products
- Non-toxic
- Control of hydrogen absorption in case of Ti alloys

Scribing templates: They are used to define the areas for exposure to chemical dissolution. The most common scribing method is to cut the mask with a sharp knife followed by careful peeling. The etch factor allowance must be included. Figure 3.7 shows NC laser scribing of masks for chemical milling of a large surface area.

Accessories: These include tanks, hooks, brackets, racks and fixtures.

TABLE 3.2
Tolerances of Different Masks

Maskant	Tolerance (μm)
Cut-and-peel masks	±180
Silk screen	±75
Photoresist	±13

TABLE 3.3
Machined Materials and Recommended Etchants in Chemicals

Etchant	Concentration	Temperature (K°)	Etch Rate (μm/min)	Etch Factor (EF = d_u/T_e)	Metal to be Machined
$FeCl_3$	12–18° Be[a]	320	20	1.5:1	Al alloys
$HCl:HNO_3:H_2O$	10:1:9	320	20–40	2:1	
$FeCl_3$	42° Be[a]	320	20	2:1	Cold rolled steel
HNO_3	10–15% (volume)	320	40	1.5:1	
$FeCl_3$	42° Be[a]	320	40	2.5:1	Cu and Cu alloys
$CuCl_2$	35° Be[a]	325	10	3:1	
HNO3	12–15% (volume)	300–320	20–40	–	Magnesium
$FeCl_3$	42° Be[a]	320	10–20	(1–3): 1	Nickel
$FeCl_3$	42° Be[a]	325	20	2:1	Stainless steel, Tin
HNO_3	10–15% (volume)	320–325	20	–	Zinc

[a] Baume specific gravity scale.

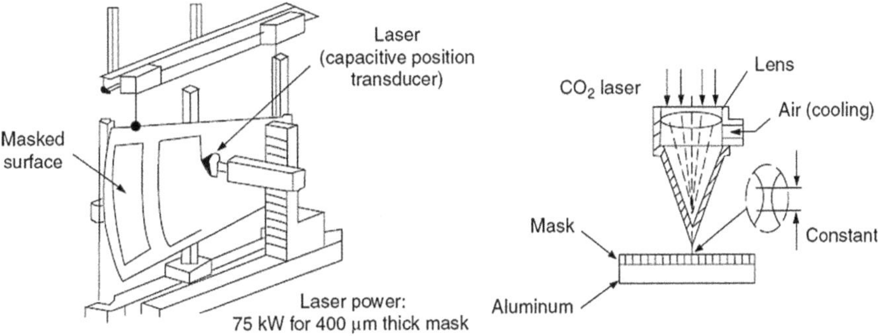

FIGURE 3.7 Laser cutting of masks for CH milling of large surfaces.

Advantages and Disadvantages of CH Milling

Advantages

- Weight reduction is possible on complex contours that are difficult to machine conventionally.
- Several parts can be machined simultaneously.
- Simultaneous material removal from all surfaces improves productivity and reduces wrapping.
- No burr formation.
- No induced stresses thus minimizing distortion and enabling machining of delicate parts.
- Low capital cost of equipment, and minor tooling cost.
- Quick implementation of design changes.
- Less skilled operator is needed.
- Low scrap rate.

Disadvantages

- Only shallow cuts are practical. Deep narrow cuts are difficult to produce.
- Handling and disposal of etchants can be troublesome.
- Masking, scribing, and stripping is repetitive, time consuming, and tedious.
- Surface imperfections if any are reproduced.
- For best results, metallurgical homogeneous surfaces are required.
- Porous castings yield uneven etched surfaces.
- Welded zones frequently etch at rates that differ from base metal.

Process Capabilities

Using fresh solutions and depending on the working conditions (etchant concentration and temperature), the etch rate ranges from 20 to 40 μm/min, Table 3.3. Etch rates are high for hard materials and low for soft ones (Metals Handbook, 1989). Generally, the high etch rate is accompanied by a low surface roughness and, hence, closer machining tolerances. Typically, surface roughness of 0.1 to 0.3 μm (R_a value), depending on the initial roughness can be obtained. However, under special conditions, surface roughness of 0.025 to 0.05 μm becomes possible (Machinability Data Handbook, 1980).

3.1.2 Photochemical Machining (Spray Etching)

Photochemical machining (PCM) (spray etching) is a variation of CH milling where the resistant mask is applied to the workpiece by photographic techniques. The two processes are quite similar because both use etchant to remove material by chemical dissolution (CD). CH milling is usually used on the three-dimensional parts originally formed by another manufacturing process such as forging and casting of irregular shapes. On the other hand, PCM is a promising method for machining foils and sheets of thicknesses ranging from 0.013 to 1.5 mm to produce accurate and micro shapes. So, the PCM process becomes a realistic alternative of shearing and punching operations performed by mechanical presses.

Additionally, a main difference between CH milling and PCM is that in CH milling, the depth of etch is controlled by the time the component is immersed in the etchant, whereas in PCM, the etch depth is controlled by the time the component is sprayed by fresh etchant through upper and lower nozzles, thus improving the performance of the PCM process by activating the etch rate and enhancing the quality. Visser et al. (1994) claimed that the etch rate of PCM is five to ten times that achieved by CH milling. Of course, in PCM highly developed expensive equipment is needed to provide high pressure/high temperature, Figure 3.8. This machine is equipped with the following units:

- System of upper and lower nozzles
- Multi-speed conveyor for serving the workpiece
- A unit for cleaning the worksheet by water and drying it with hot air
- Unit for measuring and controlling the density and concentration of etchant
- Unit for product inspection

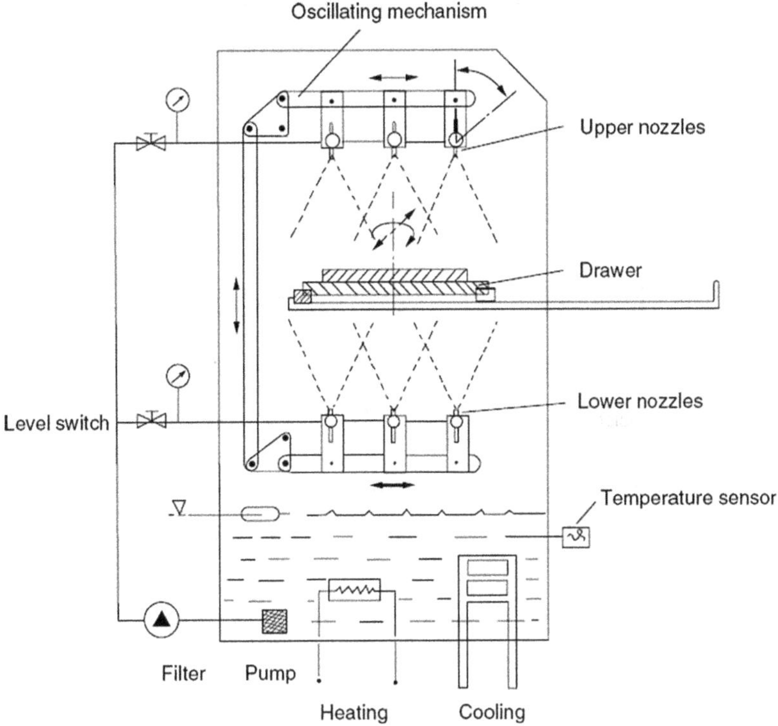

FIGURE 3.8 Schematic of PCM equipment (Visser, Junker, and Weiβinger, 1994).

In PCM, the following steps are carried out, Figure 3.9:

1. The required part shape which is considered as a primary image for the photo-tool is created by CAD. The drawing is then laser printed for accurate work such as printed circuit boards (PCB).
2. Two photographic negatives, called artwork, are produced to the actual size of the work.
3. The sheet metal is chemically cleaned and then coated with a high sensitive photoresist (called sensitive emulsion). The coating is performed by spraying, dipping, or rolling. The work is allowed to dry. The photoresist adheres to the surface protecting it during etching.
4. After coating, the work is sandwiched between both negatives (artwork), then exposed in vacuum, to an ultraviolet light. The coating is solidified in exposed areas and it is removed from the exposed by dissolving into developer.
5. The work-sheet is exposed to a powerful water jet to remove the soft photoresist. The work-sheet is then rinsed by deionized water, then dried by nitrogen gas.

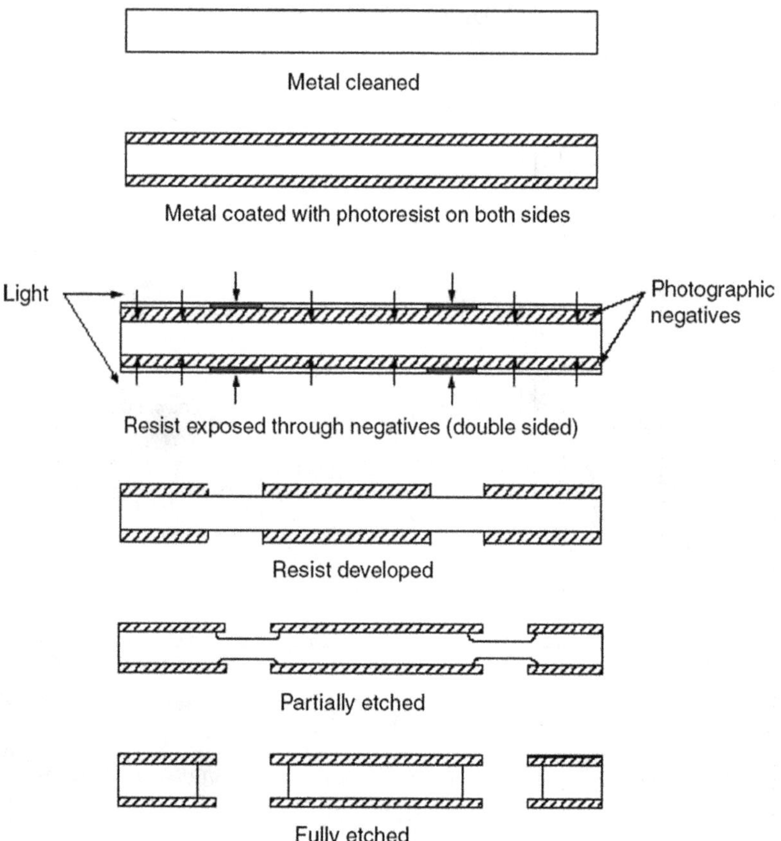

FIGURE 3.9 PCM steps.

6. The work-sheet is then spray-etched from the top and bottom. This permits the material to be etched from both sides. Thus minimizing the undercutting, reducing the machining time, and producing straighter side walls, Figure 3.9.
7. After etching, the hard photoresist is removed and the work-sheet is rinsed to avoid any reactions with suspended etchant.

Applications of PCM Process

- Aluminum, copper, zinc, steels, stainless steels, lead, nickel, titanium, molybdenum, glass, germanium, carbides, ceramics, and some plastics are photochemically machined. The process also works well on springy materials, which are difficult to punch. The materials must be flat so that they can later be bent to shape and assembled into other components. Products made by PCM are generally found in the electronics, automotive, aerospace, telecommunication, computer, and other industries. Typical products such as printed circuit boards (PCB), fine screens, flat springs etc., machined from foils are illustrated in Figure 3.10.

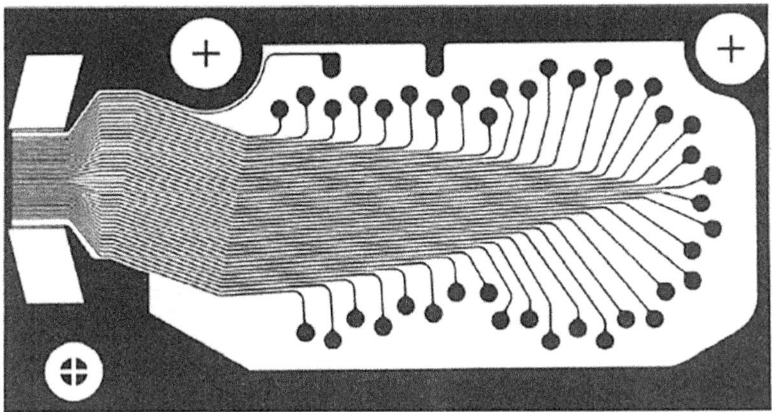

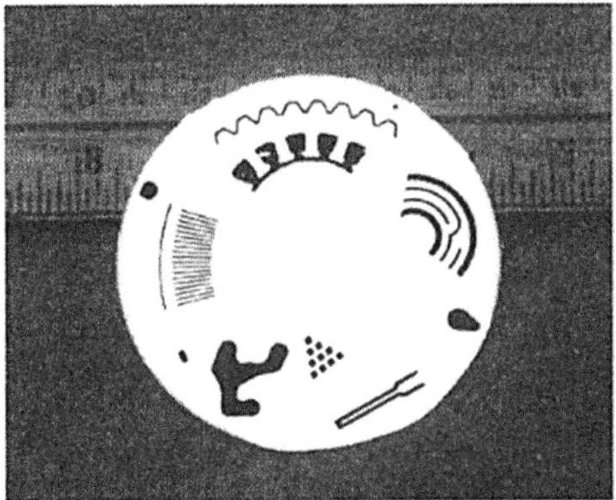

FIGURE 3.10 Typical products produced by PCM (Visser, Junker, and Weiβinger, 1994).

- Figure 3.11 illustrates a spray-etched stainless steel of 1 mm thickness, sprayed from the top and bottom, whereas Figure 3.12 shows the development of corner radius, which increases with the etching time.

Advantages and Disadvantages of PCM

Advantages

In addition to the previously mentioned advantages of CH milling, PCM is characterized by the following:

- The accuracy and etch rate are considerably greater than those realized by EC milling.
- Because tooling is made by photographic techniques, they can be easily stored, and patterns can be reproduced easily.
- Lead times are small compared to those required by processes that require hard tooling.
- Delicate and fragile parts of small thicknesses are produced by PCM without any deformation and warping.

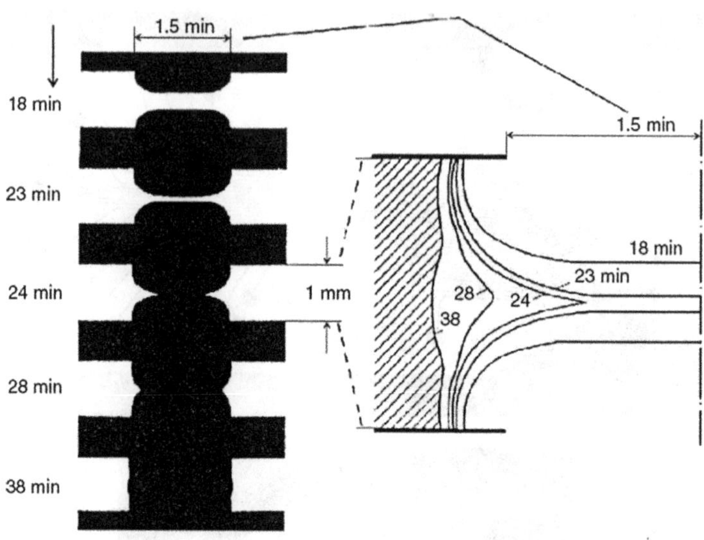

FIGURE 3.11 Cavity development with time by PCM (Visser, Junker, and Weiβinger, 1994).

Disadvantages

Apart from the disadvantages of CH milling, PCM also has the following limitations:

- PCM requires a highly skilled operator.
- PCM requires more expensive equipment.
- The machine should be protected from the corrosive action of etchants.
- The accuracy increases with the size of artwork and decreases with the depth.

3.2 ELECTROCHEMICAL MACHINES AND OPERATIONS

3.2.1 Process Characteristics and Applications

Electrochemical machining (ECM) is one of the most effective non-traditional machining processes, in which the metal removal is based on the anodic dissolution governed by Faraday's law of electrolysis (1833). According to this principle, the anodic dissociation rate in the machining (depleting) cell is directly proportional to the dc electrolyzing current and the chemical equivalent $\mathcal{E}$ of the anode material ($\mathcal{E}$ = atomic weight/valence).

In the machining cell, the workpiece is connected to the anode, while the tool is connected to the cathode of a dc source of 5 to 30 V. Both the tool and workpiece electrodes must be electrically conductive. They are separated by a gap 0.1 to 1 mm thick, into which an electrolyte (NaCl, KCl, and $NaNO_3$) is pumped rapidly to sweep away the reaction products (sludge) from the narrow machining gap. Depending on the gap thickness, a machining current of high density (20–800 A/cm^2) passes causing a high anodic dissolution rate. The shape of the cavity formed in the workpiece is the female

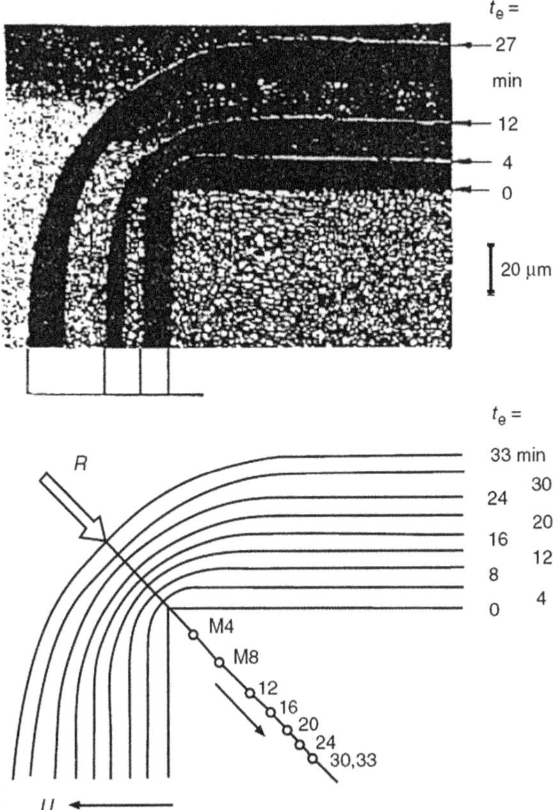

FIGURE 3.12 Corner development with time by PCM (Visser, Junker, and Weiβinger, 1994).

mating image of the tool shape. The tool advances axially toward the workpiece by means of a servomechanism at a constant feed rate V_f ranging from 0.5 to 10 mm/min, Figure 3.13. The anodic dissolution rate of the workpiece V_A adjusts itself to match with the selected tool feed ($V_f = V_A$). This matching characteristic or the self-adjusting feature of the ECM process controls and stabilizes its performance. Consequently, during machining the gap thickness attains a constant value known as the equilibrium gap.

According to the vision described above, Gusseff introduced his patent BP-335003 on EC sinking in 1930. However, the first industrial application of this patent was realized in the 1950s and 1960s. ECM is characterized by the following advantages and disadvantages.

Advantages of ECM

- Three-dimensional surfaces with complicated profiles can be easily machined in a single operation irrespective of the hardness and strength of the workpiece material.
- ECM offers a higher rate of metal removal as compared to traditional and non-traditional methods, especially when high machining currents are employed.

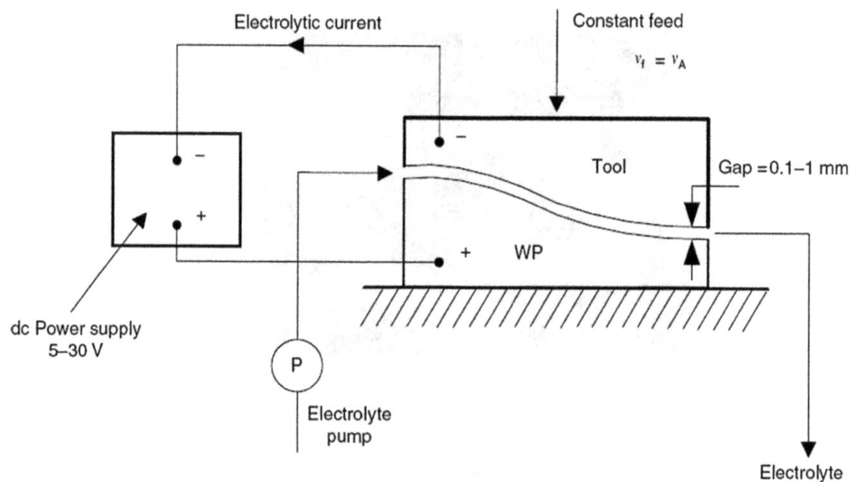

FIGURE 3.13 ECM process.

- There is no wear of the tool which permits repeatable production.
- No thermal damage or heat affected zone (HAZ).
- High surface quality and accuracy can be achieved at the highest MRR (R_a = 0.1–1.2 μm).
- Labor requirements are low.
- The surfaces produced by ECM are burr-free and free from stresses.

Disadvantages of ECM

- Non-conductive materials cannot be machined.
- Inability to machine sharp interior corners or exterior edges of less than 0.2 mm radius.
- The machine and its accessories are subjected to corrosion and rust, especially when NaCl electrolyte is used. Less corrosive but more expensive electrolytes like $NaNO_3$ can also be used.
- The endurance limit of parts produced by ECM is lowered by about 10 to 25%. In such a case shot peering is recommended to restore the fatigue strength.
- Metal removal rates are slow compared to traditional methods.
- Specific power consumption of ECM is considerably higher than that required for traditional machining.
- Cavitation channels may form which deteriorates the surface quality.
- Pumping electrolyte at high pressures into the narrow gap gives rise to large hydrostatic forces acting on the tool and workpiece, which necessitates rigid machine frame.
- The machined parts need to be cleaned and oiled immediately after machining.
- There is a danger of explosion if the hydrogen generated during machining is not safely disposed of.

- The tool and the workpiece may be damaged if arcing is initiated due to the contamination of oxides and debris in the gap, or if the tool comes into contact with the workpiece causing short circuit.

Applications of ECM

ECM has been used in a wide variety of industrial applications ranging from cavity sinking to deburring. Modifications of this process are used for turning, slotting, trepanning, and profiling in which the electrode becomes the cutting tool. Hybridization performed by integrating ECM with conventional finishing processes leads to the highly developed electrochemically assisted grinding, honing, and super finishing processes.

The process can handle a large variety of materials, limited only by their electric conductivity and not their hardness or strength. The material removal rate (MRR) is high especially for difficult-to-machine alloys. Fragile parts which are otherwise not easily machinable can be shaped by ECM. The fact that there is no tool wear in this process is advantageous, since it has a positive impact on accuracy. Moreover, a large number of components can be machined without the tool having to be replaced. Hence, ECM is well suited for mass production of complex shapes in hard and extra hard materials. The ability to machine high strength and hardened steels has led to many cost saving applications, where other processes are impractical. In the sector of electric conductive materials, both the ECM and EDM processes compete against each other. However, ECM has a special attraction due to the absence of thermal stresses and heat affected zone (HAZ). This characteristic is useful in the manufacturing of dies in hardened and tempered steel blocks (Figures 3.14 and 3.15).

3.2.2 Elements of ECM

The basic elements of the ECM process are the tool, workpiece, electrolyte, and the power supply.

1. **Tool**

   During ECM, the tool performance depends upon:

   - **Suitable choice of its material**: The tool material should be machinable, stiff, and possess high corrosion resistance, and good electric and thermal conductivity.

     ECM tools are usually made of copper, brass, and 316 stainless steel. Carbon steels are not recommended because of their low corrosive resistance, Table 3.4. Comprises properties of some metals used for such ECM tools:
   - **Tool design**: The tool is shaped not exactly to be a mirror image of the machined cavity, but its dimensions must be slightly different from the nominal dimensions of the cavity to allow for an overcut.

   The tool design must permit electrolyte flow at a rate sufficient to dissipate heat generated in order to eliminate boiling of electrolyte in the interelectrode gap. To produce smooth surfaces on the workpiece, tool deign must enable a uniform flow over the entire machining area. Ideally, flow should be laminar and free from eddies.

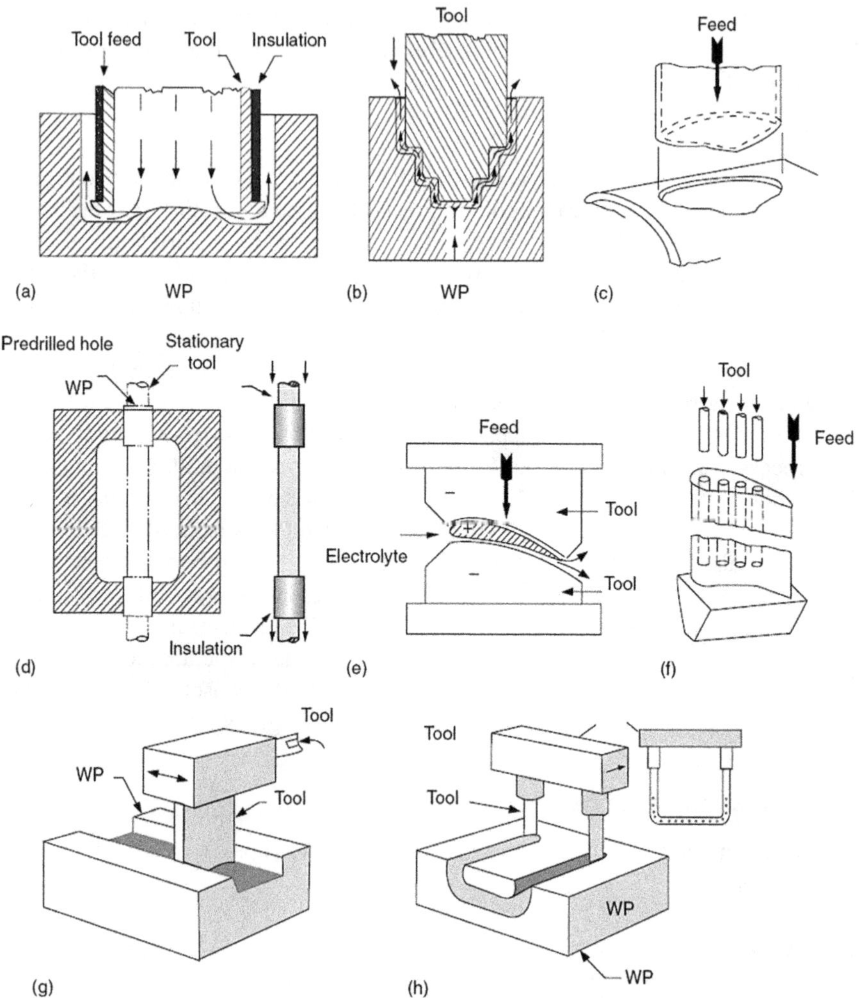

FIGURE 3.14 Typical ECM applications: (a) Hole sinking with insulated tool, (b) EC sinking of stepped through hole, (c) EC trepanning, (d) ECM of internal cavity by stationary electrode, (e) ECM of turbine blade, (f) EC deep hole drilling, (g) EC surfacing, and (h) EC hogging.

On machining complex cavities, a good designer should perform pilot tests under the same machining conditions. Accordingly, the corrections in the tool form are carried out to realize the required accuracy and surface quality.

2. **Workpiece**

   In ECM there is no restriction on the workpiece whatsoever except for it being electric conductive. The machinability is dependent only on the chemical composition of the workpiece. Carbon is passive in EC reactions; consequently, cast iron, which contains free carbon, cannot be machined satisfactorily by ECM.

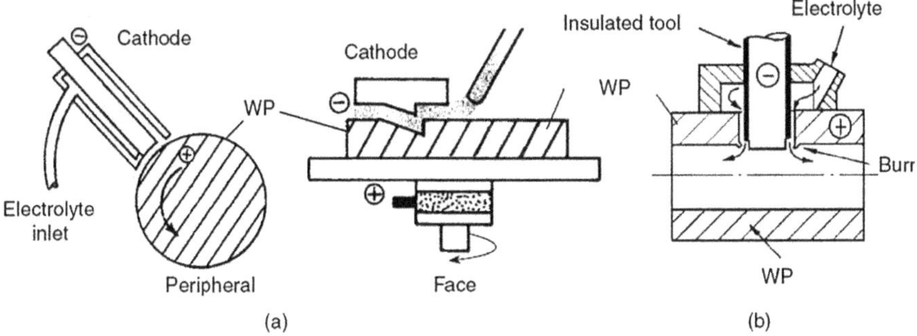

FIGURE 3.15 Other applications: (a) EC turning and (b) EC deburring.

TABLE 3.4
Relative Properties of Some Metals Used as Tools in ECM

	Tool Material			
Property	**Cu**	**Brass**	**316 Stainless Steel**	**Ti**
Electrical resistivity	1.0	4.0	53.0	48.0
Stiffness	1.1	1.0	1.9	1.1
Machinability	6.0	8.0	2.5	1.0
Thermal conductivity	25.0	7.5	1.0	2.6

Source: Reproduced from *Machining data handbook*, vol. 2, 3rd edn, Machinability Data Center, Cincinnati, OH, 1980.

3. **Electrolyte**
Electrolytes are highly conductive solutions of inorganic salts, usually NaCl, KCl, and $NaNO_3$, or their mixtures to meet multiple requirements. The main functions of the electrolytes are to:
   - Complete the electric circuit between the tool and the workpiece
   - Allow desirable reactions to occur and create conditions for anodic dissolution
   - Carry away heat generated during chemical reactions
   - Remove products of reaction (sludge) from the machining gap

Effective and efficient electrolyte should, therefore, have the following properties:
   - High electrical conductivity to ensure high current density
   - Low viscosity to ensure good flow conditions in the extremely narrow interelectrode gap
   - High specific heat and thermal conductivity to be capable of removing the heat generated from the gap

- Resistance to the formation of a passive film on the workpiece surface
- High chemical stability
- Provides high current efficiency and possesses low throwing power
- Non-toxic and non-corrosive to the machine parts
- Inexpensive and available

Electrolyte also plays an important role in the dimensional control. As shown in Figure 3.16, sodium nitrate solution is preferable, because the local metal removal rate is high at the small gap locations where both the current density and current efficiency are high. On the other hand, the local removal rate is low in the side gap where both current density and current efficiency are low. This results in smaller side gap, Figure 3.16. Several methods of supplying electrolyte to the machining gap are shown in Figure 3.17. The choice of the electrolyte supply method depends on the part geometry, machining method, required accuracy, and surface finish. Typical electrolyte conditions in the gap include a temperature of 90 to 110°C, a pressure between 10 and 20 atm, and a maximum electrolyte velocity of 25 to 50 m/s.

3.2.3 ECM Equipment (EC Sinking Machine)

This equipment comprises a dc power generator and an EC machine. The power generator should be powerful enough to supply the necessary machining current to the working gap. EC machines, equipped with power generators of current capacities ranging from 50 to 40,000 A are available in markets. The power sources supply a constant voltage ranging from 5 to 30 V. They are, generally, characterized by high power factor, high efficiency, and should be equipped with short circuit protection within a small fraction of a second to prevent catastrophic short circuits across the electrodes.

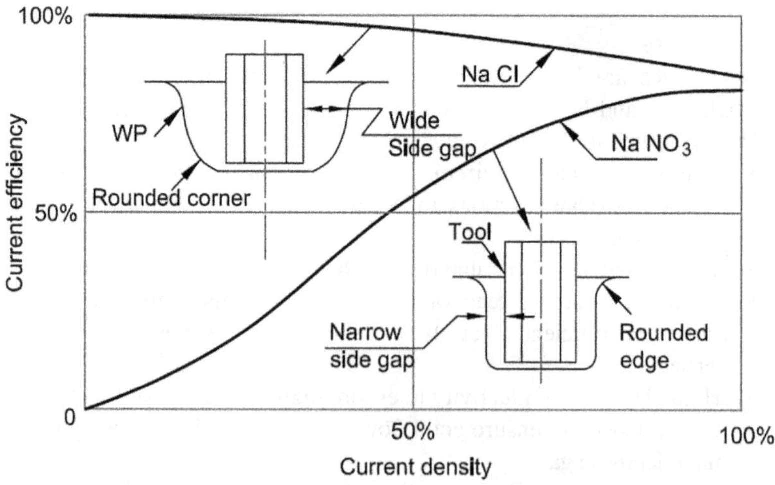

FIGURE 3.16 Current efficiency and side gap of NaCl and $NaNO_3$ (Youssef, 2016).

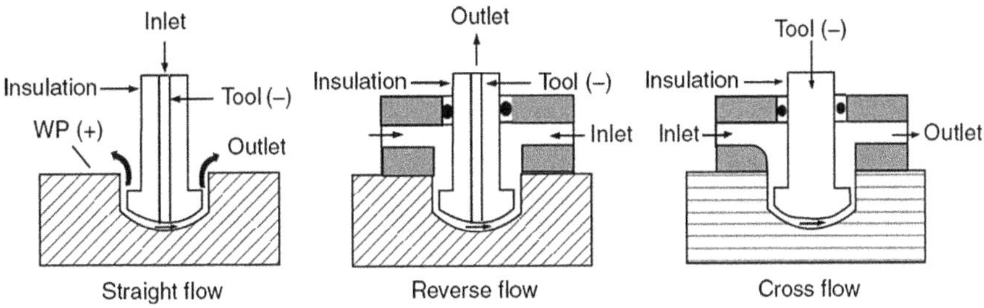

FIGURE 3.17 Methods of electrolyte feeding in ECM (El-Hofy, 2005).

Figure 3.18 illustrates schematically a typical EC sinking machine. The machine must be rigid enough to withstand the hydrodynamic pressure of the electrolyte in the machining gap which tends to separate the tool from workpiece. A servomechanism is necessary to control the tool movement in such a way that the material dissolution is balanced by the feed rate of the tool. The rate of current change is monitored and stops tool feeding when an abnormal rise in current is detected.

In contrast to conventional machine tools, EC machines are designed to stand up to corrosion attack by using non-metallic materials. For high strength and rigidity, metals with non-metallic coatings are recommended. To eliminate the danger of corrosion on other machinery, EC machines should be perfectly isolated in separate rooms in the workshop. The electrolyte-feeding unit supplies electrolyte at a given rate, pressure, and temperature. Facilities for electrolyte filtration, temperature control, and sludge removal are also included.

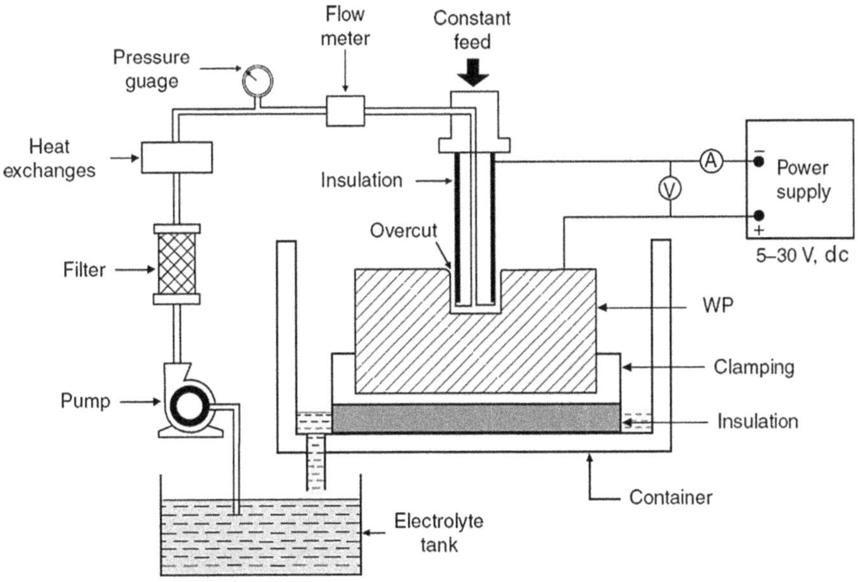

FIGURE 3.18 ECM setup (El-Hofy, 2005).

3.2.4 Process Capabilities

In ECM, the cutting rate is solely a function of ion-exchange rate irrespective of the hardness or toughness of the work material. The process provides metal removal rates in the order of 1.5 cm^3/min/1000 A. Penetration rates up to 2.5 mm/min are routinely obtained when machining carbides, and steels either hardened or not. Table 3.5 shows the electrochemical removal rate of most common metals assuming an electrolyzing current of 1000 A, and a current efficiency of 100%. The table also includes a formula to determine the chemical equivalent of alloys, necessary to calculate their removal rates (Youssef, 2005).

A well-known and unique characteristic of ECM, among all traditional and non-traditional processes, is that the accuracy and surface quality improve when applying higher removal rates (i.e. higher current densities).

A major problem of ECM is the overcut (side gap) which affects the accuracy. Roughly speaking, the side gap is governed by a complex set of parameters of which the type of electrolyte and the electrolyte flow are most crucial. It is important to have a small side gap, because the dimensional tolerance is proportional to the gap width. A typical dimensional tolerance of ECM is ±0.13 mm; however, through a proper control of the machining parameters, tight tolerance of ±0.025 mm can be achieved. It is difficult to machine internal radii smaller than 0.8 mm. A typical overcut of 0.5 mm and taper of 1 μm/mm are possible (Metals Handbook, 1989). Typical surface roughness (R_a value) of ECM ranges from 0.2 to 1 μm (0.4 to 0.8 μm is common) that decreases with increasing machining rate.

The principal tooling cost is due to the preparation of the tool electrode, which can be time consuming and costly, requiring several cut-and-try efforts except for simple shapes. There is no tool wear and the process produces stress free surfaces. The capability to cut the entire cavity in one stroke makes the process very productive, however the complicated tool from increases the tool cost.

3.2.5 ECM Allied Processes

3.2.5.1 Shaped Tube Electrolytic Machining

Shaped tube electrolytic machining (STEM) is based on the dissolution process when an electric potential difference is imposed between the anodic workpiece and a cathodic tool. Because of the presence of this electric field the electrolyte, often a sulfuric acid, causes the anode surface to be removed. After the metal ions are dissolved in the solution, they are removed by the electrolyte flow. As shown in Figure 3.19, the tool is a conducting cylinder with an insulating coating on the outside and is moved toward the workpiece at a certain feed rate while a voltage is applied across the machining gap. In this way a cylindrically shaped deep hole is obtained. STEM is, therefore, a modified variation of the ECM that uses acid electrolytes. This process is capable of producing small holes with diameters of 0.76 to 1.62 mm and a depth-to-diameter ratio of 180:1 in electrically conductive materials. It is difficult to machine such small holes using normal ECM as the insoluble precipitates produced obstruct the flow path of the electrolyte.

The machining system configuration is similar to that used in ECM. However, it has a periodically reverse polarity power supply. The cathodic tool electrode is made

TABLE 3.5
Removal Rates and Specific Removal Rates for Different Metals

Metal	ρ(g/cm³)	N (g/mol)	n	N/n (g)	Removal Rate (I = 1000 A, η = 100%) g/min	cm³/min	Spec. RR (cm³/A min)
Aluminum (Al)	2.7	27	3	9.0	5.6	2.1	0.0021
Beryllium (Be)	1.9	9	2	4.5	2.8	1.5	0.0015
Chromium (Cr)	7.2	52	2	26.0	16.2	2.3	0.0023
			3	17.3	10.8	1.5	0.0015
			6	8.7	5.4	0.8	0.0008
Cobalt (Co)	8.9	59	2	29.5	18.3	2.1	0.0021
			3	19.7	12.3	1.4	0.0014
Copper (Cu)	9.0	64	1	64.0	39.5	4.4	0.0044
			2	32.0	19.7	2.2	0.0022
Germanium (Ge)	5.3	73	4	18.3	11.2	2.1	0.0021
Gold (Au)	19.3	197	1	197.0	122.6	6.4	0.0064
			3	65.7	40.8	2.1	0.0021
Iron (Fe)	7.9	56	2	28.0	17.4	2.2	0.0022
			3	18.7	11.6	1.5	0.0015
Lead (Pb)	11.4	207	2	103.5	64.4	5.7	0.0057
			4	51.7	32.2	2.8	0.0028
Magnesium (Mg)	1.8	24	2	12.0	7.6	4.4	0.0044
Manganese (Mn)	7.5	55	2	27.5	17.1	2.3	0.0023
			3	18.3	11.3	1.5	0.0015
			4	13.8	8.5	1.2	0.0012
			6	9.2	5.7	0.8	0.0008
			7	7.8	4.9	0.7	0.0007
Molybdenum (Mo)	10.2	96	3	32.0	20.0	2.0	0.0020
			4	42.0	14.9	1.5	0.0015
			6	16.0	10.0	1.0	0.0010
Nickel (Ni)	8.9	59	2	29.5	16.2	2.1	0.0021
			3	19.7	12.2	1.4	0.0014
Platinum (Pt)	21.5	195	2	97.5	60.6	2.8	0.0028
			4	48.7	30.3	1.4	0.0014
Silver (Ag)	10.5	108	1	108.0	67.1	6.4	0.0064
Tantalum (Ta)	16.6	181	5	36.2	22.5	1.3	0.0013
Tin (Sn)	7.3	119	2	59.5	37.0	5.0	0.0050
			4	29.7	18.5	2.5	0.0025
Titanium (Ti)	4.5	48	3	16.0	10.0	2.2	0.0022
			4	12.0	7.5	1.6	0.0016
Tungsten (W)	19.3	184	6	30.7	19.0	1.0	0.0010
			8	23.0	14.3	0.7	0.0007
Vanadium (V)	6.1	51	3	17.0	10.6	1.7	0.0017
			5	10.2	6.4	1.0	0.0010
Zinc (Zn)	7.2	65	2	32.5	20.0	2.9	0.0029

$$\left(\frac{N}{n}\right)_{alloy} = \frac{1}{100}\left[x_P \frac{N_P}{n_P} + x_Q \frac{N_Q}{n_Q} + \cdots\right]^a$$

[a] This equation is used to calculate the chemical equivalent of an alloy composed of elements x_P, xq,… of atomic weights N_P, N_Q, and valences n_P, n_Q respectively.

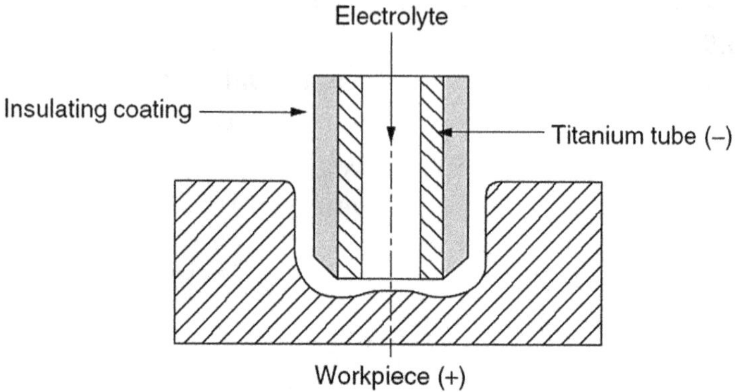

FIGURE 3.19 STEM schematic.

of titanium, its outer wall having an insulating coating to permit only frontal machining of the anodic workpiece. The normal operating voltage is 8 to 14 V dc, while the machining current reaches 600 A. It is reported that when a nitric acid electrolyte solution (15% v/v, temperature of about 20°C) is pumped at 2 to 5 bar through the gap (at 1 L/min, 10 V, tool feed rate of 2.2 mm/min) to machine a 0.58 mm diameter hole with 133 mm depth, the resulting diametral overcut is 0.265 mm, and the hole conicity is 0.01/133.

The process also uses a 10% concentration sulfuric acid to prevent the sludge from clogging the tiny cathode and ensure an even flow of electrolyte through the tube. A periodic reversal of polarity, typically at 3 to 9 s prevents the accumulation of the undissolved machining products on the cathode drill surface. The reverse voltage can be taken as 0.1 to 1 times the forward machining voltage.

Process Capabilities

Hole size:	0.5–6 mm diameter at an aspect ratio of 150
Hole tolerances :	0.5 mm diameter ±0.050 mm
	1.5 mm diameter ±0.075 mm
	60 mm diameter ±0.100 mm

Hole depth Tol. ±0.050 mm

Because the process uses acid electrolytes, its use is limited to drilling holes in stainless steel or other corrosion-resistant materials in jet engines and gas turbine parts such as:

- Turbine blade cooling holes
- Fuel nozzles
- Any holes where EDM recast is not desirable
- Starting holes for wire EDM
- Drilling holes for corrosion-resistant metals of low conventional
- Machinability
- Drilling oil passages in bearings where EDM causes cracks

Figure 3.20 shows the shape of turbulators that are machined by intermittent drill advance during STEM. The turbulators are normally used for enhancing the heat transfer in turbine engine-cooling holes.

Advantages

- The depth-to-diameter ratio can be as high as 300.
- A large number of holes (up to 200) can be drilled in the same run.
- Non-parallel holes can be machined.
- Blind holes can be drilled.
- No recast layer or metallurgical defects are produced.
- Shaped and curved holes as well as slots can be produced.

Limitations

- The process is used for corrosion-resistant metals.
- STEM is slow if single holes are to be drilled.
- Special workplace and environment are required when handling acid.
- Hazardous waste is generated.
- Complex machining and tooling systems are required.

3.2.5.2 Electrostream (Capillary) Drilling

Electrostream (ES) (capillary) drilling is a special electrochemical process used for producing fine holes that are too deep to produce by EDM and too small to drill by STEM. The cathodic tool used is made from a glass nozzle (0.025–0.50 mm diameter), which is smaller than the required hole diameter. The ES process differs from STEM, which uses a coated titanium tube as a cathodic tool. To conduct the machining current through the acid electrolyte that fills the interelectrode gap, a platinum wire electrode is fitted inside the glass nozzle, (Figure 3.21). Solutions of sulfuric, nitric, or hydrochloric acid with a concentration of 12 to 20 wt % are common electrolytes used. The type of electrolyte used depends on the workpiece. In this regard, hydrochloric solution is used for aluminum and its related alloys, while sulfuric acid

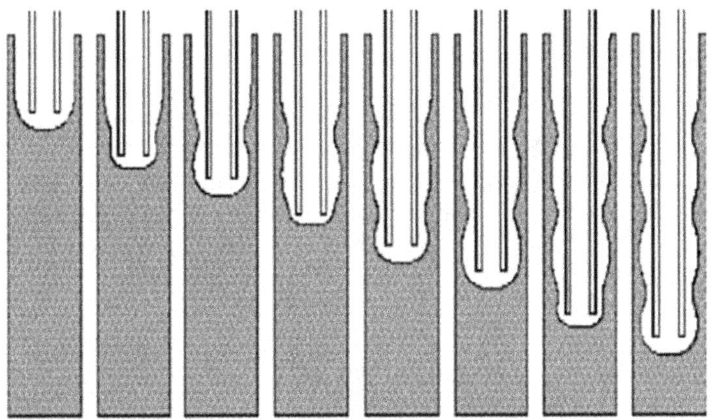

FIGURE 3.20 Turbulated cooling holes produced by STEM. (www.win.tue.nl/).

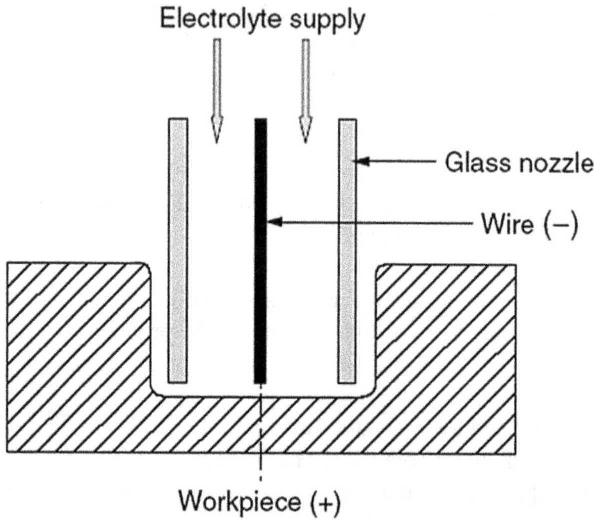

FIGURE 3.21 ES drilling schematic.

solution is recommended for Hastelloy, Inconel, Rene alloys, and carbon and stainless steels.

The electrolyte temperature is normally 40°C for sulfuric acid and 20°C for the rest. Electrolyte pressures between 275 and 400 kPa are recommended. During machining, the electrolyte stream is negatively charged and squirted against the anodic workpiece. The acid temperature, pressure, concentration, and flow rate must be carefully monitored for satisfactory machining. A gap voltage of 70 to 150 V is employed, which is ten times greater than that of normal ECM. Banard (1978) drilled small rows of cooling holes (0.127–1.27 mm diameter) in turbine blades, with depth-to-diameter ratios up to 50:1. The process can also be used in drilling side holes in inaccessible positions, Figure 3.22. Wire EDM start holes of less than 0.5 mm can also be drilled using ES.

Process Capabilities

Feed rates for ES drilling range from 0.75 to 2.5 mm/min. The feed rate depends on the material to be machined irrespective of the number of holes to be drilled simultaneously. Similar to ECM, higher removal rates are associated with larger feed rates and tool diameters. Additionally, higher removal rates have been reported for titanium alloys than for steel. Normal tolerances are within ±10% of the produced hole diameter. The normal hole depth tolerance is ±0.05 mm. These tolerance levels can further be reduced using special control and with pure metals.

Advantages

- High depth-to-diameter ratios are possible.
- Many holes can be drilled simultaneously.
- Blind and intersecting holes can be machined.
- There is an absence of recast and metallurgical defects.

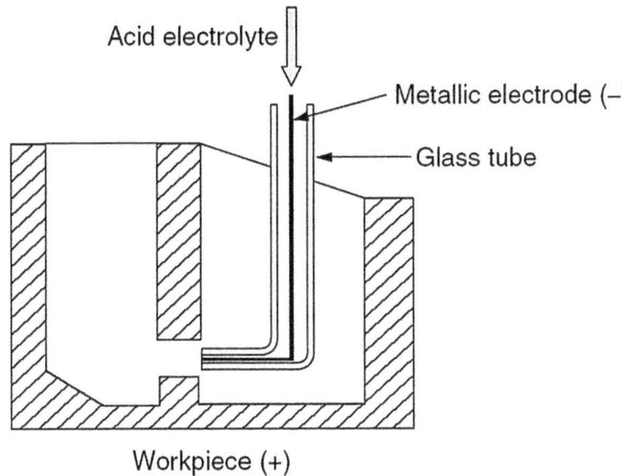

FIGURE 3.22 ES drilling in inaccessible positions.

- Powder metallurgy hard materials can be tackled.
- Burr-free holes are produced.

Limitations

- Can only be used with corrosion-resistant metals.
- Hazardous waste is generated.
- The process is slow when drilling a single hole.
- The handling of acid requires a special environment and precautions.
- Oblique entry is difficult.

3.2.5.3 Electrochemical Jet Drilling

Electrochemical jet drilling (ECJD) is mainly used for fine hole drilling at a diameter-to-depth ratio of 1:12 which is lower than that obtained in ES (capillary) drilling (1:100). As shown in Figure 3.23, the process does not require the entry of the tool as in the case of ES drilling. The process, therefore, avoids the use of fragile tooling. The jet of a dilute acid electrolyte causes dissolution, and enough room is required for the electrolyte to exit, preferably in the form of spray. A typical voltage in the range of 400 to 800 V is considered optimum. The produced hole diameter depends on the electrolyte throwing power. Generally, holes produced by ECJD are four times the diameter of the electrolyte jet. In ES capillary drilling the hole diameter/capillary diameter ratio is normally less than 2. The taper for ECJD is about 5° to 10° included angles, whereas tapers of zero are obtainable in ES capillary drilling.

3.2.5.4 Electrochemical Deburring

When machining metal components, it is necessary to cross-drill holes to interconnect bores. Hydraulic valve bodies are a typical example where many drilled passages are used to direct the fluid flow. The intersection of these bores creates burrs, which must be removed (Figure 3.24) to avoid the possibility of them breaking off and severely damaging the system. Manual removal of burrs is tedious and time

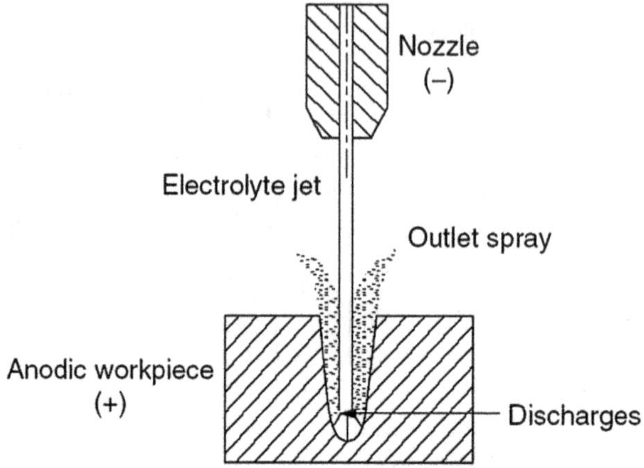

FIGURE 3.23 Electrochemical jet drilling.

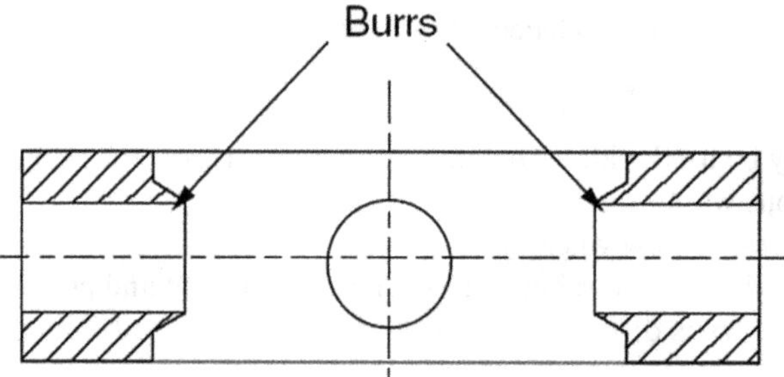

FIGURE 3.24 Burrs formed at intersections of holes.

consuming. In electrochemical deburring (ECDB), the anodic part to be deburred is placed in a fixture, which positions the cathodic electrode in close proximity to the burrs. The electrolyte is then directed, under pressure, to the gap between the cathodic deburring tool and the burr. On the application of the machining current, the burr dissolves forming a controlled radius. Since the gap between the burr and the electrode is minimal, burrs are removed at high current densities, Figure 3.25. ECDB can be applied to gears, spline shafts, milled components, drilled holes, and punched blanks. The process is particularly efficient for hydraulic system components such as spools, and sleeves of fluid distributors.

ECDB using a rotating and feeding tool electrode (Figure 3.25) enhances the deburring process by creating turbulent flow in the interelectrode gap. The spindle rotation is reversed to increase the electrolyte turbulence. Normal cycle times for deburring reported by Brown (1998) are between 30 to 45 s, after which the spindle is retracted, and the part is removed. In simple deburring when the tool is placed over the workpiece, a burr height of 0.5 mm can be removed to a radius of 0.05 to 0.2 mm leaving a maximum surface roughness of 2 to 4 μm.

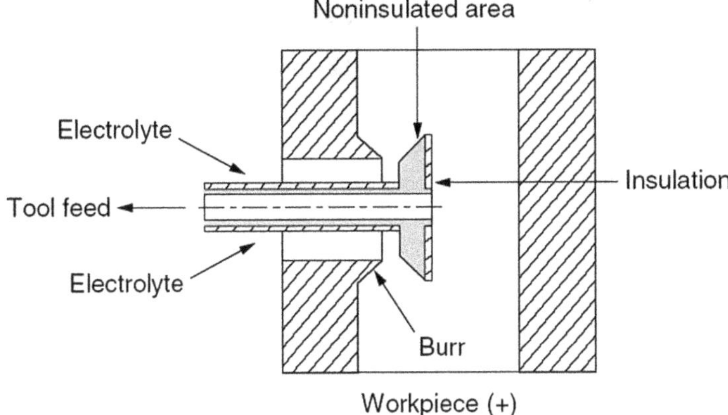

FIGURE 3.25 Hole deburring.

When burrs are removed from intersections of passages in housing, the electrolyte is directed and maintained under a pressure of 0.3 to 0.5 MPa using a special tool. That tool has as many working areas as practical so that several intersections are deburred at a time, Figure 3.26. Proper tool insulation guarantees the flow of current in areas near the burr. The deburring tool should also have a similar contour of the work part thus leaving a 0.1 to 0.3 mm interelectrode gap. Moreover, the tool tip should overlap the machined area by 1.5 to 2 mm in order to produce a proper radius.

The choice of the electrolyte plays an important role in the deburring process. Table 3.6 presents different electrolytes and the operating conditions for ECDB of some materials. ECDB power units supply a maximum current of 50 A. However,

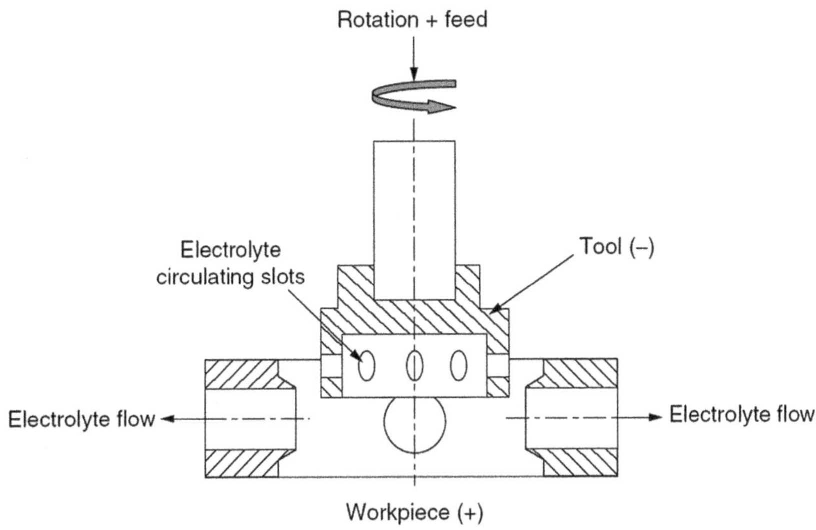

FIGURE 3.26 Electrochemical deburring using a rotating tool.

TABLE 3.6
Machining Conditions for Deburring of Different Materials

Material	Electrolyte	Applied Voltage (volt)	Current Density (A/cm²)	Time s
Carbon and low carbon steel	5–15% $NaNO_3$			
	2–2.5% $NaNO_2$			
Copper alloys	5–15% $NaNO_3$	12–24	5–10	5–100
Aluminium alloys	5–20% $NaNO_3$			
Stainless steels	5% $NaNO_3$			
	+0.5% NaCl			

Adapted from Rumyantsev, E and Davydov, A, 1984.

power units having 500 A are used to remove burrs generated by turning and facing operations on large forged parts.

Advantages

- Elimination of costly hand deburring
- Increase of product quality and reliability
- Ensures the removal of burrs at the required accuracy, uniformity,
- proper radius, and clean edge
- Reduced personnel and labor cost
- Can be automated for higher productivity

3.2.5.5 Electrochemical Polishing

Electrochemical polishing (ECP) is a special form of ECM arranged for polishing holes or cavities in electric conductive materials, Figure 3.27. Polishing parameters are similar to those for ECM, but without the feed motion. ECP generally uses a larger gap and a lower current density than ECM. This requires modestly higher voltages. Sometimes, after ECM of a cavity, the ECM electrode is allowed to dwell at the end of cut to perform a polishing action. Dwelling should be performed under carefully selected conditions and timing to prevent a bulge in the cavity. Occasionally, the ECM electrode is retracted slightly, to double or triple the cutting gap, to promote better ECP. An uninsulated electrode may also be inserted in place of cutting electrode to polish rough side walls of deep pockets.

ECP operates in the same manner as ECM; however, the operating parameters are slightly modified. The working gap is larger, typically 0.5 mm; the current density is lower, less than 25% of that of ECM; and the voltage is higher, up to 50% of that of ECM. The electrolyte flow rate is maintained at its usual high level. ECP requires modifications of electrodes and operating conditions of regular ECM equipment.

ECP removes insignificant amounts of material. The accuracy of the timing of the ON cycle controls the tolerances and surface roughness. A roughness R_a of 0.3 am can be easily attained in stainless steels and super alloys.

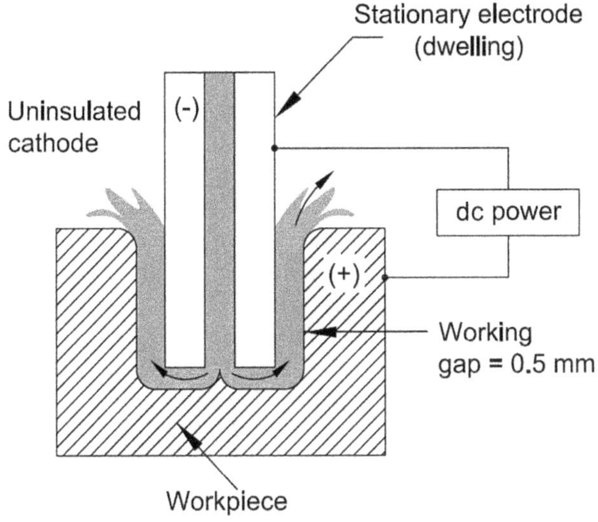

FIGURE 3.27 ECP schematic.

3.2.5.6 Electrochemical Sharpening

Electrochemical sharpening (ECS) is a special form of ECM arranged to perform sharpening or polishing using a hand tool. Figure 3.28 illustrates a schematic of a portable ECS equipment with electrolytic supply of small flow, and small current. A suction tube (not shown in the figure) picks up the used electrolyte for filtration and recirculation. Only 10 amperes or less are required for a portable ECS equipment, and 2–3 L/min flow rates are sufficient. The voltage range is similar to ECM, 8–18 V. ECS equipment have been packed in potable sets with the necessary accessories. Rapid, localized finishing to low roughness levels can be achieved on conductive materials and tools. Spot removal and recast layers or affected heat zone areas are practical finishing bench operations. ECS smooths without significant material

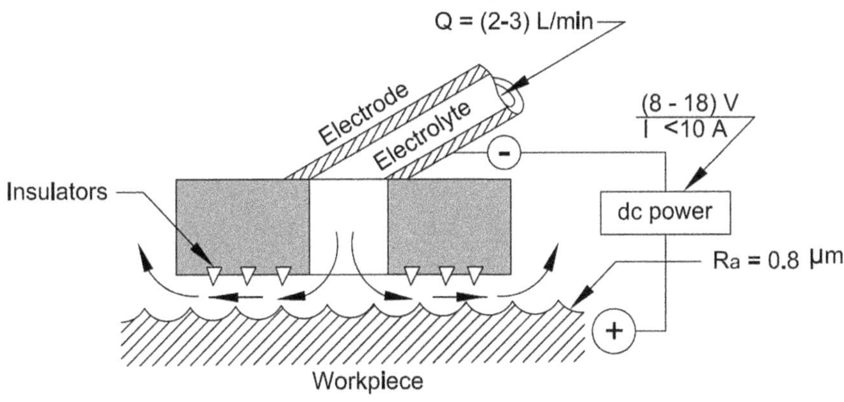

FIGURE 3.28 ECS schematic.

removal. Roughness levels of $R_a = 0.8$ μm from an initial roughness of $R_a = 1.6$ μm are possible within a few seconds of application.

3.2.5.7 Electrochemical Grinding

Electrochemical grinding (ECG) is one of the most important hybrid processes, in which metal is removed by a combination of electrochemical dissolution and mechanical abrasion. The equipment used in ECG is similar to a traditional grinding machine, except that the grinding wheel is metal-bonded with diamond or borazon abrasives; the wheel is the negative electrode and is connected to the dc supply through the spindle insulated from the machine frame, whereas the work is connected to the positive terminal of the dc supply, Figure 3.29.

A flow of electrolyte, usually sodium nitrate ($NaNO_3$) is provided in the direction of wheel rotation for the ECM phase of the operation. The wheel rotates at a surface speed of 25 to 35 m/s. The abrasives in the wheel are always non-conductive, thus they act as an insulating spacer, maintaining a separation (electrolytic gap) of 12 to 80 μm between electrodes. The abrasives also mechanically remove the reaction residue from the working gap and cut chips from the workpiece. Therefore, removal of workpiece material occurs due to the flow of current (100 to 300 A/cm^2) through the gap and the rate of decomposition is speeded up by the grinding action of the abrasive grains. With proper operation, typically 95% of material removal is due to electrolytic dissolution, and only about 5% is due to the abrasion effect of the grinding wheel. Consequently, the wear of the wheel is very low, thus eliminating or considerably reducing the need for redressing which reduces the sharpening costs by approximately 60%.

The lack of heat damage, distortions, burrs, and residual stresses in ECG is very advantageous, particularly when coupled with high MRR in addition to far less wear of the grinding wheel. That is why the process has been applied most successfully in sharpening cutting tools, die inserts, and punches made of hardened high strength steel alloys. ECG is suitable for applications similar to those for milling, grinding,

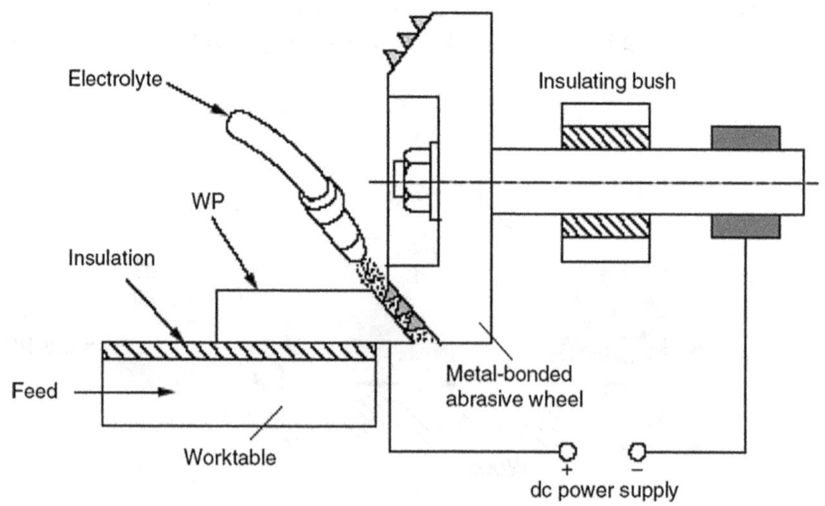

FIGURE 3.29 ECG setup (El-Hofy, 2005).

cutting-off, and sawing. It is not adaptable to cavity-sinking operation such as die making. The process offers the following specific advantages over traditional diamond wheel grinding:

- Increased MRR due to the added electrochemical effect
- Reduced tool wear and sharpening costs
- Less risk of thermal damage and distortion
- Absence of burrs
- Reduced wheel pressure which improves accuracy

However, ECG has the following disadvantages:

- Higher capital cost of the equipment
- Limited to electrically conductive materials
- Hazard due to corrosive nature of electrolyte, for that reason $NaNO_3$ of limited corrosion nature is used
- Necessity of electrolyte filtering and disposal

Some typical applications of ECG operation are:

- Sharpening of carbide cutting tools. ECG provides savings of about 75% in wheel costs and about 50% in labor costs when sharpening carbide tools (Pandey and Shan, 1980).
- It is particularly applied for grinding fragile parts such as honeycomb, thin walled tubes, and skin hypodermic needles. Figure 3.30 shows a thin walled fragile tube made of 316 stainless steel which is electrochemically ground. After machining, the tube is free of distortions and the ground edges are free of burrs. The production rate is, approximately, 12 pieces per hour.

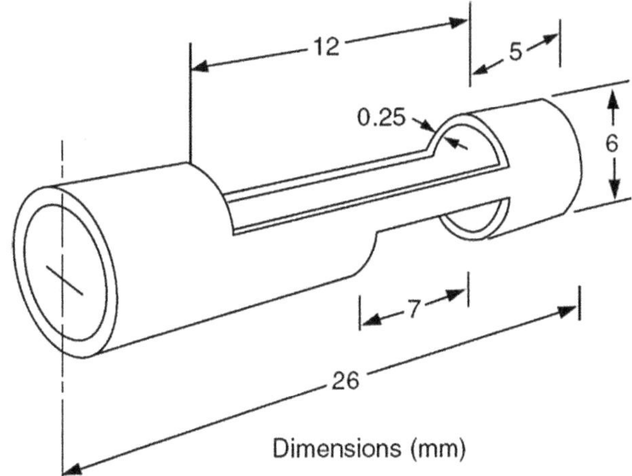

FIGURE 3.30 ECG of thin walled fragile tube made of 316 stainless steel (ASM International, 1989).

ECG machines are now available with numerical controls, thus improving accuracy, repeatability, and increased productivity. In general, metal removal rates of the ECG process of the order of 1 cm^3/min/100A are realized. Surface roughness in the range of 0.2 to 0.6 µm can be easily obtained; generally, the higher the hardness of metal or alloy, the better the finish. Typical tolerances achieved by ECG are of the order of ±10 µm. If more tight tolerances are essential, the majority of stock can be removed by ECG and a final pass of 10 to 100 µm can be taken traditionally, on the same equipment, by switching off the machining current.

3.2.5.8 Electrochemical Honing

Electrochemical honing (ECH) is the removal of conductive materials by anodic dissolution combined with mechanical abrasion from a rotating and reciprocating abrasive stick, carried on a cathode spindle, separated from the workpiece by a rapidly flowing electrolyte, Figure 3.31. The principal material removal action comes from the anodic dissolution. The abrasive sticks are dedicated to maintaining size and to clean the surface to expose fresh metal to the electrolytic action. The small gap is maintained by the non-conductive abrasives bonded to the honing sticks, which exert radial pressure against the surface to be honed. The mechanical honing action of the ECH uses speeds typical to traditional honing; however, lower pressures are used. The material removal rate is three to five times faster by ECH than by traditional honing. The advantages of ECH are most pronounced when honing hard materials, which must be electric conductive. The absence of heavy cutting forces keeps the workpiece cool, free from heat distortion, and reduces the tool wear. The electrolytic action of ECH introduces no stresses and deburrs. Production tolerances of ±13µm are achievable, and a surface roughness R_a of 0.2–0.8 is routine.

Operating Parameters of ECH

dc power supply:

- 8–30 V
- 100–3000 A (15–460 A/cm^2)

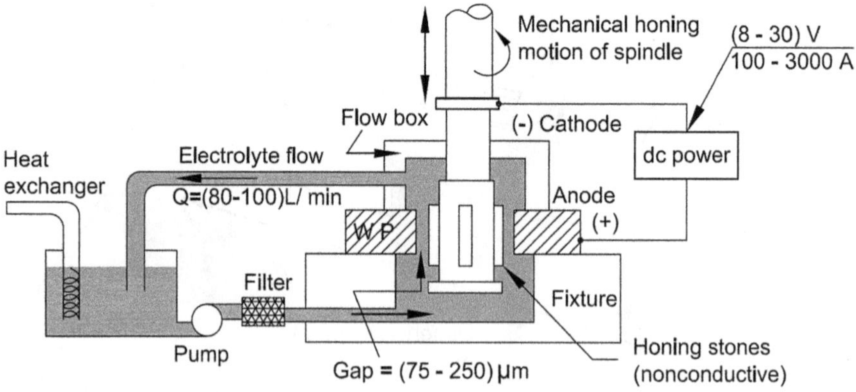

FIGURE 3.31 ECH setup (Machinability Data Center, 1980).

Electrolyte:

- $NaNO_3$(240 g/L), NaCl (120 g/L)
- Temp. = 38°C
- Pressure = 500–1000 kPa
- Flow rate = 80–100 L/min
- Gap = 75–250 am

For a controlled surface roughness and better surface appearance, the stones are sometimes allowed to cut for a few seconds after the current has been turned off. This sequence, however, will leave a light, compressive residual stress in the surface.

Commercial ECH machines are equipped with suitable pumps, tank filters, power packs, and controls for the combined material removal mode. Automatic size gauging is sometimes incorporated. Elements and tooling coming into contact with electrolyte should be fabricated from stainless steels and plastics that are resistant to the corrosiveness of the electrolyte. Sizes of ECH machines range from 10 to 150 in diameter with power packs up to 3000 amperes.

A reversed-polarity version of ECH combines honing with plating. This "Hone-Forming" uses plating solutions in place of deplating electrolytes and can achieve rapid accurate metal deposition. The mechanical honing action prepares a clean surface for the plating and sizes the holes to close tolerances. The equipment is quite similar to the ECH machine, and the acronym ECF is applied.

3.3 REVIEW QUESTIONS AND PROBLEMS

3.3.1 What are the advantages and disadvantages of ECM? Show diagrammatically the main elements of an ECM machine.

3.3.2 What are the advantages and disadvantages of PCM?

3.3.3 What are the advantages and limitations of CHM?

3.3.4 Explain what is meant by EC deburring.

3.3.5 What are the factors on which the selection of a resist (maskant) for use in CHM depends? Distinguish between cut and peel resists and photographic resists.

3.3.6 What measures should be considered to achieve maximum dimensional control in ECM?

3.3.7 Describe the PCM process and list its fields of application.

3.3.8 What are the functions of an electrolyte? What factors need to be considered while selecting it? Discuss the advantages and limitations of some common electrolytes.

3.3.9 Write down Faraday's law of electrolysis.

3.3.10 What is the "self-adjusting feature" in ECM?

3.3.11 Use a neat sketch to briefly explain the principles of the ECG process. What are the main advantages of ECG over conventional grinding?

3.3.12 On what types of works is the process of chemical machining (CHM) best suited and what are its advantages and limitations?

3.3.13 What are the specific advantages of using chemical machining over electrochemical machining? Give some practical applications of the chemical machining process.

3.3.14 Explain the difference between CHM and ECM.

3.3.15 What is the principle of electrochemical machining? What are the materials commonly used for making a tool for this method? Is there any limitation on the type of material that can be machined by ECM?

3.3.16 Write a short note on the effect of high temperature and pressure of electrolyte in the ECM process.

3.3.17 What is the main factor which limits the MRR in ECM? Explain why the surface quality of electrochemically machined holes improves with increasing feed.

3.3.18 "One of the main advantages of ECM is that no tool wear takes place." Explain this fact by considering the reactions taking place in the electrolytic gap.

3.3.19 In what ways does electrochemical grinding (ECG) differ from the ordinary grinding process?

3.3.20 What are factors affect both mechanical and EC stock removal rate in ECG?

3.3.21 What is the main application of ECG?

3.3.22 For what type of works is ECG best suited?

3.3.23 Calculate the expected machining rate and electrode feed rate when iron is electrochemically machined using copper electrode and NaCl electrolyte (ρ_s = 5.0 Ω.cm). The following data are provided:

- Valence of Fe n = 2 and 3
- Atomic weight of Fe N = 56
- Current I = 5000 Amp
- Density of iron ρ_{Fe} = 7.78 gm/cm^3
- Gap voltage E = 18 V (dc)
- Current efficiency η = 100%
- Tool work equilibrium gap h = 0.5 mm

Draw the relation between the MRR and the current I, and the relation between the tool feed rate V_f, and the interelectrode gap thickness h for both valences.

3.3.24 Repeat the same for the following anode materials: Aluminum N = 27, n = 3, and silver N = 108, n = 1.

3.3.25 Determine the theoretical metal removal in ECM of zinc workpiece at a machining current of 100 Amp, the zinc density ρ = 7.13 gm/cm^3, and of chemical equiv. = 33.

3.3.26 A Nimonic alloy of density 7.85 gm/cm^3 is machined electrochemically using NaCl (electrolyte conductivity χ = 0.2 Ω^{-1}.cm^{-1}) to remove a stock of 200 grams. If 200 Amp current and 15 V were used which caused a current density of 80 A/cm^2). Calculate the equilibrium gap, and the equilibrium feed rate.

3.3.27 Calculate the amount of Ti removed in grams in ECM, given that:
- Atomic wt. N = 47.9, valence n = 4, ρ_{Ti} = 4.56 gm/cm^3
- Machining time = 2 min
- Work area = 12 cm^2
- Current density = 250 A/cm^2, gap voltage E = 12 V
- Faraday's const. = 96500 coulombs
- Electrical resistivity $\rho_s = 4\ \Omega$.cm

Calculate the expected tool feed rate and gap thickness.

3.3.28 Determine the metal removal rate in an ECM operation on an alloy of Ni-Cr-Mo at 10000 A. The density of alloy is 8.91 g/cm^3 and its composition (%) is as follows: C 0.1, Cr 15, Fe 5.5, Mo 15.9, Ni 59.5, W 4. Neglecting carbon, which does not dissolve anodically, the atomic weights of the elements are 52.01, 52.86, 95.95, 58.69, 183.92 respectively. It may be assumed the following valences will apply in the process 2,2,3,2,6 respectively.

3.3.29 Calculate the time taken to grind a 1 mm layer from the face of a hardened steel insert 15 mm square by ECG. Assume a current eff. of 0.95, and the mass of metal removed mechanically is 10% of the total mass of the metal removed. The density of steel = 7.8 g/cm^3. Then calculate the volumetric metal removal rate as removed mechanically.

3.3.30 Mention the parameters affecting the MRR of the processes: USM, ECM, WJM.

3.3.31 Mention briefly the purpose of:
- Adding 1% of a long-chain polymer to the water in WJM
- Applying a premagnetizing current in addition to HF current in USM
- Applying suction of abrasive slurry in USM

REFERENCES

ASM International, *Metals handbook* 1989, Machining, vol. 16, Materials Park, OH.

Banard, J. 1978, Fine hole drilling using electrochemical machining.*MTDR Conf. Proc.*, pp. 503–510.

Brown, J. 1998. *Advanced machining technology handbook*. New York, NY: McGraw-Hill.

CASA Research Project <www.win.tue.nl/casa/research/casaprojects/noot.htm>.

El-Hofy, H 2005, *Advanced machining processes, nontraditional and hybrid processes*, McGraw-Hill Co., New York.

Gusseff, W 1930, *Method and apparatus of electrolytic treatment of metals*, BP, Nr. 335003.

Kalpakjian, S 1984, *Manufacturing processes for engineering materials*, Addison Wesley Publishing Co., Reading, MA.

Machinability Data Center. *Machining data handbook* 1980, vol. 2, 3rd edn, Cincinnati, OH.

Pandey, PC & Shan, HS 1980, *Modern machining processes*, Tata McGraw-Hill Co., New Delhi.

Rumyantsev, E & Davydov, A 1984, *Electrochemical machining of metals*, Mir Publishers, Moscow.

Visser, A, Junker, M & Weiβinger, D 1994, *Sprühätzen metallischer Werkstoffe*, 1st Auflage, Eugen G. Leuze Verlag, Saulgau, Württ, Germany.

Youssef, HA 2005, *Non-traditional machining processes-theory and practice*, 1st edn, El-Fath Press, Alexandria.

Youssef, HA 2016, *Machining of stainless steels and super alloys: Traditional and nontraditional techniques*, John Wiley.

4 Thermo-Electrical Non-Traditional Machining Operations and Machine Tools

4.1 ELECTRICAL DISCHARGE MACHINES AND OPERATIONS

4.1.1 Process Characteristics and Applications

Of all the NTM processes, none has gained greater industrial wide acceptance than electrical discharge machining (EDM). It is well known that when two current conducting wires are allowed to touch each other, an arc is produced. Although this phenomenon was detected by Priestly in 1790, it was not until the 1940s that a machining process based on this principle was developed by Lazerenko (1944) in Russia. The principle of EDM (also called spark erosion machining) is based on erosion of metals by spark discharges between a shaped tool electrode (usually negative) and the workpiece (usually positive).

The tool and the workpiece are separated by a small gap of 10 to 500 μm. Both are submerged or flooded with electrically non-conducting dielectric fluid. When a potential difference between the tool and the workpiece is sufficiently high, the dielectric in the gap is partially ionized, so that a transient spark discharge ignites through the fluid, at the closest points between the electrodes. Each spark of thermal power concentration, typically 10^8 W/mm^2, is capable of melting or vaporizing very small amounts of metal from the workpiece and the tool, Figure 4.1. A part of the total energy is absorbed by the tool electrode yielding some tool wear, which can be reduced to 1% or less if adequate machining conditions are carefully selected.

The instantaneous vaporization of the dielectric produces a high-pressure bubble that expands radially. The discharge ceases with the interruption of the current, and the metal is ejected leaving tiny pits or craters in the workpiece and metal globules suspended in the dielectric, Figure 4.2. Sludge of black carbon particles, formed from hydrocarbons of the dielectric, is produced in the gap and expelled by the explosive energy of the discharge, remaining in suspension until removed by filtering. Immediately following the discharge, the dielectric surrounding the channel deionizes and, once again, becomes effective as an insulator.

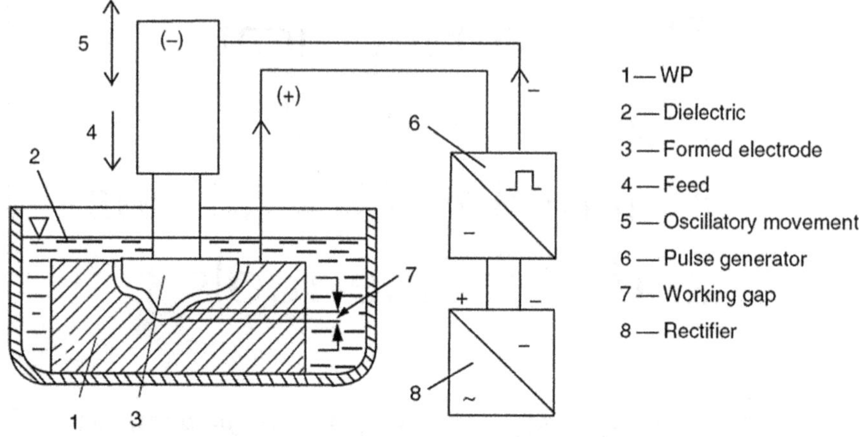

FIGURE 4.1 Concept of EDM (Düniβ, Neumann, and Schwartz, 1979).

The capacitor discharge is repeated at rates between 0.5 and 500 kHz, at a voltage between 50 and 380 V and currents from 0.1 A to 500 A. EDM is characterized by the following advantages:

- It is applicable to machine metals, alloys, and carbides irrespective of their hardness and toughness.
- Because there is no mechanical contact between the tool and the workpiece, very delicate work can be machined.
- The process is widely used to produce cavities and profiles of complex and intricate shapes accurately in extra-hard materials.
- The process leaves no burr on the edges.
- EDM electrodes can be accurately produced from machinable materials, which have a positive impact on the accuracy of machined holes and cavities.

On the other hand, the process is hampered by the following limitations:

- It cannot be used if the workpiece material is a bad electric conductor.
- On machining materials, the process produces a heat-affected zone (HAZ), Figure 4.3, which is characterized by hair cracks and thin, hard recast layer. The surface integrity (SI) is improved if other machining processes such as chemical machining, shot peeking, etc., are used subsequently to EDM. This measure is done if the product will be operating in a fatigue environment.
- EDM cannot produce sharp concerns and edges.
- The process is characterized by high specific energy.

As compared to traditional machining, a major problem of EDM is the low MRR.

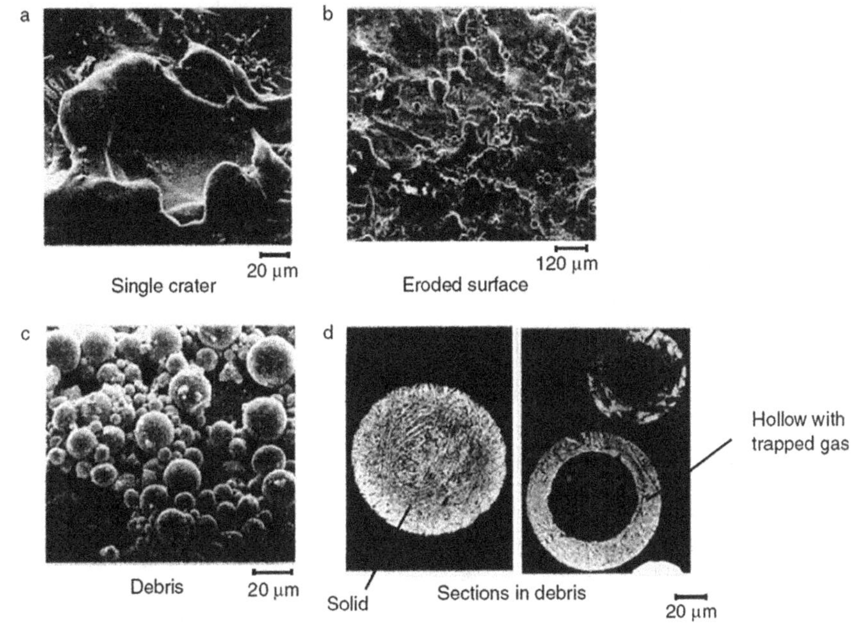

a
Single crater 20 μm

b
Eroded surface 120 μm

c
Debris 20 μm

d
Hollow with trapped gas
Solid
Sections in debris
20 μm

	Working conditions V_o = 150 V			
	Workpiece	Electrode	t_d (μs)	i_d (A)
a	Cu(–)	Cu	200	34
b	56 NiCrMoV 7	Cu	100	43
c	56 NiCrMoV 7	C	100	28
d	56 NiCrMoV 7	C	250	60

FIGURE 4.2 SEM surfaces and debris produced.

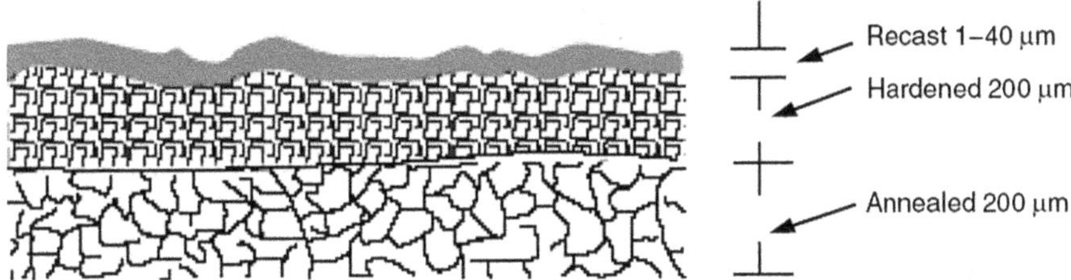

FIGURE 4.3 HAZ of surface produced by EDM.

Applications

EDM has become an indispensable process in modern manufacturing. It is used in numerous applications such as producing die cavities for automotive-body components, connecting rods, and various intricate shapes to a high degree of accuracy. About 80 to 90% of EDM work is in the manufacture of tool and die sets for the production of castings, forgings, stampings, and extrusions. Micromachining of holes, slots, texturing, and milling are also typical applications of the process. Figure 4.4 illustrates some typical products, made by EDM.

- Machining of forming molds (Machinability Data Handbook, 1980)
- Machining of fine holes in carbide plate and fuel injection made of heat-treated steel (Kalpakjian, 1984)
- Sinking of two connecting rods with electrode (Nassovia-Krupp, 1967)
- Sinking of extrusion insert plate of motor anchor with electrode (Nassovia-Krupp, 1967)
- Machining of embossing die (Nassovia-Krupp, 1967)
- Sinking of injection mold insert for light alloys (Nassovia-Krupp, 1967)
- Machining of spherical internal cavity using a specially designed mechanically rotating electrode, which is equipped with a hinged tip) (Nassovia-Krupp, 1967)

4.1.2 Sinking Machine

A typical setup of an ED sinking machine is illustrated in Figure 4.5. Most machines are of the ram type, in which the sinking head is actuated by a hydraulic cylinder.

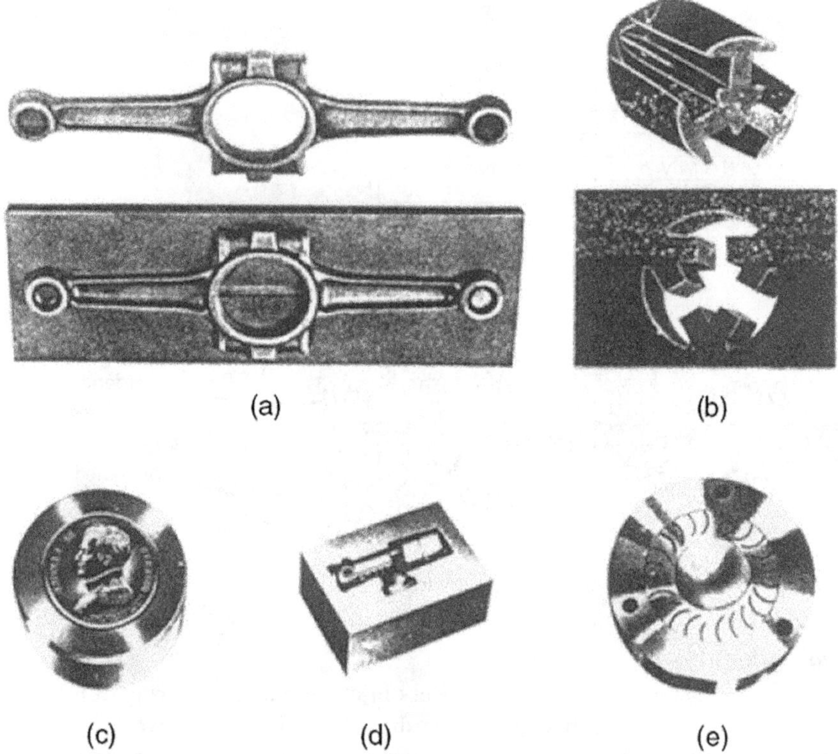

FIGURE 4.4 Typical products produced by EDM: (a) Forging die of two connecting rods, machined with one electrode, (b) extrusion insert plate of motor anchor and electrode, (c) embossing die, (d) forging die, and (e) injection mold insert for light alloys.

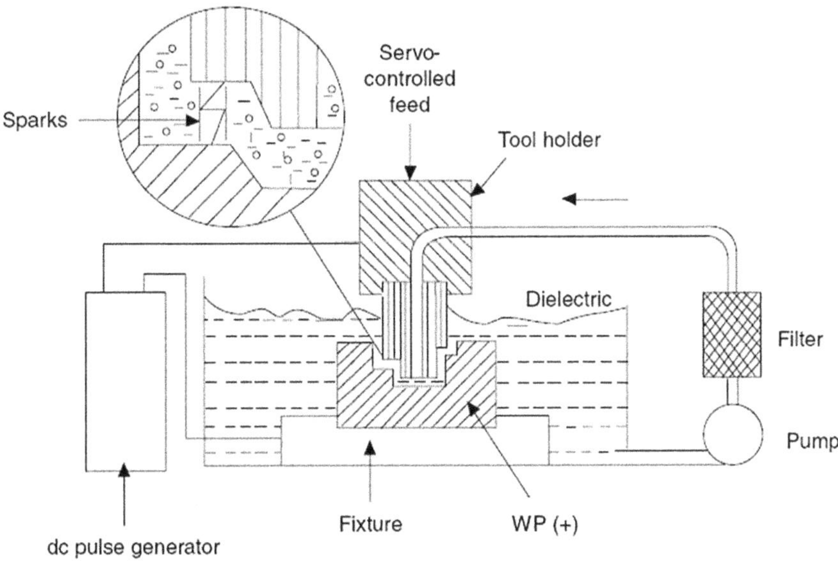

FIGURE 4.5 ED sinking machine.

In EDM, the gap between the electrode and the workpiece is critical, thus the down feed of the tool should be controlled by a servomechanism which automatically maintains a constant gap. The servo gets its input signal from the difference between the selected reference voltage and the actual voltage across the gap. This signal is amplified, and the tool is advanced by hydraulic control. A short circuit across the gap causes the servo to reverse the motion of the tool until proper control is restored.

The workpiece is clamped within the tank containing the dielectric fluid, and its movements are numerically controlled. The machine is equipped with a working table mounted on orthogonal slide ways, often provided with accurate optical scales, similar to that of a jig-boring machine. The machine is also equipped with a pump and filtering system of the dielectric fluid. The dielectric is flushed through the spark gap, supplied either through a hollow tool or from external jets, or both. It may also be supplied through holes in the workpiece.

Dielectric fluids: The dielectric fluids have four main functions. They act as an insulator between tool and work, as a spark conductor, and provide a cooling and flushing medium. The fluid must ionize to provide a channel for the spark and deionize quickly to become an insulator. A good dielectric has the following properties:

- Low viscosity to ensure effective flushing
- High flash point
- High latent heat
- A suitable dielectric strength, e.g. 180 V/25 μm
- Rapid ionization at potentials 40 to 400 V followed by rapid deionization
- Non-toxic
- Non-corrosive
- Non-expensive

The most common dielectric fluids are hydrocarbons, although kerosene and distilled and deionized water may be used in specialized applications. Polar compounds such as glycerine water (90:10) with triethylene oil as an additive have been proven to improve the MRR and decrease the tool wear as compared to kerosene.

Flushing techniques: Dielectric flushing techniques are of vital importance in EDM. At the beginning, the fresh supply of dielectric is clean and has a higher insulation strength than one containing particles (contaminated). When spark discharges commence, debris is created, and insulation strength is diminished by particles acting as stepping stones in the interelectrode gap. If too many particles are allowed to remain, a bridge is formed resulting in arcing across the gap causing damage to the tool and workpiece. Therefore, the contamination in the gap must be controlled to provide optimum conditions, Figure 4.6. In the same figure, the effect of optimum flushing on the MRR with the tool advance is schematically illustrated.

Figure 4.7 differentiates between the techniques of flushing:

- Injection flushing in which a slight taper is produced on the sides of the cavity due to lateral discharges as debris pass up the side of the tool,Figure 4.7a
- Suction flushing through which the side taper is avoided, Figure 4.7b
- Side flushing, in which a slight taper is produced on the side of the cavity at the outlet (downstream) of the dielectric, Figure 4.7c

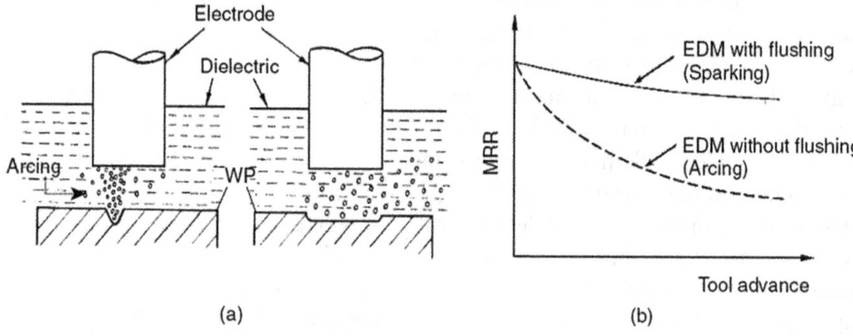

FIGURE 4.6 (a) Contamination in EDM (Lissaman and Martin, 1982) and (b) effect of flushing on MRR (Youssef, 2005).

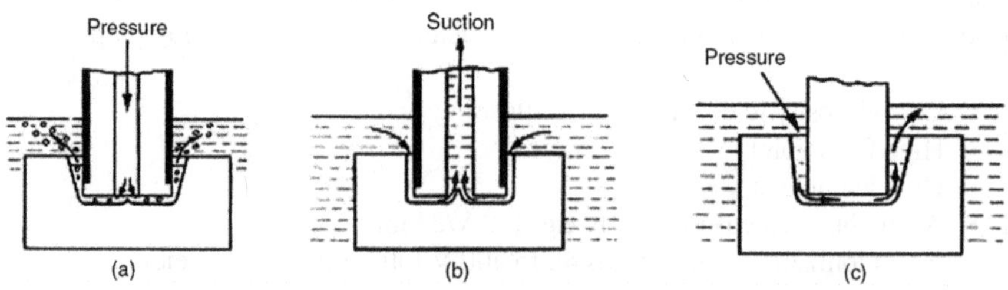

FIGURE 4.7 Dielectric flushing modes: (a) Injection flushing, (b) suction flushing, and (c) side flushing (Lissaman and Martin, 1982).

4.1.3 EDM-Spark Circuits (Power Supply Circuits)

The ED machine is equipped with a spark generating circuit which can be controlled to provide optimum conditions for a particular application. This generator should supply voltage adequate to initiate and maintain the discharge process and provide the necessary control over the process parameters such as current intensity, frequency and cycle times of discharge. The cycle time ranges from 2 to 1600 μs (Lissaman and Martin, 1982). Two main types of generators are applicable for this purpose. These are the resistance-capacitance generator (RC circuit) and the transistorized pulse generator.

4.1.3.1 Resistance-Capacitance RC Circuit

It is also called the Lazerenko circuit (1944) which is basically a relaxation oscillator. It is a simple, reliable, rigid, low-cost power source that is ordinarily used with copper or brass electrodes. It provides fine surface texture of 0.25 μm R_a, but the machining rate is slow, because the time required to charge the capacitors prevents the use of high frequencies. The relaxation circuit operates at selectively high input voltages and is difficult to operate. The reversed polarity encountered in the relaxation circuit leads to an additional tool wear.

The basic form of the RC circuit is shown in Figure 4.8a. On commencing operation, the capacitor is in the uncharged condition. Then it is charged from a dc voltage source *Vo* usually 200 to 400 V via the resistor *R* which determines the charging rate. The capacitor voltage *Vc* increases exponentially as charging proceeds, Figure 4.8b.

$$V_C = V_O\left(1 - e^{-t/RC}\right) \tag{4.1}$$

Where,

t = time, s

RC = time constants = resistance (Ω) × capacitance, Farad

When *Vc* attains the level of breakdown voltage V_s existing in the working gap, the capacitor charges across the gap eroding both the workpiece (causing material removal) and tool electrode (causing wear). The spark is not sustained because the capacitance is discharged quicker than it can recharge via the resistor,

$$t_d = 0.1 t_c \tag{4.2}$$

where,

t_c = charging time

t_d = discharging time

The cycle of charging and discharging is repeated until the cut is performed.

For maximum production rate (Barash, 1962):

$$V_S = 0.73\, V_O \tag{4.3}$$

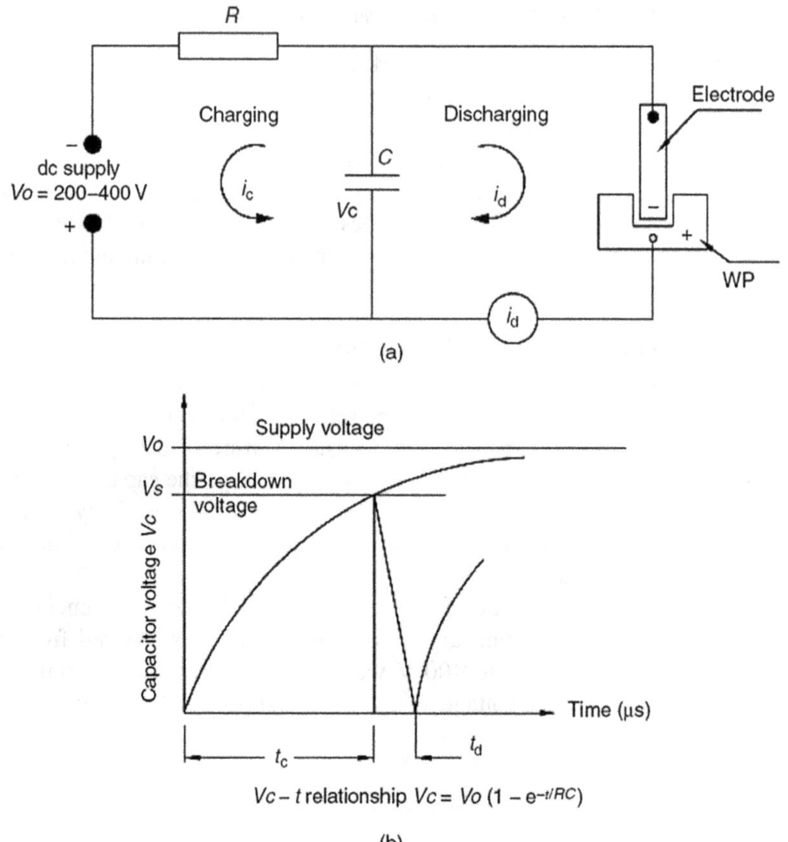

FIGURE 4.8 (a) RC circuit and (b) capacitor voltage-charging time exponential relationship.

The energy of each individual spark discharge in joule is given by

$$E_d = \frac{1}{2} C V s^2 \tag{4.4}$$

Therefore, the increase of V_o, V_s, and C leads to an increase of machining rate, however it leads to poor surface texture.

A reduction of V_s enables a smaller gap to be used, improving finish and accuracy, but reducing machining rate. High rates of machining are obtained by reducing the time constant RC in Equation 4.1 to give rapid charging. However, as R is reduced, the frequency increases and may reach a point at which deionization is prevented from taking place and arcing occurs. Arcing causes effective machining to cease and creates thermal damage to the machined surface. It follows that in the RC circuit, the machine setting for optimum performance in a given set of machining conditions involves a compromise in selecting the process parameters.

Illustrative Example 1

In an EDM operation using Lazerenko's generator, if $Vo = 250$ V, $R = 10\ \Omega$, $C = 3\ \mu F$.

If the cut is required to be performed at the maximum removal rate condition, calculate:

1. Discharge voltage
2. Charging time, t_c
3. Cycle frequency, f_r
4. Energy/individual discharge of the capacitor, E_d
5. If the dielectric used has a strength of 180 V/25 μm, estimate the expected gap thickness to realize this cut

SOLUTION

The cut is performed at maximum removal rate condition, then:

1. $V_s = 0.73\ Vo$
   $= 0.73 \times 250 = 182.5$ V

2. Charging time, t_c

$$V_S = V_O\left(1 - e_c^{-t/RC}\right)$$

$$\frac{V_S}{V_O} = 0.73 = 1 - e_c^{-\frac{t}{30}}\ (C \text{ in } \mu\text{F, and } t_c \text{ in } \mu\text{s})$$

from which,

$$t_c = 39.2\ \mu\text{s}$$

3. Cycle frequency, f_r

$$f_r = \frac{t}{t_c + t_d}$$

$$t_d = 0.1 \quad t_c = 3.9\ \mu\text{s}$$

$$t_c + t_d = 39.2 + 3.9$$

$$= 43.1\ \mu\text{s} = 43.1 \times 10^{-6}\ \text{s}$$

$$f_r = \frac{10^6}{43.1} = 23200\ \text{Hz} = 23.2\ \text{kHz}$$

4. Energy/individual discharge, E_d

$$E_d = \frac{1}{2}CVs^2 = \frac{1}{2} \times 3 \times 10^{-6} \times (182.5)^2 = 0.05\ \text{J}$$

5. Expected gap thickness, h_g

$$h_g = \frac{Vs}{180} \times 25 = \frac{182.5}{180} \times 25 = 25.5\ \mu\text{m}$$

4.1.3.2 Transistorized Pulse Generator Circuits

Among the disadvantages of the RC relaxation circuits are the interdependence (lack of control of parameters), and the restricted choice of electrode material and their high wear rate. The adoption of the transistorized pulse generators in the 1960s allowed the process parameters (frequency and energy of discharges) to be varied with a greater degree of control, in which charging takes only a small portion of the cycle. Furthermore, the voltage of these machines is reduced to the 60 to 80 V range, permitting discharge characteristics with lower current pulses of a square profile. This results in shallower and wider craters, which means better surface texture. Alternatively, when required, they provide high material removal rates at the expense of surface quality by permitting high discharge currents. Moreover, this type of generator provides considerably lower electrode wear as compared to simpler RC circuits.

In the simple form of the transistorized pulse generators, the parameters are selected and pre-adjusted according to the machining duty. The selected parameters remain constant, i.e. not influenced by the variation of working conditions in the gap during machining.

An improved circuit incorporating feedback (Isopulse generator, Charmilles) is illustrated in Figure 4.9. In such a circuit, the conditions into the spark gap are monitored by a detector unit which determines the exact moment of current flow after the ignition lag. The time base on-time then becomes effective, providing a constant discharge period. The time base for the off-time ensures a constant interval

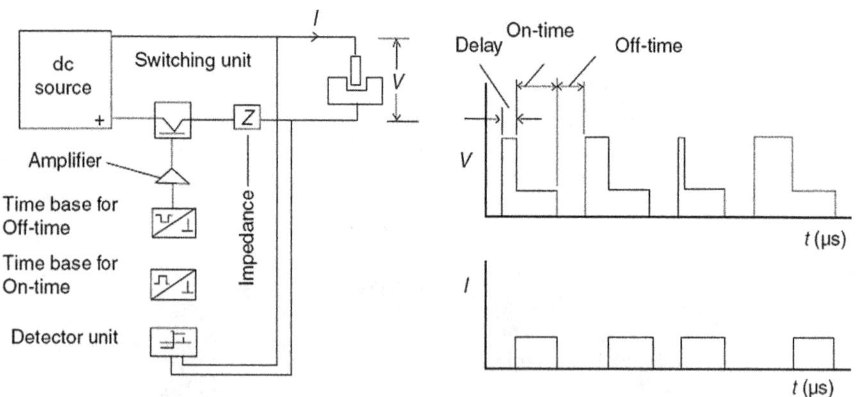

FIGURE 4.9 Pulse generator of Charmilles Technologies, Geneva, Switzerland.

of deionization and flushing away the debris by the dielectric. The following are the specifications of a typical isopulse generator, 25A, produced by Charmilles.

Power	2 kW
Open gap voltage	80 V
Discharge energy	0.18–1 J
Maximum discharge current	25 A
Discharge duration	Off-time 2–1600 μs; on-time 2–1600 μs
Achieved roughness	$R_a = 0.4$ μm

4.1.4 EDM Tool Electrodes

In ED sinking, electrodes are often the most expensive part of an EDM operation. Material fabrication, wear, and redressing costs must be carefully weighed to determine the best electrode material and EDM machine setup. The ideal electrode material should have high electric conductivity, a high melting point, be easy to fabricate, and strong enough to stand up to EDM without deformation.

Most electrodes for EDM are usually made of graphite, although brass, Cu, or Cu-W alloy may be used. These electrodes are shaped by forming, casting, and powder metallurgy or frequently by machining. EDM tool wear is an important factor as it affects the dimensional and form accuracy. It is related to melting point of the tool material involved: the higher the melting point, the lower the wear rate. Consequently, graphite electrodes have the highest wear resistance, since graphite has the highest melting point of any known material (3600°C), moreover it is low in cost and readily fabricated. Tungsten (3400°C) and W alloys are next in melting temperature, followed by molybdenum (2600°C), however these metals are expensive and difficult to fabricate. The tool wear can be minimized by reversing the polarity which depends on the tool/workpiece combination. Table 4.1 illustrates the recommended polarity for various electrode/workpiece material combinations.

TABLE 4.1
Polarity for Most Common Electrode/WP Material Combination

	Electrode Material		
WP Material	**Graphite**	**Cu**	**Cu-W**
Steel	SR	S	S
Cu	R	R	R
Cemented carbide	R	SR	SR
Al	S	S	S
Ni-base alloys	SR	S	S

Note: S – straight polarity (WP positive electrode) and R – reverse polarity (WP negative electrode).
Source: (ASM International, 1989).

The wear may attain zero value during the so-called no-wear EDM process. Work material machinable by no-wear EDM can be steels, satellites, Ni-base alloys, and aluminum. However, no-wear EDM is not recommended for machining carbides.

No-wear EDM requires pulse generators and equipment capable of attaining the following conditions:

- Reverse polarity of the tool electrode
- Low pulse frequency ranging from 0.4 to 20 kHz. Generally, about 2 kHz is recommended
- Graphite, Cu, Cu-W, or Ag-W electrodes
- High duty cycle of more than 90%
- High intensity discharge current
- Smooth control of servomechanism
- Supply voltage of not more than 80 V
- Temperature of dielectric not above 40°C, and dielectric recycled at low pressure
- Dielectric flow must not trap particles
- No capacitance across the spark gap, and no inductance in series with it

4.1.5 Process Capabilities

EDM is a slow process compared to conventional methods. It produces matte and pitted surfaces composed of small craters, which are characterized by a non-directional, randomly distributed nature due to succession of individual sparks of the process. The metal removal rates usually range from 0.1 to 600 mm^3/min. In EDM, there is a proportional relationship between the MRR and the surface roughness. High removal rates produce a very rough finish, having a molten and recast structure with poor surface integrity and low fatigue strength. The finish cuts are made at low removal rates, or the recast layer during rough cuts is removed later on by finishing EDM operations. Table 4.2 illustrates proportional interrelation between removal

TABLE 4.2
Proportional Interrelation between Volumetric Removal Rate and Surface Roughness R_a

Processing	Volumetric Removal Rate (mm^3/min)	R_a (μm)
Metal electrodes		
Gentle	0.75	<1.6
Finishing	0.75–1.5	1.6–3.2
Normal	1.5–110	3.2–6.3
Graphite electrodes (usually of +ve polarity)		
Roughing	110–400	6.3–12.5
Abusive	>400	>12.5

Source: (Youssef, 2005)

rate and surface roughness R_a when cutting steel using metal and graphite electrodes at different processing levels.

The metal removal rate (MRR) depends not only on the workpiece material but also on the machining variables, such as pulse conditions (voltage, current, and duration), electrode material and polarity, and the dielectric. The results in Figure 4.10 show the machining rate and surface roughness (R_z value) for EDM of different materials (El-Hofy, 2005). It is depicted from the same figure that the material of low melting point (A1) has the highest MRR and, hence, the highest surface roughness and vice versa (graphite).

In EDM, the surface finish varies widely as a function of the spark frequency, voltage, current, and other parameters which also control the MRR. New techniques use an oscillating electrode to provide very fine surface quality. Alternatively, bad surfaces and surface defects characterize EDM using graphite electrodes.

The size of the cavity cut by the tool electrode is larger than the tool by a side overcut between the surfaces of the electrode and workpiece. This overcut is due to the sparking in the side gap. It is equal to the length of the spark, which is essentially the constant all over the electrode areas, regardless of size or shape. Typically, overcut values vary from 10 to 300 μm, depending on the breakdown voltage and size of debris flowing in the side gap. In this respect, suction flushing is preferred, because the debris is not drawn past the side gap, thus lateral sparking is minimized leading to smaller overcut and side taper. EDM equipment manufacturers publish overcut charts for their machines; however these values should be only used as a guide for the tool designer.

Typical taper varies from 1 to 5 μm/mm per side, depending on the machining conditions, especially the flushing technique used. The minimum corner radius equals more or less the size of the overcut. Tolerance of ±50 μm can be easily achieved; however, with close control tolerance of ±5 to ±10 μm can be obtained.

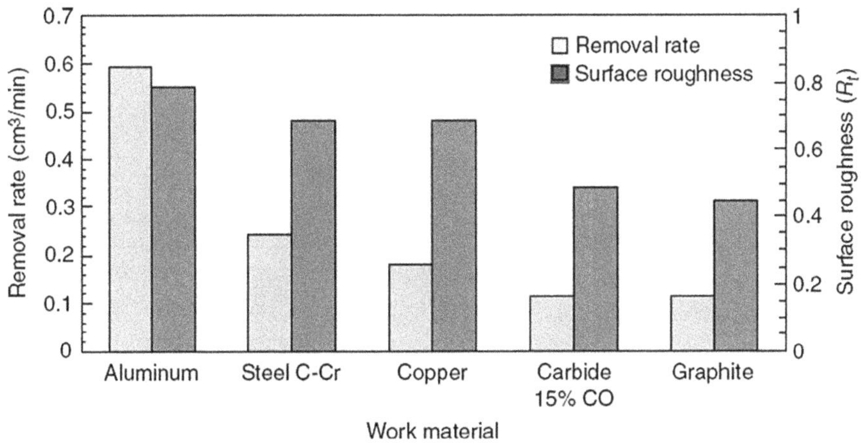

FIGURE 4.10 Machining rate and surface roughness (R_t) for EDM of different materials.

4.1.6 EDM Allied Processes

4.1.6.1 Electrical Discharge Milling (ED Milling)

Conventional ED sinking, discussed in the previous section, requires a preliminary phase for producing specially shaped electrodes. These electrodes are very expensive since they are difficult to design and manufacture, and therefore, they add more than 50% to the total machining cost of the product. Recently, a revolutionary breakthrough in the EDM realm has been achieved by adopting a new ED milling technology, which makes use of simple and cheap standard rotating pipe electrodes.

In this process, three-dimensional cavities are machined by successive sweeps of the electrode down to the desired depth, while the numerical control by means of powerful algorithms automatically compensates the electrode front wear to ensure the product accuracy along the three axes, Figure 4.11. Therefore, there is no need to manufacture the specially shaped electrodes as in the case of conventional EDM, which means a saving of time and money.

The theory of ED milling (also termed ED scanning) is shown in Figure 4.12. The thickness of the layer removed per one path ranges from 0.1 mm to several mms during rough paths and from 1 μm to 100 μm during finish paths.

Process Characteristics

Advantages of ED Milling

- Design and manufacture of electrodes is totally omitted.
- Fine shapes can be readily produced.
- Electrode wear does not need to be considered.
- The surface roughness and waviness is favorable even for large areas.
- Estimating machining time is easy.
- NC data can be directly generated from the EDM die data.
- Sharp edges and corners can be readily produced.

Limitations of ED Milling

- The removal rate may be less than that achieved by conventional EDM.
- If there is large side taper (10° or more), it is difficult to maintain side accuracy.

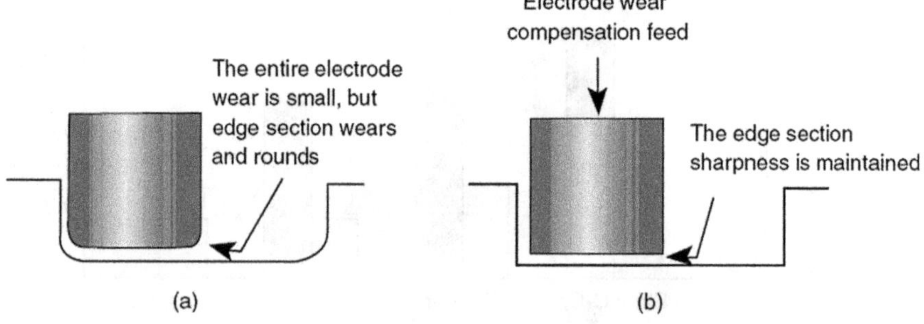

FIGURE 4.11 Compensation of electrode frontal wear in EDM milling: (a) Machining with low wear conditions (conventional method) and (b) machining with conditions providing wear (EDSCAN machining) (Mitsubishi EDSCAN Technical Data, 1997).

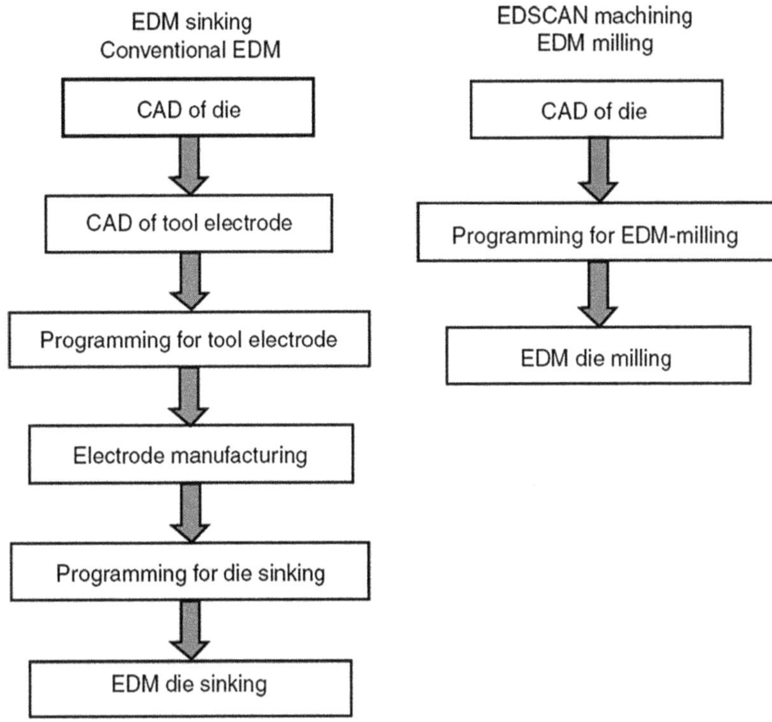

FIGURE 4.12 ED sinking and ED milling steps.

Fields of applications of ED milling: ED milling technology is particularly applicable for machining cavities with or without taper, including three-dimensional shapes. It is used notably for making molds of parts for the electrical and electronic industries, household appliances, and automotive and aeronautical industries. Another technological breakthrough of ED milling is that the process has entered the domain of micromachining. It becomes possible to produce fine and intricate shapes with sharp corners using this process.

When machining with conventional ED sinking, the process is adjusted to ensure low wear conditions of the electrode. Simultaneously, the edges wear and become round. However, when using conditions providing front wear to a certain degree on the pipe electrode, sharp edges of the electrode base can be maintained, while the electrode front wear is automatically compensated. This compensation occurs at a micro level when the depth is periodically measured to ensure final dimensions, keeping the waviness within prescribed limits rendering flat surfaces, Figure 4.13.

The path of the two-dimensional shape is created by the NC system built in the machine beforehand, to allow the target shape to be stored. Layered machining is then carried out by executing the NC path program several times till the required depth is achieved as shown in Figure 4.14b. In the Z direction, electrode wear caused by machining is automatically compensated. A Cu-W pipe electrode ($\phi = 0.2/0.12$ mm) is used. Figure 4.15 illustrates a fine part produced by ED scanning using a fine

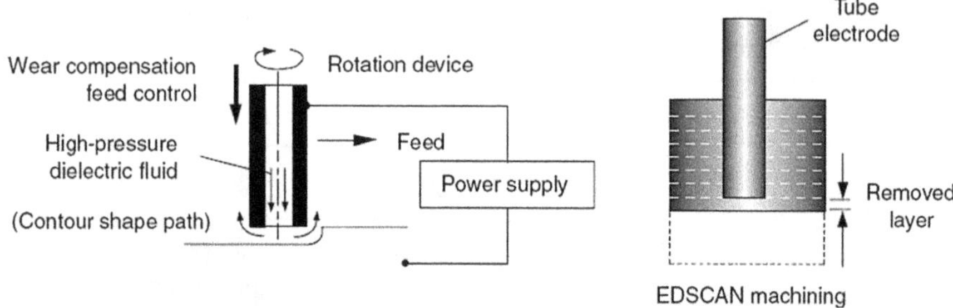

FIGURE 4.13 Theory of ED milling (Mitsubishi EDSCAN Technical Data, 1997).

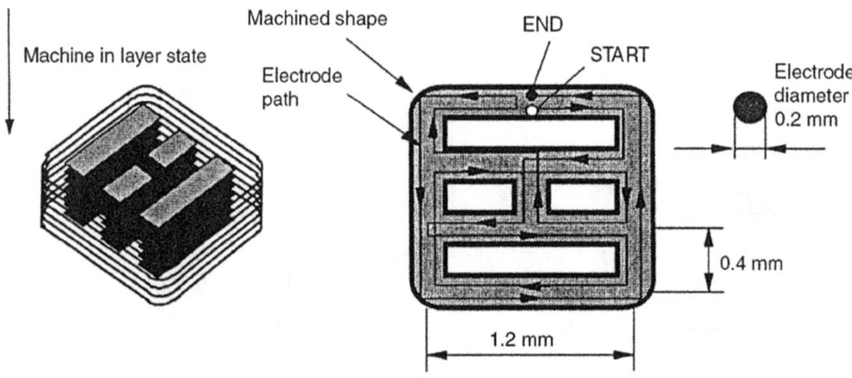

FIGURE 4.14 Micro EDM (Mitsubishi EDSCAN Technical Data, 1997).

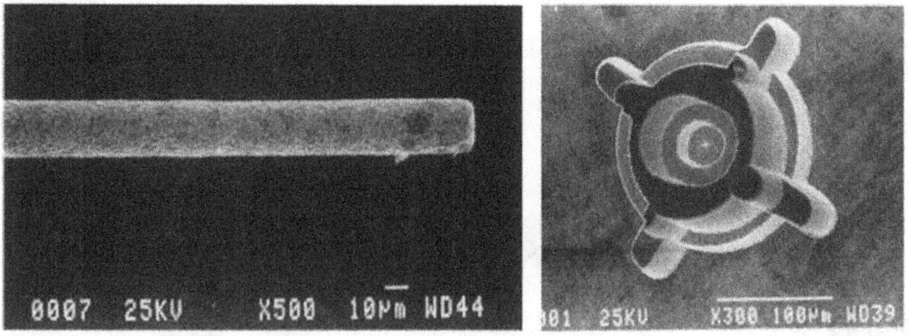

FIGURE 4.15 A fine part produced by ED milling using a graphite electrode of 10 µm diameter (Mitsubishi EDSCAN Technical Data, 1997).

graphite electrode of 10 µm diameter or less, which is dressed by a dressing unit mounted on the EDSCAN8E of a Mitsubishi machine. ED milling technology was announced by the Japanese (Mitsubishi) and Charmilles at a micro-machine exhibit in October 1996. Since then, this technology has gained attention in applications such as fabrication of micro-dies, etc.

4.1.6.2 Electrical Discharge Wire Cutting

Electrical discharge wire cutting (EDWC) is a variation of EDM which is similar to contour cutting with a band saw. A continuously moving wire travels along a prescribed path, cutting the workpiece with discharge sparks acting like cutting teeth. The tensioned wire is used only once, traveling from a take-off spool to a take-up spool while being guided to provide an accurate narrow kerf. The horizontal movement of the worktable is numerically controlled to determine the path of cut, Figure 4.16.

EDWC is used to cut plates as thick as 300 mm, used for making punches, tools, stripper plates, and extrusion dies in hard materials. It is also used to cut intricate shapes for the electronics industry. Figure 4.17 shows typical products which are cut

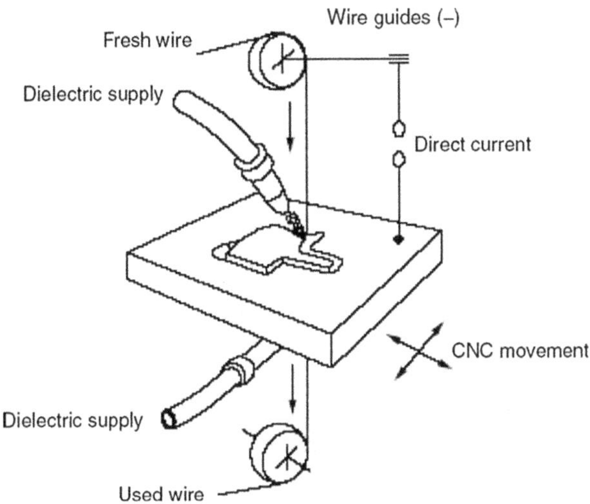

FIGURE 4.16 The EDWC process.

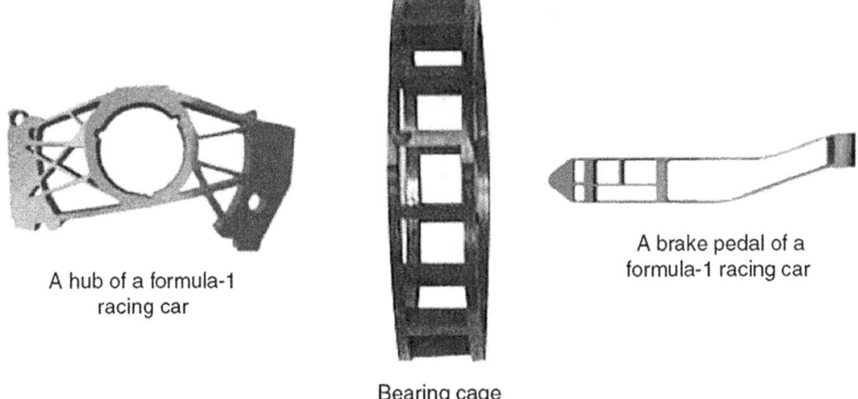

FIGURE 4.17 Typical products cut by EDWC (AGIE Charmilles Group, 2004).

by EDWC. EDWC machines are now available with CNC facility where a taper of the workpiece up to ±30°C is fully integrated.

Pulse generators: EDWC machines are only equipped with pulse generators. Peak current and on-time are the major variables controlling spark energy. Modern machines of EDWC are equipped with sophisticated Isopulse generators, Figure 4.9, where the previously mentioned variables along with off-time and spark frequency can be set independently, while mentoring the gap. The wire has a limited current capacity, so that the current rating rarely exceeds 30 A. The potential difference between wire electrode and workpiece is usually set between 50 and 60 V. Because wire electrode wear is of little importance, negative (straight) polarity is always used. Larger wire diameter can handle higher energy of sparks and, therefore, cuts at higher machining rates.

Dielectric flushing in EDWC: Effective flushing in EDWC is very important. Flushing nozzles should be as close as possible to the working gap, Figure 4.17. Workpieces of large variations in thickness are especially troublesome as they prevent effective dielectric flushing. The usual result of poor flushing is wire breakage, which may be avoided by decreasing the on-time which is accompanied by a slower cutting rate.

Deionized water is used almost exclusively as a dielectric. The low viscosity is ideal for the difficult flushing conditions found in EDWC. Additives are sometimes used as antirust compounds or ethylene-glycol base compounds to make the dielectric slippery. Light oils are sometimes used as dielectrics for EDWC. Good filtration is also important because contaminates affect the gap distance and consequently the machining accuracy and calls for arcing and wire breakage. Dielectric cooling and temperature control are essential for stable operation.

Wire electrodes: Brass is the widely used wire since it has most of the requirements needed for EDWC. It possesses high tensile strength, high electrical conductivity, and good wire drawability to close tolerances. Layered wires are also suggested and are more expensive; however, they cut faster than brass. One example is steel/copper/graphite wire, with a steel core for tensile strength, a copper layer for electrical conductivity, and graphite at surface for attaining high machining speeds. Zinc-coated brass with Mo-core is also available. Wire diameters range from 5 to 300 μm. It travels at a constant velocity ranging from 0.2 to 9 m/min.

Cutting speed: In EDWC, the cutting speed is generally given in terms of cross-sectional area cut per unit time. Typical examples are 18000 mm^2/hr for 50-mm-thick tool steel and 45000 mm^2/hr for 150-mm-thick aluminum block. This indicates a linear cutting speed of 6 mm/min and 5 mm/min respectively.

4.2 ELECTRON BEAM MACHINING EQUIPMENT AND OPERATIONS

4.2.1 Process Characteristics and Applications

The pioneering work of electron processing is related to Steigerwald (1958), who designed a prototype of electron beam equipment, which has been built by

Messer-Griessheim in Germany for welding applications. This new technology has quickly spread in industry to embrace other fields of application such as machining and surface hardening.

Electron beam machining (EBM) is a thermal NTM process that uses a beam of high-energy electrons focused to a very high power density on the workpiece surface causing rapid melting and vaporization of its material. A high voltage, typically 120 kV, is utilized to accelerate the electrons to speeds of 50 to 80% of light speed. Figure 4.18 shows the power density for different EB applications against the recommended pulse duration. The interaction of the electron beam with the workpiece produces hazardous X-rays; consequently, shielding is necessary, and the equipment should be used only by highly trained personnel. EBM can be used to machine conductive and non-conductive materials. The material properties such as density, electric and thermal conductivity, reflectivity, and melting point are generally not limiting factors. The greatest industrial use of EBM is the precision drilling of small holes ranging from 0.05 to 1 mm diameter, to a high

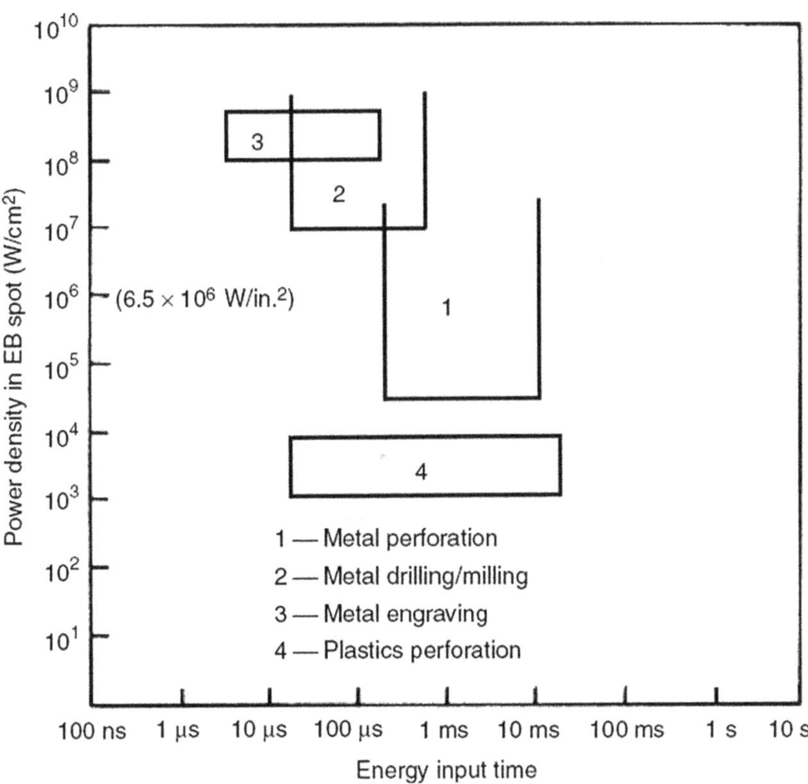

FIGURE 4.18 Power density for different EB applications against recommended pulse duration (ASM International, 1989).

degree of automation and productivity. The process has the following advantages and limitations:

Advantages of EBM

- Drilling of fine holes is possible at high rates (up to 4000 hole/s)
- Machines any material irrespective of its properties
- Micromachines economically at higher speeds than that of EDM and ECM
- Maintains high accuracy and repeatability of ±0.1 mm for position and ±5% of the diameter of the drilled hole
- Drilling parameters can easily be changed during machining even from row to row of holes
- Produces best finish compared to other process
- Provides high degree of automation and productivity
- No difficulty encountered with acute angles

Limitations of EBM

- High capital cost of equipment
- Time loss for evacuating the machining chamber
- Presence of a thin recast layer and HAZ
- Necessity for auxiliary backing material
- Need for qualified personnel to deal with CNC programming and the X-ray hazard

Applications of EBM: EBM is almost confined to drilling and slitting operations. Drilling is preferred when many small holes are to be made, or when holes are difficult to drill because of the hole geometry or material hardness. The textile and chemical industries use EB drilling as a perforating process to produce a multitude of holes for filters and screens.

Figure 4.19 shows typical components machined by the EBM process:

1. Drilling of holes starting from 20 μm diameter in metallic and non-metallic sheets
2. Machining of spinnerets of synthetic fibers, 60 μm width, 0.5 mm depth, and 1.2 mm arm length
3. Machining of vaporization mask with 60 μm trace, in a tungsten strip of 50 μm thick
4. Multi-hole drilling (1600 holes) in a tungsten strip 3.5 × 3.5 and 0.1 mm thick
5. Multi-hole drilling of cylindrical filter shell made of V2A stainless steel, 0.1 mm thick at a drilling rate of 3000 holes/s
6. Hybrid circuit engraved with 40 μm traces at a traverse speed of 5 m/s

4.2.2 Electron Beam Machining Equipment

A typical EBM equipment (also called electron beam gun) is shown schematically in Figure 4.20. The electrons are released from a heated tungsten filament. A high

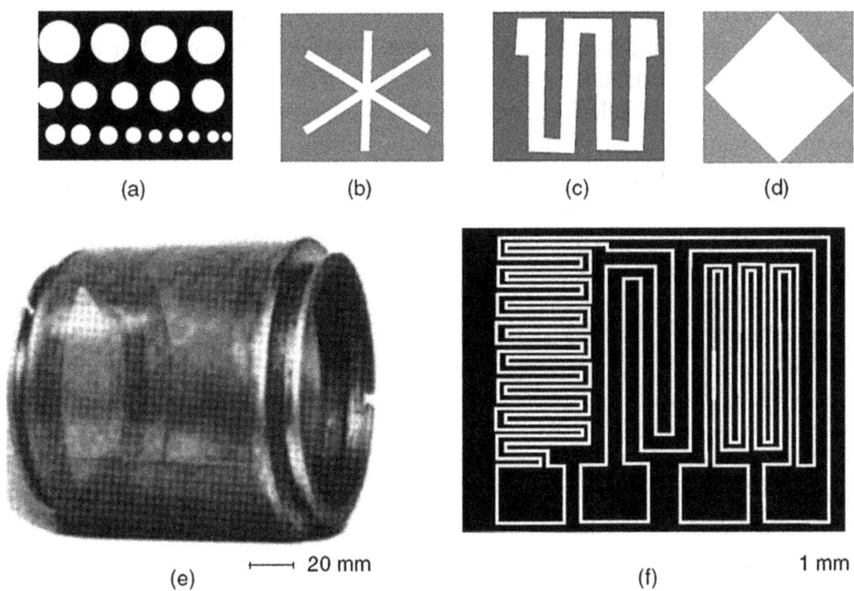

FIGURE 4.19 Typical components produced by the EBM process.

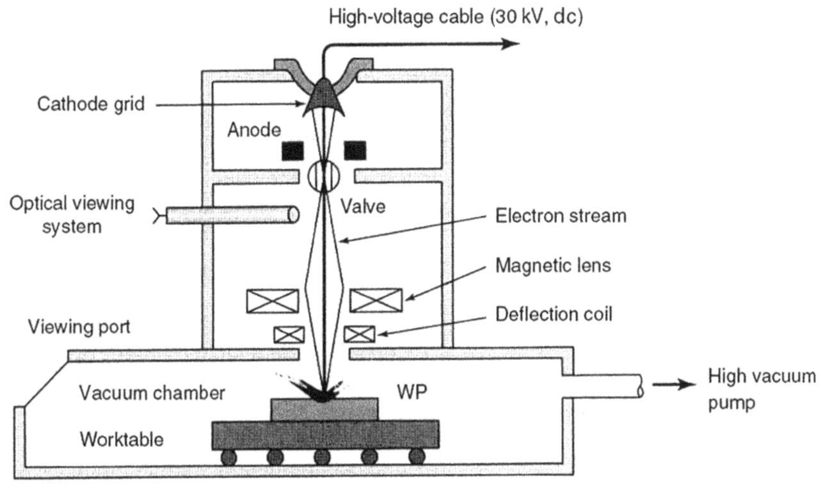

FIGURE 4.20 Electron beam gun.

potential voltage, typically 120 kV, is necessary to accelerate the electrons from the cathode (filament) toward the hollow anode, and the electrons continue their motion in vacuum toward the workpiece. A bias cup (Wehnelt electrode) located between the cathode and anode acts as a grid that controls the beam current (1–80 mA) by controlling the number of electrons. The bias cup also acts as a switch of the beam current.

A magnetic lens is intended to focus the beam on a spot of diameter ranging from 12 to 25 μm, whereas deflection coils are used to deflect the beam within an angle of no more than 5° to extend the machining range. The beam loses its speed and circularity. Through beam deflection, standard shapes and configurations similar to those shown in Figure 4.21 can be produced without workpiece manipulation. For drilling purpose, the beam is pulsed once per hole.

Another alternative, in the case of workpiece manipulation, is the beam is deflected, so that is moves in sequence with the part, thus allowing drilling while the part is moving. This is called on-the-fly drilling operation. The manipulator shown in Figure 4.22 includes a rotary axis with translation-motion capability. The work sheet is formed as a cylinder is clamped over the tensioning drum together with the backing material. A flying drilling operation is performed. As each beam pulse starts hole drilling, the beam is deflected while the drum rotates at constant speed. When the hole is completed, the pulse is turned off, and the beam instantaneously goes back to its original position to drill the next hole. This on-the-fly drilling process is continuously repeated. The translational axis is moved in synchronization with the beam to advance the workpiece under the beam. In this design, the chamber with its powered sliding door is usually a part of the manipulator. A more sophisticated multi-axis manipulator, Figure 4.23, can be used for drilling holes in complicated shapes. The workpiece is held in a chuck while the motion of the axes is computer numerically controlled. It is worth noting that the bearings must be carefully sealed to prevent them against damage from metal vapor and drilling debris.

The electron beam is confined with the workpiece in the evacuation chamber to prevent:

- Oxidation of filament and other elements
- Collision of electrons with the massive molecules of O_2 and N_2 to eliminate loss of their kinetic energy
- Contamination of metal vapor and debris

4.2.3 Process Capabilities

The parameters which affect the performance of EBM are:

- Density and thermal properties of workpiece (such as specific heat, thermal conductivity, and melting point)

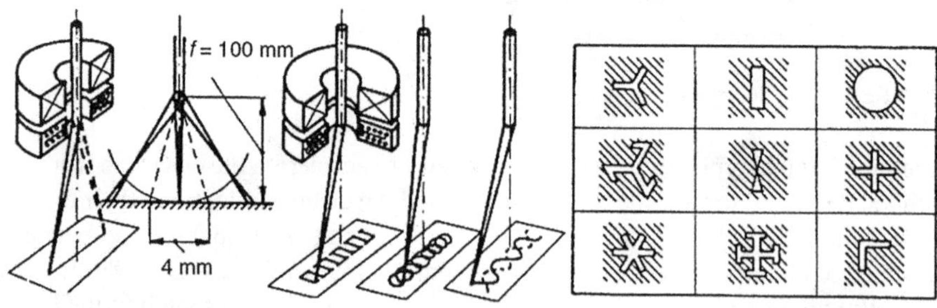

FIGURE 4.21 Standard configurations produced by EB without WP manipulation (Visser, 1968).

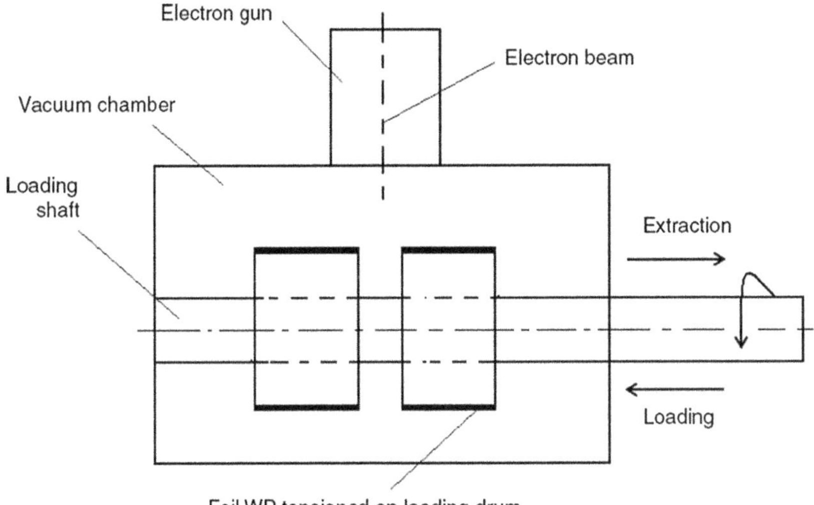

FIGURE 4.22 Rotary axis WP manipulator for the on-the-fly drilling EB operation (ASM International 1989).

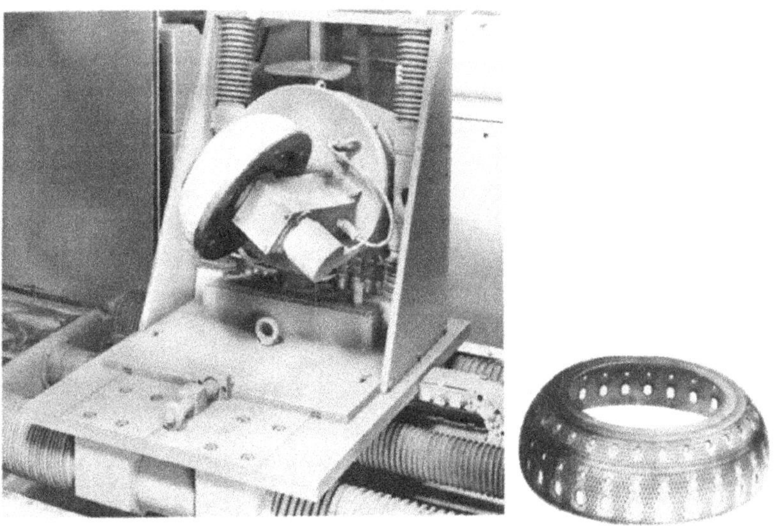

FIGURE 4.23 Multiaxis manipulator for complicated shapes (Courtesy of MG Industries/ Steigerwald).

- Accelerating voltage
- Electron beam current
- Pulse energy
- Pulse duration
- Pulse frequency
- Spot diameter
- Traverse speed of the workpiece

Equation 4.5 expresses the traverse speed V_f in terms of the machining parameters of the process, (Kaczmarek, 1976).

$$v_{\mathrm{f}} = \frac{C_{\mathrm{d}}}{d_{\mathrm{f}}}\left[0.1\frac{P_{\mathrm{e}}}{\theta_{\mathrm{m}} \cdot t_{\mathrm{l}} \cdot k_{\mathrm{t}}}\right]^2 \text{ m/s} \tag{4.5}$$

where

t_l = plate thickness (m)
P_e = power of electron beam (N m/s) = $i_b \times V_b$ = beam current × acceleration voltage
d_f = beam focusing diameter (m)
k_t = thermal conductivity (N/s °C)
C_d = coefficient of thermal diffusivity = $k_t/\rho \cdot cl$ (m^2/s)
ρ = density of WP material (kg/m^3)
cl = specific heat of WP material (Nm/kg °C)
θ_m = melting point of WP material (°C)

Table 4.3 illustrates typical parameters of the EBM process.

EBM is characterized by its minor volumetric removal rate that reaches a maximum value of 0.1 cm^3/min. The volumetric removal rate increases by increasing pulse energy (power intensity) provided the same number of pulses is used. The machinability depends on the melting point of the material to be machined. Therefore, tin and cadmium have the highest machinability, whereas W and Moare have low machinability.

In EB hole drilling, the achieved tolerance depends on the hole diameter and workpiece thickness, whereas the surface roughness depends, mainly, on the pulse energy. Depending on the working conditions, the tolerances of the drilled holes (and

TABLE 4.3
Typical Machining Parameters of EBM

Parameter	Value
Acceleration voltage (V_b)	50–150 kV
Beam current (i_b)	0.1–40 mA
Power (P_e)	1–150 kW
Pulse duration (t_i)	4–60,000 µs
Pulse frequency (f_r)	0.1–16,000 Hz
Vacuum	10^{-3} to 10^{-5}torr (1 torr = 1 mmHg)
Minimum spot (d_f)	12–25 µm
Beam deflection	6 mm
Beam intensity	10^6–10^9 W/cm^2
Tolerance	Depends on plate thickness or hole diameter, otherwise, ±10% of lateral dimension.
Roughness (R_a)	1 µm

Source: (Machinability Data Center, 1980)

slits) may attain values between ±5 and ±125 µm. The surface roughness R_a ranges from 0.2 to 6.3 µm.

Illustrative Example 2

Calculate the traverse speed *Vf* in mm/s for cutting a tungsten sheet that is 2.5 mm thick, if EB equipment of 10 kW is used. The equipment is capable of focusing the beam to a diameter of 0.5 mm. The following WP data are given: $\theta_m = 3400°C$, $C_d = 8.1 \times 10^{-5}$ m²/s, $k_t = 214$ N/s °C.

SOLUTION

$$V_f = \frac{C_d}{d_f}\left[0.1\frac{P_e}{\theta_m \cdot t_1 \cdot k_t}\right]^2 \text{ m/s}$$

$$= \frac{8.1\times10^{-5}}{0.0005}\left[0.1\frac{10\times1000}{3400\times0.0025\times214}\right]^2$$

$$= 0.049 \text{ m/s} = 49 \text{ mm/s}$$

4.3 LASER BEAM MACHINING EQUIPMENT AND OPERATIONS

4.3.1 Process Characteristics

After the age of electricity, the realm of photons began from the early 1960s. The photons of zero charge have replaced the negative electrons as a favorite tool in modern industry. The photon beam (laser) is currently used in everything from eye surgery to communication and metal processing. The earliest industrial applications in welding occurred in the 1970s while the 1980s saw the beginning of laser beam machining and micromachining.

Laser is an acronym for "Light Amplification by Stimulated Emission of Radiation." It is a highly collimated monochromatic and coherent light beam in the visible or invisible range. Laser beam machining (LBM) is a promising NTM process of machining any material irrespective of its physical and mechanical properties. It is used to cut and machine both hard and soft materials, such as steels, cast alloys, refractory materials, ceramics, tungsten, titanium, nickel, borazon (CBN), diamond, plastics, cloth, alumina, leather, woods, paper, rubber, and even glass when its surface is coated with radiation-absorbing material such as carbon. However, machining of Al, Cu, Ag, and Au is especially problematic since these metals are of high conductivity and have the tendency to reflect the applied light. But recently, yttrium aluminum garnet (YAG) with enhanced laser focusing has been suggested and used to cut such metals after treating their surfaces by oxidizing them or increasing their roughness. YAG is far superior to a CO_2 laser because it emits a shorter wavelength.

Laser is a versatile tool, useful in many areas from precision watchmaking to the heavy metalworking industry. The key of laser's effectiveness lies in its ability to deliver, in some cases, a tremendous quality of highly concentrated power as high as 10^{12} W. Tuning the laser beam makes it possible to deliver just the right concentration of power for the right amount of time to perform a specific piece of work.

It is possible to make automotive engine blocks out of aluminum with a thin, hard layer inside the cylinder by lasers, thus the engine weight is considerably reduced. As long as the beam is not obstructed, it can be used to machine inaccessible areas.

As one of its main advantages, a laser beam does not need time consumed for the evacuation of the machining area as EB; laser can operate in a transparent environment such as air, gas, vacuum, and in some cases even liquids. However, LBM is quite inefficient and cannot be considered as a mass metal removal process. A big limitation of laser drilling is that the process does not produce round and straight holes. This can, however, be overcome by rotating the workpiece as the hole is being drilled. A taper of about 1/20 is encountered. HAZ is produced in LBM, and heat-treated surfaces are also affected.

High capital and operating cost and low machining efficiency (η = 1%) prevent LBM from being competitive with other NTM techniques. Protective measures are absolutely necessary when working around laser equipment. Extreme caution should be exercised with lasers; even low power of 1 W can cause damage to the retina of the eye. In all cases, safety goggles should be used, and only authorized personnel should not be allowed to approach the laser-working zone.

Process Advantages and Limitations

Advantages

- A wide variety of metallic and non-metallic materials can be machined.
- No mechanical contact; therefore, no deformations of the workpiece and no tool wear are encountered.
- Work fixation is easily performed.
- Laser beam can travel without diffraction and can be branched to different work stations working at the same time.
- It can reach inaccessible areas on the workpiece.
- It produces microholes in difficult-to-machine and refractory materials.
- It is easy to control the beam characteristics to adapt the machining duty.
- There is no need for the time-consuming evacuation as in EBM.
- Holes can be drilled at an acute angle to surface (10°).
- The process can be automated easily.
- The operating cost is low.
- It has a narrow heat-affected zone as compared to other thermal NTM process.
- No cutting lubricants are required.

Limitations

- High equipment and maintenance cost
- Not safe in operation
- Blind holes of precise depth are difficult to achieve
- Laser produces tapered holes and hence limited thickness due to taper
- Holes produced are of limited dimensional and form accuracy, and of relatively bad surface quality
- HAZ cannot be avoided

- The process is of low efficiency and is not considered as a mass removal process compared to stamping
- Material thickness is restricted to 50 mm in the case of drilling
- Adherent materials at hole exits need to be removed
- Assist or cover gas is required

4.3.2 Types of Lasers

4.3.2.1 Pyrolithic and Photolithic Lasers

A laser beam is of high-power density especially when focused at a small spot on the surface of the workpiece. Depending on its wavelength, the beam interacts with the workpiece material either pyrolithically (thermally) or photolithically.

In the pyrolithic laser, the material removal occurs by the melting and vaporization of the material spontaneously. This laser is used mainly in applications such as cutting, drilling, welding, and surface hardening. The removal or processing rate depends on the material being machined, its thickness, and its physical and optical properties, such as the specific heat, latent heats of melting and vaporization, and the surface reflectivity. The machinability of materials increases by decreasing the above-mentioned properties. It depends also on the beam characteristics, especially its power density.

In the domain of machining, the pyrolithic laser is used to machine a wide spectrum of metallic and non-metallic engineering materials, taking into account that the thermal diffusivity does not allow heat to be transmitted beyond the machined surface.

In the photolithic laser, the material removal is not removed thermally, but it is affected by the dissociation and breaking chemical bond between the material molecules, in case its bond energy is below the photon energy of the beam. The photon energy of the beam is inversely proportional to its wavelength.

The fluorine excimer laser is a beam of ultrashort wave length (λ = 157 μm), consequently it possesses a high photon energy of 7.43 eV (1 eV = 1.6×10^{-19} J), whereas the CO_2 laser is an infrared (IR) laser beam of a long wavelength (λ = 10600 μm), which has a low photon power of 0.12 eV. It follows that the excimer laser is capable of machining plastic and Teflon photolithically, since its photon energy is greater than the chemical bond energy which ranges from 1.8 to 7 eV for most of the plastics. The CO_2 laser is not capable of machining plastics photolithically, but pyrolithically.

In photolithic material removal, three phases are necessary:

1. The ultrashort wave photons are absorbed into the surface to a depth of about 200 nm.
2. The chemical bond between molecules is broken.
3. The reaction products escape as gas and small particle ashes.

4.3.2.2 Industrial Lasers

Industrial lasers comprise, in most cases, the solid-state lasers such as neodymium: yttrium aluminum garnet (Nd:YAG), Nd:glass, ruby and gas lasers (CO_2, excimer, and He-Ne). Basically, four types prevail in metal working processes, namely the CO_2, Nd:YAG, Nd:glass, and excimer lasers. Out of these, CO_2 and YAG are considered the most applicable workhorses.

In most cases, the same laser can be applied in cutting, welding, machining, and surface treatment. It is only necessary to vary the laser beam parameters, such as power density, focus diameter, and pulse duration to suit the specified machining process. The four types of industrial lasers are described in brief.

Nd:YAG lasers: It is a single crystal of yttrium aluminum garnet doped by 1% neodymium as an active lasing material. This laser is compact, economical, its wavelength is 1060 nm, and it can operate in either the pulsed (P) or the continuous wave (CW) mode. It is characterized by relatively high efficiency, high pulsating frequency, and operates with a simple cooling system. Its pulsating frequency ranges from 1 to 10000 p/s and pulse energy 5 to 8 J/p. It has an average power output close to 1 kW.

Nd:glass laser: It is glass rod doped by 2 to 6% neodymium as the acting lasing material. This laser is often uneconomical. It has the same wavelength as the Nd:YAG laser. It operates only in the pulsed (P) mode. Due to the low thermal conductivity of glass, the pulse rate should be limited. Consequently, it is only used in drilling and welding necessitating higher energy output and low pulse frequency (1–2 p/s).

CO_2 lasers: In these lasers, the active lasing material is the CO_2 gas. However, a mixture of gases is used ($CO_{2:}$ N_2: He = 0.8: 1:7). Helium acts a coolant of the gas cavity. CO_2 lasers are characterized by their long wavelength of 10600 μm, thus the material removal depends only the thermal interaction with the workpiece. CO_2 lasers yield the highest depth-to-diameter ratio in most metals using a gas-jet assistance. These lasers are bulky but economical. There are two types of CO_2 lasers:

1. **Axial-flow CO_2 laser**: This laser can operate in pulsed (P) and continuous wave (CW) modes. In this type, the typical average output for (CW) ranges from 250 to 5000 W, while in the pulsed (P) operation, the average power output is reduced to 100 to 2000 W, and the pulse frequency ranges from 1 to 10000 p/s.
2. **Transverse-flow CO_2 laser**: It operates only in the continuous wave (CW) mode. This type is used when high power output between 2500 and 15000 W is required.

Excimer lasers: Excimer lasers are a family of pulsed lasers operating in the ultraviolet (UV) region of the spectrum. Excimer is an acronym for "excited dimmer." The beam is generated due to fast electrical discharges in a mixture of high-pressure dual gas, composed of one of the halogen gas group (F, H, Cl) and another from the rare gas group (Kr, Ar, Xe). The wavelength of the excimer laser attains a value from 157 to 351 nm, depending on the dual gas combination. Excimer lasers have low power output, so they remove the material photolithically, and have a remarkable application in machining of plastics and micromachining as previously mentioned. The main characteristics of important industrial lasers are given in Table 4.4.

TABLE 4.4
Main Characteristics of Important Industrial Lasers

		Beam Characteristics				
Laser Type	**Mode**	**λ (nm)**	**W_{ave} (W)**	**f_r (p/s)**	**d_f (μm)**	**Comments**
Nd:YAG	(P)	1,060	100–500	1–10,000	13	Compact, economical
	(CW)	1,060	10–800	–	13	
Nd:glass	(P)	1,060	1–2	0.2	25	Often uneconomical
CO_2	Axial (P)	10,600	100–2,000	400	75	High efficiency, bulky but
	(CW)	10,600	250–5,000	–	75	economical
	Transverse (CW)	10,600	2,500–15,000	–	75	
Excimer	(P)	157–531	~100	10–500	N/A	Micromachining, plastic, ceramic

Note: λ = wavelength, W_{ave} = average power, f_r = pulse/s, d_f = focus diameter, N/A = not available.

4.3.3 Laser Beam Machining Operations

As previously mentioned, industrial lasers operate either in continuous wave (CW) mode or in pulse (P) mode. Generally, CW lasers are used for processes like welding, soldering, surface hardening, etc., which require uninterrupted supply of energy for melting and phase transformation. Controlled pulse energy is desirable for cutting, drilling, marking, etc., striving at less heat distortion and the minimum possible HAZ. The applications of beam in machining are as follows:

1. **Drilling**
   Laser beam drilling is done either by percussion drilling or trepanning.
   a. *Percussion drilling* is used for relatively small holes (<1.3 mm diameter) though metal sections up to 25 mm thick are most often performed using pulsed Nd:YAG lasers because of their higher pulse energy, Figure 4.24.

   Operating Parameters
   - Pulse power: 100–250 W (average)
   - Pulse energy: up to 40 J/p depending on hole diameter and material thickness higher energy provides faster drilling rate but less hole quality
   - Pulse duration: 0.5–2 ms
   - Pulse frequency: 5–20 Hz with Nd:YAG
     100 Hz with (P) CO_2
   - Focal length: 100–250 mm, lager focal length for large hole diameters and depths
   - Focal position: Optimized, above, below, or on surface, depending on the desired results. Most often, faces at a depth of 5 to 15% of metal thickness, determined empirically for best quality as judged through roundness, taper, microcracking, and recast layer

   b. *Trepanning* is used for large holes (>1 mm diameter). It is commonly performed with both CO_2 and Nd:YAG lasers. Trepanning is essentially

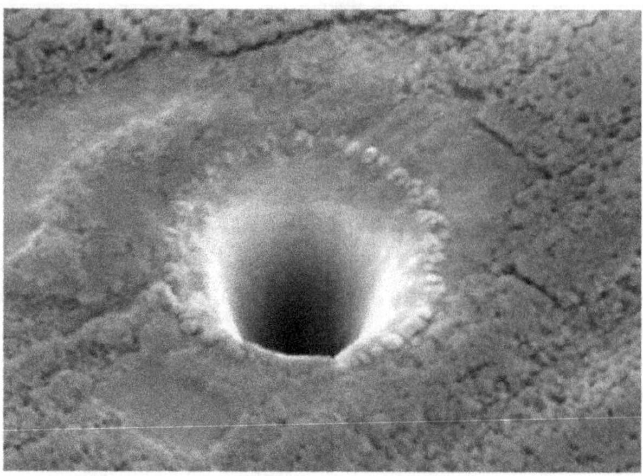

FIGURE 4.24 SEM micrograph hole drilled in 250 μm thick silicon nitride with Nd:YAG laser (Dhakal, 2007).

a machining process that can be performed by operating the laser in the (CW) or (P) modes. Trepanning requires a percussion drilled pilot hole using operating parameters for percussion drilling. As the metal thickness increases, the trepanning speed decreases.

Operating Parameters

- Pulse power: It is established at levels that maximize the trepanning rates, i.e. maximizing the part or beam translation without sacrificing the quality.
- Pulse duration: <2 ms with both pulsed CO_2 and Nd:YAG lasers.
- Pulse frequency: Lower pulse frequencies are used as metal thickness increases.
- Focal length: Similar to those used in percussion drilling. If a CO_2 laser is used, then the recommended focal length is 125 mm or less.

2. **Cutting**

Similar to trepanning, cutting can be accomplished by a laser operating in either the (CW) or (P) modes.

CO_2 laser: (CW) mode is used for thicker metal sections, while (P) mode is used for thinner metal sections.

(P) Nd:YAG lasers are used to cut thick sections of super alloys.

More often, cutting is done by a CO_2 laser because of its faster machining rate.

Operating Parameters

- Power: 250–5000 W for (CW) CO_2
  100–2000 W for (P) CO_2
  <800 W for (CW) Nd:YAG
  <500 W for (P) Nd:YAG
  Lower power is sufficient for pulsed lasers because of their higher peak instantaneous powers.

- Pulse duration:
  <0.75 ms is used for intricate cutting of thin metals. Longer duration up to 2 ms provides greater pulse energy for thicker metals.
  Similar durations for both CO_2 and Nd:YAG.
- Pulse frequency:
  200–500 Hz for (P) CO_2
  30–100 Hz for (P) Nd:YAG
  Higher frequency is used to cut thinner metals, and lower frequency to cut thicker metal sections.
- Pulse energy:
  2 J/p at longer pulse durations and lower pulse frequency for (P) CO_2
  Up to 80 J/p (pulse freq.) limited by maximum power rating for (P) Nd:YAG for thicker material.
- Focal length:
  CO_2 laser
  l_f = 6.5 mm for materials up to 6 mm thick
  = 125 mm for materials from 6 to 16 mm thick= 190 mm for materials thicker than 13 mm
  Nd:YAG lasers
  l_f = 100 mm for materials thinner than 3 mm
  = 150–250 mm for materials up to 25 mm thick
  >250 mm for materials thicker than 25 mm
  Wider kerf is obtained when using longer focal lengths. Al requires large kerf for good results.

3. **Marking**
   The laser system is used to imprint characters (letters and symbols) to a depth of 5 to 250 μm. Laser marking arrangement is made up of a low power pulsating laser system lasting only for nanoseconds, and a CNC beam scanning system. The necessary information to generate different characters is stored on the computer. Accordingly, 30 characters can be imprinted per second. Table 4.5 provides a general selection guide of industrial lasers for different applications.
4. **Gas-Assisted Laser Cutting**
   It is an important development in which a coaxial nozzle is supplemented with a continuous jet of air, O_2, or one of the inert gases (N_2, Ar, He). The selection of the gas depends upon the type of work material, its thickness, and type of cut, Table 4.6.
   In gas-assisted lasers, the gasses cutting have the following functions:
   - Providing oxidizing atmosphere (in the case of O_2 and air), thus reducing reflectivity and improving absorption of beam energy.
   - Promoting exothermic reactions (in the case of O_2), thus improving process efficiency by providing 80% of the energy needed for cutting.
   - In all cases, the pressurized gas expels vapors and molten metal from the machining zone or the bottom of the hole in the case of drilling.

TABLE 4.5
Laser Beam Selection Guide for Different Applications, Modified

Application	Type of Laser Beam
Cutting, trepanning	
Metals	(CW) CO_2, (P) CO_2, (P) Nd:YAG
Plastics	(CW) CO_2
Ceramics	(P) CO2
Drilling, percussion drilling	
Metals	(P) Nd:YAG, (P) CO_2
Plastics	Excimer
Marking, micromachining	
Metals	(P) Nd:YAG, (P) CO_2
Plastics	Excimer
Ceramics	Excimer
Welding, soldering	(CW) CO_2, (P) CO_2, (P) Nd:YAG
Surface hardening	(CW) CO_2

Source: (Kalpakjian and Schmidt, 2003)

TABLE 4.6
Gas Selection of Gas-Assisted Laser Cutting and Marking

Material	Best	Optional
Gas-assisted laser cutting		
Steels and stainless steel	O_2	N_2
Ti	He	He/Ar, CO_2
Non-ferrous alloys	O_2	Air
Nickel alloys	O_2	Air
Non-metals	O_2	Air
Composites	O_2	N_2, Ar
Plastics	O_2	Air, Ar
Wood	N_2	Air
Gas-assisted laser marking (scribing)		
Aluminum	O_2	(CW) CO_2, 1000 W, no burrs
Ceramic alumina	N_2	(P) CO_2, 50 W, 0.5 ms, no HAZ

Source: (Machinability Data Center, 1980)

O_2 is the most commonly used assisting gas for steels and most metals. When an oxide-free surface of high-quality cut is desired, an inert gas is used. Additionally, the oxide-free edges can improve the weldability. Inert gas is also used to prevent plastics and other organic materials from charring.

Operating Parameters

- Laser used: (CW) Nd:YAG, (CW) CO_2 for cutting (P) Nd:YAG for drilling
- Power: 0.25–16 kW depending on type of material and its thickness
- Feed rate: 0.25–7.5 m/min depending on the power and type of material and its thickness
- Focus diameter: 50 μm (minimum), focal point on the surface or slightly lower
- Gas pressure: 1–3 atm. for O_2
- 2–6 atm. for inert gases
- SOD: 1–15 mm for CO_2 laser, and 5 mm for Nd:YAG laser

4.3.4 LBM Equipment

Figure 4.25 shows schematically three important elements of LBM equipment, namely, a lasing material (solid state or gas), a pumping energy source required to excite the atoms of the lasing material to a higher energy level, and a mirror system. One of these mirrors is fully reflective, while the other one is partially transparent to provide the laser output (output mirror). It allows the radiant beam to either pass through or bounce back and forth repeatedly through the lasing material. To make the laser beam useful for machining, its power density should be increased by focusing, thus attaining power density values between 10^5 and 10^7 W/mm^2. The laser beam is usually delivered to the workpiece in the transverse excitation mode (TEM).

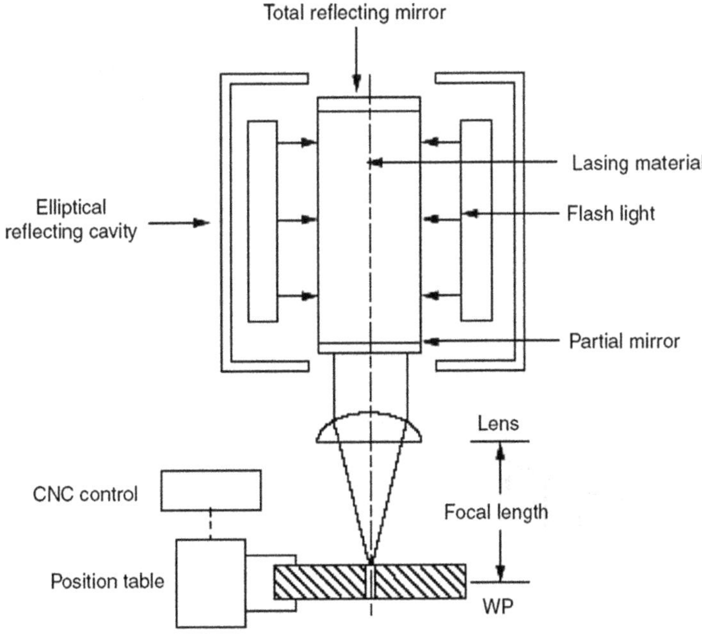

FIGURE 4.25 Schematic of LBM.

The common mode of optical configuration is TEM_{00}, which is a Gaussian output beam with lowest beam divergence, and consequently the highest power density. This mode provides the most uniform beam profile. The other widely used modes are TEM_{10}, where a broader and less intense spot is required in cutting and welding, and TEM_{01} which is useful for machining large holes and surface hardening.

To improve the process performance, most of the LBM equipment is provided with Q-switching facility to amplify the power. It provides the beam despite energy loss due to magnification, with giant power (hundreds or thousands of its normal pulsing power) acting on extra short pulse duration of a nanosecond order. Therefore, Q-switching enhances the beam capabilities regarding the removal rate and the quality of cut. The Q-switched beam is capable of evaporating the material in almost no time. In the case of Nd:YAG and CO_2 lasers operating at continuous wave (CW) mode, the Q-switching also converts the continuous wave into a train of pulsating power. Most of the new lasers are computer controlled to take advantage of their high-speed processing. During machining, motion can be provided to the workpiece, the beam, or both.

In recent LBM developments, significant progress has been made in integrating robot technology and CNC facility with lasers in a setup called a flexible machine station, Figure 4.26. A single laser beam travels to the processing locations without diffraction or loss of power, where it is divided to perform many functions simultaneously. Sometimes, the LBM equipment is provided with a laser beam torch (LBT) which may be used in cutting, welding, and surface hardening operations, Figure 4.27.

4.3.5 Applications and Capabilities

Laser and gas-assisted cutting techniques are best suited for applications demanding high accuracy, and for machining jobs in which the HAZ is to be as narrow as possible to avoid distortions and obtaining cuts of high-quality edge finish.

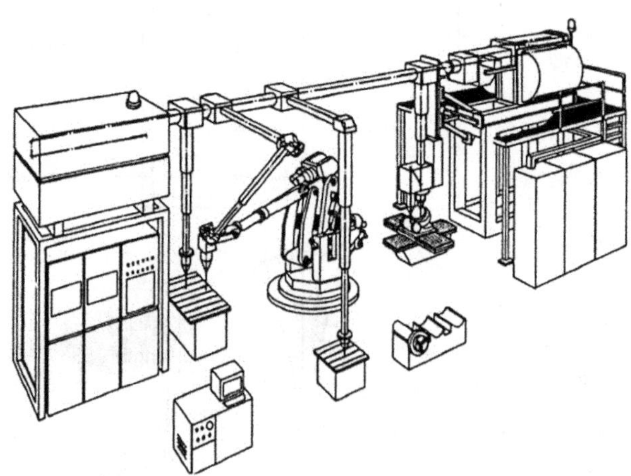

FIGURE 4.26 Flexible laser beam machine station (König, 1990).

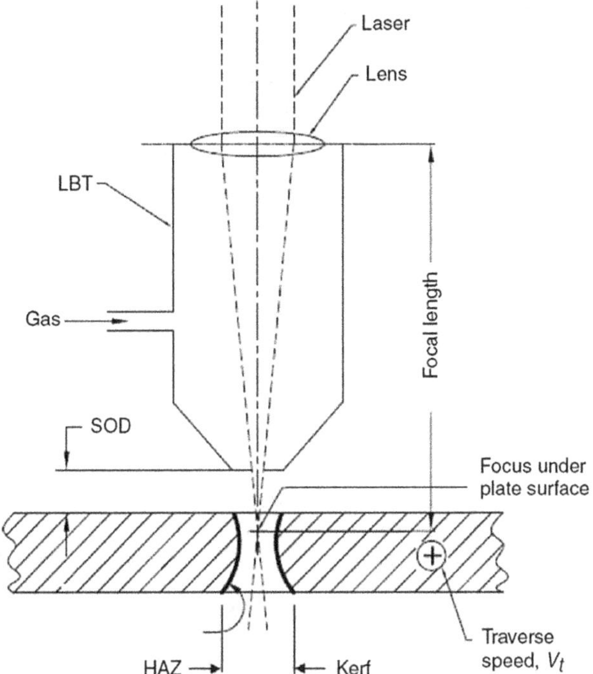

FIGURE 4.27 Laser beam torch.

Laser drilling was one of the first practical applications of laser technology. The demand for laser drilling, especially of microholes of 75 μm diameter, is increasingly emphasized.

With increasing hole diameter and depth, the ejected liquid metal deposits on the walls and the bottom, such that perfectly cylindrical holes cannot be obtained, and that is why laser is not used for producing deep holes. In industry, laser drilling is widely used for the drilling of watch jewels, diamond drawing dies, and such jobs where practically, a high level of precision is not demanded.

Transverse speeds attained using lasers are impressive. In this regard, when using a CO_2 laser capable of providing power of 0.5 kW, traverse speeds are realized:

4.5 m/min for 1-mm-thick steel sheet
0.5 m/min for 6-mm-thick steel sheet
2.0 m/min for 1.5-mm-thick stainless-steel sheet
0.9 m/min for 3-mm-thick stainless-steel sheet

Using 1 kW power, the machining speeds are doubled.

Some typical applications of LBM are listed below:

- Machining of microholes in filter screens, carburetors, and fuel injection nozzles
- Machining of miniature holes of diameter 0.1 to 0.5 mm at rates of 1 to 10 hole/s

- Micro drilling of diamond wire drawing die (50 μm), using a Q-switched microsecond pulse Nd:YAG laser, and nanosecond pulse excimer laser
- Laser drilling of rubber cups. Since there is no forces involved, then there is no deformation and the lack of stiffness is of no significance, Figure 4.28
- Scribing to widths 5 to 10 μm at speeds up to 12 m/min. Ultrashort wave excimer lasers now produce cuts 0.5 μm in width
- Trimming of flashes from plastic parts
- Marking and engraving in metallic and non-metallic workpieces
- In the aircraft turbine industry, laser drilling is used to make holes of air bleeds, air cooling, or passage of other fluids
- Lasers may be used for machining hard materials (white CI, Inconel, etc.) in combination with a traditional machining process (milling or turning). The laser is directed onto a spot in front of the turning tool in laser assisted turning (LAT), Figure 4.29

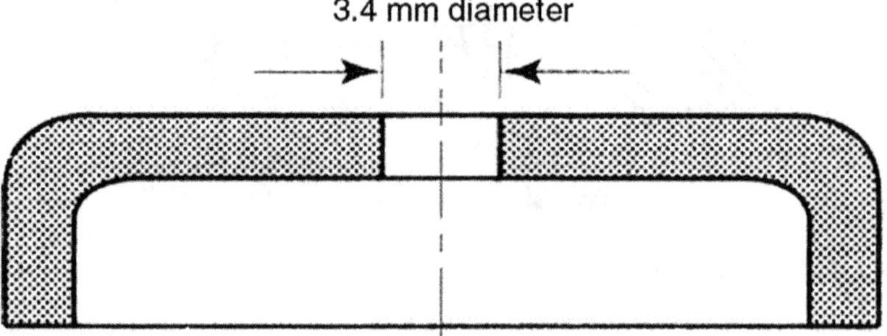

FIGURE 4.28 Laser drilling of rubber cup.

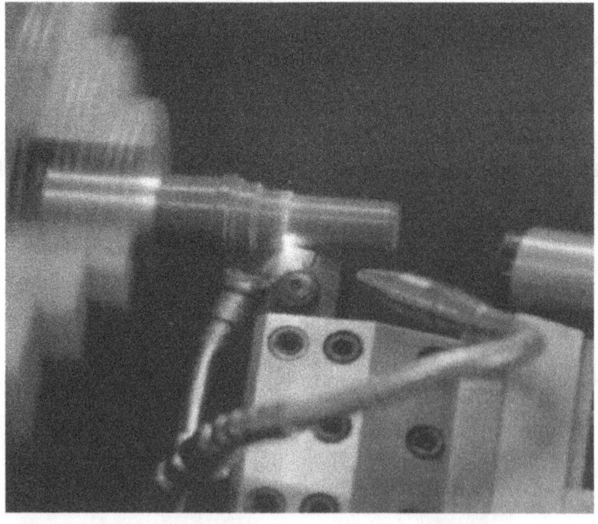

FIGURE 4.29 Laser-assisted turning.

- CO_2 lasers have recently been used for cutting of cut and peel masks, needed for CH milling of airplane wings (laser power 75 kW, and the mask 0.4 mm thick)
- Restoring of dynamic balance of high-speed rotors and shafts by removing infinitesimal small material during rotation on the dynamic balancing machine

Other applications include making holes in hypodermic needles, ceramic substrates for electronic circuits, holes in WC tool plates, and so on.

4.4 PLASMA ARC CUTTING SYSTEMS AND OPERATIONS

4.4.1 Process Characteristics

When a gas is heated to a high temperature in the order of 2000°C, its molecules separate out as atoms. If the gas temperature is raised above 3000°C, the electrons of some of the atoms dissociate and the gas becomes ionized, i.e. consisting of free electrons, positive-charged ions, and neutral atoms. This state of ionized gas is known as plasma gas and is characterized by high electrical conductivity.

The plasma arc is initiated in a confined gas-filled chamber by a high-frequency spark. The direct current from a high-voltage source sustains the arc and the plasma stream, which exits from the nozzle at sonic speed.

The source of heat generation in the plasma is due to the recombination of electrons and ions into atoms and recombination of atoms into molecules. The liberated bonding energy is responsible for increased kinetic energy of atoms and molecules formed by the recombination. The temperatures associated due to the recombination can be in the order 20000 to 30000°C. Such a temperature melts out and even vaporizes any work material subjected to machining or cutting. Plasma arc cutting (PAC) is a thermal NTM process, which was adopted in the early 1950s as an alternative method for oxy-fuel cutting of stainless steel, aluminum, and other non-ferrous metals. Recently, cutting of conductive and non-conductive materials by PAC has become much more attractive. The main attraction is that PAC is the only method which cuts faster in stainless steel than it does in mild steel.

Advantages of PAC

- The process provides smooth cuts, free of contaminants.
- It can cut exotic metals at high rates.
- The process has the least specific cutting energy among all NTM processes.

Disadvantages of PAC

- Reduced accuracy and surface quality are expected.
- The process requires high powers.
- It produces toxic fumes.
- Due to high thermal effects, the workpiece is highly distorted and HAZ of large depth reduces the fatigue resistance.

- The plasma arc produces infrared (IR) and ultraviolet (UV) radiations which cause eye injury (cataract) and loss of sleep. UV radiation leads to skin cancer. Therefore, gloves, goggles, and ear plugs should be used.

4.4.2 Plasma Arc Cutting Systems

PAC systems operate either in the transferred arc mode or non-transferred jet mode. In the transferred arc mode, Figure 4.30a, the arc is struck from the rear electrode of the plasma torch to the conductive workpiece (+ve electrode) causing temperatures as high as 33000°C. Owing to the greater efficiency of the transferred systems, PAC systems are often used in cutting of any electrically conductive material including those of high electrical and thermal conductivity that are resistant to oxy-fuel cutting as aluminum.

In the non-transferred jet mode, Figure 4.30b, the arc is struck with the torch itself. The plasma is emitted as a jet through the nozzle orifice causing a temperature rise of about 16000°C. Since the torch itself is switched as the anode, a large part of the anode heat is extracted by cooling water, and therefore not effectively used in material removal processes. Non-conductive materials that are difficult to be cut by other methods are often successfully cut by plasma non-transferred systems.

A constructional assembly of a typical transferred plasma torch is illustrated in Figure 4.31. The nozzle diameter depends on the arc current and the flow rate of the working gas. It ranges from 1.2 to 6 mm. Fine nozzles of 50 μm diameter are especially used to cut metals with kerf width of 0.1 mm and operate at low power of 1 kW. Multiple torch cuts are possible on tracer controlled cutting tables for plates up to 150-mm-thick stainless steel.

The commonly used working gases are He, Ar, H_2, N_2, or a mixture of them. The gas flow rate ranges from 0.5 to 6 m^3/h, depending on the arc power and the plate thickness.

The non-consumable electrodes are made of 2% thoriated tungsten to resist wear. Shielded plasma torches may be gas or water shielded.

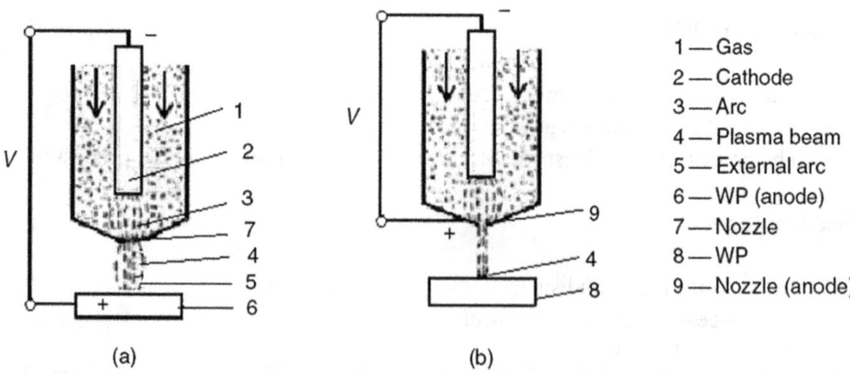

FIGURE 4.30 (a) Transferred and (b) non-transferred plasma torches.

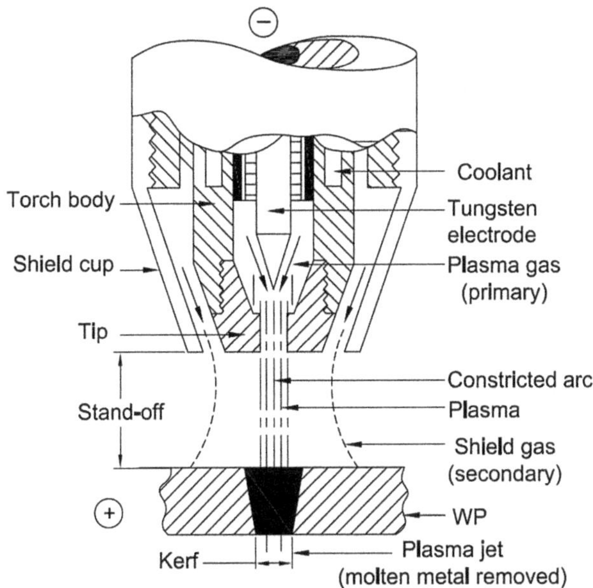

FIGURE 4.31 Constructional assembly of transferred plasma torch (Machinability Data Center, 1980).

Gas shielded plasma: During cutting aluminum, stainless steel, and mild steels, shielding gases are used in order to obtain cuts of acceptable quality. An outer gas shield (N_2 or Ar/H_2) is added around the main stream of plasma. CO_2 shield is favorable for ferrous metals. For mild steel, air or O_2 may also be used as shielding gases.

Water shielded plasma: In this case, N_2 is used as the main working gas, while the shield is a water curtain, Figure 4.32. It is reported that the cooling effect of water reduces the kerf width and improves the quality of cut; however, the cutting rate is not improved.

4.4.3 Applications and Capabilities of PAC

- PAC has a special attraction in the case of profile cutting of metals such as stainless steel and aluminum that are difficult to tackle by the oxy-fuel technique.
- Oxyacetylene flame cutting has the advantage that it cuts metals of heavier sections than PAC does. For this reason, dual operating systems (plasma/flame) are now available on the market. Dual systems have an extended application range covering all materials.
- Plasma arc is used as a non-traditional tool integrated with some of the traditional processes such as turning, milling, shaping, etc. Figure 4.33 illustrates a layout of plasma arc turning (PAT) for rough and smooth machining of traditionally difficult-to-machine materials such as Inconel, Rene 41, Hastalloy, and precipitation hardened stainless steels. According to the machined material, and degree of finish required, the cutting speeds are

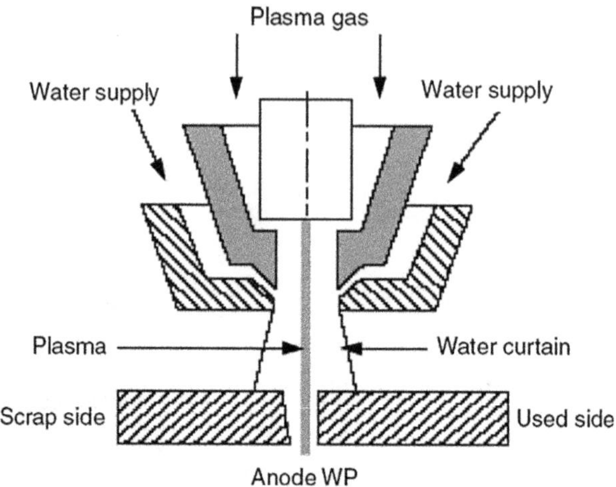

FIGURE 4.32 Water-shielded plasma.

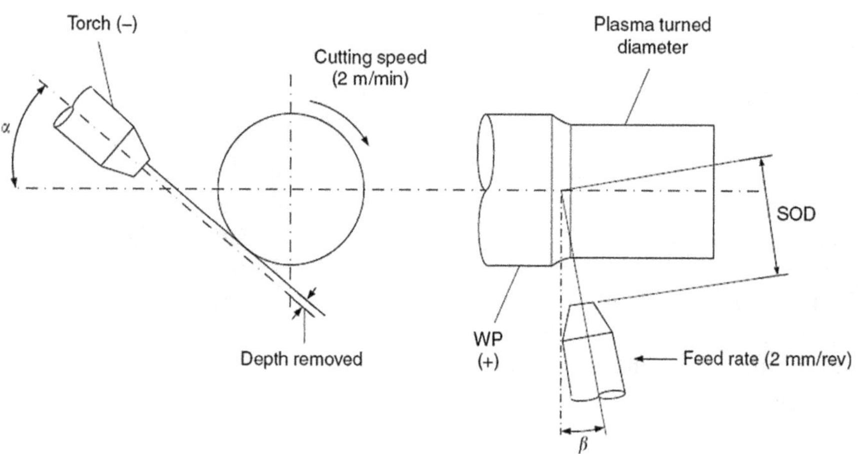

FIGURE 4.33 Plasma arc turning (PAT).

selected in a range between 10 and 100 m/min, and the nozzle feed between 1 and 5 mm/rev, depending on the required finish and the depth of cut. The nozzle is fixed at a suitable SOD.

- Underwater plasma cutting is used to reduce the plasma noise and to get rid of plasma fumes and glare. N_2 is preferred as working gas. Underwater plasma is characterized by two disadvantages, namely, the reduced cutting rate and the problems associated due to immersing the cutting torch in the water.
- PAC is also used to cut non-conductive materials such as textiles, nylon, and polypropylene with thicknesses ranging from 0.1 to 1 mm at a high traverse speed, 1000 m/min. Working gas should be Ar or Ar/H_2. A non-transferred nozzle is used in this case.

4.5 REVIEW QUESTIONS AND PROBLEMS

4.5.1 Using a neat sketch, explain the principle of material removal in EDM. Draw a typical relaxation circuit used for EDM power supply. Explain the main disadvantages of the relaxation circuits used in EDM.

4.5.2 Show diagrammatically the main elements of an EDM machine.

4.5.3 State the main functions of a dielectric used for EDM. Show the different modes of dielectric feeding to the EDM gap.

4.5.4 Explain the advantages and disadvantages of EDM.

4.5.5 Compare wire EDM and milling by EDM.

4.5.6 In an EDM operation employing a relaxation circuit, discuss the effects of charging resistance, gap setting, and capacitance on the rates of metal removal. How does this type of machine compare with EDM machine equipped with pulse generator?

4.5.7 What is the difference between sparking and arcing? Which condition leads to arcing in EDM?

4.5.8 What are the main applications of graphite electrodes in EDM?

4.5.9 Explain the term “no-wear EDM.”

4.5.10 Draw a neat sketch to show the difference between injection and suction flushing as applied in EDM. What type do you recommend in the following cases? Give your reasoning.

- Production of true cylindrical holes
- Production of forming tools and dies

4.5.11 For an EDM operating on RC circuit under optimum working conditions, determine the average power output, and the breakdown voltage V_s, given that the resistance $R = 3.2\ \Omega$; capacitance $C = 150\ \mu F$; supply voltage $Vo = 200$ V.

4.5.12 Define EBM.

4.5.13 State the important parameters that influence the MRR in EDM, LBM, and PBM.

4.5.14 What are the advantages and limitations of PAC?

4.5.15 How does a laser operate, and what does it consist of?

4.5.16 Explain why the CO2 laser is particularly effective for machining non-metals.

4.5.17 List the important advantages and limitations of LBM.

4.5.18 Make a comparison between LBM and EBM on the basis of their applications and limitations.

4.5.19 Precision engineering is a term that is used to describe manufacturing of high-quality parts with close tolerances and good surface finishes. Based on their process capabilities, make a comprehensive list of non-traditional machining processes (NTMPs) with decreasing order of quality of parts produced. Include a brief commentary on each method.

4.5.20 Make a list of non-traditional machining processes that may be suitable for the following materials: Ceramics, CI, thermoplastics, thermosets, diamond, and annealed copper.

4.5.21 Arrange the following NTMP in descending order of maximum metal rate, indicating their values in mm^3/min: USM, ECM, CHM, EDM, LBM, PAC.

4.5.22 Mention the parameters affecting SRR of the following processes: USM, ECM, WJM.

4.5.23 Draw a sketch to show the relaxation RC circuit used in EDM. What are the various materials of which electrodes are made for the EDM process and what are their advantages?

4.5.24 What are the main applications of graphite electrodes in EDM?

4.5.25 Discuss the factors influencing the choice of electrode material in EDM. Name the best electrode material for finish machining a small die made of WC by EDM process.

4.5.26 Explain briefly the function of using the dielectric fluids in EDM. What are the main types of dielectric fluids that can be applied?

4.5.27 Draw neat sketches to show three different flushing techniques used in EDM. What type do you prefer and why?

4.5.28 Draw a section through the heat-affected zone of a surface which is rough machined by EDM showing the hardness distribution and give reasons for its variation.

4.5.29 Show that for a spark machine operating on a relaxation circuit the breakdown voltage V_s is given by $V_s = V_o\,(1 - e^{-t/RC})$

where,

V_o = supply voltage
t = charging time
R = resistance of charging circuit
C = capacitance of circuit

Hence, determine for the above machine the average power output, given that the resistance $R = 3.2\ \Omega$; capacitance $C = 150\ \mu F$; supply voltage $V_o = 200$ V and the breakdown voltage $V_s = 160$ V.

4.5.30 For an RC generator, show that if t = RC, the voltage of the capacitor is 63.2% of the supply voltage after the elapse of time period t. What percentage of the supply voltage is reached by the capacitor when t = 4 RC?

4.5.31 For an RC circuit; if: $V_o = 125$ volt, $R = 3\ \Omega$, $C = 5\ \mu F$.

Calculate the breakdown voltage to achieve maximum power delivery, then determine at such condition:

- Charging time t_c
- Sparking frequency f if $t_d << t_c$
- Maximum charging current
- Charging time at the moment of breakdown
- Energy per spark

4.5.32 With the help of a sketch, describe the constructional features of an electron gun used for generating an electron beam. Explain why EBM is performed in vacuum. By what means is the EBM directed to precise location on the workpiece. What are the capabilities and limitations of EBM?

4.5.33 List the important advantages and limitations of LBM.

4.5.34 Describe an important limitation of LBM which occurs on the sidewall of holes. Could you suggest a remedy?

4.5.35 Give some important precautions that must be observed when operating laser beam equipment.

4.5.36 Give typical values of:

- depth of heat-affected zone in ECM, ... mm
- optimum break-down voltage V_s in EDM, for a source voltage V_o = 150 V, ...V
- surface roughness of ECMed surface, R_t = ... μm
- penetration rate in the case of USM of glass, ... mm/min
- gap voltage in ECM, ... V
- gap thickness in EDM, ... mm
- tool oscillation amplitude in USM, ... mm
- hole oversize in USM, ... mm
- nozzle diam. in AJM, ... mm
- SOD in AWJM, ... mm
- oscillation frequency in USM, ... Hz
- gap thickness in ECM, ... mm
- frontal gap thickness in USM, ... mm
- side gap thickness in USM, ... mm
- tool wear in ECM, ... mm
- depth of heat-affected zone in EDM, ... mm
- accelerating voltage of an EB, ... KV
- power density in EBM, ... kW/mm^2
- min. hole size produced by EB, ... μm

4.5.37 Mark true (T) or false (F) for each of the following statements:

[ ] To produce accurate cylindrical holes by EDM, injection flushing and not suction flushing should be used.

[ ] A main advantage of ECM is that tool wear is equal to zero.

[ ] While a relaxation oscillator is highly desirable in that it is simple and rugged, it is severely limited in metal removal capability.

[ ] EBM and EDM are thermal NT processes, which are used to machine only electrically conductive materials.

[ ] It is possible to set working conditions in EDM to obtain zero or minimal tool wear.

[ ] Controlled pulse circuits have the disadvantages of low metal removal rate and high tool wear.

[ ] In USM, the removal rate depends upon the mechanical properties of the WP.

[ ] In ECM, the removal rate is proportional to the valence, and inversely proportional to the atomic weight of the anode material.

[ ] In USM, the only function of the premagnetizing current is to avoid frequency doubling.

[ ] WJM is suitable for metallic WPs only.

[ ]The surface quality of electrochemically machined holes and cavities improves with increasing feed.

[ ] USM is economical in machining only electrically conductive hard materials to complex shapes with high accuracy and reasonable surface finish.

[ ] In AJM, it is always recommended to reuse the abrasive powder.

[ ] In WJM, material is removed by the mechanical action of a high-velocity stream impinging on a small area, whereby its pressure exceeds the yield strength of material.
[ ] In ECM, the passage of current through the gap results in material transfer from anode to cathode in the form of anions.
[] NaNO3 has desirable characteristics as an electrolyte and is less corrosive than NaCl. However, it has a tendency to passivate chemical reactions.
[ ] A CO2 laser is a kind of solid-state laser.
[ ] A laser is usable for metal removal only.
[ ] Mechanical shear is the main mechanism of material removal in EBM.
[ ] MRR in AJM is greater than that in AWJM.
[ ] In USM, for the same static load, the larger the tool diameter, the greater the penetration rate will be.
[ ] Complex shapes are produced in glass using EDM.
[ ] The current used in EDM is an AC.
[ ] PAM produces more accurate parts than EDM.
[ ] In EDM, it is desirable that the tool should be of positive polarity, if a short pulse duration is used.
[ ] In ECM, the only parameter that restricts an infinite penetration rate is the capacity of thc electrolyte to carry away heat without causing boiling.
[ ] Controlled pulse circuits have the disadvantages of low rate of metal removal and high tool wear.
[ ] In USM, the only function of the premagnetizing current is to avoid frequency doubling.
[ ] Graphite is most favorable in EDM electrode making.
[ ] A servo-control is used for providing feed rate in EDM.
[ ] Water-jet machining is suitable for metallic workpieces only.
[ ] ECM produces rough surfaces compared to EDM.
[ ] Mechanical shear is the main mechanism of material removal in WJM.
[ ] Laser removes metals mainly by melting.
[ ] CO_2 laser is less frequently used in material cutting.
[ ] In USM, premagnetization of the transducer is done to avoid freq. doubling and to achieve max. magnetostriction effect.
[ ] USM is economical in machining only electrically conductive hard materials to complex shapes with high accuracy and reasonable surface finish.
[ ] AJM is characterized by low capital investment and low power consumption. However, this process is restricted to brittle materials.

4.5.38 Define EDM and describe the mechanism of metal removal. Can you justify its very low efficiency in metal removal? Prove that the optimum working conditions for an RC relaxation circuit is realized if the breakdown voltage reaches 72% of the supply voltage.

4.5.39 Why has electrical discharge machining become a so widely used NTM operation?

4.5.40 A hole is to be machined in Widia by either ECM or USM. An important parameter that affects machining economy is the penetration rate. It is required to calculate the penetration rate in mm/min for both cases, provided that the machining took place under the following conditions.
Workpiece: Widia, Chem Equiv. N/n = 29.
Density $\rho = 12.8$ gm/cm^3, fracture hardness $H = 72000$ kg/cm^2

ECM	USM
Tool	
Brass of tube form ϕ 12/8 mm	Hardened steel of tube ϕ 12/8 mm
Machining Parameters	
Current. density $J = 200$ Amp/cm^2	Resonance frequency $f = 22$ kHz
Current eff. $\eta = 100\%$	Grit Mesh 320; $d_g = 50$ μm
Gap voltage = 20 V	Static stress $\sigma = 2.5$ kg/cm^2
Electrolyte spec. resist. $\rho_s = 5$ Ω.cm	Oscillation amplitude $\xi = 40$ μm

Actual penetration rate in USM can be calculated from the equation:

$$v_a = 30 f \frac{\sigma_s}{H} \sqrt{\frac{R_g \xi}{10^6}} \quad \text{mm/min}$$

R_g and ξ in μm, f in 1/sec.

Calculate then for ECM:

a. The machining current needed
b. The expected equilibrium gap h

REFERENCES

Barash, MM 1962, 'Electric spark machining', *International Journal of Machine Tool Design and Research*, vol. 2, pp. 281.

Dhakal, HN 2007, *Lecture notes on NTMPs, mechanical and marine engineering*, School of Engineering, University of Plymouth, UK.

El-Hofy, H 2005, *Advanced machining processes, nontraditional and hybrid processes*, McGraw-Hill Co., New York.

Kaczmarek, J 1976, *Principles of machining by cutting, abrasion, and erosion*, Peter Peregrines Ltd, Stevenage, Hertfordshire.

Kalpakjian, S 1984, *Manufacturing processes for engineering materials*, Addison Wesley Publishing Co, Reading, MA.

Lazarenko, BR & Lazarenko, NI 1944, *Elektrische Erosion von Metallen*, Cosenergoidat, Moskau.

Lissaman, AJ & Martin, SJ 1982, *Principles of engineering production*, Hodder and Stoughton Educational, London, UK.

Machinability Data Center *Machining data handbook* 1980, vol. 2, 3rd ednCincinnati, OH.

Nassovia-Krupp, Werkzeugmaschinenfabrik 1967, Langen/Frankfurt

Steigerwald, KH 1958, 'Materialbearbeitung mit Elektronenstahlen', *Fourth International Kongress fur Elktronenmikroskopie*, Springer, Berlin.

Unit II

Advanced Machining Technology

5 Machining of DTC Materials (Stainless Steels and Super Alloys) by Traditional and Non-Traditional Methods

5.1 INTRODUCTION

Over recent decades, engineering materials have been greatly developed. These materials such as hardened steels, stainless steels, super alloys, carbides, ceramics, and fiber-reinforced composite materials are frequently applied in modern industry. The cutting speed and the material removal rate when machining such materials using traditional methods such as turning, milling, grinding, and so on, tend to fall. Sometimes, it is difficult to machine hard materials to certain shapes using these traditional methods. It is no longer possible to find tool materials that are sufficiently hard to cut such materials. To meet these challenges, new processes with advanced methodology and tooling have to be developed. These are the non-traditional processes, which are capable of machining a wide spectrum of these difficult-to-cut materials irrespective of their hardness. The increasing use of ceramics, high-strength polymers, and composites will also necessitate the use of non-traditional methods of machining. In addition, grinding will be applied to a greater extent than in the past, with greater attention given to creep feed grinding (CFG), and the use of poly-diamond (PCD) and poly-cubic boron nitride (PCBN) (ASM International, 1989).

5.2 TRADITIONAL MACHINING OF STAINLESS STEELS

5.2.1 Types, Characteristics, and Applications of SSs

Stainless steel does not constitute a single well-defined material, but instead, depending on the alloying elements additions, comprises several families of alloys; each generally having its own characteristics, microstructure, alloying elements, and properties, that are suited to a wide range of applications.

The American Iron and Steel Institute (AISI) and the Unified National Standard (UNS) grouped these alloys by chemistry and assigned a three-digit number (AISI) or a five-digit number (UNS) that identifies stainless steels. Accordingly, stainless steels have been subdivided into five families: three basic families (*ferritic*, *martensitic*, and *austenitic*), and two derived families (*duplex* and *precipitation hardened*).

Basic (standard) alloys of stainless steels: These comprise the following three categories:

1. ***Ferritic stainless steels of AISI designations***: Series 400 [405-409-430-434-436-442-444-446], Figure 5.1.
   These have carbon levels below 0.12% (442 and 446 are at 0.2% C) and high Cr content (10.5–27%), and relatively small amounts of other alloying elements. Higher levels of Cr in alloys 442 and 446 promote their corrosion and oxidation resistance. Mo is added to 434 and 444 to improve corrosion resistance particularly in chloride-containing solution.
   These grades are magnetic and cannot be hardened by heat treatment, however, hardened by cold working, but not to the same extent as austenitic alloys. Ferritic alloys have reduced corrosion resistance compared to austenitic. They are generally not chosen for toughness. In the annealed condition, they have a yield tensile strength (YTS) of 275–350 MPa.
   The last four alloys in Figure 5.1 are free-machining ferritic alloys since they contain free-machining additives.

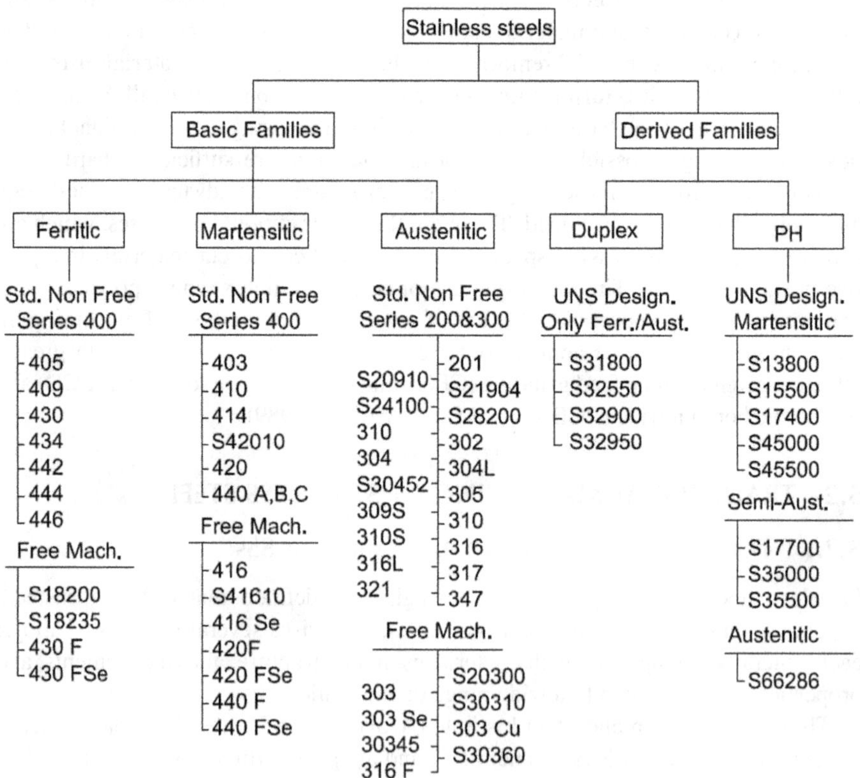

FIGURE 5.1 Basic (standard) and derived families of stainless steel alloys (Youssef, H. 2016).

Since ferritic alloys are the cheapest type of stainless steel, they should be given first consideration when SS alloy is required. Ferritic stainless steels are generally used for non-structural applications such as kitchen and restaurant equipment, automobile trims, heaters, dishwashers, and annealing baskets.

2. ***Martensitic stainless steels also of AISI designation***: Series 400 [403-410-414-416-420-422-431-440], Figure 5.1.
   Martensitic alloys have a relatively high carbon level (0.15–1.2% C) as compared to ferritic and austenitic grades, and Cr level from 11.5 to 18%. Mo (<1%) can be added to improve mechanical properties and corrosion resistance (UNS-42010). Nickel (<2.5%) can be added for the same reason, AISI (414 and 431).
   Martensitic alloys are also magnetic. They are not as corrosion resistant as the other two basic classes. In the annealed condition, the yield tensile strength (YTS) is about 275 MPa, and thus these alloys can be moderately hardened by cold working similar to ferritic alloys. However, when hardened and tempered, their YTS increases up to 1900 MPa, depending primarily on carbon content. In the annealed condition, they are machinable. The last seven alloys in Figure 5.1 are free-machining martensitic alloys, since they contain considerable amounts of free-machining additives of S or Se (minimum of 0.15% each).
   Martensitic alloys cost about 1.5 times as much as ferritic. Martensitic alloys are used for applications such as cutlery, surgical tools, instruments, valves, rivets, screws, hand tools, vegetable choppers, razor blades, riffle barrels, mining machinery, bolts, nuts, and aircraft fittings.
3. ***Austenitic stainless steels of AISI designation***: Of Series 300 [Fe-Cr-Ni], and Series 200 [Fe-Cr-Mn], Figure 5.1.
   This category contains maximum 0.15% C, a minimum of 16% Cr and sufficient Ni and/or Mn to retain an austenitic structure at all temperatures from the cryogenic region to the melting point of the alloy. These alloys cannot be hardened by heat treatment and are non-magnetic. They all exhibit excellent corrosion resistance (Mo is added for resistance to chlorides). Austenitic stainless steels are the most ductile of all SSs; hence they can be easily formed. However, with increasing cold work, their formability and ductility are reduced, and they strengthen significantly, Figure 5.2. The response of the popular 304 alloy (also known as 18-8 of composition 18% Cr and 8% Ni) to a small amount of cold work (15%) is illustrated in Table 5.1.

Corrosion resistance of austenitic alloys varies from good to excellent, depending on the alloy contents. Higher Cr, Ni, Mo, or Cu can be added to improve corrosion and oxidation resistance [316, 317, and 310,]. To prevent inter-granular corrosion after high temperature exposure, Ti or Nb is added to stabilize carbon in alloys such as 321 and 347. Carbon levels are reduced to low values to produce weldable alloys such as 304 L and 309 S.

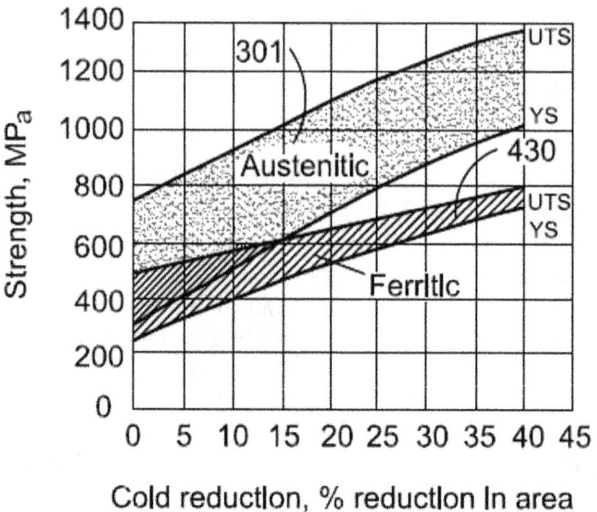

FIGURE 5.2 Comparison between cold-working rates of ferritic 430 and austenitic 301 stainless steel (ASM International, 1989).

TABLE 5.1
Yield and Tensile Strength of AISI 304 Stainless Steel at Water Quench and Cold Rolled Conditions

AISI 304	Water Quench	15% Cold Rolled
YTS (MPa)	260	805
UTS (MPa)	620	965
Elong. % (50 mm)	68	11

Source: (DeGarmo et al., 2012).

Austenitic stainless steels are classified into two groups, Figure 5.1.These are:

1. Standard group of AISI designation (Series 300), where Ni is the austenite stabilizer
2. The Mn group of AISI (Series 200), where a substantial quantity of Mn is added

The last eight alloys in Figure 5.1 are free-machining versions, since they contain free-machining additives such as S and Se.

Austenitic alloys are costly; thus, they should not be specified where less expensive ferritic or martensitic alloys would be adequate. Austenitic alloys of standard series 300 may cost twice as much as the ferrite variety due to their expensive alloying elements (Ni and Cr). Mn group (series 200) is of lower cost, but is a somewhat lower quality alloy. Austenitic alloys are used in a wide variety of applications, such as kitchenware, fittings, welded constructions, lightweight transportation equipment, furnace and heat exchanger parts, and equipment for severe chemical environments. The austenitic 304 is the most commonly used SS in the world. The excellent forming and welding characteristics make it the standard steel for many applications, architecture, and transportation. The grade 316 is the second most commonly used austenitic SS. Like 304, it has excellent forming characteristics, but the added Mo gives 316 an improved corrosion resistance, so it is usually regarded as "marine steel grade" (Youssef, 2016). Table 5.2 summarizes typical compositions of the basic alloys of SSs. The microstructure that a stainless steel attains depends primarily on its composition, in which the main alloy components Cr and Ni are most important.

Derived alloys of stainless steels: These comprise the following two categories.

1. ***Duplex alloys***: Of designations shown in Figure 5.1.
   These contain 21–29% Cr and 2.5–6.5% Ni. They are water-quenched from 960–1020°C to produce ferritic/austenitic microstructure; hence they are magnetic. Duplex alloys have a YTS of about 550 MPa in the annealed condition, which is about twice that of the standard water-quenched austenitic alloy. They generally have good ductility and toughness, along with higher resistance to both corrosion and stress-corrosion cracking than austenitic alloys. Nitrogen is added for strengthening (UNS-31803, UNS-32950), Figure 5.1. Duplex alloys do not embrace any free-machining categories.

   Typical applications of duplex alloys are in water-treatment plants and heat exchanger components. It is used in the chemical, food, medical, and papermaking industries, and in processes that include acids or chlorine, and the gas industry.

TABLE 5.2
Typical Compositions (wt%) of the Basic Ferritic, Martensitic, and Austenitic SSs

Element	Ferritic (bcc)	Martensitic (bcc)	Austenitic (fcc)
Carbon	0.08–0.2	0.15–1.2	0.03–0.25
Chromium	11–27	11.5–18	16–26
Nickel	0–1	0–2.5	3.5–22
Manganese	1–1.5	1–1.5	2(Series 200, 5.5–10)
Silicon, Molybdenum	1	1	1–1.5
Phos. & sulfur*	0–2.5	0–1	Some cases: 1.5–3
Titanium	0.075	0.075	0.075
	0–1	–	0–0.4

*In free-machining alloys, the minimum S (or Se) content is 0.15%.
Source: (DeGarmo et al., 2012).

2. ***Precipitation hardened alloys***: Of designations shown in Figure 5.1. These alloys contain Cr and Ni, along with Cu, Al, Ti, Nb, or Mo. Molybdenum is added to improve the mechanical properties and corrosion resistance. Cu is also added to enhance corrosion resistance (UNS-45000). Precipitation hardened (PH) alloys are characterized by their ability to be age-hardened to various strength levels. PH alloys can attain a YTS of up to 1700 MPa. These alloys have generally good ductility, toughness, and strength at elevated temperature, with acceptable good corrosion resistance.

Depending on their composition, these alloys may be subdivided into the following:

- Martensitic (magnetic)
- Semi-austenitic (magnetic)
- Austenitic (non-magnetic)

A better combination of strength and corrosion resistance is achieved by austenitic and semi-austenitic types than martensitic PH alloys. The most common PH alloy UNS-17400 contains Cr and Ni as do all alloys. Similar to duplex, PH alloys do not include a free-machining category; however, the alloy UNS-17400 has enhanced machining characteristics (it is machined at high speed without chattering).

The main applications of these alloys are in aircraft, aerospace, and structural components. The ultra-strength alloy 17-7 PH (UNS 17700) is frequently used in spring, pressure vessels, and where strength coupled with good corrosion resistance is required.

5.2.2 Machinability and Machinability Ratings of SSs

Ordinary stainless steels are difficult-to-machine because of their work-hardening properties and their tendency to seize during cutting. For these reasons, special free-machining alloys have been developed, within each family of the basic alloys by the addition of free-machining elements such as S, Se, Te, and so on, thus forming inclusions in these stainless steels that significantly improve their overall machining characteristics. However, the benefit of improved machinability due to the addition of these elements is not of course obtained without degrading other important properties of stainless steels such as corrosion resistance, strength, ductility, toughness, hot workability, cold formability, and weldability. The improvement in machinability must be balanced against the possible reduction of the degraded properties, especially corrosion resistance.

5.2.2.1 Free-Machining Additives of Stainless Steels

Free-machining additives form inclusions which increase tool life, allow high cutting rates, and may also affect disposability, and surface finish. Important free-machining additives are:

1. **Sulfur (S)**: The amount of sulfur addition is limited by the allowable degradation of other properties. Increasing sulfur increases tool life. Larger sulfur additions enhance disposability (ease of cut); they decrease surface finish.

2. **Selenium (Se)**: It is another commonly used free-machining agent after S in stainless steels, forming inclusions of MnSe analogous to sulfides. Sulfur is used in Europe, whereas Se is frequently used in the United States. Se is less effective than an equivalent weight percent of S in improving the overall machining characteristics of stainless steel. However, Se-bearing alloys provide a better machined surface finish than S-bearing alloys. Se-bearing stainless steels may offer improved cold formability and somewhat improved corrosion resistance compared to the corresponding S-bearing alloys.
3. **Tellurium (Te)**: It is like Se and S, because it forms inclusions similar to sulfides (MnTe). Tellurium is more effective in improving the machinability of austenitic stainless steel.
4. **Lead (Pb) and Bismuth (Bi)**: Both Pb and Bi have low solubility in stainless steel, forming metallic inclusions that improve machinability. Lead is more beneficial to machinability of austenitic stainless steels than other free-machining additives. It is also reported that the use of Pb, with or without limited S, results in better machined surface finish, corrosion resistance, and cold formability than the use of S alone. Similar effects have been attributed to the use of Bi.
5. **Phosphorous (P)**: It is added in conjunction with S or Se to enhance machinability, not by forming inclusions, but by modifying the matrix properties.

5.2.2.2 Machinability of Free- and Non-Free-Machining Stainless Steels

The free-machining alloys are found in applications where mechanical properties are second in the level of importance. Figure 5.1 illustrates the free-machining alloys as derived from the three basic stainless alloys.

1. ***Ferritic and martensitic alloys***: Free-machining ferritic alloys such as S43020 (AISI-430F), and annealed, low carbon martensitic free-machining alloys such as S41600 (AISI-416) are the easiest to machine among stainless steels. Their machinability ratings are close to those of free-machining carbon steels.
   The non-free-machining of lower Cr-ferritic alloys (AISI-405, AISI-430), and annealed low carbon, straight-Cr martensitic alloys (AISI-403, AISI-410), Figure 5.1, are generally easier to machine than most other non-free-machining alloys. The higher Cr-ferritic non-free-machining alloys, such as AISI-446, Figure 5.1, are somewhat difficult to machine than the lower Cr alloys, because of their gumminess and stringy chips.
   In martensitic alloys, the machinability decreases as the carbon content increases from (AISI-410) to (AISI-420) to (AISI-440C) for the non-free-machining alloys, Figure 5.1. With higher C-levels, there is a tendency for smaller difference in machinability between the corresponding free- and non-free-machining versions. Ni content also influences machinability by increasing the annealed hardness levels.
2. ***Austenitic alloys***: The difficulties in machining stainless steels are, in general, more specifically attributed to the austenitic stainless steels. Compared

to ferritic and martensitic alloys, typical austenitic alloys have a higher work-hardening rate, a wider spread between yield tensile strength (YTS) and ultimate tensile strength (UTS), Figure 5.2, and higher toughness and ductility. The high ductility of these steels works against them in machining. Poor chip breaking and built-up edge (BUE) at cutting face can easily occur. The thermal conductivity of austenitics is low, compared to other types of stainless steels, Table 5.3, so heat can easily build up at the tool face. Distortion or poor tolerance control during machining can be affected by the higher thermal expansion rates of these steels. In their annealed condition, these steels are not ferromagnetic.

This means that magnetic clamping devices cannot be used. The combination of these effects makes it seem that austenitics appear difficult to machine. When machining austenitic stainless steels, particularly, the non-free-machining alloys are more characterized by:

- Tools becoming hotter, tending to form a large BUE
- Forming stringier chips, with tendency to angle, causing difficult disposal
- Chattering occurs if tool rigidity is inadequate or marginal
- Producing work-hardened surfaces, which are more difficult to machine, especially if the feed rate is too low

Carbon and nitrogen can affect the work-hardening rate and increase the strength and hardness of austenitic stainless steels. Higher levels of either or both elements decrease machinability. Consequently, high-N austenitic alloys such as S20910 and S28200, are more difficult to machine than the standard lower-N austenitic alloys.

3. ***Duplex alloys***: The machinability of duplex stainless steels is generally poor. It is limited due to their high annealed strength levels. Machining of duplex causes chip hammering and generates a lot of heat, which causes plastic deformation and severe crater wear. Small entering angles are preferable. Good tool clamping and work fixation are essential.

   Table 5.4 compares the machinability (in terms of drilling depths) of duplex alloy S32950 with that of high-N austenitic alloy S20910, and a conventional austenitic alloy S31600 of standard S-content of 0.03%, and enhanced-machining version of S-content of 0.03%, Figure 5.1.The duplex

TABLE 5.3
Thermal Conductivities of Austenitic SSs as Compared to Ferritic SSs and Carbon Steel

Work Material	Thermal Conductivity (W/m.°K)
Carbon Steel	44
Austenitic SS (AISI-302)	16
Ferritic SS (AISI-430)	23

TABLE 5.4
Comparison of Machinability in Terms of Drill Penetration Depth (mm), [25 s, 45 kg Thrust Load]

Work Material	Drill Penetration Depth (mm)
High N-Aust. SS (S20910), 98 HRB	1.9–2.4
Duplex SS (S32950), 100 HRB	2.4–3.0
Standard Aust. SS (AISI-316), 79 HRB	3.2–3.7
Enhanced lower N-Aust. SS (AISI-316L), 76 HRB	3.7–4.2

S32950 has a hardness level comparable to that of the high-N austenitic alloy S20910, but provides higher machinability. However, it is not machinable as either the standard or the enhanced-machining alloy AISI-316.

4. ***PH alloy***: The machinability of PH-SSs depends on the type and hardness level of the alloy. Martensitic PH-SSs are often machined in the solution-treated condition. Most of these alloys machine comparable to, or somewhat worse than, a standard austenitic alloy such as AISI-304. The martensitic PH alloy S17400, Figure 5.1, is available in an enhanced-machining version that allows machining at higher cutting speeds with a significantly reduced tendency toward chattering. Semi-austenitic alloys S35000 and S35500 provide best machinability if supplied in over-tempered condition. Austenitic PH alloys such as S66286 machine poorly, requiring lower speeds than even the highly alloyed austenitic stainless steels. Machining in the aged condition requires even lower speeds.

5.2.2.3 Enhanced Machining Stainless Steels

It should be emphasized that the enhanced-machining versions of non-free-machining alloys provide machining performance superior to that of the corresponding standard alloys, but still do not provide the machinability of comparable free-machining alloys. However, other properties are superior to those of the corresponding free-machining alloys. Whereas free-machining alloys are only available in ferritic, martensitic, and austenitic alloys, the enhanced machining versions are available in all types of SSs.

5.2.2.4 Machinability Ratings of Stainless Steels

From the previous discussion, about 50 grades of stainless steels, comprising non-free and enhanced machining alloys are considered. They are grouped according to their machinability levels. Table 5.5 provides a proposal for ranking these grades through a ten-level machinability in descending order, based on a reference material (resulfurized and rephosphorized plain carbon steel AISI-1212). It also shows the recommended cutting speeds used. The machinability rating of the Ni-based heat-resisting super alloy (Hastelloy X) and some plain carbon and alloy steels have been considered for comparison purposes.

TABLE 5.5
Machinability Ratings for Stainless Steels and Some Selected Steel Alloys and Related CF Cutting Speeds in Descending Order

Grade (AISI Designation)	Approximate Cutting Speed (m/min)	Machinability Rating Based on 1212 (R %)
Resulf. andrephos. free cutting plain carbon steel 1212	60	100
Stainless steels: In annealed condition unless otherwise stated		
1. Mart. 416	66	110
2. 203/303/Ferr. 430F	51	85
3. Mart. 420F/Ferr. 430	40	67
4. Mart. 403/ Mart. 410	32	54
5. Mart.416 High Tens./Mart. 420/Mart. 422/Mart. 17-4 PH-H1150 (age hardened)/Standard Duplex 2205	28	50
6. 303 High Ten s./Mart.431/Mart.440FSe/Mart.15-5 PH/Mart. 15-7PH-(Custom 450)/Mart. 17-4 PH/Semi-auste. 17-7 PH	30	45
7. 302/304/304L/Pyromet, Semi-aust.AM350PH/Semi-au st.AM355 PH/Mart. 440A/Mart. 440C	24	40
8. 309/310/316/316L/317/317L/317LM/321/347/Ferr.446/ Mart. 13-8 M PH	21	36
9. 302"B"/304"B"/Mart. Custom 455 PH/Aust. Pyromet A286 PH/Duplex 7-Mo Plus	18	30
10. Mn-aust.: Nitronics 40,50,60/316"B"	13	22
Super alloy: Hastelloy X	10	17
Plain carbon and alloy steels: In annealed condition		
1020,1015 /	43	72
1050	32	54
1141	48	81
L 14 12	102	170
52100 (Ball bearing steel)	24	40

Source: Falcon Metals Group, British Stainless Steel Association, and Carpenter Technology Corporation.

5.2.3 Machining and Machining Conditions of SSs

In this section, the machining of stainless steels using different machining processes will be discussed for the following operations.

1. ***Turning***: Traditionally, HSS tools have been used for most turning operations of SSs, but carbides and carbide-coated tips are also used to realize higher removal rates. The choice of tool material depends on machining parameters, production rate, and the available power and rigidity of machines.
   Figure 5.3 illustrates the suggested tool geometry of a single-edge tool, intended for turning stainless steels, whereas Table 5.6 suggests the

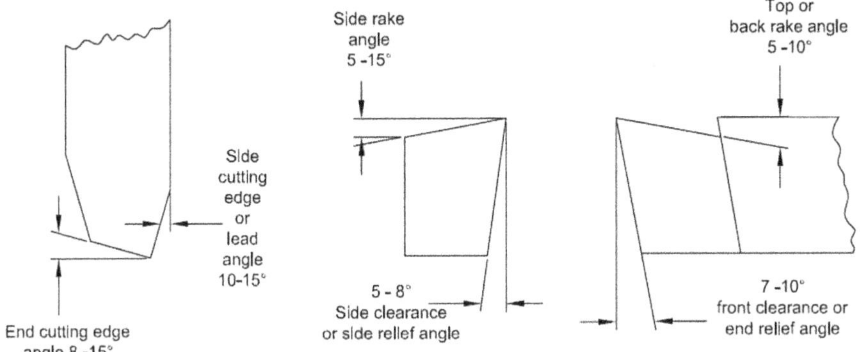

FIGURE 5.3 Suggested geometry of single-edged turning tools for machining stainless steels.

recommended cutting speeds, and depths of cut for turning SSs, using HSS and carbide tools. Too high speed can result in tool tip burning, and too low speed can result in BUE formation. Cutting fluids must be used. It is beneficial to select as large a tool as possible to provide a greater heat sink as well as a more rigid setup.

Because of their toughness, and work-hardening characteristics, austenitic stainless steels require HSS tools ground with side rakes of 10–15° to control chips and may require increased side relief angles of 5–6° to prevent rubbing and localized work-hardening.

Non-free-machining stainless steels tend to produce long stringy chips that can be very troublesome. This difficulty can be alleviated by using chip breakers that, in addition to controlling long chips, reduce friction on tool face. Chip breakers for free-machining stainless steels do not need to be as deep as those needed for the non-free-machining alloys.

Carbide tools can also be used. They allow higher speeds than HSS tools, Table 5.6. However, they require greater attention to the rigidity of the tooling and workpieces; moreover, interrupted cuts should be avoided. Speeds for coated carbides are approximately 30% higher than those for carbides.

2. ***Drilling***: In drilling, the material is mechanically pushed from the center of the drill tip (chisel edge) to the outer cutting edge. This pushing action work-hardens the material at the drill point, in turn hardens the material, causing excessive wear of the drill and hard spots in the stainless steel. This work-hardening effect especially of the austenitic grades, such as 304 and 316, is the main cause of problems associated with drilling stainless steels. Table 5.7 lists the recommended speeds and feeds for drilling different grades of stainless steels. Although standard HSS drills are suitable for drilling stainless steel, they should preferably be shorter to reduce deflection and breakage in use. Carbide-tipped drills can also be used to realize higher productivity. Figure 5.4 illustrates the proper tool geometry of an HSS drill used for drilling stainless steels. The point angle should be 140°, although smaller

TABLE 5.6
Recommended Speeds and Feeds for Single-Edged Turning of Stainless Steels

Stainless Steel Alloy	Treatment and HB	Rough R or Finish F	Speed m/min HSS	Speed m/min Carbide Brazed	Speed m/min Carbide Insert	Tool/Code* HSS	Tool/Code* Carbide
Wrought Martensitic							
Free-machining grades	Ann.-160	R	45	150	170	M2/M3	C6/C7
		F	52	170	190	M2/M3	C6/C7
Lower C/lower Cr grades	Ann.-175	R	33	120	145	M2/M3	C6/C7
		F	40	145	175	M2/M3	C6/C7
Lower C/lower Cr grades	Q&T-300	R	20	80	100	M2/M3	C6/C7
		F	25	98	120	M2/M3	C6/C7
Higher C/higher Cr grades	Ann.-240	R	20	95	105	M42/T15	C6/C7
		F	24	105	120	M42/T15	C6/C7
Higher C/higher Cr grades	Q&T-320	R	15	67	80	M42/T15	C6/C7
		F	21	80	90	M42/T15	C6/C7
Wrought Ferritic							
Free-machining grades	Ann.-160	R	45	150	165	M2/M3	C6/C7
		F	50	165	185	M2/M3	C6/C7
12–17% Cr grades	Ann.-160	R	33	137	155	M42/T15	C6/C7
		F	40	155	180	M42/T15	C6/C7
Wrought Austenitic							
Free-machining	Ann.-160	R	31	125	140	M2/M3	C2/C3
		F	36	140	155	M2/M3	C2/C3
Other grades (304, 316, 321, etc.)	Ann.-160	R	23	85	95	M42/T15	C2/C3
		F	28	97	112	M42/T15	C2/C3
Wrought Duplex	Ann.-230	R	24	53	55	M42/T15	C2/C3
		F	30	60	70	M42/T15	C2/C3
Cast Austenitic	Ann.-270	R	15	68	83	M42/T15	C2/C3
		F	20	83	100	M42/T15	C2/C3

Rough (R): DOC = 3.8 mm, Feed = 0.38 mm/rev, **Finish(F)**: DOC = 0.75 mm, Feed = 0.2 mm/rev.
Treatment: Ann. = Annealed; Q&T = Quenched and tempered.
Speeds: For coated carbides are approx. 30% higher, ***Tool Code**, See (Youssef, 2016).
Source: Adapted from British Stainless Steel Association.

angles (120°) can be used for drilling free-machining stainless steels. Wear of the drill point is an indication that a larger point angle should be used. The lip clearance should be 8–16°. It decreases as the drill diameter increases as indicated in the following table.

Drill diameter, (mm)	3	6	12	20	25 and over
Lip relief angle, (deg.)	16	14	12	10	8

TABLE 5.7
Recommended Speeds and Feeds for Drilling Stainless Steels Using HSS Drills

	Treatment	Speed	Feed mm/rev. for hole diam. of							
Stainless Steel Alloy	**and HB**	**m/min**	**1.6 mm**	**3 mm**	**6 mm**	**12 mm**	**20 mm**	**25 mm**	**50 mm**	**HSS grade***
Wrought Martensitic										
Free-machining grades	Ann.-160	36	0.025	0.075	0.15	0.255	0.33	0.4	0.635	M1,M7,M10
Lower C/lower Cr grades	Ann.-175	18	0.025	0.075	0.125	0.205	0.3	0.4	0.61	M1,M7,M10
Lower C/lower Cr grades	Q&T-300	17	0.025	0.075	0.1	0.175	0.255	0.3	0.455	M1,M7,M10
Higher C/high. Cr grades	Ann.-240	15	0.025	0.05	0.075	0.125	0.2	0.255	0.38	M42, T15
Higher C/high. Cr grades	Q&T-320	12	0.025	0.05	0.05	0.1	0.15	0.2	0.3	M42, T15
Wrought Ferritic										
Free-machining grades	Ann.-160	40	0.25	0.075	0.15	0.255	0.355	0.455	0.635	M1,M7,M10
12–17% Cr grades	Ann.-160	20	0.25	0.05	0.1	0.175	0.255	0.3	0.455	M42, T15
Wrought Austenitic										
Free-machining grades	Ann.-160	30	0.25	0.075	0.15	0.255	0.355	0.455	0.635	M1,M7,M10
Other grades (304,316,321,etc.)	Ann.-160	17	0.25	0.05	0.1	0.175	0.255	0.3	0.455	M42, T15
Wrought Duplex	Ann.-230	14	0.25	0.05	0.075	0.125	0.2	0.255	0.38	M42, T15
Cast Plain Cr	N&T-210	14	0.25	0.05	0.075	0.125	0.2	0.255	0.33	M42, T15
Cast Austenitic	Ann.-170	10	0.25	0.05	0.075	0.15	0.2	0.255	0.28	M42, T15

Ann. = Annealed Q&T = Quenched and tempered N&T = Normalized and tempered * HSS grade: See (Youssef, 2016).

Source: Adapted from: British Stainless Steel Association.

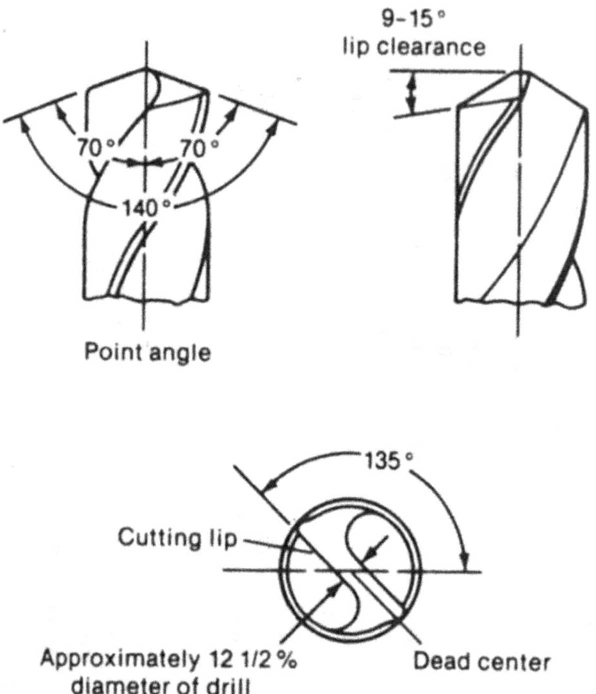

FIGURE 5.4 Recommended geometry of HSS drills for drilling stainless steels.

The web thickness at the point should be generally around 1/8 of the drill diameter, except for small diameters where thicker webs improve the strength and rigidity of the drill. This can, however, impede chip flow. Conversely, with a thinner web, although the drill has less rigidity, chip flow is easier; additionally, it is easier to start the hole. A thinner web reduces the feed force, heat generation, and work-hardening of the bottom of the hole; in other words, it enhances the drill ability.

3. ***Reaming***: The difficulties involved in reaming stainless steels are most often caused by previous operations, particularly with the non-free-machining austenitic alloys. For example, if the feed in the previous drilling operation is too light, the hole wall can be severely work-hardened and can resist cutting by the reamer.

   Straight- or spiral-fluted designs of reamers are used for cylindrical or tapered holes. Spiral fluted are preferable for reaming stainless steels as they produce less chatter, can better dispose chips from deep holes, and are capable of producing a better finish than straight-fluted reamers. For normal clockwise tool rotation, right-hand spiral tools cut more freely than left-hand spiral tools but tend to self-feed into the hole. Left-hand spiral tools have a lower tendency to self-feed which can be beneficial when precise feed rate control is needed.

   Figure 5.5 illustrates an HSS-spiral-fluted reamer. The rake angles should be between 3 and 8°, larger angles suiting austenitic stainless-steel

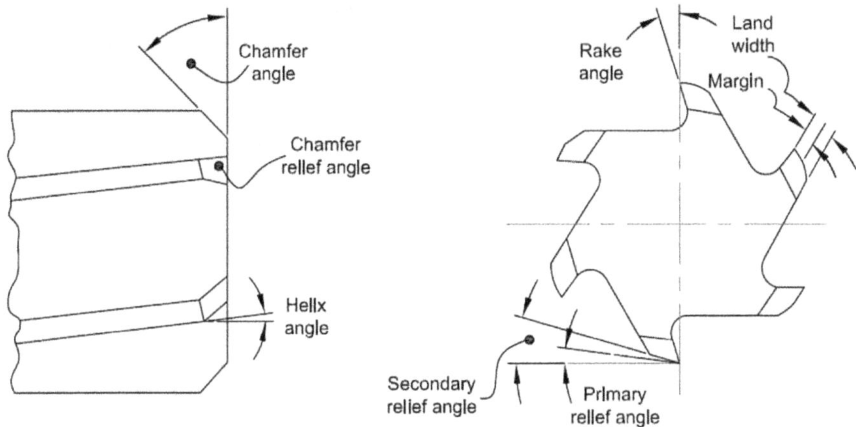

FIGURE 5.5 HSS-spiral-fluted reamer for machining stainless steels.

grades (304, 316, and so on). Margin width should be 0.13–0.38 mm for HSS tools and 0.05–0.125 mm for carbide-tipped reamers. The margin width increases within this range with increasing reamer diameter. Insufficient primary relief angle or too wide a land can cause chattering.

Table 5.8 illustrates the recommended tool geometries of HSS and carbide-tip reamers for machining stainless steels. Smooth finishes require significantly lower speeds as compared to drilling. To improve the surface finish of finally reamed holes, the suggested speeds in Table 5.7 should be reduced by 50% and the feeds by 25%. Rough holes or burned tools usually indicate that too high a speed is being used. To obtain smooth surfaces, the

TABLE 5.8
Recommended Geometries of HSS and Carbide-Tipped Reamers for Machining Stainless Steels

	Values	
Geometrical Feature	**HSS reamer**	**Carbide reamer**
Primary (working sec.) relief angle, deg.	4–5	6–12
Margin (land) width, mm	0.13–0.38	0.05–0.125*
Chamfer angle, deg.	30–35	2
Chamfer length, mm	1.5	4.8
Chamfer relief angle, deg.	4–5	NA
Rake angle (working sec.), deg.	3–8	7–10
Helix angle, deg.	0–10	5–8

* For reaming the non-free-machining, ferritic and austenitic PH grades and grades, using carbide-tipped reamers, the margin width should be increased to 0.125 to 0.25 mm.
Source: Compiled from ASM International, Metals Handbook, Vol. 16 Machining, 1989.

cutting fluid must also be kept clean. Reaming produces slivers and very fine chips which can float in the cutting fluid and damage the surface finish, especially if the machine is equipped with a circulating system.

4. ***Milling***: HSS cutters are used in milling stainless steels, although tooling with carbide inserts can also be used, particularly for alloys that are more difficult-to-machine. The smoothest finishes are obtained with helical or spiral HSS cutters running at high speed, particularly for cuts over 20 mm wide.

   Milling deep slots in stainless steel sometimes presents problems of chattering and jamming of wide chips; such difficulties are eliminated by using a staggered-teeth cutter. End milling of stainless steels is recommended using a solid-shank end mill because of its high strength. Figure 5.6 illustrates the suggested geometry of a side-milling cutter. Excessive vibration indicates that the cutter has insufficient clearance, provided that the rigidity of tooling, fixtures, and machine are adequate. Hogging-in, generally, indicates too much rake or possibly too high a cutting speed.

   Table 5.9 lists speeds and feeds in the case of peripheral end milling stainless steels using either HSS or carbide cutters. If the feed is too light, the tool will burnish the work; if too heavy, the tool life will be shortened. A roughing cut runs with heavier feeds and slower speeds, than those used for finishing cuts.

5. ***Broaching***: Broaches for stainless steels are usually made from HSS or PM-HSS (powder metallurgy high speed steel). Basically, a broach can incorporate a roughing, a semi-finished, and a final precision cut. For internal broaches, the back-off angle should be held to a minimum (2°, and not exceeding 5°). Too much back-off angle will shorten broach life due to size reduction from re-sharpening. Any nicks on cutting edges of the broach will score the surface of the work. Therefore, careful handling is very important.

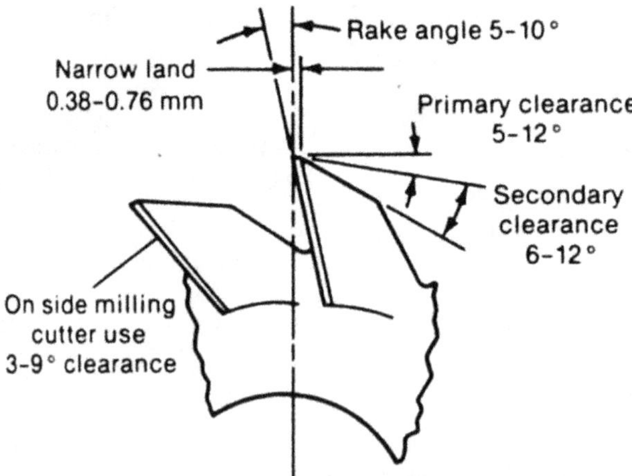

FIGURE 5.6 Recommended geometry of milling cutters for machining stainless steels.

TABLE 5.9
Nominal End-Milling Parameters for Machining Wrought Stainless Steels

Stainless Steel Alloy	Treatment and HB	Speed m/min	Feed mm/Tooth for Cutter Diam. 6 mm	13 mm	19 mm	25–50 mm	Tool Material HSS: M2, M7 Carb.: C2, C6
Wrought Martensitic							
Free-machining grades	Ann.-160	30	0.03	0.06	0.11	0.12	M2, M7
		90	0.03	0.06	0.13	0.17	C6
Lower C/lower Cr grades	Ann.-175	34	0.05	0.075	0.12	0.15	M2, M7
		107	0.025	0.05	0.12	0.15	C6
Lower C/lower Cr grades	Q&T-300	27	0.025	0.05	0.12	0.15	M2, M7
		82	0.025	0.05	0.12	0.15	C6
Higher C/higher Cr grades	Ann.-240	23	0.025	0.05	0.12	0.15	M2, M7
		72	0.025	0.05	0.12	0.15	C6
Higher C/higher Cr grades	Q&T-300	20	0.025	0.05	0.12	0.15	M2, M7
		68	0.025	0.05	0.12	0.15	C6
Wrought Ferritic							
Free-machining grades	Ann.-160	44	0.03	0.06	0.1	0.12	M2, M7
		122	0.03	0.06	0.13	0.17	C6
12–17% Cr grades	Ann.-160	32	0.05	0.075	0.13	0.15	M2, M7
		100	0.025	0.05	0.1	0.15	C6
Wrought Austenitic							
Free-machining grades	Ann.-160	32	0.03	0.06	0.11	0.14	M2, .M7
		100	0.03	0.06	0.11	0.14	C7

(Continued)

TABLE 5.9 (CONTINUED)
Nominal End-Milling Parameters for Machining Wrought Stainless Steels

Stainless Steel Alloy	Treatment and HB	Speed m/min	Feed mm/Tooth for Cutter Diam.				Tool Material HSS: M2, M7 Carb.: C2, C6
			6 mm	13 mm	19 mm	25–50 mm	
Other graded 304, 316, 321, etc.	Ann.-160	24	0.05	0.075	0.13	0.15	M2. M7
		82	0.025	0.05	0.075	0.13	C6
Wrought Duplex	Ann.-230	23	0.05	0.075	0.13	0.15	M2, M7
		79	0.025	0.05	0.075	0.13	C2
Wrought PH Alloys	Ann.-210	22	0.025	0.05	0.07	0.1	M2, M7
		62	0.025	0.05	0.07	0.12	C2

Ann. = Annealed, Q&T = Quenched and tempered.
Optimum values of speeds and feeds may be higher or lower, depending on machining conditions.
Tool material: See (Youssef, 2016).
Source: Adapted from ASM International, Metals Handbook, Vol. 16 Machining, 1989.

Table 5.10 lists the nominal feeds and speeds for broaching free- and non-free-machining stainless steels using powder HSS tools (Carpenter-Trademark of AK steel Corp.). When higher hardness is required, pieces should be first broached and then heat treated. Sulfo-chlorinated oils diluted with paraffin oil, rather than emulsifiable fluids, are suggested.

6. ***Grinding***: Corundum (Al_2O_3) wheels are most commonly used for stainless steels. However, carborundum (SiC) wheels can also be used for special

TABLE 5.10
Nominal Broaching Parameters of Free- and Non-Free-Machining Stainless Steels Using HSS Broaches

Stainless Steel Alloy			Speed m/min	Super-elev. mm/tooth	HSS.* Broach
Martensitic Alloys					
	Non-free	410	8	0.1	
	Free-mach.	416	10	0.1	
	Non-free	420	6	0.075	
	Free-mach.	420F	8	0.075	
	Non-free	440A, 440B	6	0.05	
	Non-free	440C	4	0.05	
	Free-mach.	440F	6	0.05	
Ferritic Alloys					
	Non-free	430	8	0.075	
	Free-mach.	430F	12	0.1	
	Non-free	443	8	0.075	
Austenitic Alloys					Recommended broaches are M42, or T15
Non-free	302, 304,	316	6	0.1	
	Free-mach.	303	8	0.085	
	Free-mach.	203	8	0.085	
	Non-free	S24110	4	0.075	
	Non-free	S24904	4	0.075	
	Non-free	S20910	4	0.075	
Duplex Alloy: Annealed, 230 HB			5	0.075	
PH Alloys:					
Martensitic-Custom			3	0.05	
455-Annealed			4	0.05	
Custom 455-Aged			4	0.05	
Semi-Aust.-Pyromet 350-355,			3	0.05	
Over-temp.					
-Pyromet 350-355, Aged					

The listed speeds and deeds are conservative recommendations.
Higher values may be attainable depending on machining environment.
HSS Broach*: See (Youssef, 2016).
Source: Compiled Machining Data Handbook, Machinability Data Center,Metcut, Research Associates, Inc. Vol. 2, 3rd Ed. Cincinnati, OH. 1980 and Carpenter Technology Corporation.

applications, but at a reduced wheel life. Medium density wheels of hardness grades H to L are generally selected for grinding stainless steels, although harder wheels are more suited to thread grinding. Grit sizes commonly used are 46, 54, or 60; finer grits can be used to produce a finer finish. Vitrified and resinoid bonded wheels are mostly used.
Typical wheel speeds are 1500–2000 m/min. For surface grinding, table speeds are 15–30 m/min with a down feed of up to 0.05 mm/pass for rough grinding, and 0.013 mm/pass for finishing, and a cross-feed of 1.3 to 13 mm/pass. Because of lower thermal conductivity of stainless steels, an efficient coolant is necessary. Conventional water-soluble fluids generally provide lower GW-life than heavy-duty sulfo-chlorinated oils.

5.3 TRADITIONAL MACHINING OF SUPER ALLOYS

5.3.1 Types, Characteristics, and Applications of SAs

Super alloys are a relatively new class of materials which exhibit high mechanical strength, ductility, creep strength at high operating temperatures, high fatigue strength, and typically superior resistance to corrosion and oxidation even at elevated temperatures. These features make super alloys ideal for applications in aircraft, submarines, nuclear reactors, dies for hot working of metals, and petrochemical equipment.

These alloys are usually classified into three main groups which are Fe-based, Ni-based, and Co-based alloys. The physical, mechanical, and machining behavior of each group varies considerably due to the chemical compositions of the alloy and the metallurgical processing it receives during manufacturing. Super alloys are also classified into wrought, cast, and powder metallurgy (PM) super alloys. The wrought may be solid-solution-strengthened, or precipitation hardened alloys. Representative listing of super alloys and compositions, emphasizing alloys developed and used in the United States, are given in Tables 5.11–5.14. These are generally identified by trade names or by special numbering systems.

Super alloys generally contain Ni, Cr, Co, Mo, and Fe as major alloying elements. Others are Al, W, Ti, and so on. The role of these alloying elements is to enhance the characteristics of super alloys in the following manner:

- Ni stabilizes alloy structure and properties at high temperatures.
- Co, Mo, and W increase strength at elevated temperature.
- Cr, Al, Si enhance resistance to oxidation and provide high temperature corrosion.
- C increases creep strength.

Super alloys are typically characterized by an austenitic face-centered crystalline structure. They are classified into three grades which are Fe-base, Ni-base, and Co-base alloys, as demonstrated in Tables 5.11–5.14.

Fe-base alloys: These alloys are sometimes designated as Fe-Ni-base alloys. They have been developed from austenitic steels. This group is typically the easiest to be

TABLE 5.11
Nominal Compositions of Commonly Used Solid-Solution Strengthened Wrought Super Alloys Covered in This Chapter and Their UNS Designation

Super Alloy (Trade Name)	UNS No.	Composition, wt%										
		Cr	Ni	Co	Mo	W	Nb	Ti	Al	Fe	C	Other
Fe-base Alloys:												
M-155 (Multimet)	R30155	21.0	20.0	20.0	3.0	2.5	1.0	-	-	32.2	0.15	0.15 N, 0.2 La, 0.02 Zr
Haynes 556	-	22.0	21.0	20.0	3.0	2.5	0.1	-	0.3	29.0	0.10	0.5 Ta, 0.02 La, 0.002 Zr
19-9 DL	K63198	19.0	9.0	-	1.25	1.25	0.4	0.3	-	66.8	0.30	1.10 Mn, 0.60 Si
Incoloy 800	N08800	21.0	32.5	-	-	-	-	0.38	0.38	45.7	0.05	
Incoloy 801	N08801	20.5	32.0	-	-	-	-	1.13	-	46.3	0.05	
Incoloy 802	-	21.0	32.5	-	-	-	-	0.75	0.58	44.8	0.35	
Ni-base Alloys:												
Haynes 214	-	16.0	76.5	-	-	-	-	-	4.5	3.0	0.03	0.25 Cu
Inconel 600	N06600	15.0	76.0	-	-	-	-	-	-	8.0	0.08	<0.25 Cu
Inconel 625	N06625	21.5	61.0	-	9.0	-	3.6	0.2	0.2	2.5	0.05	
Hastelloy X	N06002	22.0	49.0	<1.5	9.0	0.6	-	-	2.0	18.8	0.15	
Nimonic 75	-	19.5	75.0	-	-	-	-	0.4	0.15	2.5	0.12	
Co-base Alloys:												
AiResist 213	-	19.0	<0.5	65.0	-	4.5	-	-	3.5	<0.5	0.17	6.5 Ta, 0.15 Ze, 0.02 Zr
Haynes 25 (L605)	R30605	20.0	10.0	50.0	-	15.0	-	-	-	3.0	0.10	1.5 Mn
Haynes 188	R30188	22.0	22.0	37.0	-	14.5	-	-	-	<3.0	0.10	0.9 La
S-816	R30816	20.0	20.0	42.0	4.0	4.0	4.0	-	-	4.0	0.38	0.3 Ti
MP-159	-	19.0	25.5	35.7	7.5	-	0.6	-	0.2	9.0	-	

TABLE 5.12
Nominal Compositions of Commonly Used Precipitation Hardened Wrought Super Alloys Covered in this Chapter and their UNS Designation

Super Alloy (Trade Name)	UNS No.	Composition, wt%										
		Cr	Ni	Co	Mo	W	Nb	Ti	Al	Fe	C	Other
Fe-base Alloys:												
A-286 (pyromet)	K63198	15.0	26.0	-	1.25	-	-	2.0	0.2	55.2	0.04	0.005 B, 0.3 V
Discaloy	K66220	14.0	26.0	-	3.0	-	-	1.7	0.25	55.0	0.06	0.15 Si
Incoloy 903	-	<0.1	38.0	15.0	0.1	-	3.0	1.4	0.7	41.0	0.04	0.40 Si
Incoloy 907	-	-	38.4	13.0	-	-	4.7	1.5	0.03	42.0	0.01	0.01 B, (0.5 max)
Incoloy 909	-	-	38.0	13.0	-	-	4.7	1.5	0.03	42.0	0.01	
V-57	-	14.8	27.0	-	1.25	-	-	3.0	0.25	48.6	<0.08	
Ni-base Alloys:												
Astroloy	-	15.0	56.5	15.0	5.25	-	-	3.5	4.4	<0.3	0.06	0.03 B, 0.06 Zr
Inconel 100	-	10.0	60.0	15.0	3.0	-	-	4.7	5.5	<0.6	0.15	1.0 V, 0.06 Zr, 0.015 B
IN-100	-	10.0	60.0	15.0	3.0	-	-	4.7	5.5	<0.6	0.15	0.06 Zr, 1.0 V
Inconel 718	N07718	19.0	52.5	-	3.0	-	5.1	0.9	0.5	18.5	<0.08	<0.15 Cu
M 252	-	19.0	56.5	10.0	10.0	-	-	2.6	1.0	<0.75	0.15	0.005 B
Nimonic 80A	N07080	19.5	73.0	1.0	-	-	-	2.2	1.4	1.5	0.05	<0.1 Cu
Nimonic 90	N07090	19.5	55.5	18.0	-	-	-	2.4	1.4	1.5	0.06	0.01 B
René 41	N07041	19.0	55.0	11.0	10.0	-	-	3.1	1.5	<0.3	0.09	0.01 B, 0. 05 Zr
René 95	-	14.0	61.0	8.0	3.5	3.5	3.5	2.5	3.5	<0.3	0.16	0.005 B
Udimet 500	N07500	19.0	48.0	19.0	4.0	-	-	3.0	3.0	<0.4	0.8	0.03 B
Udimet 700	-	15.0	53.0	18.5	5.0	-	-	3.4	4.3	<1.0	0.07	0.006 B, 0.09 Zr
Waspaloy	N07001	19.5	57.0	13.5	4.3	-	-	3.0	1.4	<2.0	0.07	

Co-base alloys are not included in this group.

TABLE 5.13
Nominal Compositions of Ni-Base and Co-Base Cast Super Alloys Covered in This Chapter and Their UNS Designation

Super Alloy	Composition, wt%											
(Trade Name)	C	Ni	Cr	Co	Mo	Fe	Al	B	Ti	Ta	W	Zr	Others
Ni-base Alloys:													
B-1900	0.1	64.0	8.0	10.0	6.0	-	6.0	0.015	1.0	4.0	...	0.10	1.5 Hf
CMSX-4	-	bal.	6.5	9.0	0.6	-	5.6	-	1.0	6.5	6.0	-	-
CMSX-10	-	bal.	1.8–4.0	1.5–9.0	0.25–2.0	-	5.0–7.0	-	0.1–1.2	7.0–10.0	3.5–7.5	-	-
Hastelloy X	0.1	50.0	21.0	1.0	9.0	18.0	-	-	-	-	1.0	-	-
Inconel 100	0.18	60.5	10.0	15.0	3.0	-	5.5	0.01	5.0	-	-	0.06	1.0 V
Inconel 718	0.04	53	19.0	-	3.0	18.0	0.5	-	0.9	-	-	-	0.1 Cu, 5 Nb
MAR-M246	0.15	60.0	9.0	10.0	2.5	-	5.5	0.015	1.5	1.5	10.0	0.05	-
René 41	0.09	55.0	19.0	11.0	10.0	10.0	-	1.5	0.01	3.1	-	-	-
Udimet 700	1.0	53.5	15.0	18.5	5.2	-	4.2	0.03	3.5	-	-	-	-
WAX-20(DS)	0.2	72.0	-	-	-	-	6.5	-	-	-	20.0	1.5	-
Co-base Alloys:													
AiResist 13	0.45	-	21.0	62.0	-	-	3.4	-	-	2.0	11.0	-	0.1 Y
Haynes 21	0.25	3.0	27.0	64.0	5.0	1.0	-	-	-	-	-	-	-
MAR-M302	0.85	-	21.5	58.0	-	0.5	-	0.005	-	9.0	10.0	0.2	-
MAR-M509	0.6	10.0	23.5	54.5	-	-	-	-	0.2	3.5	7.0	0.5	-
NASA Co-W-Re	0.4	-	3.0	67.5	-	-	-	-	0.1	-	25.0	1.0	2 Re
X-40 (Stellite31)	0.5	10.0	22.0	57.5	-	1.5	-	-	-	-	7.5	-	0.5 Mn, 0.5 S

TABLE 5.14
Composition of Mechanically Alloyed Super Alloys Covered in This Chapter and Their UNS Designation

Alloy	BHN	Order of Machinability Rating	Composition, wt%										
			Ni	Fe	Cr	Al	Ti	Ta	W	Mo	Zr	B	Y_2O_3
Fe-base Alloy:													
Incoloy MA 956	270	1	-	74.5	20	4.5	0.5	-	-	-	-	-	0.5
Ni-base Alloy:													
Inconol MA 754	277	2	88.60	-	20.0	0.3	0.5	-	-	-	-	-	0.6
Inconol MA 6000	450	3	68.74	-	15.0	4.5	2.5	2.0	4.0	2.0	0.15	0.01	1.1

machined, partly because it has the poorest hot strength properties of all groups. It consists of an iron-base with larger amounts of Ni and Cr than typical types of stainless steels. This grade includes: Incoloy, Discaloy, and crucible A-286, Table 5.11 and 5.12. Some of Fe-base alloys have very low thermal coefficient of expansion (Incoloy 909), which make them especially suited to shafts, rings, and casings. At lower temperatures, and depending on strength needs for an application, Fe-base alloys find some use more than Co-base and Ni-base alloys. Table 5.14 presents the composition of the so-called mechanically alloyed super alloys (produced through PM techniques). Incoloy MA 956 is an Fe-base alloy with no Ni content at all.

The features and specific applications of some important Fe-base super alloys are listed as follows:

1. ***A-286*** is a precipitation hardened alloy. It is designed for applications requiring high strength and good corrosion resistance at temperatures up to 700°C. This alloy offers high ductility in notched sections. It is useful in high temperature application such as jet engines, superchargers, blades, and afterburner parts.
2. ***Incoloy800*** is a solid solution strengthened Fe-base super alloy of moderate strength and good resistance to oxidation and carburization at elevated temperatures. It is particularly useful for high temperature equipment in the petrochemical industry, such as heat exchangers, process piping, and furnace components.
3. ***N-155*** is also a solid solution strengthened Fe-base alloy, having good ductility, excellent corrosion resistance, and can be readily fabricated and machined. It is used in numerous aircraft applications such as tail cones and tailpipes, exhaust manifolds, combustion burners, and turbine blades.

Ni-base alloys: These constitute the largest group of super alloys and are generally very difficult and demanding to machine. They are mainly used in applications requiring high corrosion resistance or high strength at elevated temperatures. They currently constitute over 50% of the weight of advanced aircraft engines. Alloys of this grade contain 38–76% Ni, and up to 28% of each of the elements Cr, Co, and Mo. Ni-base super alloys are basically of four types. Solid solution strengthened (Table 5.11), precipitation hardened (Table 5.12), cast (Table 5.13), and oxide-dispersed strengthened (ODS), Table 5.14. Table 5.15 illustrates how the attractive properties of Ni-base alloys are achieved through the addition of the shown alloying elements, Seco Tools, 2014).

The features and specific application of some commonly used Ni-base alloys are given below:

1. ***Inconel 718*** is a recently developed precipitation hardened Ni-base alloy, containing significant amounts of iron, niobium, and molybdenum along with lesser amounts of Al and Ti. It is designed to display exceptionally high yield and creep rupture properties at a temperature up to 700°C. It has excellent weldability as compared to Ni-base alloys hardened by Al and Ti. Application examples include plane engines, nuclear activation furnaces, rocket engines, furnaces, and military warships.

TABLE 5.15
Enhancement of Properties of Ni-Base Super Alloys through the Addition of the Shown Elements

Addition of Alloying Elements	Leading to
Cr, Fe, Mo, W, Ta	Higher strength
Al, Ti	Higher temperature toughness
Al, Cr, Ta	Higher oxidation resistance
B, C, Zr	Higher creep resistance
Hf	For intermediate temperature ductility, and prevents oxide flaking

Source: (Seco Tools, 2014).

2. ***Hastelloy X*** is a solid solution strengthened Ni-base alloy that possesses exceptional strength and oxidation resistance up to 1200°C. It is found to be exceptionally resistant to stress cracking in petrochemical applications. The alloy has excellent forming and welding characteristics. It is recommended especially for use in furnace applications.
3. ***Nimonic 90*** is a precipitation hardened Ni-base alloy of extra-high mechanical properties along with corrosion resistance. It is used in the aerospace industry and as springs operating at high temperatures.
4. ***René 41*** is a high temperature and high strength Ni-base alloy. It possesses good oxidation resistance at high temperatures up to 800°C. It is a useful alloy in gas turbine, aircraft, and marine applications.
5. ***Waspalloy*** is a precipitation hardened, Ni-base alloy, which possesses excellent corrosion resistance and is used in elevated temperature applications (900°C), for example gas turbines and aircraft jet equipment.

Table 5.16 shows the mechanical properties of some very common Ni-base super alloys at elevated temperatures (870°C). It is remarkable that the ratio of (YTS/UTS) attains very high value compared to steels and alloy steels. Such a ratio is highly evaluated by the designers for applications requiring high strength at high operation temperatures, such as aircraft applications.

Co-base alloys: These super alloys are not as strong as Ni-base alloys, but they display superior hot corrosion resistance, and retain their strength at higher temperatures as compared to Ni-base alloys. Co-alloys show superior thermal fatigue resistance and weldability over Ni-base alloys. Co-base super alloys generally contain 35 to 67% Co, 19 to 30% Cr, and up to 35% Ni. Common alloys of this group are:

1. ***Haynes 25 (L-605)*** combines good formability and excellent high temperature properties. This alloy is resistant to oxidation and carburization up to 1050°C. Haynes 25 has a good service in many jet engine parts. Some include turbine blades, combustion chambers, afterburners, and turbine rings.
2. ***AiResist 213*** is a cast Co-base alloy, which exhibits good cast ability and excellent resistance to thermal fatigue and oxidation.

TABLE 5.16
Classification and Grouping of Super Alloys as Based on Recommended Cutting Speeds (For Carbide Tools) in Descending Order

Trade Name of Super Alloy	Alloy Condition	UTS MPa	YTS MPa	Ratio YTS/UTS	Elong.% 50 mm
Astroloy	PH-Wrought	770	690	0.90	25
Hastelloy X	SS-Wrought	225	180	0.80	50
IN-100	Cast	885	695	0.79	6
Inconel 625	SS-Wrought	285	275	0.97	125
Inconel 718	PH-Wrought	340	330	0.97	88
MAR-M200	Cast	840	760	0.90	4
René 41	PH-Wrought	620	550	0.89	19
Udimet 700	PH-Wrought	690	635	0.92	27
Waspaloy	PH-Wrought	525	515	0.98	35

SS = Solid Solution Strengthened, PH = Precipitation Hardened.
Source: (Kalpakjian and Schmid, 2003).

5.3.2 Machinability and Machinability Rating of Super Alloys

5.3.2.1 Machinability Aspects of Super Alloys

Super alloys are generally classified as having poor machinability. The Fe-base alloys, which have descended from stainless steels, usually machine more easily than the Ni-base and Co-base super alloys under similar conditions of processing and heat treatment. However, the Fe-base alloys do present chip breaking problems, which often require special tool geometries. The Ni-base and Co-base alloys have several characteristics in common that contribute to high machining costs.

All super alloys have, in general, high strength at high temperatures and produce segmented chips during cutting, hence creating high dynamic forces. An increase in high temperature strength of these alloys makes them harder and stiffer at the cutting temperature, thus increasing forces at the cutting edge during machining and consequently promoting chipping or deformation of the tool edge. Poor heat conductivity and high hardness generate high temperatures during machining. The strength and work-hardening characteristics of super alloys create an extremely abrasive environment of the cutting edge. Accordingly, when machining super alloys, carbide inserts should have good edge toughness, and when coated, a good adhesion of coating to the substrate should be secured to provide good resistance to ablation and plastic deformation. In general, tool inserts of sharp edges and positive rakes are to be used.

To summarize, the main factors affecting the machining characteristics of super alloys are that they:

- Exhibit austenitic matrix which promotes rapid work-hardening during machining
- Retain strength at high temperatures, where less efficient HSS tools begin to soften

- Possess usually high dynamic shear strength
- Contain in their microstructure hard carbides that make them abrasive
- Possess low thermal conductivity, which leads to high cutting edge temperature
- Form a tough continuous chip, which leads to BUE formation
- Produce abrasive carbides in their microstructures
- Are reactive with cutting tool materials under atmospheric conditions

Super alloys constitute a wide spectrum of alloys. Consequently, it is impossible to quote one set of cutting recommendations for the entire class, and the machining behavior can vary greatly even within the same alloy group. In fact, the same material can have numerous machining recommendations depending on its heat treatment conditions.

The super alloy as a raw material could be provided as cast, wrought and forged (bar and plate stocks), and in sintered forms. Forged materials usually have a finer grain size than in castings. Forgings generally possess higher strength, improved grain flow, better fatigue and fracture resistance, as compared to cast super alloys. They are, however, more abrasive with greater tendency to deform the tool during machining. Machining with reduced speeds and increased feeds leads to a reduction in the potential of work-hardening and notching of the cutting edge. In casting, the opposite applies, i.e. applying higher speeds and lower feeds can be beneficial. Casting makes machining more difficult and can cause notch wear on the insert. This necessitates harder and wear resistant insert grades to be used than for forgings. Casting alloys are intended for parts requiring less strength and are suitable for the production of near-net-shape (NNS) components such as turbine blades. More complicated and near-net-shape components can be produced using the powder metallurgy technique.

5.3.2.2 Machinability Rating of Super Alloys

One of the main criteria to assess machinability is the tool life and the related cutting speed to affect a predetermined allowable wear mark that terminates the tool life. Another important criterion for assessment of the machinability is the power consumption. A material of good machinability requires lower power consumption, and consequently lower specific cutting energy.

1. ***Machinability as based on tool life and nominal cutting speeds***: As usual, the turning process has always been selected to determine the machinability of materials. The machinability of a large number of super alloys as based on tool life and nominal cutting speeds, based on the data delivered by different sources such asASM International (1989), are arranged in Table 5.17, in descending order. Accordingly, these super alloys fall into 13 different categories that may be classified into three main groups, namely, easy-, medium-, and hard-to-machine groups.

   The easy-to-machine group consists of five categories, listed in Table 5.5. These are the cast Fe-base categories (Fe-C1, Fe-C2, Fe-C3), and the wrought Fe-base category (Fe-W), along with the wrought Ni-base category (Ni-W5), representing only one alloy (TD-Nickel-90% Ni and 2%

TABLE 5.17
Classification and Grouping of Super Alloys as Based on Recommended Cutting Speeds (for Carbide Tools) in Descending Order

No.	Category and Group	BHN	Cutting Speed m/min (Carbides)	Super Alloy
	Easy-to-machine:			
1	Fe-C1 [Ann]	135–185	76	ASTM A 297 Grade HC
2	Fe-C2 [Ann or N]	135–185	60	ASTM A 351 Grades HK30, HK40, HT30
3	Ni-W5 [R]	180–200	60	TD-Nickel
4	Fe-C3 [C]	160–200	53	Other types specified in **3.3**Table 3.3
5	Fe-W [ST or ST-Ag]	180–320	30–58	A-286, Discaloy, [Incoloy 800, 801, 802], N-155, V-57, W-545, 16-25-6, 19-9DL, Incoloy MA 956 (PM)
	Medium-to-machine:			
6	Ni-W4 [Ann. or ST, Cd or Ag]	140–310	15–35	[Hastelloy B, B-2, (C-276), G, S, X], [Incoloy 804, 825], [Inconel 600, 601], refractory 26, Udimet 630
7	Ni-W1[Ann. or ST, ST-Ag]	200–400	15–30	Haynes 263, Incoloy 901, [Inconel 617, 625,702,706,718,722,X-750,751], M252, [Nimonic 75, 80], Waspaloy, Inconel MA754 (PM)
8	Co-W [ST, ST-Ag]	180–320	15–27	AiResist 213, [Haynes 25 (L605), 188], J-1570, [MAR-M905, M918], S-816, V-36
9	Ni-C1 [C or C-Ag]	200–375	14–26	[Hastelloy B, C], ASTM A297 (Grades HW, HX), ASTM A608 (Grades XW50, HX50)
10	Ni-W2 [ST, ST-Ag]	225–400	15–24	Astroloy, IN-106, Inconel700, [Nimonic 90, 95], [René 41, 63], [Udimet 500, 700, 710]
	Hard-to-machine:			
11	Co-C [C or C-Ag]	220–425	9–18	[AiResist 13, 215], [HS 6, 21, 25, 31 (X-40)], [MAR-M302, M322, M509, NASA Co-W-Re, W1-52, X-45]
12	Ni-C2 [C or C-Ag]	250–425	9–18	B-1900, IN-100, [IN-738, 792], [Inconel 713 C, 718], M252, [MAR-M200, M246, M421]
13	Ni-W3 [ST, ST-Ag]	275–475	9–15	[René 77, 95], Inconel MA 6000 (PM)

ST: Solution Treated, Ann.: Annealed, N: Normalized, C: Cast, W: Wrought, C-Ag: Cast and Aged, ST-Ag: Solution Treated and Aged, R: Rolled, Cd: Cold drawn.
Source: Adapted from Machining Data Handbook, Machinability Data Center Metcut, Research Associates, Inc. Vol. 2, 3rd Ed. Cincinnati, OH. 1980.

ThO_2). The medium-to-machine group also comprises five categories. These are the wrought Ni-base categories (Ni-W4, Ni-W1, Ni-W2) and the cast Ni-base category (Ni-C1), along with the wrought Co-base category (Co-W). The hard-to-machine group comprises three categories which are the cast Co-base category (Co-C), the cast Ni-base category (Ni-C2), and finally, the wrought Ni-base category (Ni-W3).The mechanically alloyed versions (PM) of super alloys can be machined using practices appropriate for wrought alloys of similar composition and hardness. As shown in Table 5.17, Ni-base alloy MA 754 (277 BHN) has lower machinability than Inconel 718 and Fe-base alloy A-286. The Fe-base alloy MA 956 of approximately the same hardness (270 BHN) as MA 754 appears to be more machinable than MA 754. Table 5.18 lists the nominal speeds and feeds for turning these alloys.

From the above, it is depicted that the easiest-to-machine from the super alloys is the Fe-base alloys such as A-286, Discaloy, N-155, etc., as well as the alloy TD-Nickel, while the hardest-to machine are the Ni-base and Co-base super alloys, such as IN-100, Réne 77, AiResist 13, 215, and NASA Co-W-Re, Table 5.5.

Similarly, Machinability Data Center (1980) and others (ASM International, 1989; High Performance Alloys, Inc., and All Metals & Forge Group), tested many metals and alloys, and compared cutting speeds to those obtained when machining a reference alloy, AISI-1212, a resulfurized and rephosphorized plain carbon steel under the same cutting conditions; AISI-1212 got a score of 1. Materials scoring above 1 are easier-to-machine; on the other hand, materials scoring less than 1 are more difficult-to-machine, Table 5.19.

In such cases, it is highly recommended to include the material hardness (e.g. in BHN), because the material hardness has a considerable effect on the machinability rate (MR). So, if a certain super alloy of a hardness value H1 has a rating MR1, a certain metal working, or heat treatment processing, which has raised its hardness number from H1 to H2, then it acquires a machinability rating MR2, which can be evaluated by:

$$\mathrm{MR2} = \frac{\mathrm{H1}}{\mathrm{H2}}\mathrm{MR1}$$

Referring to Table 5.19, Inconel 901 has received such a treatment that increased the hardness from 200 to 300 BHN. Before the treatment, it had a rating of MR1 = 0.2; then the expected new rating MR2 will be:

$$\mathrm{MR2} = \frac{200}{300} \times 0.2 = 0.13$$

2. ***Machinability as based on specific cutting energy***:
Sandvik Coromant (2013) provided data regarding the specific cutting energy $k_{s.1}$, as well as the related chip thickness exponents of basic types of super alloys, Table 5.20. Figure 5.7 illustrates the global values of $k_{s.1}$ for Fe-, Ni-, and Co-base alloys, along with those of some commonly used austenitic stainless steels for comparison.

TABLE 5.18
Nominal Speed and Feeds for Turning Mechanically Alloyed Super Alloys

			HSS Tools			Index. Carbide Tools		
Alloy	**BHN**	**DOC mm**	**v m/min**	**f mm/rev**	**Tool Material**	**v m/min**	**f mm/rev**	**Tool Material**
Fe-base								
Incoloy MA956	270	Rough 6.5	12–15	0.75	T-15, M-36	50–75	0.5	C-6
		Finish 1.3	18–21	0.25	T-15, M-36	75–90	0.2	C-8
Ni-base								
Incoloy MA754	277	Rough 6.5	3–6	0.25	T-15, M-36	12–18	0.25	C-2
		Finish 1.3	5–6	0.20	T-15, M-36	15–30	0.20	C-2
Inconel MA6000	450	Rough 2.0	3–4	0.10	T-15, M-36	9–18	0.2	C-2
		Finish 0.2	3–4	0.12	T-15, M-36	9–18	0.12	C-2

Tool material: Refer to (Youssef, 2016).

Source: Adapted from Machining Data Handbook, Machinability Data Center Metcut, Research Associates, Inc. Vol. 2, 3rd Ed. Cincinnati, OH. 1980.

TABLE 5.19
Machinability Rating (MR) in Turning of Commonly Used Super Alloys, as Based on Resulfurized and Rephosphorized Plain Carbon Steel AISI-1212 as a Reference Material, Tool Material HSS

Alloy Grade	v (m/min)	MR	Alloy Grade	BHN	MR
AISI-1212	60	1.0			1.0
Super Alloys:			**Super Alloys:**		
A-286	17	0.28	Discaloy	135	0.4
Haynes 25(L-605)	5	0.09	Hastelloy B	200	0.12
Inconel 600	15	0.22	Hastelloy C	170	0.20
Inconel 625	6	0.12	IN-100	320	0.09
Inconel 718	6	0.12	Haynes 31-Cast	-	0.06
Inconel X-750	6	0.12	Inconel X	360	0.15
Waspaloy	14	0.2	Inconel 718-Cast	290	0.09
MP35N	14	0.2	Inconel 702	225	0.11
MP159	14	0.2	Inconel 901	200	0.20
Hastelloy C-276	12	0.18	Inconel 901	300	**0.13***
Hastelloy X	14	0.2			
René 41	5	0.09			

Source: Compiled from: High Performance Alloys, Inc. Tipton, IN 46072 USA, and All Metals & Forge Group. LLC Fairfield, New Jersey 07004 USA.

TABLE 5.20
Specific Cutting Energy of Super Alloys

Super Alloy	Processing	Heat Treatment	BHN	Spec. Cutting Energy $k_{s.1}$(N/mm²)	Thickness Exponent z
Fe-base alloy	Not specified	Annealed	200	2400	0.25
		Aged	280	2500	0.25
Ni-base alloy	Wrought	Annealed	250	2650	0.25
		Aged	350	2900	0.25
	Cast	Not specified	320	3000	0.25
Co-base	Wrought	Annealed	200	2700	0.25
		Aged	300	3000	0.25
	Cast	Not specified	320	3100	0.25

Source: (Sandvik Coromant, 2013).

5.3.3 Machining and Machining Conditions of Super Alloys

Traditional machining has found much use in super alloys because they provide much higher metal removal rates than those achieved using non-traditional methods that will be dealt with afterwards. Single-point tool turning is the most frequently used machining process for super alloys.

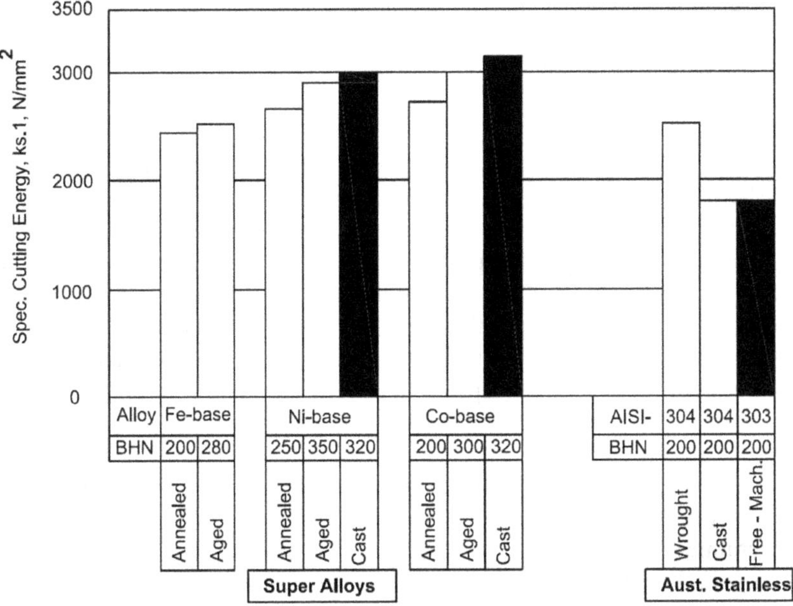

FIGURE 5.7 Global values of k_s for Fe-, Ni-, and Co-base alloys, along with those of some commonly used austenitic stainless steels for comparison (Youssef, 2016).

Challenges and machining guidelines for super alloys: Due to low thermal conductivity of super alloys, temperature during machining can be as high as 1000 to 1300°C, thus causing crater wear and severe plastic deformation of the cutting tool. Therefore, resistance against crater wear is an important tool property requirement for machining super alloys. Plastic deformation, on the other hand, can blunt the cutting edge, thereby increasing the cutting forces.

The chemical reactivity of these alloys facilitates BUE formation and coating delamination, which severely degrades the cutting tool leading to short tool life. An ideal cutting tool should exhibit chemical inertness when machining such alloys. The hard abrasives inter-metallic compounds in the microstructure cause severe abrasive wear of the tool tip. Heat generated during machining can potentially alter the microstructure. The chip produced when machining super alloys is tough and continuous and requires superior chip breaker geometry.

As general guidelines, when machining super alloys, high feed rate and depth of cut should always be maintained to minimize work-hardening. The tool is never allowed to dwell to avoid the possibility of work-hardening and problems in subsequent processes. Except with ceramic tooling, a generous quantity of coolant is recommended to reduce high temperature and consequently rapid tool wear.

The following is a practical guide for machining high temperature super alloys. Machine alloys in the softest state possible:

- Use a positive rake insert
- Use sharp edges

- Use strong geometry
- Use a rigid setup
- Prevent part deflection
- Use a high lead angle

A positive rake cutting edge is recommended for semi-finishing and finishing operations whenever possible. Positive rake geometry minimizes work-hardening of the machined surface by shearing the chip away from the workpiece in an efficient way in addition to minimizing built-up edge. Dull or improperly ground edges increase cutting forces during machining, causing metal build-up, tearing and deflection of the work material. Machining with a rigid setup prevents vibration and subsequent chatter that deteriorates surface finish and can cause tool fracture. Tighter tolerance can be maintained with rigid setups. Deflection of the work material should be prevented, especially when machining thin-walled components or parts.

1. ***Turning***: All types of cutting tools are applicable in turning operations of super alloys. Carbide tools are usually used, although coated carbides, ceramics, cubic boron nitride (CBN), and even HSS are also used. A carbide tool of grade C-2 is frequently selected for roughing, while C-3 is frequently dedicated for finishing, Table 5.21.
   To realize a high production rate when machining the harder Ni-base (wrought or cast), and Co-base, strengthened and aged (ST-Ag) alloys, HSS and carbide tooling should be replaced by CBN or ceramics. CBN realizes a cutting speed ranging from 120–185 m/min, while ceramics may be used, without cooling, for higher cutting speeds. Table 5.22 summarizes the nominal speed ranges for rough and finish turning of super alloys using CBN and coated carbide tools. Figure 5.8 shows an HSS turning tool for super alloys (wrought and cast) of BHN ranging from 140 to 475.
2. ***Drilling***: The high cutting forces when drilling super alloys necessitate maximum rigidity of the tool and workpiece. In terms of tool design or selection, the most important requirement is that the drills be as short and rigid as possible of heavy web thickness, Figure 5.9. Both HSS and carbide drills are used when drilling super alloys.
   Carbide-tipped-twist drills are more suitable for drilling the Ni-W3 category, Table 5.17, and the mechanically alloyed Ni-base alloy MA 6000, Table 5.14. These alloys present high resistance if machined by conventional HSS drills. A carbide drill is preferred when drilling alloys in the Ni-C2 and Co-C categories, if they exhibit hardness of 320 to 425 BHN, and 250 to 425 BHN, respectively. Figure 5.9 shows the recommended tool geometry for a twist drill used for drilling super alloys. Nominal speeds and feeds along with the preferred drill material for drilling various types of super alloys are shown in Table 5.23.

   When drilling deep holes, for example, of an aspect ratio of 8:1, the cutting speeds should be reduced by 40% and the feed rates by 20%. Depending on the type of super alloy, cutting speeds ranging from 2 to 18 m/min are used when drilling with an HSS drill. For the more difficult-to-cut super

TABLE 5.21
Recommended Speeds, Feeds, and Cutting Tools for Cylindrical Turning of Medium- and Difficult-To-Cut Super Alloys

				HSS			Indexable Carbides			Borazon	
Group of super Alloy	Condition	BHN	DOC mm	v m/min	f mm/rev	Mat.	v m/min	f mm/rev	Mat	v m/min	f mm/rev
Ni-W1	Ann. or ST	200–300	0.8	8	0.13	T-15, M-42	30	0.13	C-3	-	-
			6	6	0.18	T-15, M-42 T-15, M-42	24	0.18	C-2	-	-
			-	-	-		18	0.40	C-2	-	-
	ST-Ag	300–400	0.8	8	0.13	T-15, M-42	29	0.13	C-3	185	0.08
			2.5	5	0.18	T-15, M-42	23	0.18	C-2	150	0.13
			5.0	-	-	-	15	0.40	C-2	135	0.13
Co-W	ST	180–230	0.8	8	0.13	T-15, M-42	27	0.13	C-3	-	-
			2.5	6	0.18	T-15, M-42	21	0.18	C-2	-	-
			5.0	-	-	-	17	0.25	C-2	-	-
	St-Ag	270–320	0.8	6	0.13	T-15, M-42	24	0.13	C-3	-	-
			2.5	5	0.18	T-15, M-42	20	0.18	C-2	-	-
			5.0	-	-	-	15	0.25	C-2	-	-
Ni-C1	C or C-Ag	200–375	0.8	5	0.13	T-15, M-42	24	0.13	C-3	-	-
			2.5	3.6	0.18	T-15, M-42	21	0.18	C-2	-	-
			5.0	-	-	-	17	0.25	C-2	-	-
Co-C	C or C-Ag	220–290	0.8	6	0.13	T-15, M-42	18	0.13	C-3	-	-
			2.5	3.6	0.18	T-15, M-42	15	0.18	C-2	-	-
			5.0	-	-	-	-	-	-	-	-
	C or C-Ag	220–425	0.8	3.6	0.13	T-15, M-42	14	0.13	C-3	185	0.08
			2.5	3	0.18	T-15, M-42	9	0.13	C-2	135	0.13
			5.0	-	-	-	-	-	-	120	0.13

(Continued)

TABLE 5.21 (CONTINUED)
Recommended Speeds, Feeds, and Cutting Tools for Cylindrical Turning of Medium- and Difficult-To-Cut Super Alloys

				HSS			Indexable Carbides			Borazon	
Group of super Alloy	Condition	BHN	DOC mm	v m/min	f mm/rev	Mat.	v m/min	f mm/rev	Mat	v m/min	f mm/rev
Ni-C2	C or C-Ag	250–300	0.8	5	0.13	T-15, M-42	18	0.13	C-3	-	-
			2.5	3.6	0.13	T-15, M-42	14	0.18	C-2	-	-
			5.0	-	-	-	11	0.25	C-2	-	-
	C or C-Ag	220–425	0.8	3.6	0.13	T-15, M-42	15	0.13	C-3	185	0.08
			2.5	3	0.13	T-15, M-42	11	0.18	C-2	135	0.13
			5.0	-	-	-	9	0.25	C-2	120	0.13
Ni-W3	ST	275–390	0.8	5.0	0.13	T-15, M-42	15	0.13	C-3	-	-
			2.5	3.6	0.18	T-15, M-42	14	0.18	C-2	-	-
			5.0	-	-	-	11	0.25	C-2	-	-
	ST-Ag	400–475	0.8	3.6	0.13	T-15, M-42	15	0.13	C-3	185	0.08
			2.5	3.0	0.18	T-15, M-42	12	0.18	C-2	135	0.13
			5.0	-	-	-	9	0.25	C-2	120	0.13

NB: The abbreviations as in Table 5.17.

Source: (Youssef, 2016).

TABLE 5.22
Speed Range for Rough and Finish Turning of Super Alloys Using CBN and Coated Carbide Tools

		Roughing		Finishing	
		CBN	Coated	CBN	Coated
Super Alloy Group as Defined in Table 5.17	BHN	v (m/min)		v (m/min)	
Easy-to-cut	135–210	-	70–100	-	100–145
Med.- and Hard-to-cut	200–475	120–135	-	160–185	-

The cutting speed decreases as the hardness of the alloy increases.

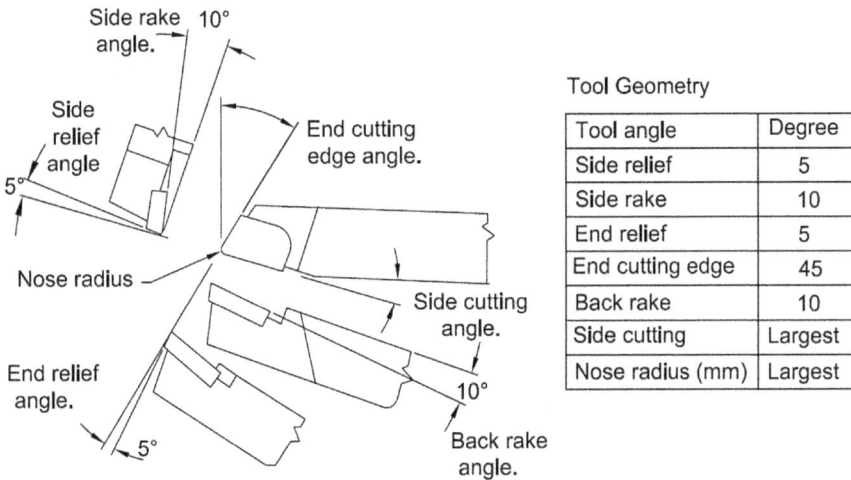

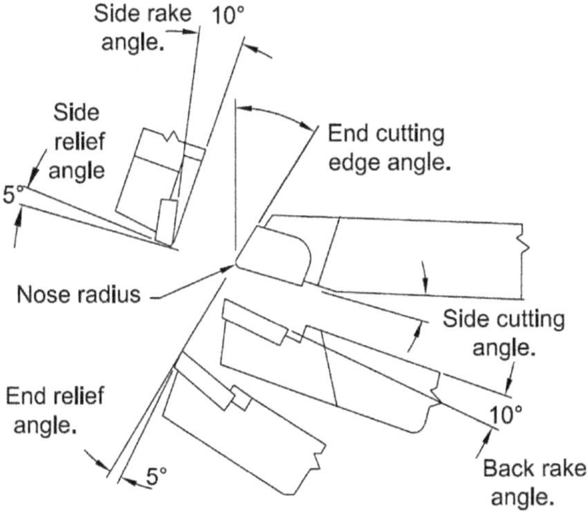

Tool Geometry

Tool angle	Degree
Side relief	5
Side rake	10
End relief	5
End cutting edge	45
Back rake	10
Side cutting	Largest
Nose radius (mm)	Largest

FIGURE 5.8 HSS-turning tool for super alloys (wrought and cast) of BHN ranging from 140 to 475 (Youssef, 2016).

alloys, such as René 41 and Haynes 25, speeds as low as 2 m/min may be required. Speeds lower than 3 m/min are seldom used because in this case the shear action is poor and the drills are liable to fail by chipping.

When carbide drills are used, speeds are usually two to three times as fast. A steady rate of feed is also important when drilling super alloys that work-harden readily. Hand feeding is sometimes used; however, it is not recommended for drilling super alloys.

3. ***Reaming***: For reaming super alloys, speeds of less than 3 m/min are seldom used because cutting edges are likely to chip. The feed must be great enough to maintain cutting action with a practical size of chip thickness. The reaming allowance is especially critical with super alloys. Removing an excessive amount of stock overloads the reamers, while insufficient stock will induce burnishing, causing the alloy to work-harden. For super alloys,

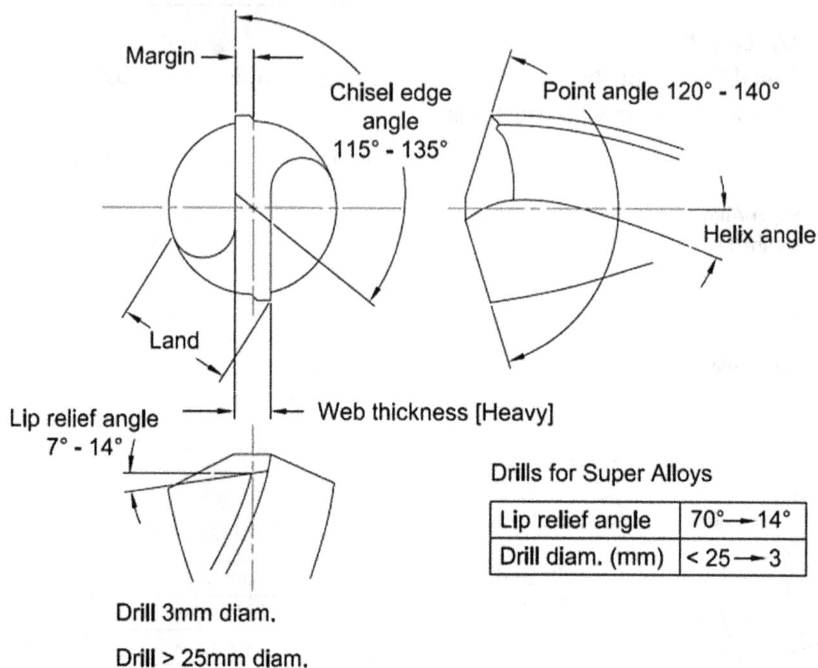

FIGURE 5.9 Recommended tool geometry for a twist drill used for drilling super alloys (Youssef, 2016).

the optimum amount of reaming allowance varies with the hole size, but 0.13 mm on the radius should be a minimum, even for the smallest holes (ASM International, 1989).

It is generally preferable to use carbide-tipped reamers, although HSS reamers are also used for easy-to-cut super alloys. Carbide-tipped reamers should be used when reaming hard-to-cut super alloys and mechanically alloyed MA6000 products. The recommended speeds when using carbide-tipped reamers range from 3 to 14 m/min depending on the machinability of the super alloy. If HSS reamers are used, only 2/3 of these values are recommended (Machining Data Handbook, 1980).

4. ***Milling***: Climb milling of super alloys is generally preferred than conventional. It requires milling machines equipped with backlash eliminator, however, cuts deeper than 1.5 mm are seldom attempted with climb milling of super alloys, because it is virtually impossible to attain the required rigidity (Seco Tools, 2014). It is common practice to employ a fairly low cutting speed in combination with a moderately high feed/tooth (no less than 0.1 mm/tooth) to prevent work-hardening of the material. The flank wear should not exceed 0.2–0.3 mm; otherwise, the chances of catastrophic failure increases rapidly. For milling super alloys, two issues of cutter design must be specially considered. First, the tooth strength must be greater than that required for milling steel or cast iron, and second, relief angles must be large enough to

TABLE 5.23
Nominal Speeds and Feeds along with the Preferred Drill Material for Drilling Various Types of Super Alloys

Super Alloy Machinability Group	Condition	BHN	Speed v m/min	Feed mm/rev Nominal Diam. (mm)						Tool Material
				3	6	12	18	25	50	
Fe-C1	Ann.	135–185	18	0.05	0.1	0.18	0.25	0.35	0.4	M-1, M-7, M-10
Fe-C2	Ann. or N	135–185	16	0.05	0.1	0.18	0.25	0.30	0.4	M-1, M-7, M-10
Ni-W5	R	180–200	16	0.05	0.1	0.18	0.25	0.4	0.45	M-1, M-7, M-10
Fe-C3	C	160–210	14	0.05	0.08	0.12	0.2	0.25	0.3	M-1, M-7, M-10
Fe-W	ST	180–230	8	0.05	0.1	0.15	0.2	0.25	-	T-15, M-42
	ST-Ag	250–320	6	0.05	0.1	0.15	0.2	0.2	-	T-15, M-42
Ni-W4	Ann. or ST	140–220	6	0.05	0.075	0.075	0.1	0.1	-	M-1, M-7, M-10
	Cd or Ag	240–310	5	0.05	0.075	0.075	0.1	0.1	-	M-1, M-7, M-10
Ni-W1	Ann. or ST	200–300	6	0.05	0.075	0.075	0.1	-	-	T-15, M-42
	ST-Ag	300–400	5	0.05	0.075	0.075	0.1	-	-	T-15, M-42
Co-W	ST	180–230	6	0.05	0.075	0.075	0.1	-	-	T-15, M-42
	ST-Ag	270–330	5	0.05	0.075	0.075	0.1	-	-	T-15, M-42
Ni-C1	C or C-Ag	200–375	2	0.05	0.075	0.075	0.1	-	-	T-15, M-42
Ni-W2	ST	225–300	5	0.05	0.075	0.075	0.1	-	-	T-15, M-42
	ST-Ag	300–400	3.6	0.05	0.075	0.075	0.1	-	-	T-15, M-42
Co-C	C or C-Ag	220–290	2.4	0.05	0.075	0.075	0.1	0.15	-	T-15, M-42
	C or C-Ag	290–425	5	0.025	0.05	0.075	0.1	-	-	C-2
Ni-C2	C or C-Ag	250–320	2.4	0.05	0.075	0.075	0.1	-	-	T-15, M-42
	C or C-Ag	320–425	5	0.025	0.05	0.075	0.1	-	-	C-2
Ni-W3	ST	275–390	6	0.25	0.05	0.075	0.1	-	-	C-2
	ST-Ag	400–475	5	0.025	0.05	0.075	0.1	-	-	C-2

Source: (Youssef, 2016).

prevent a rubbing action and consequent work-hardening of the super alloy being cut. Regardless of the cutter material, and excluding small cutters, inserted blades are used on nearly all cutters, because even under the most favorable machining conditions, the tool life of the cutting edge is short.

Mechanical methods of securing the blades on the cutter body are preferred, because replacement of chipped or broken blades (not brazed) is easier. Figure 5.10 illustrates an HSS face-milling cutter for milling wrought and cast super alloys of BHN ranging from 200 to 475 m/min.

Nominal speeds and feeds for face milling of super alloys are given in Table 5.24. The parameters listed in this table assume that the milling operations are conducted under optimum conditions, regarding adequate rigidity of the setup, optimum tool geometry, and plentiful supply of coolant.

5. ***Broaching***: Successful broaching of super alloys requires the following important considerations:
   - Broach design that provides ample strength and clearance for swarf
   - Rigid machine combined with adequate power
   - Rigid tool and workpiece setup
   - Avoidance of cutting edge rubbing against the workpiece
   - Careful selection of cutting oil

Tool angles and gullet shape are important design issues due to the behavior of super alloys in shearing and chip formation. The pitch of the teeth should be approximately 25% more than that for broaching plain-carbon and low-alloy steels in order to provide the necessary greater chip clearance.

The large pitch also will decrease the total load by reducing the number of teeth in engagement. Nominal speeds and super-elevations for broaching using HSS broaches are illustrated in Table 5.25.

Figure 5.11 illustrates the tool geometry of three different HSS broach designs. The first is a standard broach for super alloys, Figure 5.11a. The second and the third are specially designed broaches for A-286, Figure 5.11b, while the last one for broaching René 41, Figure 5.11c. All of which are made from high-speed steel.

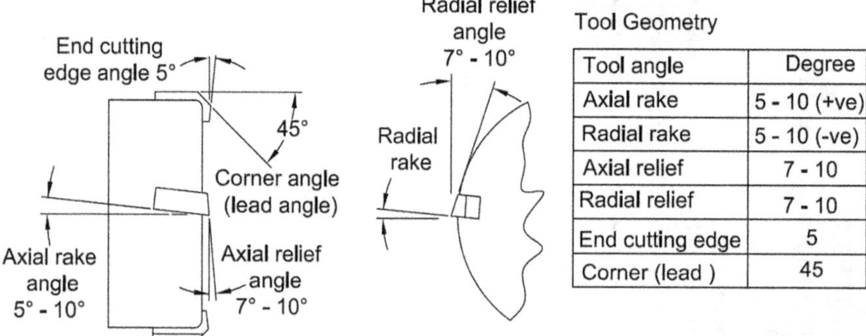

Tool angle	Degree
Axial rake	5 - 10 (+ve)
Radial rake	5 - 10 (-ve)
Axial relief	7 - 10
Radial relief	7 - 10
End cutting edge	5
Corner (lead)	45

FIGURE 5.10 HSS face-milling cutter for milling wrought and cast super alloys of BHN ranging from 200 to 475 m/min (Youssef, H., 2016).

TABLE 5.24
Recommended Speeds, Feeds, and Cutting Tools for Face Milling of Medium- and Difficult-To-Super Alloys

				HSS			Uncoated Index. Carbides		
Group of Super Alloy	**Condition**	**BHN**	**DOC mm**	**v mm/min**	**f mm/tooth**	**Mat.**	**v mm/min**	**f mm/tooth**	**Mat.**
Ni-W1	Ann. or ST	200–300	1	8	0.10	T-15, M-42	-	-	-
			4	6	0.15		-	-	-
			8	-	-		-	-	-
	ST-Ag	300–400	1	6	0.07	T-15, M-42	-	-	-
			4	5	0.13		-	-	-
			8	-	-		-	-	-
Co-W	ST	180–230	1	9	0.05	T-15, M-42	21	0.13	C-2
			4	8	0.07		20	0.13	C-2
			8	6	0.1		-	-	-
	ST-Ag	270–320	1	5	0.05	T-15, M-42	18	0.13	C-2
			4	3	0.07		17	0.15	C-2
			8	2.5	0.10		-	-	-
Ni-C1	C or C-A	200–375	1	8	0.10	T-15, M-42	23	0.13	C-2
			4	6	0.15		21	0.15	C-2
			8	5	0.20		-	-	-
Ni-W2	ST	225–300	1	6	0.1	T-15, M-42	-	-	-
			4	5	0.15		-	-	-
			8	-	-		-	-	-
	ST-Ag	300–400	1	5	0.07	T-15, M-42	-	-	-
			4	3	0.13		-	-	-
			8	-	-		-	-	-

(Continued)

TABLE 5.24 (CONTINUED)
Recommended Speeds, Feeds, and Cutting Tools for Face Milling of Medium- and Difficult-To-Super Alloys

				HSS			Uncoated Index. Carbides		
Group of Super Alloy	Condition	BHN	DOC mm	v mm/min	f mm/tooth	Mat.	v mm/min	f mm/tooth	Mat.
Co-C	C or C-Ag	220–290	1	5	0.05	T-15, M-42	15	0.13	C-2
			4	3	0.07		14	0.15	C-2
			8	2.5	0.1		-	-	-
	C or C-Ag	290–425	1	3.5	0.05	T-15, M-42	11	0.13	C-2
			4	2.5	0.05		8	0.15	C-2
			8	2.0	0.07		-	-	-
Ni-C2	C or C-Ag	250–300	1	6	0.05	T-15, M-42	-	-	-
			4	3.5	0.05		-	-	-
			8	2.8	0.07		-	-	-
	C or C-Ag	300–425	1	5	0.05	T-15, M-42	-	-	-
			4	3	0.05		-	-	-
			8	2.5	0.07		-	-	-
Ni-W3	ST	275–390	1	5	0.05	T-15, M-42	-	-	-
			4	3	0.07		-	-	-
			8	-	-		-	-	-
	ST-Ag	290–425	1	3.5	0.05	T-15, M-42	-	-	-
			4	2.5	0.07		-	-	-
			8	-	-		-	-	-

The abbreviations as in Table 5.17.
Source: (Youssef, 2016).

TABLE 5.25
Nominal Cutting Speed and Super Elevation (Chip Load) in Broaching of Medium- and Difficult-To-Super Alloys, Using HSS Broaches

Group of Super Alloy	Condition	BHN	Speed m/min	Super-elev. mm/tooth	HSS Grade
Ni-W1	Ann. or ST	200–300	2.5	0.05	T-15, M-42
	ST-Ag	300–400	2.0	0.05	T-15, M-42
Co-W	ST	180–230	2.5	0.05	T-15, M-42
	ST-Ag	270–320	2.0	0.05	T-15, M-42
Ni-C1	C or C-Ag	200–375	2.5	0.05	T-15, M-42
Ni-W2	ST	225–300	2.5	0.05	T-15, M-42
	ST-Ag	300–400	2.0	0.05	T-15, M-42
Co C	C or C-Ag	220–290	2. 5	0.05	T-15, M-42
	C or C-Ag	290–425	2.0	0.05	T-15, M-42
Ni-C2	C or C-Ag	250–320	2.5	0.05	T-15, M-42
	C or C-Ag	320–425	2.0	0.05	T-15, M-42
Ni-W3	ST	275–390	2.0	0.05	T-15, M-42

NB: Abbreviations as in Table 5.17.
Source: (Youssef, 2016).

6. ***Grinding***: Super alloys are generally more difficult and costly to grind than low-alloy steels. Specifically, Ni-base and Co-base super alloys are sensitive to the grinding heat, metallurgical alterations, and microcracking that occur within a considerable thickness, resulting in a deleterious effect of the surface integrity of the component. Therefore, machining parameters during grinding of super alloys should be carefully selected to achieve an optimum surface integrity.
   Corundum wheels are selected for most super alloys, although, for some precision applications, CBN grinding wheels are used. Depending on the type of grinding operation, most super alloys are ground using medium-hard wheels (F to L). For surface grinding, it is recommended to use a grinding wheel (H), for internal grinding (F/G), while for external cylindrical grinding, a grinding wheel of the grade (JK/L) is preferred. Medium wheel structure numbers from 7 to 10 are recommended for grinding super alloys, depending on the ductility of the super alloy. Wheel designations are provided in Table 5.26 and Table 5.27 without the structure number. Vitrified bonded grinding wheels (designated by V) are most commonly used for grinding super alloys. However, resinoid-bonded wheels (B) are used for high speed grinding wheels (≥1400 m/min).Table 5.26 illustrates the recommended machining condition when surface grinding all types of super allays using vitrified bond wheels, while Table 5.27 illustrates the same, but for external cylindrical grounding.

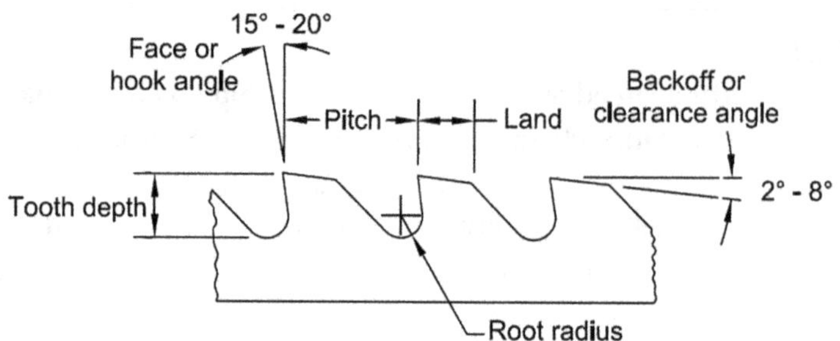

a - Standard broach for super alloys

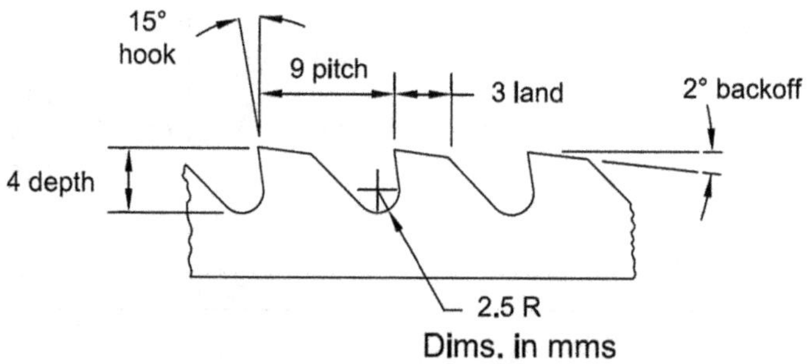

b - Broach for A - 286

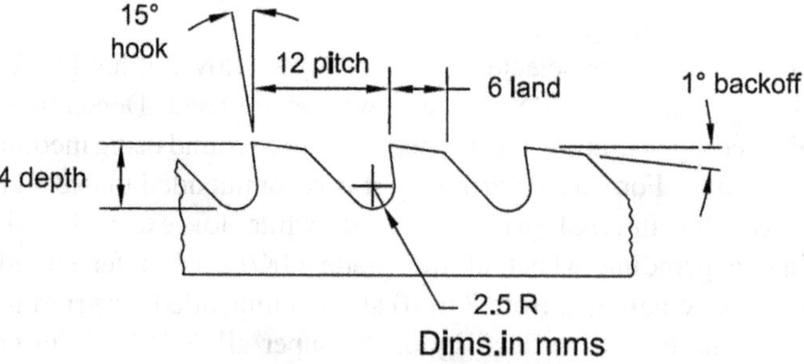

c - Broach for Rene´41

FIGURE 5.11 Tool geometry of three different HSS-broach designs for super alloys (Machinability Data Center, 1980 and ASM International, 1989).

TABLE 5.26
Recommended Machining Condition for Surface Grinding of Super Alloys

Work Material	BHN	GW Designation	GW speed vw m/min	Table speed vs m/min	Cross-feed: Fraction of GW width/pass
All types of super alloys	140–475	**A-46-H-V**	900–1200	15–30	1/12

Down feeds: 25μm, and final finishing feed = 12 μm; Grinding fluid: water-base soluble-oil emulsion or sulfurized oil.

Source: Adapted from Machining Data Handbook, Machinability Data Center Metcut, Research Associates, Inc. Vol. 2, 3rd Ed. Cincinnati, OH. 1980.

TABLE 5.27
Recommended Machining Conditions for External Cylindrical Grinding of Super Alloys

Work Material	BHN	GW Designation	GW speed vs m/min	In-feed μm/pass for	
				Roughing	Finishing
All types of super alloys	140–475	**A-60-J-V**	900–1200	25	5

Work speed: 15–30 m/min, traverse feed rate: 0.2 GW-width/rev (roughing,. 0.1 GW-width/rev (finishing); grinding fluid: Water-base soluble-oil emulsion or sulfurized oil.

Source: Adapted from Machining Data Handbook, Machinability Data Center Metcut, Research Associates, Inc. Vol. 2, 3rd Ed. Cincinnati, OH. 1980.

5.4 NON-TRADITIONAL MACHINING OF STAINLESS STEELS AND SUPER ALLOYS

While the majority of stainless steels and super alloys are machined traditionally, non-traditional techniques are used when justifiable. Such justification involves cost savings when machining alloys at the extremes of toughness and hardness, or when machining intricate shapes. This section briefly describes, as based on the available data in literature, specialized company information and handbooks on some of the non-traditional machining processes which have been applied successfully to stainless steels and super alloys. Also, some hybrid processes are considered important and promising for machining these materials.

5.4.1 Machining of Stainless Steels and Super Alloys by Mechanical Techniques

Jet machining processes: Although AJM is best suited to hard materials, it has been used to de-burr and clean stainless steel and super alloys. One advantage of using jet machining processes is the fact that they are not thermal processes.

Table 5.28 shows the machined surface finish of a soft austenitic SS AISI-316 processed by AJM. The starting surface had been ground to $R_a = 0.47$ μm.

WJM is also successfully used to cut SS alloys of series 300, and 400 of stocks ranging from 0.25 to 100 mm. AWJ not only possesses the versatility of WJ but also extends the applications to harder and denser workpieces. The addition of abrasives allows the cutting of difficult-to-cut materials such as super alloys, stainless steels, composites, and ceramics. Table 5.29 lists traverse cutting rates of some SSs processed by AWJM (Schwartz, 1985). Figure 5.12 predicts the effect of SS plate thickness on the cutting rate (traverse speed) when using AWJM Youssef, 2005). Figure 5.13 illustrates a 760-mm-diameter turbine wheel which was machined with an AWJ from a solid, 45-mm-thick disk of Inconel. The objective was to remove

TABLE 5.28
Surface Roughness for Annealed SS AISI-316 in AJM

Abrasive	Grit Size (μm)	R_a (μm)
Al_2O_3	25	0.25–0.53
	50	0.38–0.96
SiC	20	0.3–0.5
	50	0.43–0.86
Glass beads	50	0.30–0.96

Starting surface had been ground to $R_a = 0.47$ μm.
Source: Adapted from Machining Data Handbook, Machinability Data Center Metcut, Research Associates, Inc. Vol. 2, 3rd Ed. Cincinnati, OH. 1980.

TABLE 5.29
Traverse Cutting Rates for Some SS Alloys Using AWJM

Stainless Steel Alloy	Machining Condition	Plate Thickness mm	Cutting Rate mm/min
PH-S 15500 (15 Cr-5Ni)	Pressure: 310 MPa, with 60 mesh garnet	3	230–380
		64	13–25
Aust.-S31600 (bar stock)		76 (diam.)	13–50
Mart.-S17400 (630)	Pressure: 200 MPa, with 60 mesh garnet	25	50

Source: (Schwartz, 1985).

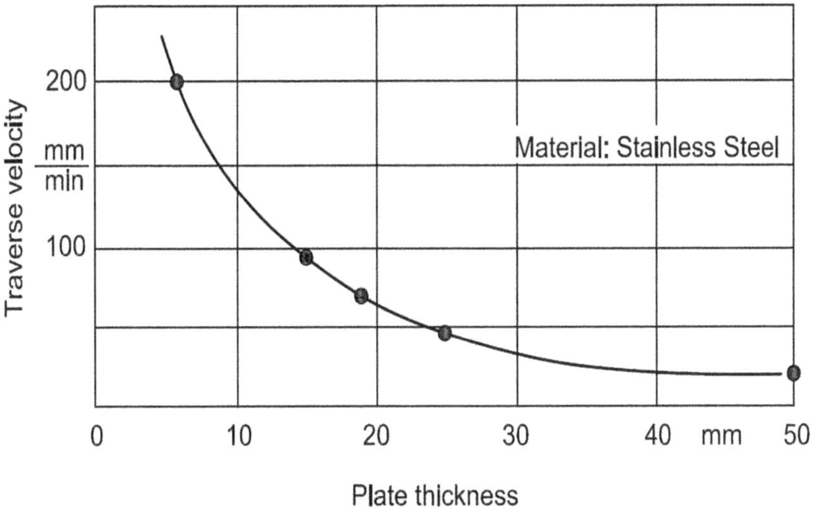

FIGURE 5.12 Effect of SS plate thickness on the traverse speed in AWJM (Youssef, 2005).

FIGURE 5.13 AWJM of an Inconel disk to produce integral turbine wheel blade (ASM International, 1989).

the material from between the turbine blades. Final shaping has been performed by ECM, where the total machining time was 48 hours.

Ultrasonic machining (USM) of stainless steels and super alloys: Generally, USM is not recommended to machine all types of stainless steels and super alloys, since such a process is intended to machine only hard and brittle materials. However, it has been reviewed in literature that Neppiras and Foskett (1958) tried to cut SS 304 ultrasonically, where they used boron carbide B4C (mesh 100). They reported very low MRR of 3 mm^3/min. Stainless steels and super alloys, when machined ultrasonically, acquire an index 2, as based on soda glass of index 100, provided the same machining condition. Since that time, no encouraging technical data were found in literature related to USM of stainless steels and super alloys. However, recently (2003), Park Myung Ho Samcheok National Univ., Korea developed an advanced USM technology for Inconel, using 60 kHz and 75 kHz high frequency transducers, of amplitudes of about 8 and 4 μm, respectively. Such high frequencies and low amplitudes secure highly efficient and precise USM of Inconel.

Abrasive flow machining of stainless steels and super alloys: Abrasive flow machining (AFM) was initially developed for the critical de-burring of aircraft valve bodies and components. Other applications include finishing impellers, integrally bladed rotors (IBRs), compressor wheels, turbine disks and gears. Milled surfaces of an IBR, as finished by AFM, need only 15 to 30 min operation time. Their surface finishes are considerably improved while eliminating hours of hand finishing, Figure 5.14.

Other important applications are:

a. Cross-drilled and intersecting holes that present a major problem for conventional de-burring methods are easily handled by AFM media. De-burring of hole intersections in stainless steel part, Figure 5.15a.
   Operating Conditions
   Grit type SiC, 700 grit
   Extrusion pressure 30 bar
   Process time 1.5 min
   No. of strokes 6
   Pieces/fixture 1
   Surface roughness $R_a = 0.4$ μm

FIGURE 5.14 Milled surfaces of an aircraft IBR finished by AFM (ASM International, 1989).

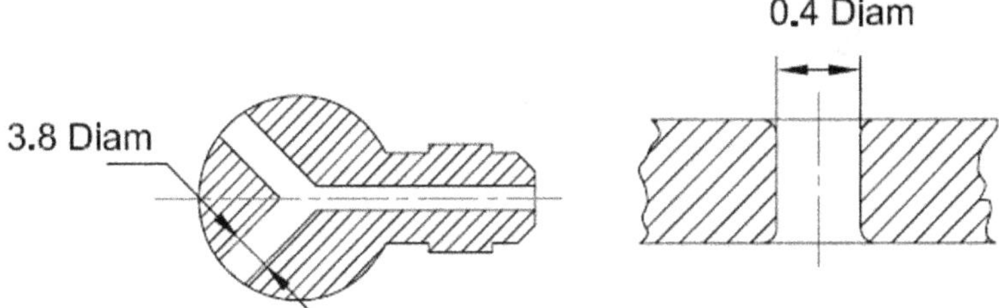

NB : All dimensions in mm.

a - Stainless steel **b - Cast Ni-base super alloy**

De-burring of hole intersection Removal recast residues

FIGURE 5.15 Processing of stainless steel and Ni-base super alloy using AFM (Machinability Data Center, 1980).

b. Removal recast residues from small holes of cast Ni-base super alloy Figure 5.15b.
   Operating Conditions
   Grit type SiC, 220 grit
   Extrusion pressure 32 bar
   Process time, controlled by volume, not strokes and time
   Pieces/fixture 1
   Surface roughness $R_a = 0.8$ μm

Viscous medium used to finish both parts was the Dynetics D080. The surface finish depends mainly on the abrasive grit size, which ranges from 20 to 700.

5.4.2 Machining SSs and SAs by Electrochemical and Chemical Techniques

Electrochemical machining (ECM): The major advantages of ECM are its process specific characteristics of high material removal rate in combination with almost no tool wear. ECM is specifically used in large batch size production and represents a viable alternative manufacturing technology for turbo-machinery components. In addition, high material removal rates can be realized while achieving good workpiece surface quality without the occurrence of white layers, heat-affected zones or strain hardening (Klocke and König, 2007).

Table 5.30 illustrates the recommended electrolytes (along with their concentration and inlet temperatures), used for ECM of some SS alloys. Table 5.31 shows theoretical removal rates for SS alloys, as calculated from the basic Faraday's equation.

Electrochemical machining is not a highly accurate process. The distribution of the current lines leads to rounding the corners and edges. Thus, sharp corners cannot

TABLE 5.30
Recommended Electrolytes Used for ECM of SSs

Stainless Steel Alloy	Electrolyte	Conc. (g/lit.)	Inlet Temp. (°C)
Type 410	NaCl or	36	27
(Martensitic)	NaCl+$NaNO_3$	192–216	46
Type 302 (Austenitic)	NaCl+NaF	30–32	38
Type 303 (Austenitic)	NaCl+$NaNO_3$	120–140	21
Type 316 (Austenitic)	NaCl	120	38
517400 [Custom 630]	NaCl or	96–120	27
(enh., PH, Mart.)	$NaNO_3$	240	38
Pyromet A-286	Na NO_3	240	38

Source: Adapted from Machining Data Handbook, Machinability Data Center Metcut, Research Associates, Inc. Vol. 2, 3rd Ed. Cincinnati, OH. 1980.

TABLE 5.31
Theoretical Removal Rates for ECM of Custom 630, and Pyromet A-286 SS Alloys

Stainless Steel Alloy	Theor. RR for 150 A/cm^2
S17400 [Custom 630]	2.0 cm^3/min
Pyromet A-286	1.9 cm^3/min

Source: Adapted from Machining Data Handbook, Machinability Data Center Metcut, Research Associates, Inc. Vol. 2, 3rd Ed. Cincinnati, OH. 1980.

be produced by ECM. Tolerances of about 0.12 are generally held to be typical for ECM of super alloys and stainless steels, while dimensional accuracies of 15 μm or less have been claimed under special shielding and masking of the tool cathode to direct the current flow only to the areas to be machined.

Table 5.32 shows the theoretical removal rates for ECM of some selected types of super alloys and stainless steels, as calculated according to Faraday's law using a current of 1000 Amp, and assuming 100% current efficiency which is realized if NaC1 electrolyte is used. The valences of alloying elements illustrated in the table are assumed to control the chemical dissolution of the machined alloys. Table 5.33 provides the recommended electrolytes along with their recommended concentrations and inlet temperatures, used for ECM of some typical super alloys and stainless steels. In the majority of applications, simple NaC1 electrolyte is to be selected first, then more complex electrolytes are only to be tried if necessary. NaCl is highly corrosive and non-expensive (costs 30% of $NaNO_3$). Electrolyte concentrations should be kept as low as compatible with productivity so as to reduce accumulations of salt

TABLE 5.32
Theoretical Metal Removal Rates for ECM of Super Alloys and Stainless Steels as Calculated by Faraday's Law Assuming Current Eff. η = 100%, Using a Current of 1000 Amp

Work Material	MRR, based on I = 1000 Amp (cm^3/min), η = 100%	Assumed Valences (n)
Super Alloys:		
A-286*	1.92	Al=3, Nb=3, Co=2, Cr=3, Cu=2, Fe=2,
M 252	1.80	Mn=3, Mo=4, Ni=2, Ti=4, W=6, V=5,
René41	1.77	C=0, Si=0
Udimet 500	1.80	MRR = (60/F) $(N/n)_{eq.}$ I. $1/\rho$ cm^3/min
Udimet 700	1.77	F = 96487 A.s/mol
Haynes 25 (L605)	1.75	$(N/n)_{eq}$ = Chemical equivalent of anode
		ρ = Anode density
Stainless Steels:		
17-4PH(UNS 17400)	2.02	

*A-286 is the Pyromet Stainless Steel.

deposits on equipment and tooling. The temperature control should be within ±1°C of the recommended value, in order to attain consistent electrolyte conductivity at the inlet side of the electrolytic cell. The cell open circuit voltage is set to attain the desired current density that matches the feed rate. The typical voltage setting for ECM of super alloys and stainless-steel ranges from 10 to 25 V.

It should be emphasized that electrolytes play an important role in the dimensional control of the produced holes and cavities. NaCl, for example, yields much less accurate components than nitrates ($NaNO_3$), the latter having far better dimensional control due its current efficiency/current density characteristic (see Figure 3.16).

Actual MRRs are less deviated in most cases from the Faraday's theoretically calculated values listed in Table 5.32. Such deviations are related to the not correctly assumed valences of the work material and the current efficiency which attains values not necessarily 100%, as considered in Faraday's law.

Figure 5.16 shows the relation between the open circuit voltage and the current density for two selected frontal equilibrium gaps of 0.5 and 0.25 mm, respectively, when ECM of an Ni-base super alloy René 41 for minimum starting voltage ΔE = 2.6 V. Using lower gap is preferred. Figure 5.17 shows an airfoil, directly machined on a turned Inconel 718 compressor disk to eliminate the mechanical joining of the airfoils on the disk which realizes the highest rigidity. A multi-axis NC-ECM machine is used to rough and finish the airfoils out of a turned hub. After the finish run, ECM yields a consistent surface finish and satisfactory part tolerances; accordingly, no further processing (machining or polishing) of the airfoils is required.

Table 5.34 provides typical MRR and the relevant operating conditions for ECM of stainless steels and super alloys as compared to Ti-base alloy Ti-6Al-4V.

TABLE 5.33
Electrolyte Selection Guide (Composition, Concentration, and Inlet Temperature) for ECM of Super Alloys and Stainless Steels

Work Material	Electrolyte: Composition	Electrolyte: Concentration (g/L)	Electrolyte: Inlet temperature (°C)	Remarks
Super Alloys:				
Fe-base:				
A-286(W)	$NaNO_3$	240	38	$NaClO_3$ for specialized applications
Ni-base:				
M252(W)	NaCl	240	41	ST (not Ag) gives better results
Waspaloy (W)	NaCl	120	35	
Astroloy (W)	NaCl	240	24	
	$NaNO_3$ or	120 to 240	38	Cuts easily and rapidly
	NaCl + NaF	90+30	38	
Inconel 700(W)	NaCl	120	38	
Inconel 706 (W)	NaCl	96	24	
Inconel 718 (W)	NaCl or	108–120	35 to 38	ST and Ag give better results
	$NaNO_3$	216 to 240	35 to 38	
René 41 (W)	NaCl	258 (or 120)	24 (or 35)	
Udimet 500 (W)	NaCl	120	38	
Udimet 700 (W)	NaCl	120	38	
Inconel x (W)	NaCl + $NaNO_3$	240+60	38	
IN-100(C)	NaCl	120	38	
René80 (C)	NaCl	120	38	
René 125 (C)	NaCl	120	38	
René 95 (C)	$NaNO_3$ or	216 to 240	38	
	NaCl	120	38	
Co-base:				
MAR-M509	NaCl	240	32 to 52	
HS-21	NaCl	120	38	
Haynes 25(L605)	NaCl + $NaNO_3$	240+30	38	
HS-31(X-40)	NaCl + NaF	30+2.4	38	
Stainless steels:				
302 (Aust.)	NaCl + NaF	30 + 2.4	38	
303 (Aust.)	NaCl + $NaNO_3$	120 + 20	21	
316 (Aust.)	NaCl	120	38	
410 (Marten.)	NaCl or	96	27	
	NaCl + $NaNO_3$	192 + 120	115	
17-4 PH (Custom 630, UNS 17400)	NaCl or	192 to 120	26	
	$NaNO_3$	240	38	

(W): Wrought, (C): Cast, ST: Strengthened, Ag: Aged.

Source: Adapted from Machining Data Handbook, Machinability Data Center Metcut, Research Associates, Inc. Vol. 2, 3rd Ed. Cincinnati, OH. 1980.

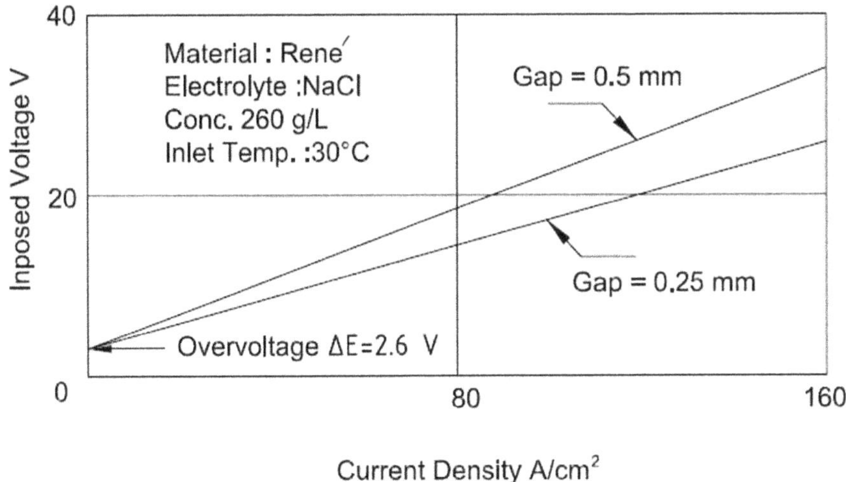

FIGURE 5.16 Relation between the open circuit voltage and the current density for two selected frontal equilibrium gaps when ECM of René 41 (Bellows, 1967).

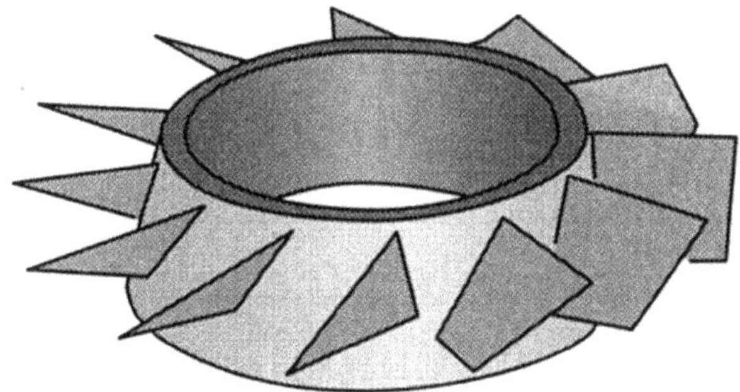

FIGURE 5.17 Airfoil directly machined on a turned Inconel 718 compressor disk (Klocke et al., 2013).

Some other examples are provided to illustrate the application of ECM in machining components made from stainless steels and super alloys, Figure 5.18 and Table 5.35:

Figure 5.18a: ECM of turbine blade made from either an Fe-base super alloy A-286, or a Ni-base Waspaloy, using a machine that provides 2000 A machining current, and an OCV of up to 20 V. The tool electrode used was made from Cu-W alloy. For both materials, the current starts at 100 A and ends at 150 A. The electrolyte was NaNO3 (of concentration 260 g/L) for the first blade, whereas it was NaCl (of concentration 200 g/L) for the second blade. Other working conditions are shown in Table 5.35.

Figure 5.18b: Jet engine blade made from Ni-based alloy Udimet 700 was provided by small size deep holes using STEM. Acid electrolyte H_2SO_4 (10% concentration) was used. Periodic voltage reversal was performed.

TABLE 5.34
Typical MRRs and Relevant Operating Conditions for ECM of Stainless Steels and Super Alloys

Work Material	Electrolyte Type	Electrolyte Concentration (g/L)	Min. starting voltage (ΔE)	MRR, Based on I = 1000A, (cm^3/min)
Super Alloys:				
Inconel 718 (ST-Ag)	NaCl	120	3.3	1.44
René 95 (ST-Ag)	$NaNO_3$	270	4.3	1.69
300M (Tempered)	NaCl	120	1.1	2.16
MAR-M509 (C)	NaCl	120	1.2	1.57
Astroloy	$NaNO_3$	240	4.0	2.05
Stainless Steel:				
17-4 PH (ST-Ag)	$NaNO_3$	270	3.6	1.41
Ti Alloy:				
Ti-6Al-4V (Ann.)	NaCl	120	3.8	1.64

(C) : Cast, ST-Ag: Solution treated and aged, Ann.: Annealed.

Figure 5.18c: This illustrates an Inconel 718 rotor in which multiple small cavities are to be machined electrochemically. Milling of such small cavities conventionally is costly because small milling cutters must be used. Seven hours are needed to machine 54 cavities. With ECM, however, these cavities can be machined either singularly or in groups.

Figure 5.18d: This shows the cross-section of a nozzle made from AISI-316 stainless tube. The EC machine can provide a maximum current of 500 A at 24 V, dc. A Cu electrode is fed into one end of the tube to machine the hole to a conical shape, with a current beginning at 20 and ending at 310 A. The open-circuit voltage was 17 V. The electrolyte was NaCl (concentration 120 g/L, inlet temperature 27°C). A feed rate of 6 mm/min has been realized. A cutting time of 8 minutes produced an excellent surface quality of $R_a = 0.12–0.25$ μm.

After this operation, the part is inverted, and another Cu electrode was used to machine the radius at the other end of the nozzle. The current began at 20 A, and ended at 220 A. The tool feed was 2.5 mm/min, the cutting time was 2 minutes, producing a similar surface finish as before.

Filtration of electrolytes to the 50 μm level or less is most desirable. Placement of filters immediately ahead of the electrode is good practice. The volume of metal hydroxide or metal hydrate is considerable. In case of salt electrolytes, it can be 100–500 times the volume of the metal removal. For super alloys and stainless steels, the volumetric ECM sludge ratio is about 200 times, which is considerably less than that for Ti-base alloys which is about 500 times the volume of metal removal. Settling and filtration to remove these chips (sludge) is highly recommended to avoid short circuiting (arcing) which may destroy both the work and the tool electrode. Sludge filtration, however, necessitates environmentally satisfactory measures.

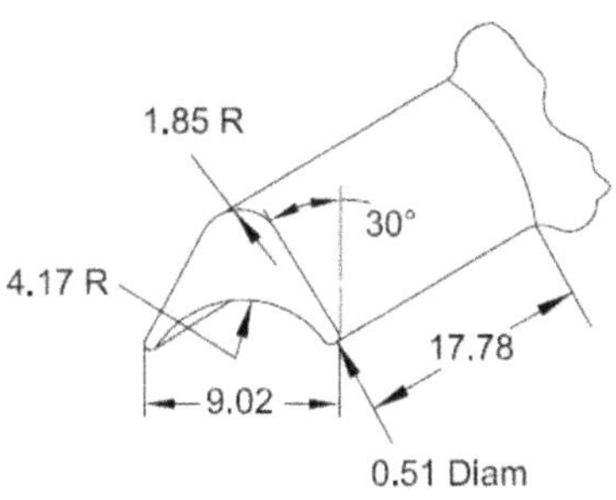

a - Turbine blade

Material : A-286 or Waspaloy
Operation : ECM

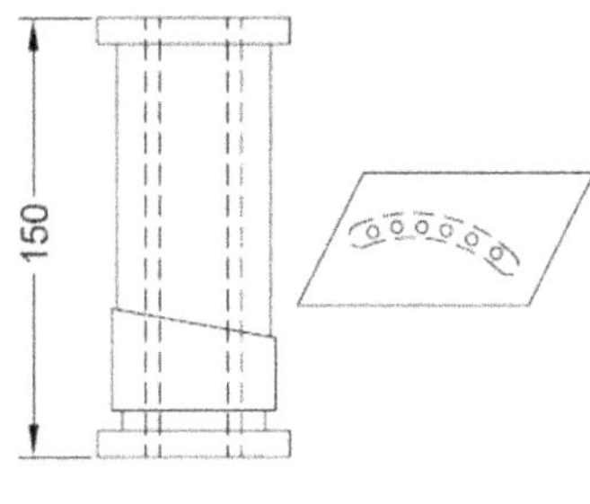

b - Jet engine blade

Material : Udimet 700
Operation : STEM

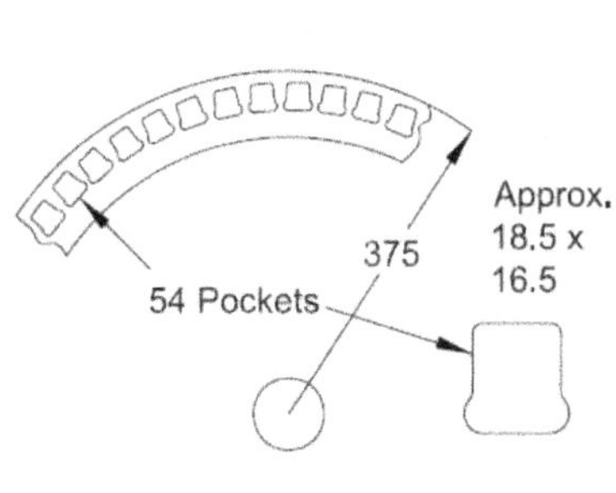

c - Turbine disk pockets

Material : Inconel 718
Operation : ECM

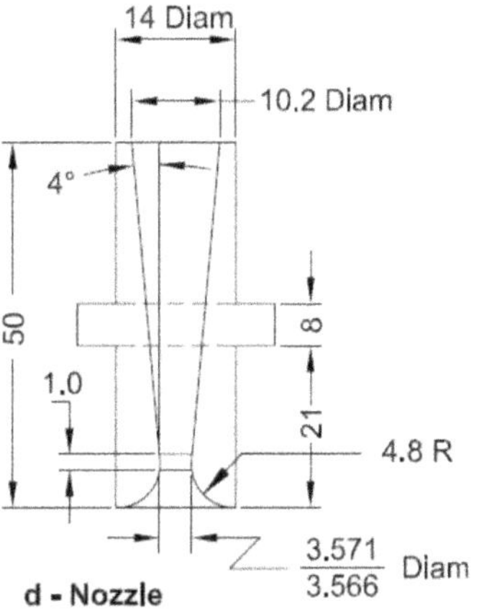

d - Nozzle

Material : Stainless 316
Operation : ECM

NB : All dimensions in mm.

FIGURE 5.18 Application of ECM and STEM in machining aero-engine components made from stainless steels and super alloys (Youssef, 2016).

Shaped tube electrolytic machining (STEM) of stainless steel and super alloys: STEM is used to drill round and shaped holes in difficult-to-cut materials such as stainless steels and super alloys. Holes ranging from 0.5 to 6 mm diameter and as deep as 600 mm can be produced with length-to-diameter ratios up to 300. Over 100 holes/machine stroke are practical.

Holes in the following stainless steels and super alloys have been drilled using STEM with H2SO4 acid electrolyte of 10% concentration.

- **Stainless steels**: AISI-304, AISI-321, AISI-414
- **Super alloys**: Udimet 500, 700, 710

TABLE 5.35
Working Conditions for ECM And STEM of Stainless Steel and Super Alloy Components Shown in Figure 5.17

	Component				
Working Conditions	a		b	c	d
Operation	ECM	ECM	STEM	ECM	ECM
Material of Component	A 286	Waspaloy	Udimet 700	Inconel 718	AISI-316
Electrolyte					
Type	$NaNO_3$	NaCl	H_2SO_4	NaCl	NaCl
Concentration (g/L)	260	200	10%	120	120
Inlet Temp. (°C)	42	32	35	38	27
Inlet Pr. (bar)	9–14	9–14	1	15–17	6
Flow Rate (L/min)	8	10	2	10	8
Current (A)					
Start	100	100	3	130	20
End	150	150	10	630	310
OCV (V)	11	12	9*	18	17
Feed Rate (mm/min)	7.5	8	1.25	1.5	6
Tolerance (μm)	±100	±100	±50	±75	4° taper
Surf. Roughn. R_a **(μm)**	0.4–0.7	0.2–0.5	1.5–3	0.7–1	0.12–0.25

***Periodic reversal in STEM.**
Source: (Youssef, 2016).

Stellite

IN-100, -102, -738
Inconel 625, 718, X-750, 825
René 41, 80, 95, 100
Haynes 25, 181
HS-31(X-40)

Hastelloy C, X
Greek Ascoloy

Electrochemical grinding (ECG) of stainless steels and super alloys: In ECG, the wheels must have insulating grits (SiC and some forms of Borazon cannot be used because they are electrically conductive). This is why the wheels in Table 5.36 are corundum. Also, NaCl electrolytes are rarely used in ECG because they are strongly corrosive to machine components.

Table 5.36 lists the recommended parameters of ECG of super alloys and stainless steels. The maximum current density must be low enough to prevent overheating the low conductive materials. For comparison, the same parameters are listed for Cu alloys, Al alloys, and carbon steels. It is depicted from the table that greater

TABLE 5.36
Recommended Parameters for ECG of Super Alloys and Stainless Steels

Work Material	Abrasives of GW	Electrolyte/Concent.	g/l H_2O	Max. Current Density A/cm^2
Super Alloy:	Al_2O_3			
A-286 (Pyromet)		$NaNO_3$	120–140	116
Hastelloy X		$NaNO_3$	120–140	116
M 252		$NaNO_3$	120–140	116
Udimet 500,700		$NaNO_3$ or NaCl	110–120	116
Waspaloy		$NaNO_3$	120–140	116
Inconels		$NaNO_3$	120–140	116
René 41		$NaNO_3$	180–230	78
René 80		$NaNO_3$	120–140	78
HS-31(X-40)		$NaNO_3$+NaCl	60–80	78
Stellite		$NaNO_3$	210–240	78
Stainless Steels:		$NaNO_3$	180–200	78
Other Alloys:				
Cu alloys		$NaNO_3$ or KNO_3	180–200	233
Al alloys		$NaNO_3$	120–140	233
Steels:				
Low carbon		KNO_3: KNO_2 (9:1)	60–120	155
High carbon		$NaNO_3$	120–180	155

NB: SiC and CBN not used for ECG since they are electric conductive.

Source: Adapted from Machining Data Handbook, Machinability Data Center Metcut, Research Associates, Inc. Vol. 2, 3rd Ed. Cincinnati, OH. 1980.

current densities are allowed for the latter alloys than stainless steels and super alloys. Corundum grinding wheels and NaNO3 electrolytes are recommended for most cases of ECG.

Chemical milling (CH milling) of SSs and SAs: The disadvantages of CH milling include low cutting rates and the fact that masked areas will be under-cut (etch factor EF) by the corroding solutions. The corrosive effect is less serious in the case of stainless steels; however, hydrogen embitterment may be a problem with hardened martensitic stainless alloys and inter-granular corrosion may occur, depending on the chemical composition of that type. Table 5.37 illustrates the main parameters for CH milling of austenitic and martensitic stainless alloys. The etching rate of martensitic alloys is considerably lower than that of austenitic alloys. Table 5.38 shows the same parameters for CH milling of some selected super alloys.

The gentle chemical action of CH milling does not introduce stress into the workpiece, since it removes the material molecule-by-molecule, resulting in surfaces free from residual stresses. Surface roughness is influenced by the initial workpiece roughness condition. Too violent agitation can lead to uneven cut or grooving. Tables 5.37 and 5.38 show the surface roughness of stainless steels and super alloys attained by CH milling, respectively.

TABLE 5.37
Parameters of CHM of Austenitic and Martensitic SSs

Stainless Alloy	Etchant	Concent.	Temp.	Etch Rate µm/min	Maskant	Etch Factor (–)	Depth Tol. ± µm	R_a µm
Austenitic	$FeCl_3$ or HCl:HNO3	42°Be*	54°C	20–130	Polyvinyl chloride	1.5–2.0	100	1.6
Martensitic	$FeCl_3$ or HCl:HNO3	52°Be*	54°C	6	Polyvinyl chloride	-	100	3.2

Be*: Baumé spec. gravity scale.

Source: Adapted from Machining Data Handbook, Machinability Data Center Metcut, Research Associates, Inc. Vol. 2, 3rd Ed. Cincinnati, OH. 1980.

TABLE 5.38
Parameters of CHM of Super Alloys

Super Alloy	Etchant	Concent.	Temp.	Etch Rate µm/min	Maskant	Etch Factor (–)	Depth Tol. ± µm	R_a µm
Co-base alloy	HCl:HNO3: $FeCl_3$	-	60°C	10–38	-	-	-	1–3.8
Inconel	$FeCl_3$ or	42°Be*	54°C	13–38	Polyethelene	-	-	1–3.8
	HCl:HNO3	42°Be*	54°C	13–38	Polyethelene	-	-	1–3.8
Nimonic	$FeCl_3$ or	42°Be*	49°C	13–38	Polyethelene	1–3	51	1–3.8
	$FeCl_3$:HNO3: HCl	-	49°C	13–38	Polyethelene	1–3	51	1–3.8

Be*: Baumé spec. gravity scale.

Source: Adapted from Machining Data Handbook, Machinability Data Center Metcut, Research Associates, Inc. Vol. 2, 3rd Ed. Cincinnati, OH. 1980.

Photochemical machining (spray etching) of SSs and SAs: The etchability of a metal or an alloy mainly depends on its chemical composition. Table 5.39 lists the etchability ratings of some stainless steels and super alloys, machined by PCM. Copper, brass, aluminum, and magnesium have a good rating. Stainless steels and super alloys such as Inconels and Hastelloy B have a fair to good rating. Udimet alloys have a poor to fair rating. Hastelloy C and René 41 along with W, Ti, Nb, and Ta have a poor etchability rating.

Etchants for PCM of stainless steels and super alloys are listed in Table 5.39, together with their operating characteristics. Ferric chloride (FeC13) solutions are used for PCM of a wide variety of metals and alloys and therefore it became the most widely used etchant in the PCM industry. Sodium hydroxide is extensively used with Al and Al alloys. Etchant compositions can be adjusted to meet the requirements of specific applications, and proprietary additives can be included to control foaming or wetting characteristics, increase or decrease etching rate, or make etching more uniform. Etching machines are made from materials (such as polyvinyl chlorides and titanium) that can withstand corrosion from ferric chloride and other etchants. Etchant temperature must be maintained below 55°C to avoid distortion of plastics used in machine construction.

Visser et al. (1994) investigated the effect of spray jet pressure on the etching speed and the etchant concentration on the surface roughness. Figure 5.19 illustrates the effect of jet pressure and temperature of sprayed etchant (acidic solution of concentration of 3.8 mol/lit.) on the etching speed, when machining martensitic stainless steel AISI-420, whereas Figure 5.20 visualizes the effect of etchant concentration on the surface roughness of austenitic alloy AISI-304. Figure 5.21 shows a proportional relationship between the volume of produced cavity in stainless 304 and the pulse charge Li (Visser, 1966).

TABLE 5.39
Etchability Ratings of Some Selected Super Alloys and Stainless Steels Machined by PCM

Etchability Rating	Super Alloys and Stainless Steel Grades (Etchability in Descending Order)
Good	Copper, brass, aluminum, magnesium, …
Good to Fair	AISI 215, 301, 302, 304, 305, 316, 321, 347
	PH 15-7, PH 17-7
	AISI 410, 420, 430
	Inconel alloys (e.g. Ni, 15% Cr, 7% Fe)
	Hastelloy B (Ni, 28% Mo, 5% Fe, 2.5% Co, 1% Cr, 0.5%V, 0.05% C)
Fair to Poor	Udimet alloys (e.g. Ni, 42% Fe, 12.5% Cr, 2.7% Ti)
Poor	Hastelloy C (Ni, 15% Mo, 14% Cr, 5% Fe, 3% W, 2.5% Co, 0.08% C)
	René 41 (Ni, 19% Cr, 11% Co, 10% Mo, 3% Ti, 1.5% Al)

Source: Adapted from Machining Data Handbook, Machinability Data Center Metcut, Research Associates, Inc. Vol. 2, 3rd Ed. Cincinnati, OH. 1980.

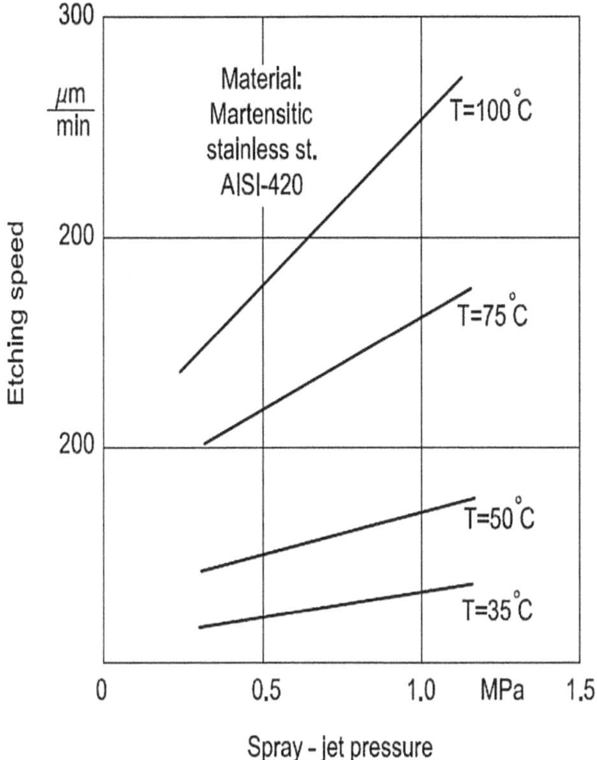

FIGURE 5.19 Effect of jet pressure and temperature in spray etching on etching rate of martensitic stainless AISI-420 (Visser et al., 1994).

5.4.3 Thermoelectric Machining of Stainless Steels and Super Alloys

Electric discharge machining (EDM) of stainless steels and super alloys: The reasons for choosing EDM over traditional processes are that the productivity is not limited by the hardness or strength of the workpiece; and complex features, or high aspect ratios of holes and cavities, can be readily machined. The turbo-machinery materials therefore specifically machined by EDM consist of the super alloys Inconel 738, Inconel 939, CMSX-4, MAR-M002, MAR- M247, Udimet 720, Nimonic 105, Nimonic 713, etc. The main areas for application of EDM include the drilling of cooling holes and die sinking of slots, pockets, and grooves Table 5.40.

When machining stainless steels with EDM, Cu electrodes are generally used with reversed polarity, and kerosene as dielectric fluid. In the case of using pulse generators, the recommended peak currents range from 2 to 12 A, pulse duration 50–200 µs, pulse-off 50–200 µs, and the duty factor 0.5. When machining stainless steel, minimum wear of electrode is realized if long pulse duration is provided. Data indicated that the fatigue life of SS alloys such as 304 and 410, machined by EDM, can be significantly reduced compared to traditional machining.

Electrical discharge milling (ED milling) of SSs and SAs: ED milling is used notably for making molds of parts for electrical and electronic industries, household

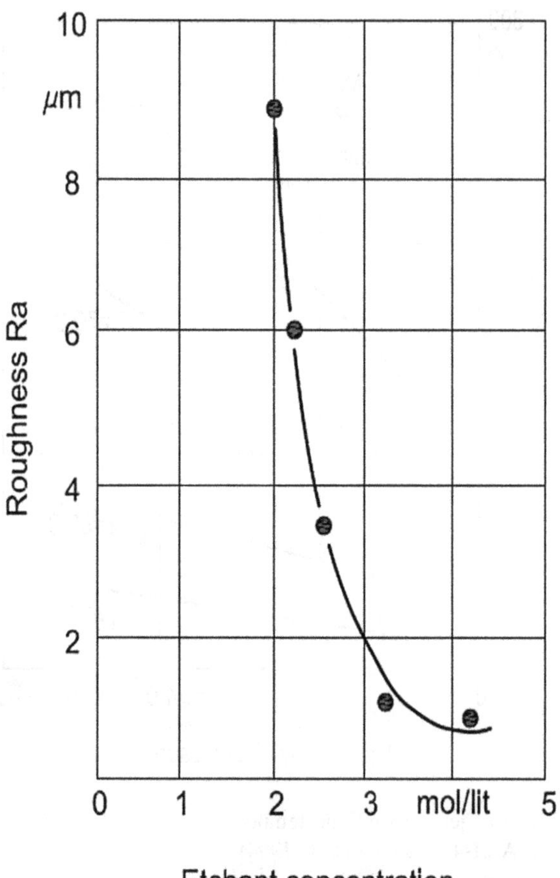

FIGURE 5.20 Effect of etchant concentration on the surface roughness of austenitic alloy AISI-304 (Visser et al., 1994).

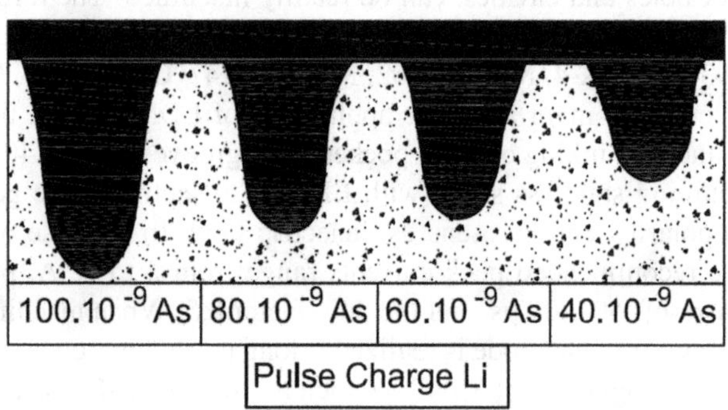

FIGURE 5.21 Proportional relationship between the volume of produced cavity in stainless 304 and the pulse charge Li.

TABLE 5.40
Etchants for PCM of Plates Made from Some Selected Super Alloys and Stainless Steels

Plate Material	Etchant Formulation	Temperature (°C)
Super Alloys:		
HyMu 80, 800		
(80% Ni, 4% Mo, Fe)	42Bé, $FeCl_3$: HC1 (9:1)	43–49
Inconels (Ni, Cr, Fe)	42Bé, $FeCl_3$	54
Nimonics (ca. 80% Ni, 20% Cr)	42Bé, $FeCl_3$: HNO3:HCl	49
Stainless Steels:		
Mo-Free Stainless Steel	35–48 Bé, $FeCl_3$	35–55
Mo Stainless Steel	36–42 Bé, $FeCl_3$ with HNO_3- Addition	35–55

Source: Adapted from Machining Data Handbook, Machinability Data Center Metcut, Research Associates, Inc. Vol. 2, 3rd Ed. Cincinnati, OH. 1980.

appliances, and the automotive and aeronautical components made from stainless steels and super alloys. Another technological breakthrough of ED milling is that the process has entered the domain micromachining, where it is possible to produce fine and intricate shapes with sharp corners.

Electron beam machining of SSs and SAs: Operating parameters for drilling holes and cutting slots in different alloys of stainless steels and super alloys are shown in Table 5.41 and Table 5.42, respectively.

TABLE 5.41
Parameters for Drilling Holes in Stainless Steels and Super Alloys with EBM

WP Thickn. (mm)	Hole Diam. (mm)	Drilling Time (s)	Accel. Voltage (kV)	Aver. Beam Current (μA)	Pulse Width (μs)	Pulse Freq. (Hz)
Ferritic and martensitic stainless-steel alloys:						
0.25	0.013	<1	130	60	4	3000
Other stainless alloys (austenitic):						
1.0	0.13	<1	140	100	80	50
2.0	0.13	10	140	100	80	50
2.5	0.13	10	140	100	80	50
6.4	0.5–1.0	180	145	4000	2100	12.5
Hastelloy (Ni-base alloy):						
10	2.5	70	130	5000	5300	100

Source: Adapted from Machining Data Handbook, Machinability Data Center Metcut, Research Associates, Inc. Vol. 2, 3rd Ed. Cincinnati, OH. 1980.

TABLE 5.42
Parameters for Cutting Slots in Stainless Steels with EBM

WP Thickn. (mm)	Slot Width (mm)	Rate of Cut (mm/min)	Accel. Voltage (kV)	Aver. Beam current (μA)	Pulse Width (μs)	Pulse Freq. (Hz)
0.05	0.05	100	130	20	4	50
0.18	0.10	50	130	50	80	50
1.57	0.2	1.25	140	120	80	50

Source: Adapted from Machining Data Handbook, Machinability Data Center Metcut, Research Associates, Inc. Vol. 2, 3rd Ed. Cincinnati, OH. 1980.

Cylindrical, conical, and barrel-shaped holes of various diameters can be drilled with consistent accuracy at rates of several thousand of holes per second. EB-drilled holes in super alloy turbine blade at angles of 60–90° to profile chord can be easily machined. Holes of inclination angle of 15° are possible. The largest diameter and depth of holes that can be accurately drilled by EB are 1.5 mm and 10 mm, respectively, and the aspect ratio is typically 1:1 to 15:1. Rectangle slots of 0.2 × 6 mm in 1.6-mm-thick SS plates are produced in 5 min using 140 kV, 120 μA, pulse duration of 80 μs, and frequency of 50 Hz. The traverse speed is inversely proportional to the work thickness (Table 5.42).

Laser beam machining of stainless steels and super alloys: Approximately 50% of all industrial laser material processing applications are laser drilling operations. It is possible to drill hundreds/thousands of cooling holes with high precision and of variable diameter and shape in multi-material blades of complex shape.

Table 5.43 and Table 5.44 provide the machining parameters and lasers for drilling and slotting operations performed in various stainless steels. It is predicted that non-satisfactory results are achieved when cutting austenitic alloys with laser due to less fluidity of the molten metal as compared to other SS alloys, Table 5.44. Also, it

TABLE 5.43
Parameters for Drilling 0.12 Mm Diameter Holes in SS 304 Using Nd:YAG (1.06 μm) Laser

WP Thickness mm	Assisting Gas	Drilling Time (s)	Lamp Current (A)	Average Power (W)	Max. Thickness (mm)
3.0	Oxygen	88	34	31	4.8
3.0	Argon	221	34	31	4.8

Source: Adapted from Machining Data Handbook, Machinability Data Center Metcut, Research Associates, Inc. Vol. 2, 3rd Ed. Cincinnati, OH. 1980.

TABLE 5.44
Parameters for Gas-Assisted CO_2 (1.06 μm) Laser Cutting Various Types of SS

Plate Thickness, mm	Assisting Gas	Cutting Rate, m/min
500 W Laser:		
0.3	Oxygen	3.71
1.0	Oxygen	1.65
3.2	Oxygen	0.89
1000 W Laser:		
0.5	Oxygen	19.00
0.8	Oxygen	16.50
1.6	Oxygen	11.40
3.2	Oxygen	5.08
1250 W Laser:		
1.0 (Austenitic)	Oxygen	8.89
3.2 (Austenitic)	Oxygen	3.05
3.2 (Austenitic)	Air	1.52
5.2 (Ferr./Mart.)	Oxygen	1.78

Source: (Saunders, 1984).

can be predicted that higher cutting rates are realized when O_2 is used as assisting gas. The traverse cutting speed V_t (up to 20 m/min for gas-assisted laser cutting of stainless steel) is a measure of machining efficiency. It depends on laser power P, and plate thickness t (Table 5.44). These are correlated for a certain material by the following equation.

$$v_t \propto \frac{P}{t}$$

In general, pulsed laser systems are applied for laser drilling processes, where selection of the pulse duration depends on the hole characteristics and material being processed. In the field of laser drilling of turbine components, pulse duration is normally of the order of nano- or milliseconds. The required pulse energy basically depends on the exact chemical composition of the material, the material thickness, and the desired hole diameter and shape.

Advanced designs of Nd:YAG lasers have cut Ni-base super alloys up to 50 mm thick at speeds greater than EDM. Table 5.45 illustrates the (percussion drilling) PD time(s) when LBM of Inconel 718 plates of different thicknesses using Nd:YAG of pulse energy of 10 J/pulse at three different power levels takes place. It is depicted from this table that the drilling time of thin sheets (up to 2.5 mm) does not depend on the power level. Therefore, small power levels are highly recommended for drilling thin plates, and that is the same reason why LBM is preferred for drilling and slotting of small sheets.

TABLE 5.45

Percussion Drilling Time in Seconds of Inconel 718 Of Different Plate Thicknesses Using Nd:YAG Laser of Pulse Energy of 10 J/Pulse at Various Power Levels

Nd:YAG Laser of	Drilling time in seconds for plate thickness (s) in mm of Inconel 718					
Average Power Level	s = 2.5 mm	s = 5 mm	s = 10 mm	s = 15 mm	s = 20 mm	s = 25 mm
75 W	0.5	3	15	37	65	95
150 W	0.5	2	5	17	33	55
250 W	0.5	1	4	12	23	38

Source: Adapted from Machining Data Handbook, Machinability Data Center Metcut, Research Associates, Inc. Vol. 2, 3rd Ed. Cincinnati, OH. 1980.

Plasma arc cutting of stainless steels and super alloys: This process is characterized by its highest cutting rate and lowest specific cutting energy when cutting stainless and exotic materials as compared to other NTMPs. The gas used for the arc (primary gas) may be N_2, H_2, Ar, or various admixtures. Compressed air may also be used to increase cutting rates, depending on the thickness of SS plate, as shown in Table 5.46. However, using air as primary gas calls for oxidizing the cut surfaces.

TABLE 5.46

Comparison of PAC Cutting Rates of SSs Using Ar/H_2 or Air as Primary Gases

	Cutting Rate, m/min		
Thickness, mm	Ar/H_2	Air	Remarks
5	5.0	5.0	Increased cutting rates if air is used.
10	2.6	3.4	
15	1.5	1.8	
20	1.1	1.2	
25	0.8	0.85	
30	0.65	0.6	
35	0.5	0.4	
40	0.4	0.3	
45	0.35	0.25	
50	0.3	0.2	
60	0.3	0.2	

Source: (Holden, 1985).

For stainless steels, CO_2 may be used with N_2 as a primary gas (Table 5.47 and Table 5.48). A water curtain may be used in place of the shielding gas or may be injected into the plasma stream to produce a cleaner cut with a reduced bevel, and narrow kerf, but without improvement of the cutting rate. However, nozzle life can be improved by cooling action of the water.

Table 5.49 shows also the cutting speeds for Al and carbon steel. It can be depicted from this table that when using PAC, Al has the best machinability rating, followed by stainless steels, then carbon steel.

TABLE 5.47
Cutting Rate and Current Selection of PAC of SS

Plate Thickness, mm	Cutting Speed, m/min	Power Selection Amperage A
6	1.78	105
	2.54	140
13	0.51	135
	1.02	190
	2.54	270
	3.81	700
25	0.51	210
	0.76	270
	2.03	540
	2.79	1000
38	0.25	280
	0.51	420
	1.02	620
	1.78	1000
51	0.13	320
	0.25	610
	1.02	950
64	0.13	410
	0.25	550
	0.51	820
76	0.13	510
	0.25	675
	0.51	1020
89	0.25	730
	0.51	1110
102	0.13	675
	0.25	900
114	0.13	900
127	0.076	1100
140	0.076	1100

Source: (Bagley, 1969).

TABLE 5.48
Cutting Speeds and Machining Conditions of PAC for SSs

Machining Conditions	Plate Thickness, mm	Cutting Speed, m/min Best/Max.
Amperage Selection: 100 A	6	1.25/2.54
Prim. gas N_2 (1.55 m^3/h, 2.07 bar)	13	0.51/0.76
Second. gas CO_2 (5.8 m^3/h, 2.75 bar)	25	0.23/0.28
Amperage Selection: 200 A	6	1.65/3.43
Prim. gas N_2 (1.95 m^3/h, 2.07 bar)	13	1.27/1.78
Second. gas CO_2 (5.8 m^3/h, 2.75 bar)	25	0.51/0.66
	38	0.30/0.40
Amperage Selection: 400 A	13	1.91/3.05
Prim. gas N_2 (1.4 m^3/h, 1.40 bar)	25	1.02/1.40
Second. gas CO_2 (5.8 m^3/h, 2.75 bar)	38	0.64/0.97
	64	0.30/0.38
	76	0.20/0.25

Source: Courtesy of Thermal-Dynamic Corp.

TABLE 5.49
Cutting Speeds and Machining Conditions of Water-Injection PAC for SS, Al, and Carbon Steel

		Cutting Speed, m/min		
	Plate Thickness,	Carbon St.	SS	Al
Machining Conditions	mm	Best/Max.	Best/Max.	Best/Max.
Amperage selection: 300 A	6	1.5/2.8	1.9/3.3	2.2/3.7
Prim. gas N_2 (2.1 m^3/h, 2.07 bar)	13	1.0/1.5	1.3/1.8	1.4/2.0
Water injection at 30 to 60 lit/hr	25	0.5/0.6	0.6/0.9	1.0/1.5
	38	0.3/0.4	0.4/0.5	0.5/0.6

Source: Courtesy of Thermal-Dynamic Corp.

5.5 REVIEW QUESTIONS

5.5.1 What makes stainless steel stainless?
5.5.2 What is the difference between 304 and 316 stainless steel?
5.5.3 Is stainless steel magnetic?
5.5.4 Can stainless steel be welded? What does the "L" designation mean?
5.5.5 Can stainless steel be hardened?
5.5.6 What is the "annealed" condition of SS?

5.5.7 What are AISI specifications of stainless steel?
5.5.8 Can stainless steel be machined?
5.5.9 Define and describe the basic (standard) alloys of SSs?
5.5.10 Give reasons why free-machining alloys are not currently available in the duplex or PH-SSs?
5.5.11 Differentiate between S and Se as free-machining additives in SS.
5.5.12 Define super alloys.
5.5.13 Classify super alloys in their basic groups.
5.5.14 What is the difference between cutting stainless steel with a laser or a water jet?
5.5.15 Does abrasive water jet cutting cost the same as laser cutting?
5.5.16 Can you cut titanium?
5.5.17 How thick of a material does abrasive water jet cut?
5.5.18 Does abrasive water jet cut hardened metals?
5.5.19 What are the limitations of water jet cutting?
5.5.20 What kind of abrasive is used in abrasive water jet cutting?
5.5.21 What types of materials can be laser cut?
5.5.22 Can you describe abrasive water jet cutting?
5.5.23 What are some of the various alloy additions that are used to improve the machinability of stainless steels?
5.5.24 Differentiate between S and Se as free-machining additives for stainless steels, regarding machinability, surface finish, cold formability, weldability, and corrosion resistance.
5.5.25 What do you understand by a free-machining stainless steel? What elements are usually added to make SS, and explain how they make the steel free-cutting?
5.5.26 Why do ferritic SSs and austenitic SSs not respond to quench, good weldability, corrosive environment, and good machinability?
5.5.27 What are the special properties of SSs that make them difficult to machine?
5.5.28 What are general rules to be considered when machining SSs conventionally?
5.5.29 When do you recommend non-traditional machining conditions for machining SSs instead of traditional machining?
5.5.30 Why should ferritic SSs be given first consideration when selecting an SS?
5.5.31 What are the tool materials commonly used when machining stainless steels?
5.5.32 Describe the three classes of PH-SSs.
5.5.33 Holes of 6 mm diameter and 25 mm depth are to be drilled into AISI-304. Estimate the recommended cutting speed (m/min), and the recommended feed (mm/rev). Calculate the spindle speed (rpm), feed (mm/min), drilling time (min), and consumed power if the specific cutting energy of the SS = 3700N/mm^2. Check whether the recommended values are feasible on commercially available equipment.
5.5.34 Describe how the machinability can be rapidly assessed using a simple drill-penetration test.

5.5.35 Mark [T] for true statements and [F] for false statements.

[ ] Austenitic SSs are non-magnetic and hardenable by heat treatment.

[ ] Martensitic alloy is one class of PH-SSs.

[ ] Ferritic alloy is one type of PH-SS.

[ ] Ferritic grades of SSs are only hardenable by cold working.

[ ] HSS and carbide tools are most widely used in machining SSs.

[ ] HSS tools are seldom used to machine SAs. They may be only used for interrupted cuts.

[ ] HSS broaches are used for broaching SAs and SSs.

[ ] HSS drills and milling cutters cannot be used for machining SAs.

5.5.36 What are the main factors that generally affect the machining characteristics of super alloys?

5.5.37 What are the practical guidelines to be observed when machining super alloys?

5.5.38 What are considerations to be observed when broaching super alloys?

5.5.39 Differentiate between super alloys and refractory metals. Why is the latter not widely used in aircraft applications?

5.5.40 Numerate three different types of each basic group of super alloys.

5.5.41 Define the machinability of a material. What are the assessments used for its evaluation?

5.5.42 What are the important techniques that enhance the machinability of DTC materials?

5.5.43 Provide your justification why the tool wear during UAM is lower compared to traditional machining.

REFERENCES

All Metals & Forge Group. LLC Fairfield, New Jersey 07004 USA, viewed 30 Nov 2011, <http://steelforge/machinability ratings.html>.

ASM International 1989, *Machining Vol. 16, metals handbook*, ASM International, Materials Park, OH.

Bagley, JA 1969, *Plasma arc cutting, technical paper MR69-578*, Society of Manufacturing Engineers, Dearborn, MI, p. 23.

Bellows, G 1967, *ECM machinability data and ratings*. Technical Paper, SME, Metcut, Dearborn.

British Stainless Steel Association <www.bssa.org.uk/topics.php.? article=194-2007/2012>.

Carpenter Technology Corporation Philadelphia, PA 19103 USA, <www. Cartech.Com/techarticles.aspx?id=1578>.

DeGarmo, EP, Black, JT & Kohser, RA 2012, *Materials and processes in manufacturing*, 11th edn, John Wiley & Sons, Inc.

Falcon Metals Group, <www.falconmetals.com/stainlessandalloys/ stainlesssteel/stainlessmachinabilityratings.htm>.

High Performance Alloys, Inc. Tipton, IN 46072 USA, viewed 30 Nov 2011, <http:www.hpalloy.com/alloysdescriptions/Machinability Ratings. Html>.

Holden, S 1985, *The plasma cutting of SS*, vol 13, no. 74, Stainless Steel Industry, p. 12–25.

Kalpakjian, S & Schmid, SR 2003, *Manufacturing processes for engineering materials*, 4th edn, Prentice Hall, Inc., Upper Saddle River, NJ.

Klocke, F & König, W 2007, *Fertigungsverfahren 3: Abtragen, Generierenund Lasermaterialbearbeitung*, Springer, Berlin. ISBN 3-540-23492-6.

Klocke, F, Zeis, M, Klink, A, & Veselovac, D 2013, Technological and economical comparison of roughing strategies via milling, sinking-EDM, wire-EDM and ECM for titanium- and nickel-based blisks. *CIRP JMST*, vol. 6, no. 3, pp. 198–203.

Machining Data Handbook 1980, *Machinability data center*, vol. 2, 3rd edn. Metcut Research Associates, Inc., Cincinnati, OH.

Neppiras, EA & Foskett, RD 1958, *Ultraschall-Metallbearbeitung: Prinzip und Apparatur*, Philips Rundschau, Heft 2, p. 37–68.

Sandvik Coromant 2013, *Workpiece materials-ISO-S HRSA titanium.*

Saunders, RJ 1984, *Laser metalworking, Metal Progress*, p. 51.

Schwartz, BL 1985, 'Principles and applications of AWJ-cutting'. High Productivity Machining Materials and Processes, *ASM*, p. 291–298.

Seco Tools 2014, *Technical guide, turning difficult-to-machine alloys.*

Thermal Dynamics, Ontario, Canada 91761.

Visser, A 1966, 'Werkstoffabtrag mittels Elektronenstrahl', Dissertation, T.H. Braunschweig, Germany.

Visser, A, Junker, M & Weissinger, D 1994, *Sprühätzen mittallischer Werkstoffe*, 1st edn, Eugen G. Leuze Verlag, Bad Saulgau, Germany.

Youssef, H 2005, *Nontraditional machining processes – theory and practice*, Alfath Publisher, Alexandria, Egypt.

Youssef, H 2016, *Machining of stainless steels and super alloys: Traditional and nontraditional techniques*, JohnWiley & Sons.

6 Machining of DTC Materials (Ceramics and Composites) by Traditional and Non-Traditional Methods

6.1 INTRODUCTION

Due to their wide range of attractive properties, non-metallic materials have always played a significant role in manufacturing. More recently, the non-metallic materials family has greatly expanded from natural materials to include listing of plastics, elastomers, ceramics, and composites. Most of these are manufactured, so a wide variety of properties and characteristics can be obtained. New materials and variations are being created on a continuous basis, and their uses and applications are widely expanding. Scientists and material engineers now refer to a *material revolution* as these new materials are developed, and research efforts have brought their cost into competition with metallic engineering materials. New products have been designed to utilize the new properties, and existing products are continually reevaluated for the possibility of material substitute. As the design requirements of products continue to push the limits of traditional materials, the role of manufactured non-metallic materials will continue to expand.

In this chapter, the machining and machinability aspects of two DTC materials of non-metallic materials are considered. These are ceramics and composites. Both have an expanding use in industrial applications. This chapter provides a solution for most of problems facing production engineers and relates to both traditional and non-traditional machining domains.

6.2 MACHINING OF CERAMIC MATERIALS

6.2.1 Ceramic as a Promising Engineering Material

Metals are usually considered to be the most important class of engineering materials. However, it is of interest to note that ceramics are actually more abundant and widely used. Included in this category are clay products (e.g. bricks and pottery), glass, cement, and more modern ceramic materials such as tungsten carbide and cubic boron nitride. This is the class of materials whose machinability will be discussed in this section. The importance of ceramics as engineering materials derives from their abundance in nature and their mechanical and physical properties, which are quite different from those of metals.

Ceramic has traditionally been defined as an inorganic, non-metallic solid that is prepared from powdered materials, fabricated into products through the application of heat and displays characteristic properties such as hardness, strength, stiffness, brittleness, and of low thermal and electrical conductivity. Ceramic materials are strong in compression, weak in shearing and tension. They withstand chemical erosion that occurs in an acidic or caustic environment. In many cases, they withstand erosion from the acid and bases applied to them. Ceramics generally can withstand very high temperatures ranging from 1,000°C to 1,600°C. However, their properties vary widely. For example, porcelain is widely used to make electrical insulators, but some ceramic compounds are superconductors.

The word ceramic comes from the Greek word *keramikos*, which means "pottery." Ceramics are typically crystalline in nature and are compounds formed between metallic and non-metallic elements such as aluminum and oxygen (alumina-Al_2O_3), calcium and oxygen (calcia-CaO), and silicon and nitrogen (silicon nitride-Si_3N_4). The atoms in ceramic materials are held together by a chemical bond. Briefly though, the two most common chemical bonds for ceramic materials are covalent and ionic. Covalent and ionic bonds are much stronger than in metallic bonds and, generally speaking, this is why ceramics are brittle and metals are ductile.

Among the oldest raw materials of ceramics is clay. It is of a fine-grained structure of more complex compound, hydrous aluminum silicate ($Al_2Si_2O_5(OH)_4$), a white clay mineral known as *kaolinite*, consisting of silicate of aluminum with alternating weakly bonded layers of silicon and aluminum ions, Figure 6.1. When added to kaolinite, water attaches to the layers, making them slippery and giving the wet clay its hydro-plasticity that makes it formable.

Other major raw materials for ceramics that are found in nature are flint (rock of fine-grained silica, SiO_2), and feldspar (a group of crystalline minerals consisting of aluminum silicates, potassium, calcium, or sodium). In their natural state, these raw materials contain impurities, which have to be removed prior to further processing into useful products of reliable performance. The modern ceramic materials, which

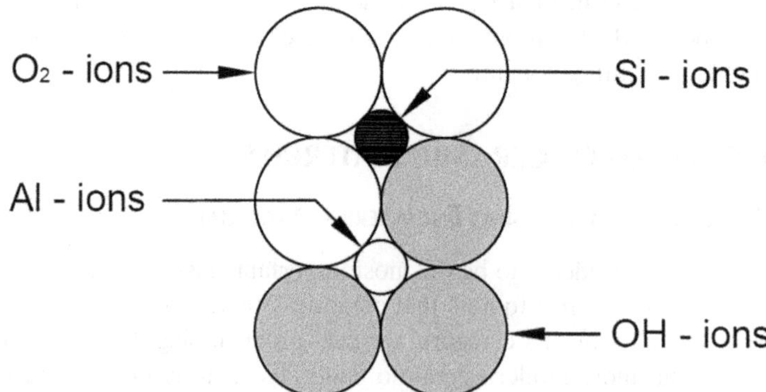

FIGURE 6.1 Hydrous aluminum silicate ($Al_2Si_2O_5(OH)_4$), a clay mineral known as kaolinite.

are classified as advanced ceramics, include silicon carbide and tungsten carbide. Both are valued for their abrasion resistance, and hence find use in applications such as tooling in the metal working industry and wear plates of crushing equipment in mining operations. Advanced ceramics are also used in the medicine, electrical, and electronics industries.

6.2.2 Types, Characteristics, Classification, and Applications of Ceramics

6.2.2.1 Types, Characteristics, and Classification

There are various classification systems of ceramic materials, which may be attributed to one of two principal categories. These are the composition base system, Figure 6.2, as oxides, silicates, nitrides, carbides, sulfides, fluorides, etc., and the application base system, Figure 6.3, as traditional and advanced ceramics.

The earliest ceramics, called traditional, were pottery objects made from clay and then glazed and fired to create a colored surface. These were generally used as domestic and art products. In the 20th century, advanced ceramic materials were developed, which found far wider applications in engineering than traditional porcelain and pottery. Table 6.8 shows some classifications and properties of advanced ceramics and metallic materials. The advanced ceramics are classified as structural ceramics (i.e. engineering ceramics) and functional ceramics. Functional ceramics are generally employed as a part of electronic components due to their inherent physical features, such as electric, magnetic, dielectric, ferroelectric, optical, or other properties, which play an active role in the electronic industry. Engineering ceramics are applied as structural components in most engineering industries. Compared with other engineering materials (e.g. metals or polymers), engineering ceramics offer numerous enhancements in performance, durability, reliability, chemical stability, hardness, mechanical strength at elevated temperatures, wear resistance, and thermal resistance.

Table 6.1 also visualizes the advantages and disadvantages of ceramics as compared to metals. The advantages include higher melting points, stiffer, harder, more

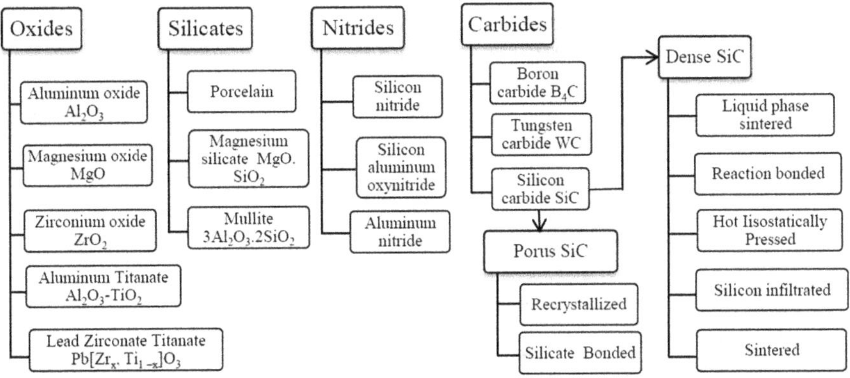

FIGURE 6.2 General classification of ceramics composition based (From Encyclopedia Britannica. Britannica.com).

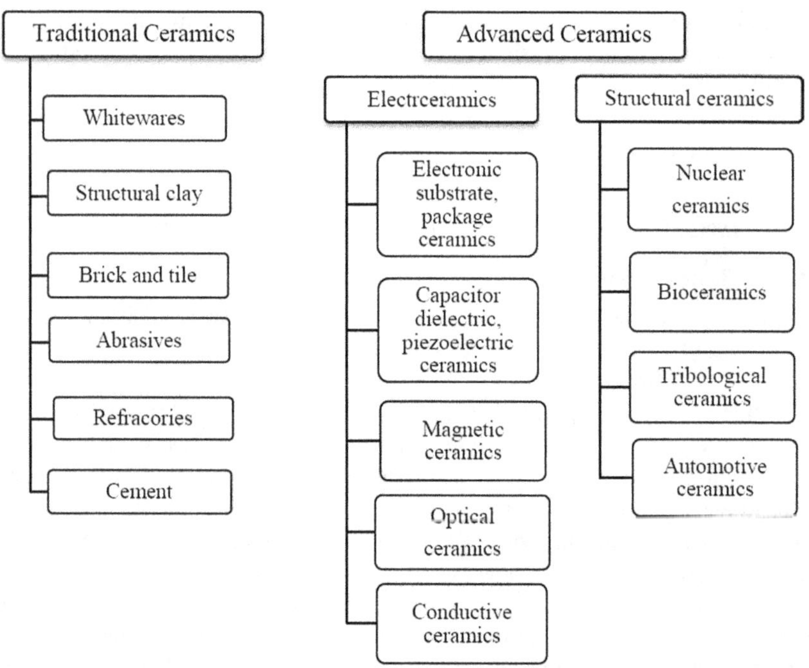

FIGURE 6.3 General classification of ceramics application based (From Encyclopedia Britannica. Britannica.com).

TABLE 6.1
Properties of Some Ceramic and Metallic Materials

Material	Melting Point (°C)	Density (g/cm³)	Elastic Modulus (GPa)[a]	Hardness (Mohs)[b]
Oxide ceramics:				
Alumina, Al_2O_3	2054	3.97	380	9
Beryllta, BeO	2578	3.01	370	8
Zirconia, ZrO_2	2710	5.68	210	8
Non-oxide ceramics:				
Boron carbide, B_4C	2350	2.5	280	9
Silicon carbide, SiC	2830	3.16	400	9
Silicon nitride, Si_3N_4	1900	3.17	310	9
Metals:				
Aluminum	660	2.70	70	3
Plain carbon steel	1515	7.86	205	5

[a] The elastic modulus, measured in units of pressure (1 GPa = 109 Pa) indicates the stiffness of a material when it is subjected to a load.

The larger the value, the stiffer the material.

[b] Number on the Mohs hardness scale range from 1 for talc (a very soft material) to 10 for diamond (the hardest known substance).

Source: (Samant and Dahotre, 2009).

resistant to wear and corrosion, and maintaining strength at high temperatures. A significant disadvantage of ceramics is their brittleness: Ceramics easily shatter and break in the case of impact and dynamic loads.

Ceramics have a unique combination of mechanical, physical, and chemical properties:

- High strength
- Hardness
- Low density
- High stiffness (modulus of elasticity)
- Good tribological properties (e.g. excellent resistance to different types of wear)
- Very low electric conductivity
- Very low thermal conductivity
- High refractoriness and thermal stability
- Good corrosion resistance

Grain size has a major influence on the strength and other properties of ceramics: the finer the grain size, the higher the strength and toughness. Hence fine ceramics acquire vital importance in specific and engineering applications. On the other hand, traditional or coarse ceramics are used in manufacturing of products such as white ware, dinnerware, tiles, bricks, and pottery. The material strengths of ceramics have been improved considerably over the last few decades. These developments have already led to increased use of ceramic materials in applications where metallic components achieve only unsatisfactory service lives, owing to their corrosive wear or inadequate temperature resistance. The most commonly used engineering ceramics are: Alumina (Al_2O_3), zirconia (ZrO_2), aluminum nitride (AlN), silicon nitride (Si_3N_4), silicon carbide (SiC), reaction bonded silicon nitride (RBSN), and hot pressed silicon nitride (HPSN). Some of their applications are listed in Table 6.2.

TABLE 6.2
Performance Advantages of Some Commonly Used Engineering Ceramics

Application	Performance Advantages	Ceramics
Wear Parts: Seals, bearing, vales, nozzle	High hardness, low friction	SiC, Al_2O_3
Cutting tools	High strength, hardness	Si_3N_4
Heat Engines: Diesel components, gas turbines	Thermal insulation, high temperature strength, fuel economy	ZrO_2, SiC, Si_3N_4
Medical Implants: Hips, teeth, joints	Biocompatibility, surface bond to tissue, corrosion resistance	Hydroxyapatite, Bioglass, Al_2O_3, ZrO_2
Construction: Highways, bridges, buildings	Improved durability, low overall cost	Advanced cements and concrete

Source: (Samant and Dahotre, 2009).

Types of advanced ceramics (highly refined ceramics): These are:

1. **Oxides** (typical examples include alumina and zirconia)
   Alumina (also called corundum, emery, aluminum oxide, Al_2O_3)
   It is the most widely used either in pure form or as a raw material mixed with other oxides. It has high hardness and moderate strength. In its natural form, it contains varying amount of impurities making it of non-uniform properties of unreliable behavior. Al_2O_3, as well SiC and many other ceramics, is now almost totally manufactured synthetically to control the quality. Synthetic alumina (first made in 1893) is obtained by fusion of molten bauxite (aluminum oxide ore), iron filings, and coke in electric furnaces. The material is then crushed, and mesh graded. Parts made from alumina are cold pressed and sintered to produce white ceramics. Their properties are enhanced by minor additions of other ceramics, such as titanium oxide and titanium carbide. The products are known as mullite (spinel), which are used as refractory materials for high temperature applications.
   Zirconia (zirconium oxide, ZrO_2)
   It is white in color, has good toughness, good resistance to thermal shocks and corrosion, low thermal conductivity, and low coefficient of friction. A more recent development is partially stabilized zirconia (PSZ), which performs more reliably than zirconia due to higher strength and toughness. Typical applications of PSZ include dies for hot forming. Another important characteristic of PSZ is that its thermal conductivity is about one-third that of other ceramics. Consequently, it is suitable for heat engine components, such as cylinder liners and valve bushing. A new further development is transformation-toughened-zirconia (TTZ), which has higher toughness than that of PSZ.
2. **Carbides** (typical examples include WC, TiC, and SiC)
   Tungsten carbide consists of WC powder with cobalt (Co) as binder. Co has a major influence on the properties of this ceramic. Toughness increases with Co content, whereas both hardness and strength decrease.
   Titanium carbide consists of TiC powder with nickel, and molybdenum as binder. It is not as tough as WC, but harder.
   Silicon carbide (SiC) has good wear, thermal shock, and corrosion resistance. It has low coefficient of friction and retains its strength at elevated temperature. It is suitable for high temperature components in heat engines and is also used as an abrasive. Synthetic SiC is made from silica sand, coke, and small amounts of nail and saw dust, in a similar way to that of synthetic Al_2O_3.
3. **Nitrides** (typical examples include CBN, TiN, and Si_3N_4)
   Cubic boron nitride (CBN) is the second hardest known substance after diamond. It does not exist in nature and was first made synthetically in the 1970s, with similar techniques as synthetic diamond.
   Titanium nitride (TiN) is extensively used as coatings on cutting tools to improve tool life due to its low coefficient of friction.

Silicon nitride (Si_3N_4) has high creep resistance at elevated temperatures, low thermal expansion, and high thermal conductivity. Hence, it resists thermal shocks. Therefore, it is suitable for high temperature applications, such as in automotive engines and gas turbine components, bearings, and components for the paper industry.

Sialon and cermets are used as materials for cutting tools and cutting tips for machining DTC materials.

4. **Silica**

   It is abundant in nature as a polymorphic material; that is, it can have different crystal structures. Most glasses contain more than 50% silica. The most common form of silica is quartz, which is a hard-abrasive hexagonal crystal. It is used extensively as oscillating crystal in communications, since it exhibits piezo-electric characteristics. Silicates are products of the reaction of silica with oxides of aluminum, calcium, magnesium, potassium, sodium, and iron. Examples include clay, asbestos, mica, and silicate glasses. Silica is used in high temperature applications, and non-structural applications such as catalyst converters, and heat exchanger components.

6.2.2.2 Fields of Applications

Due to the high hardness, higher strength, high temperature strength, high strength to weight ratio, biocompatibility, lower thermal and electrical conductivity, superior chemical stability as well as wear resistance, ceramics materials have become increasingly popular and have found their applications not only in the cutting tool industry but also in the biomedical and aerospace industries. These properties meet the demands of the manufacturing of high-quality micro-systems as the mechanical components are constantly exposed to high temperatures and mechanical loads. Biomedicine has a strong demand for tough and stable bio-inert ceramics that is met by nano-structured ZrO_2 or Al_2O_3. These ceramics are used for the fabrication of dental implants and restorations, ace tabular cups, and femoral heads for total hip replacement, bone fillers, and scaffolds for tissue engineering.

Therefore, ceramics are successfully used in the different branches of industries including:

Mechanical engineering: Cutting tools and dies, abrasives, precise instrument parts, molten metal filter, turbine engine components, low weight components for rotary equipment, wearing parts, bearings, seals

Aerospace: Fuel systems and valves, power units, low weight components, fuel cells, thermal protection systems, turbine engine components, combustors bearings, seals, and structures

Automotive: Heat engines, turbines, catalytic converters, drive train components, fixed boundary recuperators, fuel injection components, turbocharger rotors, low heat rejection diesels, water pump seal

Defense industry: Tank power trains, submarine shaft seals, improved armors, propulsion systems, ground support vehicles, military weapon systems, military aircraft (airframe and engine), wear-resistant precision bearings

Biological, chemical processing engineering: Artificial teeth, bones and joints, catalysts and igniters, heart valves, heat exchanger, reformers recuperators, refractories, and nozzle

Electrical, magnetic engineering: Memory element, varistor sensor, resistance heating element, integrated circuit substrate, multilayer capacitors

Nuclear industry: Nuclear fuel, nuclear fuel cladding, control materials, moderating materials

Oil industry: Bearings flow control valves, pumps, refinery heater, blast sleeves

Electric power generation: Bearings ceramic gas turbines, high temperature components, fuel cells (solid oxide), filters

Optical engineering: Laser diode, optical communication cable, heat-resistant translucent, porcelain, light-emitting diode

Thermal engineering: Electrode materials, heat sink for electronic parts, high-temperature industrial furnace lining

Some examples of commonly used ceramics are alumina, aluminum nitride, boron carbide, silicon nitride, zirconia, sialon, titanium carbide, titanium nitride, zirconium nitride, and silicon carbide, etc. Properties and applications of some ceramics are provided in Table 6.3. Nevertheless, the properties that have made ceramic materials one of the most desirable engineering materials also hinder their machining characteristics. The main drawback of ceramics lies in their cost and the complex manufacturing cycle, essentially during the finishing step.

6.2.3 Fabrication Techniques of Crystalline Ceramics

The high-performance material properties of ceramics make them attractive for uses but also make them difficult to fabricate. Moreover, the damage caused by machining may affect the performance of the final ceramic components. Therefore, cost-effective machining without significant reduction of the outstanding material properties is a crucial step for ceramic fabrication. Over the past few decades, numerous efforts have been performed to improve the process quality and several advanced machining techniques with reduced deleterious effects on material properties have been developed (Tuersley et al., 1994).

Figure 6.4 shows a typical route of forming engineering ceramics. The first step is powder synthesis and treatment. The second step involves the consolidation of the treated powder into an expected shape that is known as the "green body." The green body typically contains about 50 vol % porosity and is extremely weak, and machinable. The last step utilizes heat, or heat and pressure combined, to bond the individual powder particles, remove the free space and porosity in the compact via mass diffusion, and create a fully dense and well-bounded ceramic component (Kutz, 2002).

This process is known as sintering or firing. The sintered ceramics can be machined but it is very difficult. Figure 6.5 also summarizes the most commonly used techniques for machining of engineering ceramics in the green body form and in the sintered form. The machining of ceramics is very expensive and time-consuming operation representing from 50 to 90% of the total cost of the part.

TABLE 6.3
Application and Performance Properties of Advanced Ceramics Materials

Applications	Performance Properties	Ceramics
Wear parts: Seals, bearings, valves, fuel nozzles, aerospace industry, cutting tool inserts, automotive brakes, prosthetic products, piezoceramic sensors, biomedical implants, mold-dies, heat engines, next generation computer memories	High hardness, lower friction, high thermal conductivity, high stiffness, and low density	SiC, Al_2O_3
Cutting tools, gas turbine impeller manufacturing	High strength, high hardness, thermal shock, and oxidation resistance	Si_3N_4
Heat engines: Diesel engines components, gas turbines	Thermal insulation, high temperature strength, fuel economy, exceptional high fracture resistance, good corrosion resistance	ZrO_2, SiC, Si_3N_4
Medical implants: Hip joint, teeth, other joints	Biocompatibility, machined surfaces' bond to tissue, corrosion resistance	Hydroxyapatite, bioglass, Al_2O_3, ZrO_2
Ballistic applications, shielding in nuclear fission reactors, bearings, dies, cutting tools, extrusion nozzles, seals and rings	Excellent hardness, wear resistance, fracture toughness properties, low density, high compressive strength, high elastic modulus	B_4C
Construction: Highways, bridges, buildings	Improved durability, low overall cost	Advanced cements and concrete

Source: (Lu et al., 1999; Vikulin et al., 2004).

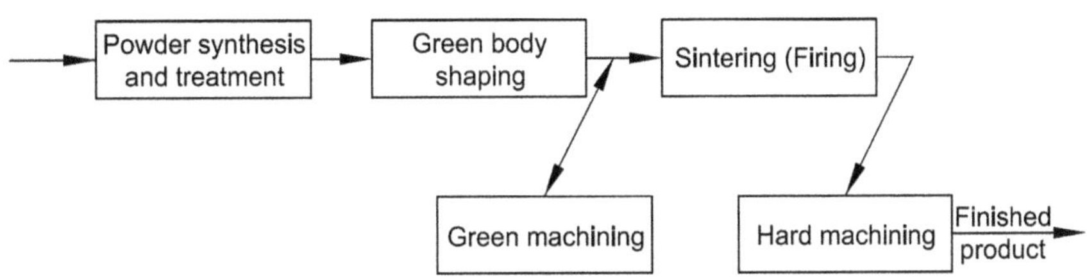

FIGURE 6.4 Simplified route of forming engineering ceramics.

6.2.3.1 Processing Techniques and Shaping of Green Bodies

Generally, the processing involves: Crushing or grinding the raw materials into very fine particles; mixing the particles with additives to impart certain desirable characteristics; and shaping, drying, and sintering the powder. Crushing of the raw materials is generally done in ball mills, either dry or wet. Wet crushing is more effective, because it keeps particles together and prevents the fine powder elevated in air. The ground particles are then mixed with additives. The ceramic green body can be directly fabricated to the final shape before sintering. The desired shapes can be easily formed due to the low mechanical strength of weakly bonded ceramic powders in the "green body" phase. The fabrication techniques under this category include: Pressing, extrusion, slip casting and gel casting, injection molding, and tape casting (Figure 6.5).

Drying and firing: After ceramic has been shaped, the next step is to dry and fire the part to give it the proper strength. Drying is a critical stage, because of the tendency for the part to warp or crack from variations in the moisture content and thickness and complexity of the part shape. Control of atmospheric humidity and temperature is important to reduce warping and cracking. Loss of moisture results in shrinkage of the part by as much as about 20% of its original moist size. The dried part (called green) can be machined relatively easily to bring it closer to the final shape; however, it must be handled carefully.

Firing (sintering) involves heating the part to an elevated temperature in a controlled environment, similar to sintering in powder metallurgy. Some shrinkage occurs during firing. Firing gives ceramic its strength and hardening. The enhancement of properties results due to:

- Development of strong bond between complex oxide particles in ceramic
- Reduced porosity

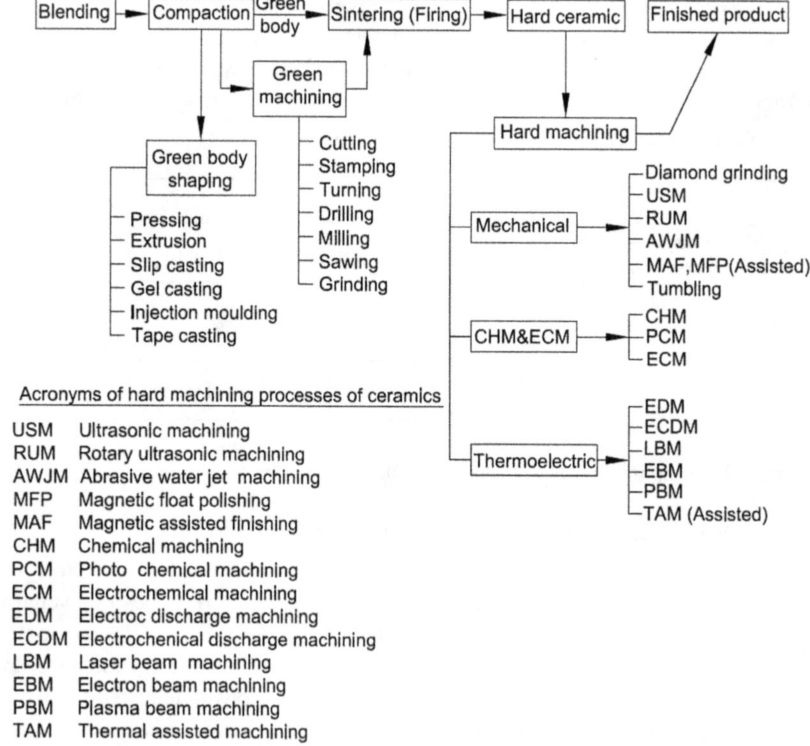

FIGURE 6.5 Detailed route of forming and machining of engineering ceramics.

Dimensions and shrinkage: The primary goal, when forming a green body, is to come as close as possible to the final dimensions and shapes, since it is costly to post-machine hard and brittle sintered ceramics. Basically, the green body is formed larger than the final dimensions due to a reduction in volume (known as shrinkage) during binder removal, drying, pre-sintering and sintering. Table 6.4 lists the typical shrinkage percentages of some advanced ceramics. Therefore, molds are usually over-dimensioned with respect to the desired geometries, in order to compensate for the shrinkages. However, the tolerances of sintered ceramic parts via industrial standard processes are approximately ±2%.

6.2.3.2 Green Machining Processes of Green and Pre-Sintered Ceramics

The machining of a ceramic in the pre-sintered state is called green machining. Sintered ceramics are very hard and therefore their machining is an expensive, difficult, and time-consuming process. Therefore, ceramic parts are conventionally machined before their final sintering stage either in the "green" (non-sintered powder) compact state or in the pre-sintered "bisque" state. The low mechanical strength of a green body gives green machining an extremely high material removal rate with relatively low tool wear. Titanium nitride (TiN)-coated high-speed steel tools, tungsten carbide tools, and polycrystalline diamond (PCD) tools are used. The material removal rate (MRR) is 10 cm^3/min, which is similar or even higher than the MRR of tool and die steels (SubsTech). Many machine tools have been used in green machining, such as computer numerical control (CNC) milling, lathes, and drilling equipment, cut-off saws, and surface grinders. Laser beams and electron beams have also been applied to drill micro-through-holes in green ceramic electronic substrates.

The glass-ceramic can be machined using conventional HSS tools although it is recommended to use carbide tools for more extended tool life. Furthermore, it is recommended to use water-soluble coolants designed for sawing and grinding glass or ceramics (Präzisions Glas & Optik GmbH, 2020).

Typical machining conditions for green machining are given below:

Sawing: Use a carbide grit blade at a band speed of 30 m/min. An alternative is a silicon carbide or diamond cut-off wheel.

Turning: Cutting speed 10–15 m/min; Feed rate 0.05–0.13 mm/rev; Depth of cut 4–6.4 mm

TABLE 6.4
Typical Shrinkages of Some Advanced Ceramics

Ceramic	Shrinkage Vol. %
Sintered silicon carbide (SiC)	18–20
Alumina (Al_2O_3)	17–20
Zirconia (ZrO_2)	25–32

Source: Adapted from Brevier Technical Ceramics.

Milling: Cutting speed 6–11 m/min; Chip load 0.05 mm/tooth; Depth of cut 3.8–5.1 mm

Drilling:

Drill Size (mm)	6	12	19	25	50
Spindle Speed (rpm)	300	250	200	100	50
Feed Rate (mm/rev)	0.13	0.18	0.25	0.30	0.35

Grinding: Diamond, silicon carbide, or aluminum oxide grinding wheels can be used.

6.2.3.3 Hard Machining Processes of Sintered Ceramics

After firing, additional operations may be performed:

- To give the part its final shape
- To remove surface flaws
- To improve surface finish and dimensional accuracy

In order to meet the tighter requirement of dimensional precision and surface finish quality as well as the variety of geometries, post-sintering-machining techniques should be employed. However, the extremely high hardness of sintered ceramics makes this process slow and costly (hence why it is called hard machining).

The operations that are commonly used in hard machining are:

- Diamond grinding, honing, lapping
- USM, RUM
- AWJM
- MAF, MAP (magnetically assisted)
- Tumbling
- CHM, PCM
- ECM
- EDM
- ECDM
- LBM, EBM, PBM
- TAM (thermally assisted)

The most used techniques in this subcategory include diamond grinding, ultrasonic and rotary ultrasonic machining and water jet machining. The choice of the process is critical in view of the brittle nature of ceramics and additional costs involved in these processes. The effect of these processes on the properties of the product must also be considered; because of notch sensitivity, the finer the finish, the higher the part strength. To improve the appearance and strength, ceramic products are often coated with a glaze material, which forms a glassy coating after firing.

1. **Diamond Grinding**

Diamond grinding is considered to be the most desirable and reliable technique for ceramic machining. This technique is capable of producing complex profiles and fine surface finish. Unfortunately, the major drawbacks of the technique are the extremely long machining time and the high machining cost, amounting to 60–90% of final costs (Samant and Dahotre, 2009). Plastic deformation and surface residual stresses after machining also degrade the final quality. However, it has to be employed as the final grinding step for good surface finish and high dimensional precision. Grinding is the most widely used method of machining of ceramics in the sintered state. The material removal rate (MRR) of grinding ceramics reaches 10 mm^3/min. The grinding zone is continuously flushed with a fluid coolant, which cools the grinding zone, lubricates the contact between the wheel and the part surfaces, and removes the micro-chips (debris) produced in the grinding process. Resin-bond wheels with either synthetic or natural diamond of different grit size pressed at different concentrations in polymer (resin) matrices are commonly used for grinding ceramics.

Ceramic materials are more brittle than metals and show very little plastic deformation under load up to the point of fracture. Therefore, the metal removal mechanism during abrasive machining involves brittle fracture, crack formation, separation, and spalling of the material. Another model assumes that the ceramic is softened by the local temperature at the machining zone. Under such conditions, ceramic becomes plastically deformed and can be machined in a similar matter to other materials. Ductile regime grinding is used to describe the material removal mechanism under suitable conditions. There is a transition depth at which the material removal mechanism switches from the ductile to the brittle regime. Such a depth is a function of the material properties governing plastic deformation and fracture. Ductile mode grinding enhances the surface quality, but it is very slow and costly. In reality a combination of the two modes, ductile and brittle, depending on the type of ceramic and the machining conditions, is involved during grinding of ceramic materials (El-Taybany et al., 2019; Abdelkawy et al., 2018).

The application of high-speed grinding of advanced ceramics either improved the surface quality or increased the machining efficiency. This was because the high wheel velocity reduced the undeformed chip thickness, thus resulting in a decreased normal grinding force. The temperature measured in the grinding zone was in the range of 150 to 300°C; not considerably high, provided that the coolant supply was sufficiently effective. Nevertheless, to realize the potential of the high-speed grinding technology, great care had to be taken on the grinding-induced vibration. The reduction of the grinding-induced vibration required a highly rigid machine, appropriate preparation of grinding wheels, and accurate dynamic balancing of the wheels used. Coolant supply was another key factor in the high-speed grinding of advanced ceramics. It was extremely important that the coolant could effectively enter the grinding zone when grinding advanced ceramics at relatively high speeds (Huang, 2009).

Electrolytic in-process dressing (ELID) is currently used for grinding hard ceramics. The metallic bond grinding wheel is made anodic and spaced 0.1 mm from a secondary cathodic electrode of graphite, stainless steel, or copper, as shown in Figure 6.6. A power source provides pulsed dc. During the grinding stage, an oxide layer is formed over the bond that prevents the grinding chips from adhering to the wheel. As the grain becomes worn, the insulating oxide layer also becomes worn. Fresh oxidation of the wheel occurs in a self-regulating fashion, providing continuous exposure of fresh cutting points. High accuracy, good surface finish, and low subsurface damage are all achieved. Most ELID applications lie within ceramics, optical materials, and bearing steel. Surface finishes of 0.011–0.36 μm R_a are possible. ELID ensures the following advantages (Ioan et al., 2016):

- Elimination of dressing time
- Elimination of conventional dressing tools
- Elimination of physical or mechanical damage to the dressed grains
- Extremely smooth surfaces are achieved by the use of ultrafine grains
- High precision parts are achieved with excellent process efficiency

2. **Ultrasonic Machining (USM) of Ceramics**

Although the principle of ultrasonic machining was recognized in 1927, the first useful description of the USM technique wasn't given in industry literature until about 1940. Since then, ultrasonic machining has attracted a great deal of attention and has found its way into industry on a relatively wide scale. By 1953–1954, the first ultrasonic machine tools, mostly based on drilling and milling machines, had been built. By about 1960, ultrasonic machine tools of various types and sizes for a variety of purposes had been seen, and some models had begun to come into regular production.

Ultrasonic machining (USM) of ceramics is the machining method using the action of a slurry containing abrasive particles flowing between the workpiece and a tool vibrating at an ultrasonic frequency. The tool vibrates

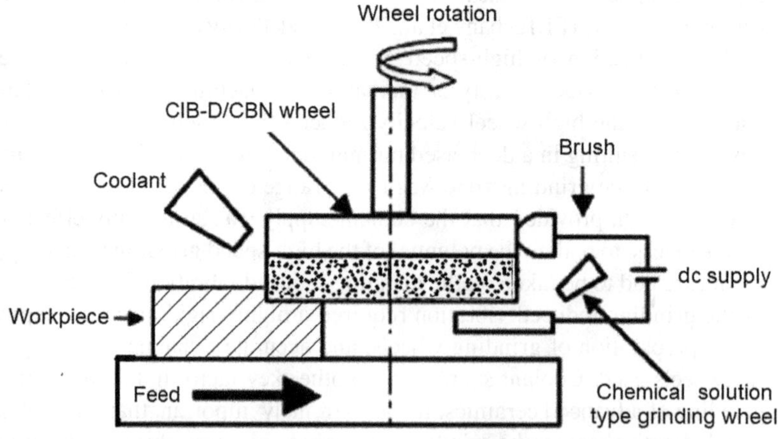

FIGURE 6.6 ELID process of grinding ceramics (El-Hofy, 2018).

at a frequency of 19~25 the amplitude of vibration 13–50 μm. During the operation, the tool is pressed to the workpiece at a constant load. As the tool vibrates, the abrasive particles (grits) dispersed in the slurry strike the ceramic workpiece and remove small ceramic debris fracturing from the surface. Conventional USM is characterized by low material removal rates up to 50 mm^3/min. USM is commonly used for drilling operation.

In ceramic applications, USM provides a number of advantages compared to conventional machining techniques. Both conductive and non-conductive materials can be machined, and complex three-dimensional contours can be machined as quickly as simple shapes. Additionally, the process does not produce a heat-affected zone or cause any chemical/electrical alterations on the workpiece surface, and a shallow, compressive residual stress generated on the workpiece surface can increase the high-cycle fatigue strength of the machined part. Ultrasonic machining could reduce the damage in machining due to the low normal forces compared with diamond grinding.

Disadvantages of USM are low accuracy and high tool wear. In USM, the slurry has to be fed to and removed from the gap between the tool and the workpiece. As a result, the material removal rate slows considerably and even stops as the penetration depth increases. The slurry can also wear the wall of the machined hole as it passes back toward the surface, which limits accuracy, particularly for small holes. Additionally, the abrasive slurry "machines" the tool itself, which causes considerable tool wear and, in turn, makes it difficult to hold close tolerances.

3. **Rotary Ultrasonic Machining**

Rotary ultrasonic machining (RUM) provides a fast, high-quality machining method for many ceramic and glass applications. RUM is a hybrid machining process that combines the material removal mechanisms of diamond grinding with ultrasonic machining (USM), resulting in higher material removal rates (MRR) than those obtained by either diamond grinding or USM alone. Experiments with calcium aluminum silicate and magnesia-stabilized zirconia have shown that the MRR obtained with RUM is six to ten times higher than that of a conventional grinding process under similar conditions, and it is about ten times faster than USM. It is also easier to drill deep holes with RUM than with USM, and the hole accuracy is improved. Other advantages of this process include a superior surface finish and a low tool pressure. However, RUM does not have the advantage of producing non-circular and complicated shapes as USM.

In RUM, a tool is rotating and simultaneously vibrating at an ultrasonic frequency. The tool is continuously fed toward the ceramic workpiece causing abrasive action performed by the rotating-vibrating diamond grits. A cutting fluid is continuously flowing through the core of the tool to the machining zone, thus performing cooling and removing the debris produced by the grinding process. RUM is much more effective than conventional ultrasonic machining. One of the major differences between USM and RUM is that USM uses a soft tool, such as stainless steel, brass, or mild steel, and a slurry loaded with hard abrasive particles, while in RUM the hard-abrasive particles are diamond and are bonded on the tools. Another major difference is that the RUM tool rotates and vibrates simultaneously, while the USM tool only vibrates. These differences enable RUM to provide both speed and accuracy advantages in ceramic and glass machining operations. The RUM material removal rate is up to 500 mm^3/min. A variety of tool shapes are used for rotary ultrasonic machining, and ceramic and technical glass machining applications typically use either a diamond-impregnated or electroplated tool. Diamond-impregnated tools are more durable, but electroplated tools are less expensive, so the selection depends on the particular application.

During RUM of advanced ceramics, the MRR increases with increasing the ultrasonic power and amplitude of tool vibration, rotational speed, and grain size. The surface roughness or hole clearance tends to increase with the increase of vibration amplitude and abrasive grit size but decrease with high applied static load. The reported coolant types have no significant effect on the MRR and surface roughness but provide better performance at certain pressure. Wear occurs on both the end face and lateral face of the tool in RUM of advanced ceramics. The tool wear in terms of the diamond grain dislodgment on the end face is more serious than the lateral face. Mechanical impact and high temperature may contribute to the weakening of the interfaces between diamond grains and metal bond. Edge chipping is an unavoidable phenomena in drilling hard and brittle materials, but the edge chipping thickness could be decreased by increasing the support length (Esah Hamzah et al., 2008). In rotary ultrasonic core drilling of a ceramic material (92% alumina), Jiao et al. (2006) concluded that the cutting force is one of the important output variables in RUM. The spindle speed and the feed rate have significant effects on the cutting force; higher spindle speed and lower feed rate result in a smaller cutting force. Only feed rate has significant effects on the MRR. The feed rate, the spindle speed, and the grit size have significant effects on surface roughness.

4. **Abrasive Water Jet Machining**
   Abrasive water jet machining (AWJM) employs a high-pressure water jet seeded with abrasive particles impinging on the surface of the workpiece to remove material. The main advantage of this technique is the relatively high cut quality, low heat-affected zone, and low tool wear. However, the machining rate is relatively low since ceramics have a similar hardness to the seeded particles. The typical cutting speed is lower than 50mm/min for ceramic matrixes and plastic deformation cannot be avoided (Savrun and Taya, 1988).
5. **Magnetic Assisted Finishing (MAF) and Magnetic Float Polishing (MFP)**
   The principle of these will be considered in Chapter 7.
6. **Tumbling**
   Tumbling is intended to remove sharp edges and grinding marks from the sintered parts.
7. **Chemical Machining (CHM) and Photochemical Machining (PCM)**

Chemical machining utilizes etchants to attack the material and remove small amounts from the surface. This technique has been developed to chemical machining and photochemical machining, etc. Sharp corners, deep cavities, and porous workpieces cannot be easily machined by these techniques. Also, the material removal rate is relatively low. For photochemical machining of alumina ceramic, the etching depth can be lower than 200 μm and the etching rate can be 3.2 μm/min (Makino and Sato, 1987). Chemical-mechanical machining is generally used in surface patterning in semi-conductor (silicon) and micro-electro-mechanical systems (MEMS), in which a softened layer on the material surface is produced under a chemical reaction and then the mechanical machining is used to generate the desired pattern on the surface. However, high costs, low machining rate, and the multi-step limit this technique within the micro-machining field (Park et al., 2002). Additionally, considering its contributions to environmental hazards, chemical machining is not considered as an environmentally friendly process.

8. **Electrochemical Machining (ECM)**

   Electrochemical machining is characterized as reverse electroplating. This method is limited to electrically conductive materials only. Therefore, it is commonly used for machining metals with excellent electrical conductivity. The low heat and stress mean that little damage is caused. However, the high power costs and the unsuitability to generate sharp corners are two major drawbacks of this technique.

   Moreover, the hard machining of sintered ceramics employs the following thermoelectric non-traditional machining processes, in which the electrical energy, continuous or in a pulsed form, is used to achieve material removal.

9. **Electrical Discharge Machining (EDM)**

   Electrical discharge machining is a non-abrasion process, which keeps a gap between the tool-piece electrode and the workpiece surface at approximately 40 μm. The material is removed by high energy plasma produced by the electrical discharge. The surface roughness R_a is typically 4 μm and the material removal rate is 0.6mm^3/min. The thermal damage on the ceramic surface may be caused due to heat generation in discharging (Lee and Lau, 1991). Another application of EDM can be found in the drilling of micro-holes, which are widely used in industry. Common examples include spinnerets holes, inkjet printer nozzles, blades cooling channels, drug delivery orifices, and diesel fuel injection nozzles.

   Currently, there is a high demand for environmentally friendly and safe manufacturing processes. This has become a goal of many companies, especially after the introduction of the ISO 14000 standard of the environmental management system. Considering that EDM has a small environmental impact, this process can be improved to a fully environmentally friendly manufacturing process by eliminating the liquid from the process.

   Ceramic materials can be conductive, semi-conductive, and non-conductive, which are all able to be machined by the EDM technique. This technique is not affected by material hardness; it is, however, restricted to conductive material which should have an electrical resistivity of less than 100 Ω cm for the process to be successful. Yoo et al. (2015) have successfully doped yttrium nitrate (YN) with SiC and showed that the EDM process can be used for machining these ceramics. Nevertheless, it has been reported that doping usually negatively affects the mechanical properties of the ceramics. Apart from this, the "assisting electrode method" was suggested in the literature by Shin et al. (1998). In the assisting electrode method, a conductive layer is applied on the surface of the non-conductive ceramics as shown in Figure 6.7. Electric sparks generate high temperature, which forces the molecules of dielectric oil hydrocarbons and workpiece material to crack, which in turn enables the binding of carbon to certain elements of the ceramic. Since the carbon compounds are conductive, new discharges allow the machining of the deposited conductive layer together with the workpiece material, which was initially under the conductive layer. Shin et al. (1998) also suggested incorporating a condenser between the tool and the workpiece in order to increase the MRR and the machining stability.

Finally, EDM has become an effective manufacturing process for machining both conductive and non-conductive ceramics. Some of the major challenges reported in the literature are:

- One of the important criteria for successful machining of ceramics is the selection of an assisting electrode. These include the selection of coating material, appropriate coating thickness, and ways of creating these coatings. Since there is a wide variety of assisting electrodes, it is difficult to effectively choose the kind of assisting electrode that would fit a particular area of application.
- Another important challenge of EDM of ceramics is the need for a modified pulse generator specifically designed for conducting EDM on ceramics.
- The major challenge is to understand the physics of the process and to develop physics-based modeling for the EDM of ceramics. Thermal fractures should be considered during the modeling to understand the crack formation in the surface and sub-surfaces.

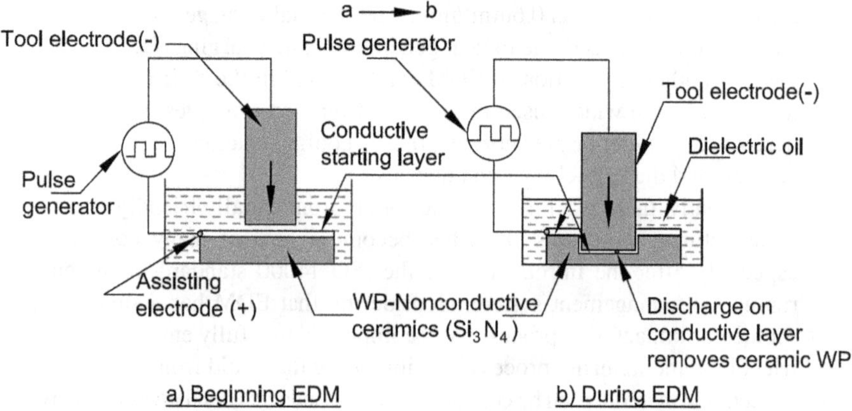

FIGURE 6.7 Basic principle of EDM of non-conducting ceramics with an assisting electrode (Gotoh et al., 2016).

- Analytical and numerical modeling of the material removal mechanisms during EDM and micro-EDM of ceramics is of prime importance to broaden the application of EDM usage in ceramic machining.
- The future research trend should focus on solving the associated problems either by developing newer hybrid machining processes or by incorporating novel ideas to improve the existing process and creating new processes for machining ceramics.

EDM versus Micro-EDM

EDM is widely used in tool- and mold-making; micro-EDM with its much lower discharge energies has been successfully applied to micro-machining of high-accuracy parts. The precision manufacturing of high aspect ratio micro geometries such as deep micro bores relies on stable process conditions in the discharge gap. Even though the physical principles of the micro-EDM are similar to those of the macro-EDM and both use spark erosion, micro-EDM is not just an adoption of the EDM to micron level. There are significant differences in the size of the tool used, the fabrication method of micro-sized tools, the power supply of discharge energy, movement resolution of machine tools' axes, gap control and flushing techniques, and in the processing techniques (Masuzawa, 2001). The most important difference between macro-EDM and micro-EDM is the dimension of the plasma channel. For macro-EDM, the size of the plasma channel is much smaller than the electrode size, while the size is comparable for micro-EDM (Tibbles, 2005). Because of small electrodes, the maximum energy that can be reached is limited in micro-EDM, since excessive discharge energy can lead to electrode burn (or wire rupture in wire EDM) (Lim et al., 2007). As a result, for each discharge, the electrode wear in micro-EDM is proportionally higher than conventional EDM. Also, the flushing of debris is more difficult in micro-EDM because of small gap size, high dielectric viscosity, and a higher pressure drop in micro-volumes (Jahan et al., 2014).

The micro-EDM process requires small energies of 10^{-6}–10^{-7} J for every discharge of 40–100 V and high frequencies of greater than 200 Hz (Jahan et al., 2014). For micro-EDM, discharge durations of less than 1 μs and discharge energies of less than 100 μJ are common. This allows the machining of geometrical features with diameter of less than 5 μm and a depth of less than 1 μm. In such conditions, the achievable surface roughness is lower than 1 μm (Schubert, 2015). The precision and accuracy of the final products are much higher in micro-EDM. The crater sizes in micro-EDM are also much smaller than that of conventional EDM (Katz and Tibbles. 2005). Figure 6.8 shows the comparison of the crater size between conventional EDM and micro-EDM (Uhlmann, 2005).

New Hybrid Micro-EDM Ultrasonic Technology of Electrically Non-Conductive Ceramics

New hybrid technology approaches, such as ultrasonic or low frequency superposition, significantly raise the process stability and speed. The micro-EDM process with ultrasonic vibration assistance—directly applied to the workpiece and indirectly applied high-intensity ultrasonic to the dielectric—in metallic materials as well as in the machining of electrically non-conductive ceramic materials has been

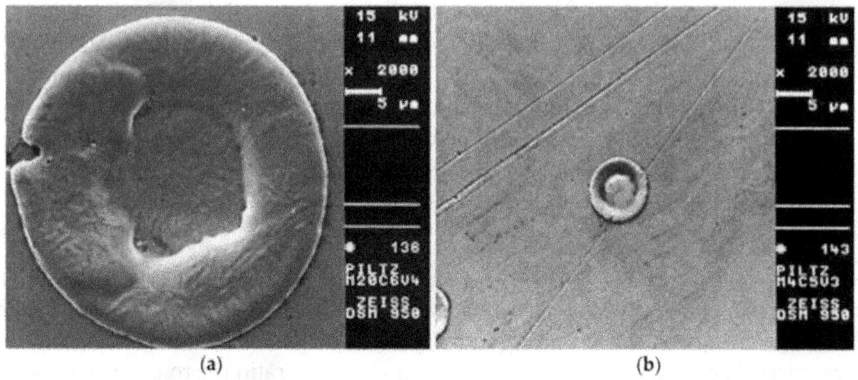

FIGURE 6.8 Comparison between crater dimensions in (a) conventional EDM (left) and (b) micro-EDM (right) (Uhlmann et al., 2005).

investigated. Using ultrasonically aided micro-EDM, the process speed can be raised by up to 40%, enabling bores of less than 90 μm in diameter with aspect ratios >40 for metallic materials.

10. **Electrochemical Discharge Machining (ECDM)**

The electrochemical discharge machining (ECDM) process is a complex combination of the electrochemical machining (ECM) and electrical discharge machining (EDM) processes. This process has very good potential in the area of micro-machining non-conductive hard and brittle materials such as ceramic, glass, quartz, and Pyrex. The ECDM process involves melting and chemical etching of the workpiece due to high electrical energy discharged on the tip of the electrode during electrolysis.[1] The literature reveals that combined metal removal rate in ECDM can be five to 50 times over EDM and ECM with decreased electrode tool wear. However, the maximum machined depth is lower than 2 mm and the mass removal rate could not exceed 2.5 mg/min. Most importantly, the machined material shows the tendency to crack due to the thermal shocks caused by the heat generated at high voltages (Chak and Rao, 2007).

This technique requires two electrodes. One is the tool electrode, which is used to produce desired machined shape, and the other is the counter electrode or auxiliary electrode made as anode. The workpiece and counter electrode (anode) are immersed in an electrolyte solution (typically sodium hydroxide or potassium hydroxide). The tool electrode (cathode) is kept 2–3 mm dipped in the electrolyte. The counter electrode or anode is a large size dummy electrode in general, which is kept at a distance of about 25–50 mm away from the tool electrode. Electrolysis starts when a voltage is supplied by a direct current (dc) power source between the tool electrode and counter electrode. A schematic of the process is shown in Figure 6.9.

The surface area of the tool electrode submerged in the electrolyte is kept very small compared to the counter electrode (anode). This results

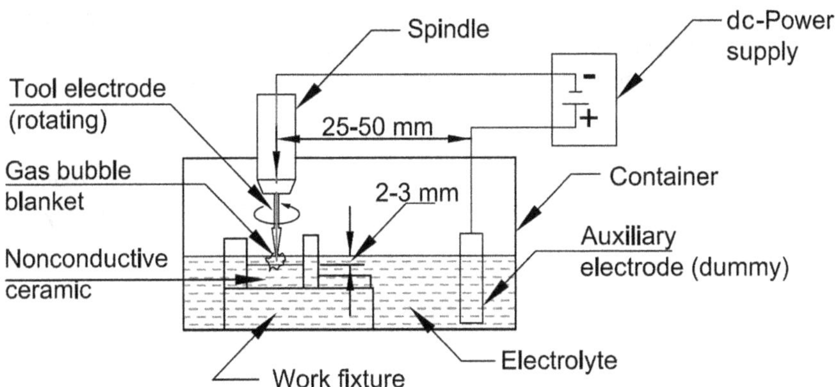

FIGURE 6.9 Schematic of the ECDM process (Gupta et al., 2015).

in high current density at the cathode. Rapid production of hydrogen gas bubbles takes place at the cathode due to ohmic heating of the electrolyte solution. Surrounding electrolyte insulates the immersed tool electrode (blanketing effect) by a gas film due to bubble coalescence. Gas film plays a key role in machining during ECDM. A stable and dense gas film controls the machining process. Sparking action takes place between the tool and the workpiece when the current density of the tool exceeds the critical value (typically around 1 A) and applied voltage also becomes more than the critical voltage (approximate 25 V). The critical value of current and voltage depends on the geometric shape of the tool point and concentration of the used electrolyte. As the workpiece lies in close proximity (typically within 20 μm) to the tool electrode, material removal takes place by melting and etching of the workpiece. This material removal process mechanism is known as ECDM.

The ECDM process has limitations such as surface defects due to localized overheating, which requires a suitable arrangement of intermittent cooling such as pulse power supply. Low material removal rate (MRR), low depth of penetration, overcut and taper in the hole are major problems in ECDM. These may be due to the lack of flow of electrolyte at tool point during machining. However, these problems can be minimized by a proper design of tool point of tool electrode. Tool electrode may be given a suitable movement in order to get better results of machining such as rotation. Tools are also vibrated ultrasonically in the direction of tool feed movement. This tool movement results in reduction of taper, overcut, and better surface quality of sidewalls. The tool electrode has a significant role in the machining. Tools of pointed ends are preferred as they concentrate sparking action at the tool point. This results in better machining performances.

Materials for both electrodes should be properly selected. The size of the counter electrode is kept as large as possible and material of this electrode must not react with the electrolyte used. The electrodes made from stainless steel or nickel are preferred because these materials have excellent corrosion resistance, chemical resistance to electrolytes, and minimum tool wear. Tool wear is negligible in ECDM, which occurs due to chemical etching and anodic dissolution. The sparking at the tool tip may cause some tool wear.

The current density being less on the counter electrode limits its anodic dissolution. A circular hollow pipe or ring can be used as anode surrounding the tool electrode. Such a type of electrode can maintain a constant inter-electrode resistance.

Operational features of the tool electrode such as geometry, rotation, and roughness greatly influence efficiency and accuracy of the ECDM process. The surface roughness of the tool is an index of quality of gas film produced during electrolysis. This influences thickness and stability of the gas film. Geometry of the cathode tip is a measure of dimensional accuracy of the machining cavity produced. Hence, tool design is important for the ECDM process for improving machining efficiency and accuracy.

Zheng et al. (2007) investigated that the flat sidewall tool, as shown in Figure 6.10, improves the quality of gas film while the drilling process is occurring in comparison to the cylindrical tool. This results in an increase in machining efficiency and accuracy.

Han et al. (2008) studied the effect of using a side insulated tool electrode as shown in Figure 6.11 in micro-machining of micro channels. Investigations show that this kind of tool decreases the stray electrolysis, minimizes the fluctuation of current peak values, and increases the

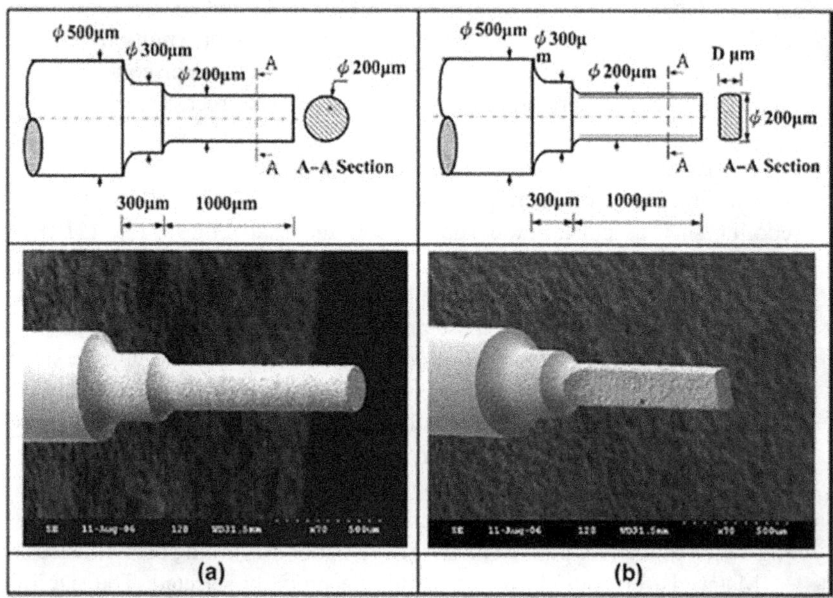

FIGURE 6.10 Different sidewalls of the tool: (a) Cylindrical tool and (b) flat sidewall flat front tool (Zheng et al., 2007).

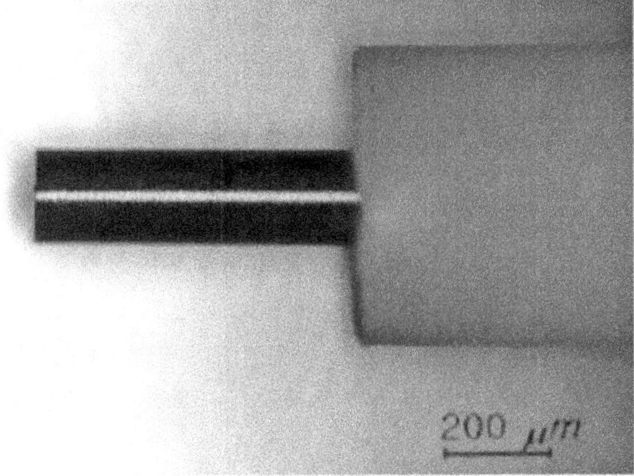

FIGURE 6.11 Side insulated tool electrode (Han et al., 2008).

discharge current uniformity of the ECDM process. The above phenomenon results in better geometrical accuracy on the surface of micro channels.

Side insulated tool electrode depicts a stable large gas bubble, which covers the tool point completely. Figure 6.12 shows the scanning electron microscopy (SEM) image of the micro channel machined by the above two types of electrodes. Machining by conventional non-insulated tool electrode lacks the uniformity in width and surface roughness as compared to the side insulated tool electrode. Han et al. (2009) investigated that the electrolyte with ultrasonic vibration causes the adequate flow of electrolyte between the tool and workpiece. The acoustic pressure distribution developed due to ultrasonic energy modifies the bubble layer. Thus, uniform spark is generated (Figure 6.13). Consequently, machining depth increases during hole drilling in the ECDM process.

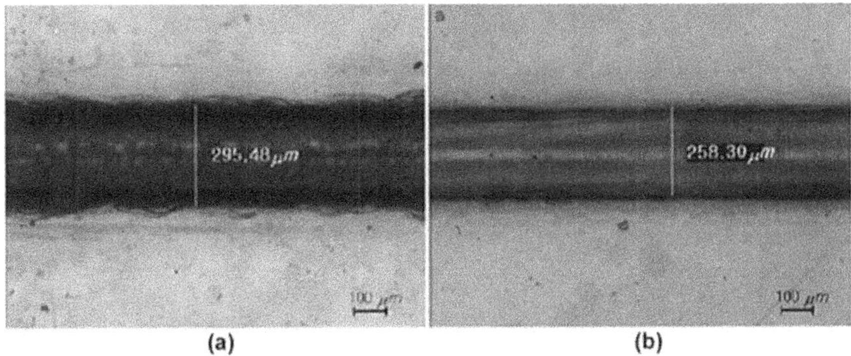

FIGURE 6.12 Micro channel fabrication using: (a) Conventional electrode and (b) side insulated electrode (Han et al., 2008).

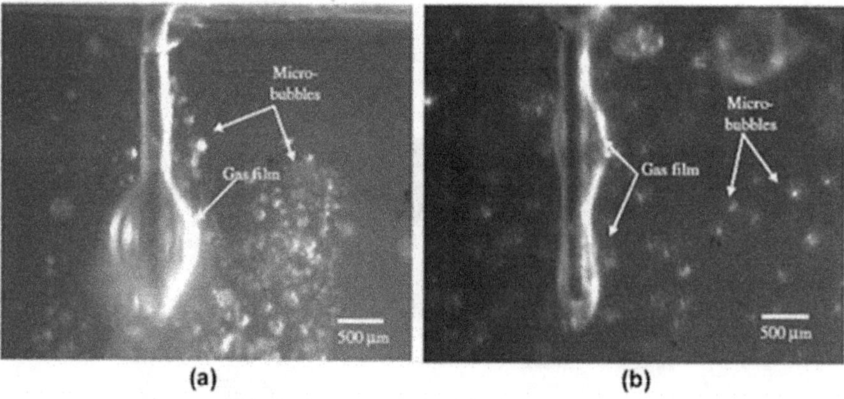

FIGURE 6.13 Difference in gas film formation: (a) Without ultrasonic and (b) with ultrasonic effects (Han et al., 2009).

11. **Laser Machining**
    In laser machining, high power light energy is focused onto the workpiece surface and the local material is hence heated, melted, dissociated, decomposed, evaporated, and/or ablated from the surface. The machining quality is jointly governed by the laser parameters employed and inherent material properties. But it is almost independent of material hardness and brittleness properties. The technique represents a possible alternative for processing ceramic materials due to a number of advantages, such as no tool wear and cutting forces, high machining quality, as well as high flexibility and automation, etc. The economic comparison of laser machining with other techniques of hard machining of ceramics is listed in Table 6.5.

Type of Lasers Used for Hard Machining of Ceramics

To date, several lasers have been used in machining of engineering ceramics, including CO_2, Nd:YAG, excimer, fiber, diode, and ultra-short (i.e. picosecond and femtosecond) pulsed lasers. Each laser has its own emitting wavelength and machining

TABLE 6.5
Economic Comparisons of Different Machining Techniques

Machining Process	Capital Investment	Tooling/ Fixtures	Power Requirements	Tool Wear
Diamond grinding	Low	Low	Low	Low
Ultrasonic machining	Low	Low	Low	Medium
Electrical-discharge machining	Medium	High	Low	High
Laser machining	Medium	Low	Very low	N/A

Source: (Pandey and Shan, 1980).

applications. Although some of these lasers can be operated in both continuous wave (CW) and pulsed mode (PM); PM lasers are considered to be preferred for ceramic machining due to the more effective control of process parameters (Islam and Campbell, 1993). For macro-machining in industrial applications, infrared (IR) lasers are more suitable than ultra-short and ultraviolet lasers due to their high average power output causing a high material removal rate. Ultra-short and ultraviolet lasers are generally used in micro-machining and surface treatment. Basically, CO_2, Nd:YAG, and fiber lasers have been widely employed in ceramic macro-machining.

CO_2 lasers: The CO_2 laser emits the light at a wavelength of 10.6 μm in the far infrared spectrum. CO_2 lasers have contributed to over 40% of industrial lasers (Steen, 2003).The overall efficiency of electrical-optical energy conversion is in the range of 10–30% and a beam power greater than 10 kW can be achieved. However, the beam quality degrades with increasing output power. Furthermore, CO_2 lasers are usually operated in CW mode, but PM can also be achieved by modulating the pump source up to 5 kHz. Therefore, CO_2 lasers are generally considered as one of general-purpose lasers for most material processing, such as heat treatment, cutting, welding, cladding, etc. Fast axial gas flow, slow axial gas flow, perpendicular direction of gas flow to the laser optical path, perpendicular direction of electrical discharge to the optical path, and sealed-off are five techniques applied in CO_2 laser systems. Sealed-off CO_2 lasers are the lowest cost lasers in the CO_2 laser family, but their output powers are also limited and lower. Periodic maintenance is required (i.e. refilling pre-mixed gas) in order to maintain the laser output power. This is the major drawback of CO_2 lasers. In addition, the wavelength emitted by CO_2 lasers cannot be delivered by optical fibers, which limits the combination with robots in industrial applications.

Nd:YAG lasers: NG:YAG lasers are solid-state lasers that use 1–2% dopants (i.e. Neodymium (Nd3+)) dispersed in a crystalline matrix (i.e. yttrium aluminum garnet (YAG) with the chemical composition $Y_3Al_5O_{12}$) as the gain medium. Pump sources are krypton or xenon flash lamps, or recently laser diodes. The wavelength of an Nd:YAG laser is emitted at 1.06 μm in the near-infrared spectrum. The power output of CW Nd:YAG lasers are lower by some kilowatts. But the Q-switch Nd:YAG lasers can generate short pulses with peak power in megawatts, repetition rates up to 100 kHz, and pulse durations between 15–400 ns. The average output power of a pulsed Nd:YAG laser is generally lower than 1 kW. The output laser beam (at 1.06 μm wavelength) can be coupled into an optical fiber for delivery. Therefore, it is suitable to combine with robots for more flexible processing in industries. Nd:YAG lasers have been widely used for cutting, welding, cladding, and drilling of metallic or non-metallic materials. The drawback of Nd:YAG lasers is the relatively low beam quality at high output powers (M2<6), causing a large spot size, low power density, and short focal depth. Therefore, high power Nd:YAG lasers are not suitable in precision processing and thick-section machining applications. Furthermore, maintenance is also essential to Nd:YAG lasers.

Fiber lasers: Although the first fiber laser was proposed in early 1960s, their practical applications were limited to laboratory demonstrations until the late 1990s. In the last decade, some unique advantages in material processing have appeared as the advanced and commercial double-clad high-power fiber lasers became available. In high power fiber lasers, a multi-clad silica-based optical fiber is used as the gain medium, which is doped with rare-earth ions, such as erbium (emitted laser at 1.55 μm wavelength), ytterbium (emitted laser at 1.07 μm wavelength), and neodymium (emitted laser at 1.06 μm wavelength). The beam quality of fiber lasers is perfect (M2<1.1), even at high powers (up to some kilowatts). The market leader among high power fiber laser manufactures—IPG Photonics Corp.—provides kilowatt class fiber lasers up to 50 kW, operating in CW or modulated modes up to 20 kHz, with wall-pump efficiencies greater than 30%. High power fiber laser as high power, high brightness, high efficiency, high flexibility (fiber-delivered beam and robot manipulation), small spot size, high beam quality, maintenance-free, high compactness, and high durability, make fiber lasers more attractive in future laser industrial applications.

All the wavelengths of the above-mentioned lasers are in the IR spectrum. Nowadays, IR lasers can partially replace mechanical material removal techniques in several engineering applications due to their unique feature. Generally, the material removal rate and the finish quality mainly depend on the employed laser parameters, such as laser power, spot size, focal plane position, feed rate, pulse repetition rate, pulse duration, etc., and rely on the thermal and optical properties of the material as well. This makes hard and brittle materials with low thermal conductivity, such as engineering ceramics, suitable to be machined by IR lasers. Although the laser cutting/machining techniques have been employed in industrial applications for many years and have become mature, laser cutting/machining of engineering ceramics still face some challenges. The major challenge is to achieve high quality (without crack and rough surface finish) and high removal rate. Therefore, many studies have been performed and several techniques developed over the past decades, in order to improve the process quality and machining rate of laser cutting/machining of engineering ceramics.

Crack is the most critical quality factor since it may cause fracture and catastrophic failure in brittle ceramic materials. Therefore, most of the studies were first focused on crack-free cutting. Previous studies indicated that high laser power with a low cutting speed can reduce the thermal-stress level and subsequently, reduce the possibility of fracture in CW laser cutting (Li and Sheng, 1995). Lu et al. (1999) specified an empirical equation to identify the boundary of operating conditions on fracture initiations in CW CO_2 laser cutting of alumina up to 2 mm thickness.

$$P \geq 1.78x(10)_{11}(t)2.41v \tag{6.1}$$

Where P is the minimum laser power to avoid crack formation (in W), t is the workpiece thickness (in m), and v is the feed rate (in m/s).

An attractive new technology known as the laser-controlled fracture cutting technique is implemented using a CO_2 PM laser. In order to separate brittle materials (e.g. ceramic or glass), this technique was developed where the incident laser energy generates localized mechanical stresses that separate the material by controllable crack extension (Tsai and Chen, 2003a). This technique requires a much lower laser power output than other laser cutting/machining techniques as mentioned above for the same workpiece thickness. Laser-controlled fracture cutting was first proposed by Lumley (1969), and now has become the major technique to separate flat glass in the LCD industry. The early investigations generally focused on the fracture technique with a coolant jet, e.g. Kondratenko. U.S. Patent (1997), Unger and Wittenbecher (1998), by which the crack propagation was unstable, and the fracture control was very difficult due to the completely tensile stress at the crack tip (cooling). The later studies paid more attention to laser-controlled fracture without coolant jet (Figure 6.14) since it can produce a compressive stress dominated region between the crack tip and the laser spot (heating), by which the fracture extension was stabilized (Tsai and Liou, 2003). Most importantly, Tsai's work (Tsai and Liou, 2003) demonstrates a significant advantage of laser-controlled fracture cutting of ceramics, i.e. the very smooth cut surface (R_t = 26 μm) and fewer defects.

Tsai and Chen (2003b) further achieved controlled fracture cutting of 10-mm-thick mullite alumina by synchronously applying an Nd:YAG laser and a CO_2 laser. The focused Nd:YAG laser was used to scribe a groove-crack on the substrate surface and then the defocused CO_2 laser was employed to release the thermal stress (Figure 6.15). The thermal-stress concentrates on the tip of the groove-crack and makes it extend through the substrate with the defocused laser beam moving along the path scribed by the previous focused laser beam. The tensile stress was induced at the groove-crack edge so that the crack unstably propagated along the thickness (z direction). But the fracture extension in the transverse direction (x direction) was stable due to the compressive stress between the laser beam and the crack tip. The fracture surfaces were very smooth (R_a = 2 μm). The dual beam can control the fracture trajectory along the cutting path more accurately than the single beam.

12. **Electron Beam Machining (EBM)**
    In electron beam machining, high-velocity electrons are directed toward the workpiece, heating or melting/vaporizing the material. It can be used for fine cutting of a wide variety of materials, with a high cut quality. Unfortunately, the need of a vacuum chamber and the beam defocusing at high speeds are two major drawbacks of this technique. Hence this technique is not suitable for large-scale and high-speed machining.

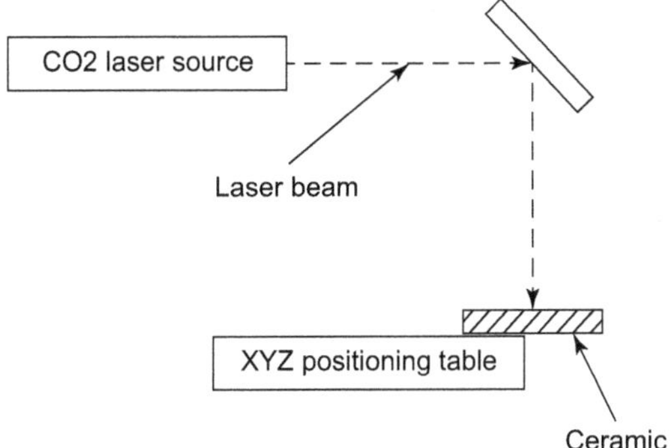

FIGURE 6.14 Schematic of laser-controlled fracture cutting and the system (Tsai and Liou, 2003).

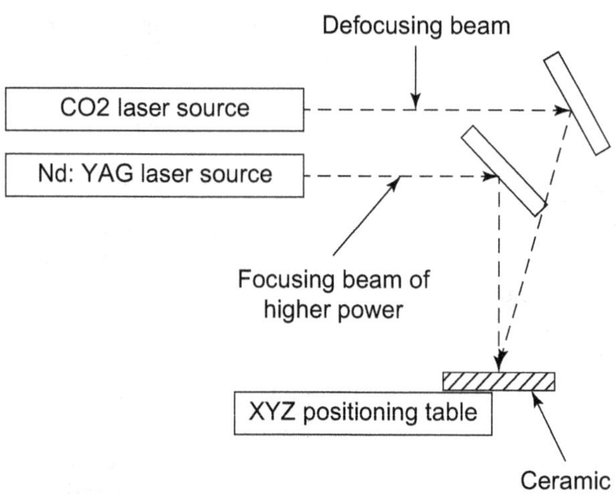

FIGURE 6.15 Configuration of the dual-beam controlled fracture cutting system (Tsai and Chen, 2003a).

13. **Plasma Arc Machining (PAM)**
    Plasma arcs induced by ionized gas can be used for machining. The generated high intensity plasma provides sufficiently high temperatures leading to melting/vaporizing the material and ejects molten material away from the processed region. The narrow-cut kerf and good surface finish are two significant advantages of this technique (Samant and Dahotre, 2009). However, similar to electron beam machining, the need of vacuum ambient limits the application of plasma arc to large-scale machining.

6.3 MACHINING OF COMPOSITE MATERIALS

6.3.1 Types, Characteristics, and Applications of Composites

Composite materials are currently replacing common engineering materials in many applications due to their high strength to weight ratio, high specific stiffness, improved fatigue and creep resistance, allowing a weight reduction in the finished parts. Composites are widely used in wind energy systems, machine tools, sports goods (golf clubs, tennis rackets, bicycles, arrows, surfboards, and skateboards), and biomedical products. Fiber reinforced polymer (FRP) composites and particularly those involving carbon fiber (CFRP) are the most widely used composite material in military and commercial aerospace systems. Figure 6.16 shows the use of fiber reinforced polymer composites in Airbus A380.

A composite material is a combination of reinforcement and matrix materials that results in better properties than each individual component that retains its separate chemical, physical, and mechanical properties. The matrix can be a polymer, metal, or ceramic. Polymers have low strength and stiffness, metals have intermediate strength and stiffness but high ductility, and ceramics have high strength and stiffness but are brittle. The matrix maintains the fibers in the proper orientation and spacing, protects

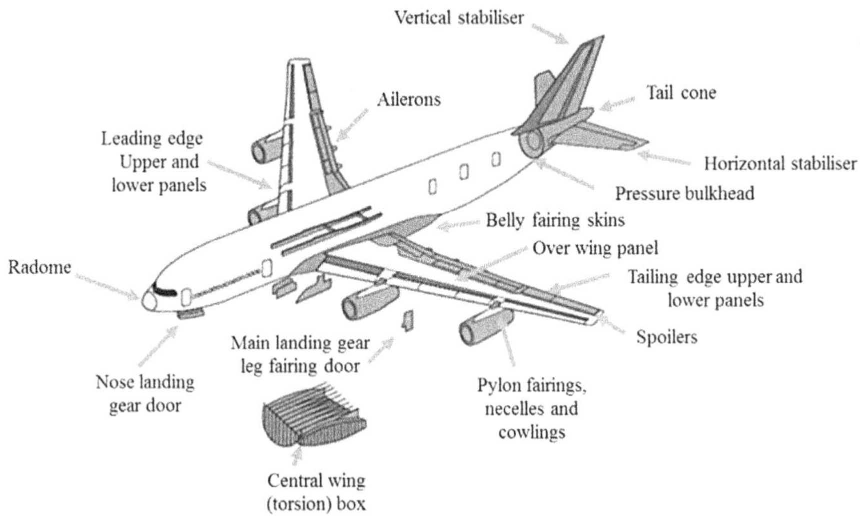

FIGURE 6.16 Use of fiber reinforced polymer composites in the Airbus A380.

fibers from abrasion, chemical reactions, and the environment, and distributes the load. It determines the service temperature of a composite material, transmits loads to the fibers through shear loading at the interface, provides good adhesion to the fiber surface, and provides compatible stress strain behavior with fibers.

The reinforcing phase can be glass, graphite, boron, aramids, and various other oxides, carbides, and nitrides. It is usually stronger, harder, and stiffer than the matrix, provides the strength and stiffness for the composite material and adds chemical, thermal, and electrical properties. Typical fibers include glass, aramid, and carbon, which may be continuous or discontinuous. When more than one type of fiber is used in a matrix then the composite is known as hybrid. Reinforcing materials are used in the form of particulates, whiskers, continuous fibers, and short fibers (Sheikh Ahmed Jamal, 2009). Composites are classified according to matrix material into polymer matrix composites (PMC), metal matrix composites (MMC), and ceramic matrix composite (CMC) composites. In general, metal and ceramic matrix composites are much more expensive than polymer matrix composites. However, they have much better thermal stability than PMC, a requirement for high temperatures applications. Table 6.6 shows the effect of composite attributes on their mechanical properties and machining behavior.

6.3.2 Traditional Machining and Machinability of Composites

Composites are produced to near-net shape. Turning, milling, trimming, sawing, or grinding operations are used for machining composites to the final shape at the required accuracy and surface quality. Due to the toughness and abrasive nature of composites, there is a need for harder and longer lasting cutting tools. The mechanisms involved while cutting composite materials are different from those observed when cutting homogeneous materials. Since composites are not homogeneous, their

TABLE 6.6
Effect of Composite Attributes on Mechanical and Machining Properties

Attribute	Properties	Comments on Machining
Fiber	High strength, high modulus	Abrasiveness of fiber increases with strength
Fiber length	–	Small pieces of fiber delaminate easier and present machining difficulties
Fiber diameter	Increasing diameter decreases tensile strength	While tensile strength reduces with diameter, cutting forces are expected to increase
Matrix	Toughness	–
% volume of fibers	Improves mechanical properties	Adversely affects machinability
Fiber layout: Unidirectional or fiber weave	Affects the degree of anisotropy of properties	Delamination is usually severe in unidirectional types

Source: Kennametal incorporation.

machining characteristics are dependent on the tool path relative to the direction of the reinforcing fibers. The knowledge of the properties of the single component (matrix or reinforcement) is not sufficient to predict the behavior of the composite material during cutting. The homogeneity of reinforcement distribution, its alignment in one direction, and the crystalline structure of the matrix are important issues. The interaction between the strength of the matrix at the interface affects both the quality of the machined surface and the cutting tool wear. Composite materials are more difficult to machine than metals mainly because they are anisotropic, non-homogeneous, and their reinforcing fibers are very abrasive. During machining, defects are introduced into the workpiece, and tools wear rapidly.

Traditional machining methods such as drilling, turning, sawing, routing, and grinding can be applied to composite materials using appropriate tool design and operating condition. Figure 6.17 shows the different factors affecting machinability of composite materials.

Drilling: Drilling is the most common composite machining operation, since many holes must be drilled in order to install mechanical fasteners. During composite drilling, the thrust force increases steadily until a constant value corresponding to steady drilling through the thickness of the laminate is reached and is followed by a sharp drop as the tool exits the opposite side, Figure 6.18. The drilling torque increases rapidly until the cutting edges of the tool are completely engaged and then increases linearly until a maximum value is reached, followed by a slight drop after hole completion, Figure 6.19. The high difference between the cutting torque T_c and the maximum torque T_m is attributable to high frictional forces between the lands of the drill and the wall of the hole. As drilling progresses, the tool is in contact with the side of the hole over an increasing area so that frictional forces at the interface create increasingly higher resistant torque. After complete penetration has occurred,

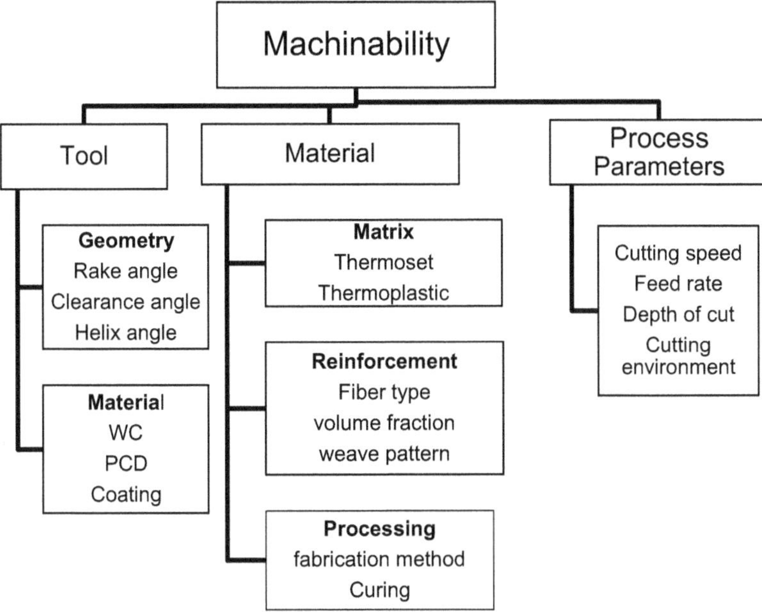

FIGURE 6.17 Factors affecting machining performance of composite materials.

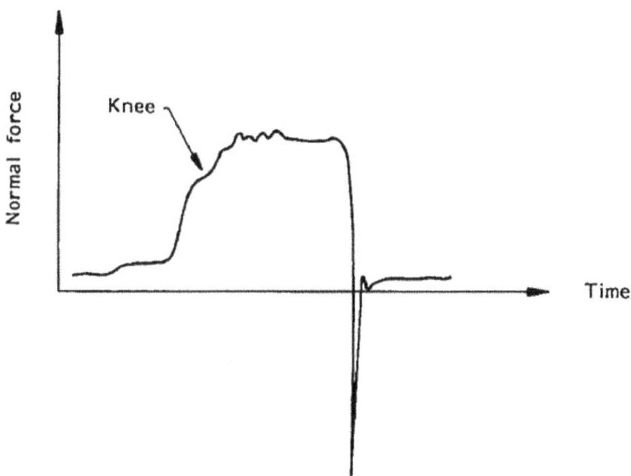

FIGURE 6.18 Typical axial force history during drilling of composites (Abrate and Walton, 1992a).

only a small decrease in torque is observed which indicates that friction is the major contribution to total torque. Increasing the feed rate and point angle of the drill, respectively, produces a corresponding increase and decrease in torque levels respectively (Abrate and Walton, 1992a). Different drill point designs used for drilling composites are described by Sheikh Ahmed Jamal (2009). Table 6.7 presents typical machining parameters for drilling composites.

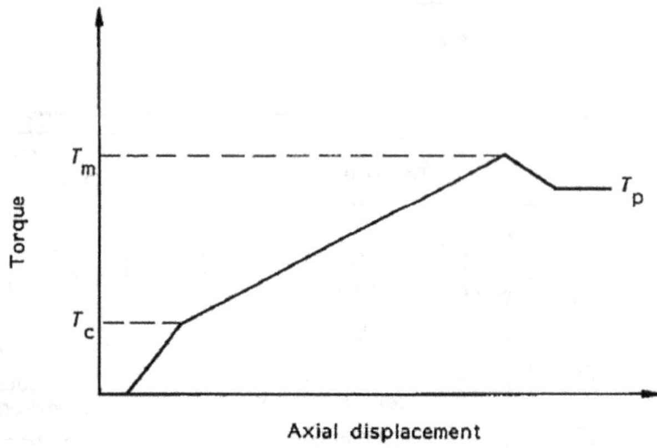

FIGURE 6.19 Variation of torque during drilling: T_c: Cutting torque; T_m: Maximum torque; T_p: Torque after penetration (Abrate and Walton, 1992a).

TABLE 6.7
Typical Machining Parameters for Drilling Composites

Workpiece Material	Tool Material	Hole Diameter, mm	Material Thickness, mm	Cutting Speed, m/min	Feed Rate, mm/rev
UD graphite-epoxy	Carbide	4.85–7.92	0–12.7	42.7	0.0254–0.0508
		4.85–7.92	12.7–19.1	33.5	0.0254
	PCD	4.85–7.92	0–12.7	61	0.0508–0.0889
		4.85–7.92	12.7–19.1	51.8	0.0508–0.0889
UD graphite-epoxy	Carbide	4.85–7.92	0–12.7	61	0.0254–0.0508
		4.85–7.92	12.7–19.1	42.7	0.0254
	PCD	4.85–7.92	0–12.7	68.6	0.0508–0.0889
		4.85–7.92	12.7–19.1	61	0.0508–0.0889
Graphite-epoxy	Carbide	4.85	6.35	60.9	0.0254
Glass-epoxy	HSS	3	10	33	0.05
		8	1.2	40.2	20–460 mm/min
	Carbide	3	10	33	0.05
Boron-epoxy	PCD	6.35	2.0	91–182	25.4 mm/min
		6.35	25.4	91–182	25.4 mm/min
MMC	PCD/Carbide	6	19.2	15–75	0.05
Kevlar-epoxy	Carbide	5.6	–	158	0.05

Source: (Abrate and Walton, 1992a).

Delamination occurs when drilling FRPs due to the large thrust force which acts normal to the ply. This force increases with feed rate and tends to separate the plies in a laminate by interlaminate cracking during conventional push drilling, Figure 6.20. There is a direct relationship between drill wear and delamination since tool wear increases the feed force. There is a critical feed rate below which delamination will

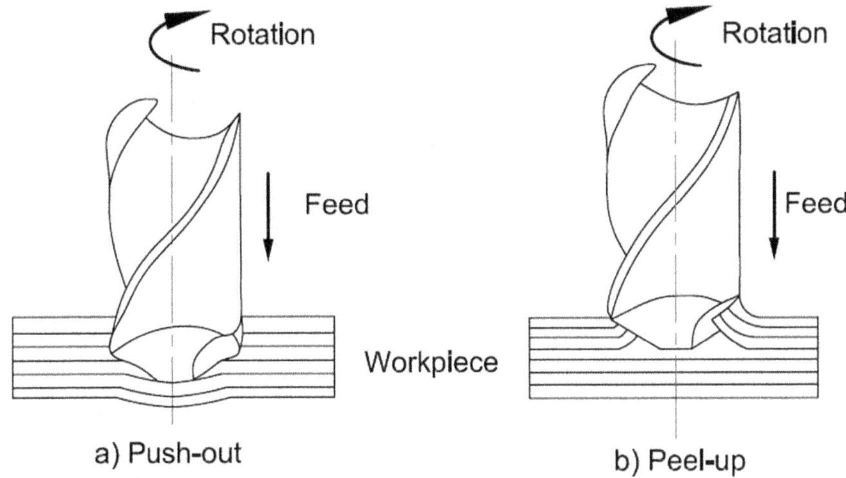

FIGURE 6.20 Push-out and peel-up delamination in drilling.

not occur. In order to avoid delamination, the selection of proper feed rate is therefore essential. The extent of delamination is measured by the delamination factor F_d as shown in Figure 6.21. Reducing delamination in drilling can be achieved by reducing or distributing the thrust force component through the proper scheduling of the feed rate in a drilling cycle, proper selection of the drill point geometry, and the use of core or saw drills for distributing the feed force (Babu et al., 2016).

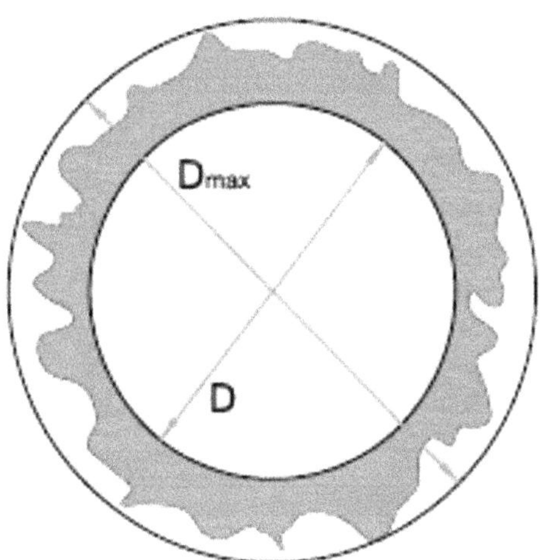

D: Hole diameter
D_{max}: Maximum delamination diameter
Delamination factor $F_d = D_{max}/D$

FIGURE 6.21 Evaluation of conventional delamination factor.

$$F_d = D_{\max} / D$$

One-dimensional delamination factor may not be the correct method of assessing the delamination. The reason for this can be explained by Figure 6.21. Figures 6.22a and 6.22b have the same F_d values and D_{max} is the same in both cases; however, the specimen in Figure 6.24b is more prone to failure at lower load application than that in Figure 6.22a, as it is more severely damaged. A two-dimensional delamination factor (F_a) is schematically shown in Figure 6.23. A_{nom} is the nominal drill area (Babu et al., 2016).

$$F_a = \left(A_d / A_{nom}\right)\%$$

where A_d is the total area of the drilled hole and delamination, and A_{nom} is the nominal hole area. Other drilling defects can be seen in Figure 6.24.

Figure 6.25 shows the orbital drilling where the tool is rotating around its axis and at the same time orbiting around the axis of the drilled hole. The cutting edge intermittently comes in contact with the inside wall of the hole. The cutter diameter is less than the diameter of produced hole. Orbital drilling has the following advantages:

- Delamination free holes
- Lower cutting temperature
- Easy chip evacuation
- One tool for different diameters
- Efficient cooling of cutter and hole surface
- Reduced risk of matrix melting

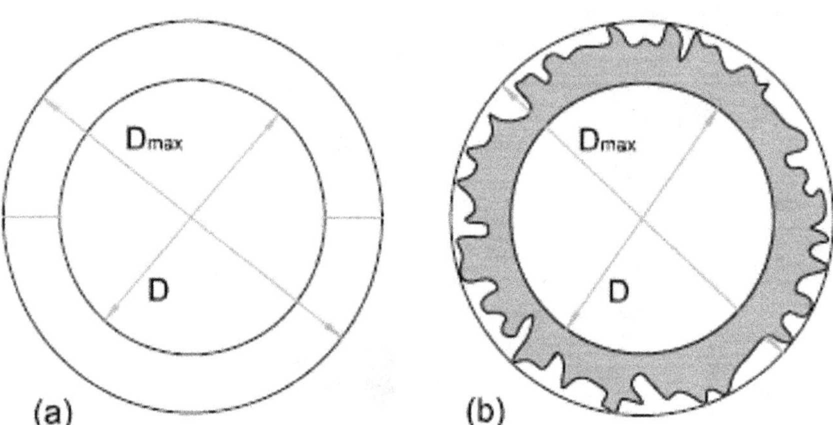

FIGURE 6.22 Diagram of the damage: (a) Fine crack and (b) severe crack with damaged area.

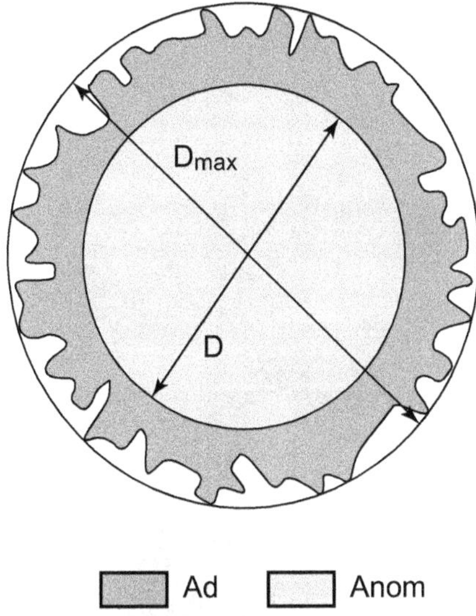

D : Hole diameter
D_{max} : Maximum delamination diameter
Ad : Area of delamiation
Anom : Area of drilled hole
Fa : (Ad/Anom) %

FIGURE 6.23 Evaluation of two-dimensional delamination factor.

Temperature generation at the cutting point causes tool wear and softening to the matrix material which, in turn, promote delamination. Therefore, high thermal conductivity through the cutting tools and through the tool cooling is recommended.

Turning: Turning is governed by fiber orientation, fiber content, cutting speed, and to a smaller degree by feed rate. Generally, tool-wear increases with the increase in fiber content, cutting speed, and feed rate. Tool wear is less when turning CFRP than GFRP composites. Less tool wear occurs when using PCD and diamond-coated tools than uncoated carbides. This trend occurs due to the effective heat dissipation capabilities of carbon fibers in CFRP and diamond in PCD which lowers the temperature at the cutting region and reduces tool wear.

The cutting forces decrease with the increase of cutting speed up to a transitional cutting speed beyond which further increase in cutting speed increases the cutting forces. The viscous behavior of polymers at higher cutting speeds may explain the rise in cutting forces. The transitional speed occurs around the melting temperature of thermoplastics and decomposition temperature of thermosets (300°C). The transition speed increases with the increase of thermal conductivity of the cutting tool material. The surface roughness in turning is closely related to the fiber orientation (Sheikh Ahmed Jamal, 2009).

Milling and trimming: Cutting tools of PCD, diamond-coated carbides, and solid carbide burr tools provide low tool wear in milling and trimming of composites.

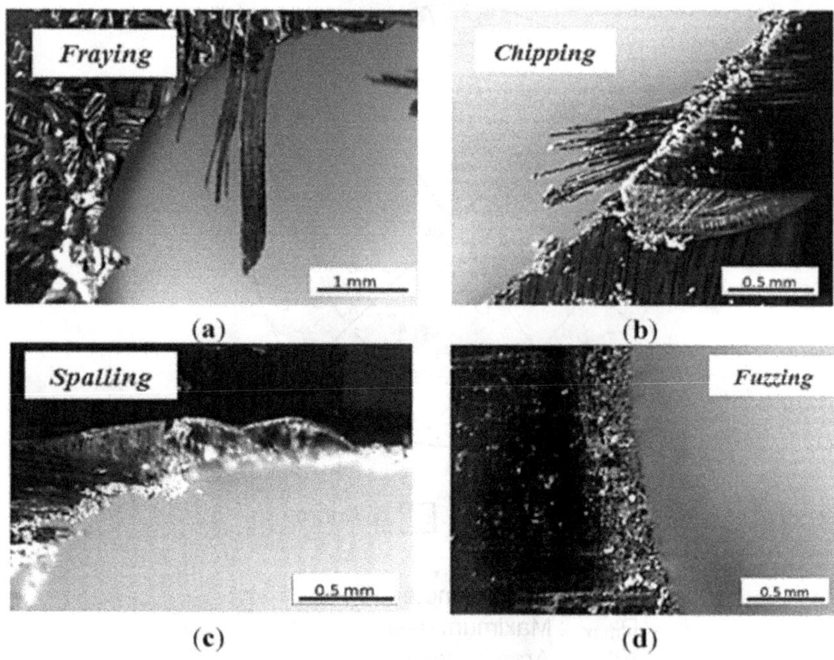

FIGURE 6.24 Different defects related with hole quality found after drilling: (a) Fraying; (b) chipping; (c) spalling; and (d) fuzzing (Feito et al., 2014).

Because of the intermittent nature of the milling and trimming processes, cutting tools perform better at smaller values of equivalent chip thickness which significantly reduces the impact forces on the cutting edge. Less tool wear and delamination are achieved by using smaller radial depth of cut, smaller feed rates, and higher cutting speeds, which is a typical combination of finishing operations. The extent of delamination is also influenced by the surface ply orientation. Therefore, delamination can be reduced by designing laminates with 0° and (0°/90°) surface plies (El-Hofy et al., 2011).

Standard type end mills generate cutting forces in one direction. With a positive helix cutter, this has the tendency to lift the workpiece while causing damage to the top edge as shown in Figure 6.26. On the other hand, the compression-style router generates cutting forces into the top and bottom surfaces of the workpiece. These forces stabilize the cut while eliminating damage to the workpiece edges.

Delamination can be reduced by proper selection of cutting tool geometry and cutting configuration. Straight flute cutters and double spiral compression tools are most suitable because they eliminate the axial force component acting normal to the ply. Down milling (climb) is more suitable than up milling (conventional) for reducing delamination and improving surface roughness. Solid carbide router bits shown in Figure 6.27 represent a group of cutting tools available for milling and trimming of composite panels.

Sawing: Diamond-plated saw blades are used for both straight sawing and contour sawing of graphite-epoxy materials. Composites with aramid fiber reinforcement up

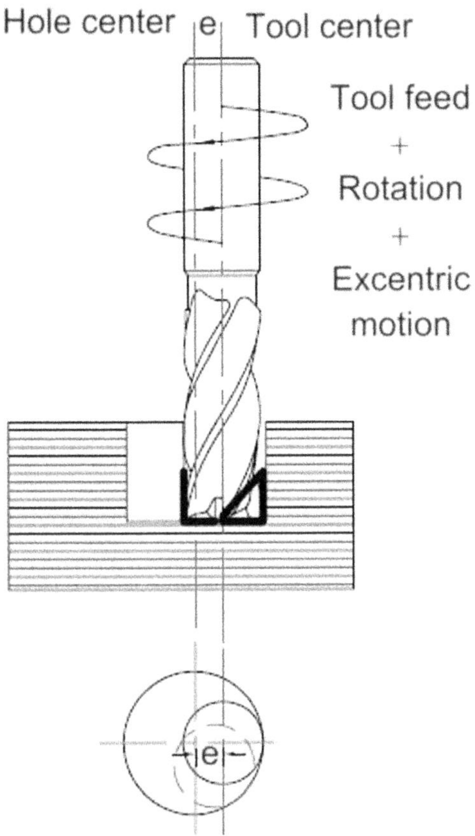

FIGURE 6.25 Orbital drilling.

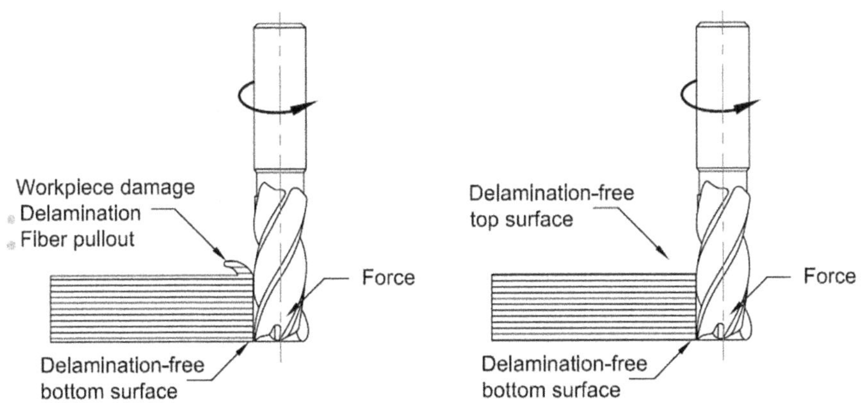

FIGURE 6.26 Standard end milling and compression end (Kennametal incorporation).

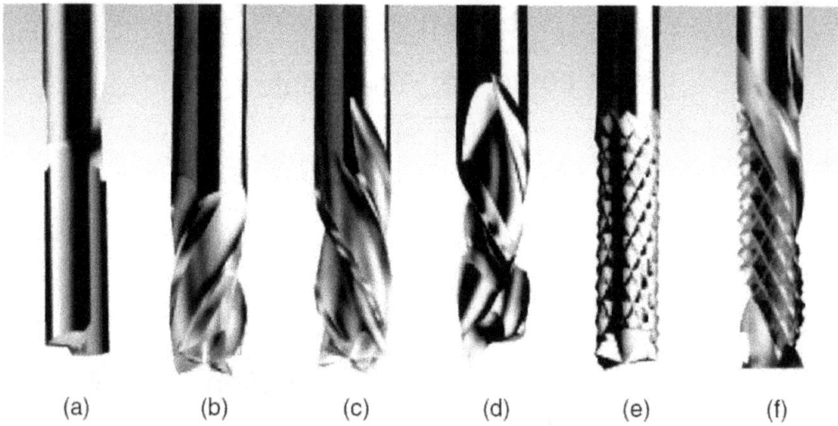

FIGURE 6.27 Geometry of cutting tools for composite materials: (a) Straight flute, (b) up cut helical tool, (c) down cut helical tool, (d) double spiral compression tool I burr tool, (f) fluted burr tool (Courtesy of ONSRUD tool company).

to 6 mm thick can be cut with band saws and saber saws with only slight fuzzing at the exit side. However, the blades are run in reverse, Figure 6.28, such that the heel of the tooth enters the composite first, which gives a more efficient shearing action. The edges of the teeth should be honed to remove sharp edges which tend to pull the Kevlar fibers and create fuzz. Boron-epoxy and boron-aluminum composites can be cut to close tolerances using diamond-coated circular saws or band saws. However, when boron-epoxy is bonded to titanium, flood coolant is required to prevent damage to the epoxy matrix. Table 6.8 shows typical sawing parameters for composite materials (Abrate and Walton, 1992a).

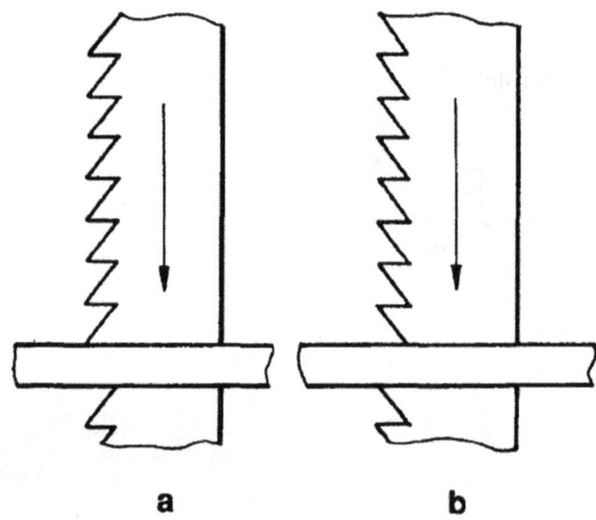

FIGURE 6.28 Sawing of aramid laminates: (a) Conventional cutting, (b) reverse cutting (Abrate and Walton, 1992a).

TABLE 6.8
Typical Sawing Parameters for Composite Materials

Material	Saw Type	Saw Material	Cut Thickness, mm	Cutting Speed, m/s	Feed Rate, mm/min
Graphite-epoxy	Circular	HSS	4	2–12	50.8
	Band saw	PCD	0–25.4	15–24	50.8
Boron-epoxy	Circular[a]	PCD	2.0	30.48	254
	Band saw[a]	PCD	25.4	15.24	50.8
Boron-aluminum (Annealed)	Circular[a]	PCD	1.52	28.5	152
Boron-aluminum (Annealed)	Band saw[a]	PCD	12.7	28.5	25.4

[a] with coolant.
Source: (Abrate and Walton, 1992a).

Grinding: Grinding is a major finishing process that received much attention by the scientific community. Abrasive machining, to the contrary, is a relatively new practice and has received very little attention. Abrasive machining has particular aspects that make it somewhat different than grinding. Differences between the two processes fall mainly in three process parameters: Depth of cut, wheel or cutter diameter (and the resulting velocity), and workpiece feed rates. Table 6.9 lists some of the common ranges of parameters for grinding and abrasive machining. Abrasive cutters have small diameters as compared to grinding wheels and are used for more massive material removal. In these cutters, flutes and coolant holes are incorporated for ease of debris removal and cooling (Sheikh Ahmed Jamal, 2009). The cutting speed and the workpiece feed are relatively lower, but the depth of cut is higher. As a result, abrasive machining equivalent chip thickness is considerably larger, the material removal rate is higher, and the surface roughness is higher than those for grinding.

TABLE 6.9
Comparison between Typical Ranges of Process Parameters in Abrasive Machining and Grinding

Parameter Range	Abrasive Machining	Grinding
Typical cutter/wheel diameter (mm)	6–25	Up to 1.0
Cutting speed (mm/min)	100–500	1500–5000
Workpiece feed rate (m/min)	0.25–1.0	5–50
Material removal rate (cm^3/min)	10–100	1.0×10^{-5}–2.5×10^{-4}

Source: (Sheikh Ahmed Jamal, 2009).

6.3.3 Non-Traditional Machining and Machinability of Composites

In order to overcome the difficulties associated with conventional cutting of composites, non-traditional machining processes such as abrasive water jet machining (AWJM), laser beam machining (LBM), and electro discharge machining (EDM) are introduced.

AWJ machining: With pure water, aramid-epoxy or glass-epoxy laminates up to 6.35 mm thick can be cut, while for graphite-epoxy the upper limit is about 0.15 mm (Sprow, 1987). The performance is significantly improved when abrasive particles are added. In that case, graphite-epoxy laminates up to 10 mm thick can be cut and for a thickness of 3.17 mm, a cutting speed of 1 m/min can be used which is nearly four times that of traditional methods (Howarth, 1990).

Adding abrasives to water je machining (WJM) makes AWJM suitable for machining ceramic, glass, and composite materials that crack and fragment under jet impact causing brittle erosion. AWJM is a suitable choice for machining composite materials due to the following advantages:

- Lack of thermal damage
- No tool wear
- Small cutting forces
- High flexibility
- Reduced material wastage
- Increased productivity

The maximum erosion occurs at impact angle near 90° for brittle materials. High jet pressure causes entrance surface chipping, delamination, internal cracking, and fiber pull-out. Better dimensional accuracy is achieved at a smaller stand-off distance. Moreover, reducing the loss in strength of the cut material is obtained at low water jet pressure, high feed rate, and large stand-off distance. Table 6.10 presents a typical AWJ traverse rate in mm/s for through cutting of different composites (pressure 345 Mpa, nozzle diameter 0.299 mm, tube diameter 0.762 mm, garnet mesh 80).

Compared to LBM and conventional machining, the following can be considered:

1. The performance of AWJM is superior compared to LBM and water jet machining (WJM) as far as delamination is concerned.
2. When AWJM is compared with LBM and conventional machining processes, no heat-affected zone is produced.
3. AWJM offers better cut quality than CO_2 LBM in terms of small out-of-roundness error and reduced hole-conicity and less reduction in tensile strength as compared to CO_2 laser machining.
4. Fatigue tests showed that damage accumulation in conventional drilling (CD) is higher than that of AWJM machined ones.
5. The endurance limit for FRP plates drilled using AWJM was 10% higher as compared to CD.

Laser beam machining: One of the major advantages of lasers is that there is no contact between the tool and the workpiece, eliminating problems associated with chatter and vibration and allowing for the machining of small or thin components

TABLE 6.10
Typical AWJ Traverse Rate in mm/s for through Cutting of Different Composites (Pressure 345 Mpa, Nozzle Diameter 0.299 mm, Tube Diameter 0.762 mm, Garnet Mesh 80)

	Thickness (mm)						
Material	**0.79**	**1.60**	**3.18**	**6.36**	**12.7**	**19.1**	**50.8**
Ceramic matrix composites							
SiC fiber in SiC		1.1	0.6	0.45			
ZrO_2-MgO			0.8	0.7			
Al_2O_3COCrAly (80%/20%)			0.95	0.65			
Al_2O_3COCrAly (60%/40%)			0.95	0.65			
Al_2O_3/SiC (7.5%) SiC Abrasives			2.7	1.4			
SiC/TiB_2 (15%)			0.29	0.15			
Metal matrix composites							
Mg.B_4C (15%)	70	30	15	10		4	
Al/SiC (15%)	70		17	10	5		
Al/SiC (25%)				9.5	5		
Al/mullite (5%)	75	35	20	12	7.5	5	2.5
Al/Al_2O_3 (15%)	65	28	15	8	4		
Organic matrix composites							
Carbon/carbon composites	42	32	22	13	7.7	4	0.85
Epoxy/glass composites	105	95	76	42	17	12	5
Graphite/epoxy composites	74	63	52	40	17	10	4.2

Source: (Abrate and Walton, 1992b).

without requiring mechanical force. FRP composites are laminated with different fibers orientation bound together in a polymer matrix. The quality of machining composite is affected by laser parameters, fiber orientations, and machining environment. Each constituent retains its own thermal properties. Table 6.11 shows typical thermal properties of matrix and fiber materials.

1. Vaporization of fiber (3300°C) and matrix (350–500°C) occur at different temperatures.
2. The heat conduction parallel to the fiber axis is faster than in the transverse directions, which results in a non-uniform HAZ size as a result of the different fiber orientations.
3. The difference in thermal conductivities 50.00 (W/m/K) for graphite and 0.2 (W/m/K) for polyester makes it more difficult to achieve uniform, high-quality laser beam machining.
4. The heat conduction along the fibers is much faster than that in the polymer matrix.
5. The time that elapses before the vaporization of fibers is larger than that for the resin, so a great amount of the heat is absorbed by the matrix.

TABLE 6.11
Typical Thermal Properties of Matrix and Fiber Materials

Material	Conductivity, W/m K	Heat Capacity, J/kg K	Vaporization Temperature, °C	Thermal Diffusivity, $(cm^2/s) \times 10^3$	Density, g/cm^3
Epoxy	0.10	1100	400–440	0.76	1.20
Polymer	0.20	1200	350–500	1.33	1.25
Aramid fiber	0.05	1420	950	0.24	1.44
Graphite fiber	50	710	3300	380.66	1.85

Source: (Sheikh Ahmed Jamal, 2009).

Proper selection of the machining parameters can reduce the thermal damage in the workpiece. Figure 6.29 shows the extent of HAZ during laser cutting of CFRP using different laser types. The major problem of laser beam machining is the formation of HAZ which can be reduced by the following:

- Short wavelength of pulsed laser radiation
- High feed rate which decreases energy per unit length
- Increased spacing between adjacent laser beams scan traces
- Small laser power
- High pressure assisting gas

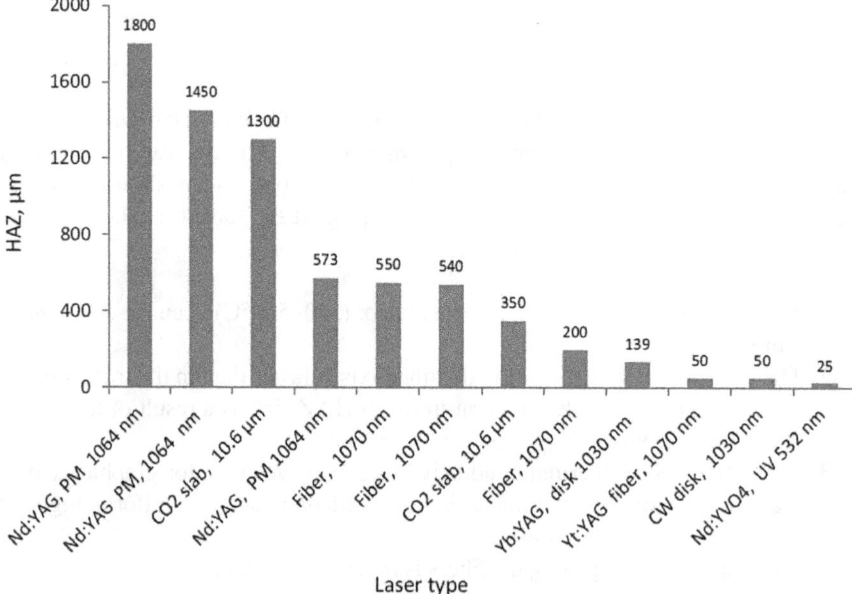

FIGURE 6.29 Heat-affected zone in CFRP composites by different laser types (El-Hofy and El-Hofy, 2018).

Table 6.12 shows typical traverse speed and specific energy for laser cutting of composites. Both HAZ and kerf width decreased and the machinability improved by decreasing the specific energy that occurs when using larger laser power to cut large CFRP thickness at high cutting speed, Figure 6.30. The maximum feed rate depends on the released mean power while the minimum spot overlap depends on the pulse energy. Narrow kerf width can be obtained at reduced spot overlap or increased cutting speed to its limit.

EDM: Machining composites by EDM is difficult, because they are inhomogeneous substances consisting of electrically conductive high tensile fiber materials and an electrical non-conductive matrix material that is usually made from a plastic or epoxy resin. EDM was applied to metal matrix composites consisting of silicon carbide whiskers in an aluminum matrix (Sic/Al) with 15% and 25% volume fraction reinforcement, respectively (Ramulu et al., 1989). The machining time for material with 25% fiber reinforcement is almost double that for material with 15% fibers.

EDM was used for die sinking of graphite-epoxy laminates since the graphite fibers are electrically conductive using voltage of 100 V and peak discharge currents of 0.5–5 A. At high currents, high temperatures are produced which cause severe melting of the composite surface, thermal expansion of the graphite fibers in the lateral direction, and debonding between fibers and the matrix. Melted matrix material smeared over the conductive graphite fibers lowers the conductivity, leading to a reduction of the MMR and degradation of surface quality. Severe distortion of the fibers into a hexagonal shape was observed. Therefore, low currents must be used (Lau et al., 1990).

TABLE 6.12
Typical Traverse Speed and Specific Energy for Laser Cutting of Composites

Material	Thickness, mm	Cutting Speed, mm/s	Power, W	Specific Energy, J/mm²
Glass-epoxy	3.2	5	250	15.625
	1.6	250	1200	3.0
	1.6	86.7	450	32.33
Glass-polyethylene	3.2	15	250	5.2
	4.8	30	250	1.7
Boron-epoxy	8	2.67	15000	702.2
SiC-aluminum	0.38	30	150	13.16
Aramid-epoxy	2	16	500	15.63
		133	500	1.88
Graphite-epoxy	0.5	38.1	400	21
Kevlar-epoxy	1	38.1	400	10.5
Kevlar-epoxy	2.36	89.7	300	1.42
Glass ceramic-SiC fibers	0.9	22	910	49.96
			1400	70.7

Source: (Abrate and Walton, 1992b).

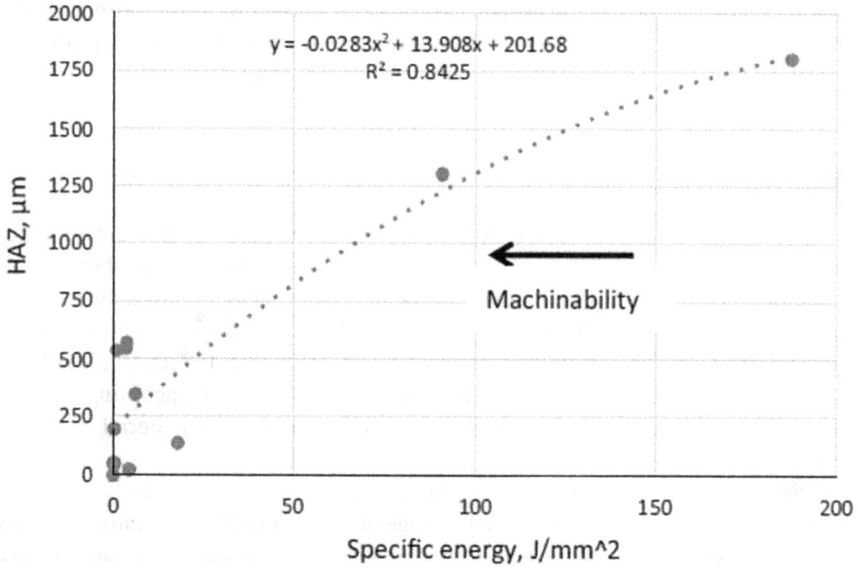

FIGURE 6.30 Effect of specific energy on HAZ and machinability of CFRP types (El-Hofy and El-Hofy, 2018).

Electric discharge wire cutting (EDWC) was used for ceramic materials. Values of cutting rates in the 13–35 mm^2 min range are reported for a sintered tungsten carbide with 15% cobalt with thicknesses ranging from 5 to 100 mm. Average peak to valley surface roughness was 7–9 μm. For die sinking of Sic/graphite conductive ceramics, MRR in the range of 0.5–7 mm^3 min were obtained for ceramic composites (König et al., 1988). With high currents, the workpiece can actually fracture during EDM (Nakamura et al., 1989).

6.4 REVIEW QUESTIONS

6.4.1 Define what is meant by ceramics.

6.4.2 What are the main properties of ceramic materials?

6.4.3 Based on their application, what are the main types of ceramics?

6.4.4 Based on their chemical composition, what are the main types of ceramics?

6.4.5 Compare ceramic materials and metals regarding melting point, density, elastic modulus, and hardness?

6.4.6 Explain how non-conductive ceramics can be machined using grinding, RUM, and EDM.

6.4.7 It is required to machine the following traditionally and non-traditionally:

a) Non-conductive ceramics
b) Conductive ceramics
c) Carbon fiber reinforced polymer (CFRP)

You are requested to give all possible traditional and non-traditional machining processes you know. Then select four from these processes (two

TM and two NTM) for each material, and use neat sketches to describe the concept of each process. What are the operating parameters of each process? What are their advantages and limitations?

REFERENCES

Abdelkawy, A, Hossam, M & El-Hofy, H 2018, 'Mathematical model of thrust force for rotary ultrasonic drilling of brittle materials based on the ductile-to-brittle transition phenomenon', *The International Journal of Advanced Manufacturing Technology*. Doi:10.1007/s00170-018-2943-4

Abrate, S, & Walton, DA 1992a, 'Machining of composite materials. Part I: traditional methods', *Composites Manufacturing*, vol. 3, no. 2, pp. 75–83. Doi:10.1016/0956-7143(92)90119-f.

Abrate, S & Walton, DA 1992b, 'Machining of composite materials. Part II: non-traditional methods', *Composites Manufacturing*, vol. 3, no. 2, pp. 85–94.

Babu, J, Sunny, T, Paul, NA, Mohan, KP, Philip, J, & Davim, JP 2016, Assessment of delamination in composite materials: a review. *Proceedings of the Institution of Mechanical Engineers, Part B: Journal of Engineering Manufacture*, vol. 230, no. 11, pp. 1990–2003. Doi:10.1177/0954405415619343.

Brevier technical ceramics: dimensions and shrinkage. Viewed 11 March 2011, <http://www.keramverband.de/brevier_engl/4/1/4_1_5.htm>.

Chak, SK & Rao, PV 2007, 'Trepanning of Al_2O_3 by electro-chemical discharge machining (ECDM) process using abrasive electrode with pulsed dc supply', *International Journal of Machine Tools and Manufacture*, vol. 47, pp. 2601–2070.

El-Hofy, H 2018, *Fundamentals of machining processes: conventional and nonconventional processes*, 3rd edn, Taylor and Francis Ltd.

El-Hofy, MH & El-Hofy, H 2018, 'Laser beam machining of carbon fiber reinforced composites: a review', *The International Journal of Advanced Manufacturing Technology*. Doi:10.1007/s00170-018-2978-6.

El-Taybany, Y & El-Hofy, H 2019, 'Mathematical model for cutting force in ultrasonic-assisted milling of soda-lime glass', *The International Journal of Advanced Manufacturing Technology*, vol. 103, pp. 3953–3968. Doi:10.1007/s00170-019-03399-6.

Encyclopedia Britannica. <Britannica.com>.

Feito, N, Díaz-Álvarez, J, Díaz-Álvarez, A, Cantero, J & Miguélez, M 2014, Experimental analysis of the influence of drill point angle and wear on the drilling of woven CFRPs. *Materials*, vol. 7, no. 6, pp. 4258–4271.doi:10.3390/ma7064258.

Gotoh H, Tani T & Mohri N 2016, 'EDM of insulating ceramics by electrical conductive surface layer control', *Procedia CIRP*, vol. 42, pp. 201–205. Doi:10.1016/j.procir.2016.02.271.

Gupta, PK, Dvivedi, A & Kumar, P 2014, 'Developments on electrochemical discharge machining: a review of experimental investigations on tool electrode process parameters', *Proceedings of the Institution of Mechanical Engineers, Part B: Journal of Engineering Manufacture*, vol. 229, no. 6, pp. 910–920. Doi:10.1177/0954405414534834.

Hamzah,Esah, Izmanand, S, & Khoo, CY, 2008, A review on the rotary ultrasonic machining of advanced ceramics. *JurnalMekanikal*, vol. 25, pp. 9–23.

Han, M, Min, BK, & Lee, SJ 2008, 'Modeling gas film formation in electro chemical discharge machining processes using a side-insulated electrode', *Journal of Micromechanics and Microengineering*, vol. 18, pp. 045019, 8 p.

Han, MS, Min, BK, & Lee, SJ 2009, 'Geometric improvement of electrochemical discharge micro-drilling using an ultrasonic-vibrated electrolyte', *Journal of Micromechanics and Microengineering*, vol. 19, pp. 065004 (8p.).

Howarth, SG & Strong, AB 1990, 'Edge effects with waterjet and laser beam cutting of advanced composite materials', *35th International SAME Symposium*, April, 2–5.

Huang, H 2009, 'High speed grinding of advanced ceramics: a review', *Key Engineering Materials*, vol. 404, pp. 11–22.doi:10.4028/www.scientific.net/kem.404.1

Ioan, D., Marinescu, Mike P. Hitchiner, EckartUhlmann, W. Brian Rowe & Ichiro Inasaki 2016, *Handbook of machining with grinding wheels*, 2nd edn, CRC Press: Boca Raton, FL.

Islam, MU., & Campbelln, G 1993, 'Laser machining of ceramics: a review', *Material and Manufacturing Processes*, vol. 8, no. 6, pp. 611–630.

Jahan, M, Rahman, M & Wong, Y 2014, *Micro-electrical Discharge Machining (micro-EDM): Processes, Varieties, and Applications*, Springer, Singapore.

Jiao Y, Liu WJ, Pei ZJ, et al. Study on edge chipping in rotary ultrasonic machining of ceramics: an integration of designed experiments and finite element method analysis. *J Manuf Sci Eng* 2005; 127: 752–758.

Kalpakjian, S & Schmid, SR 2003, *Manufacturing processes for engineering materials*, 4th edn, Prentice Hall.

Katz, Z, & Tibbles, C 2005, 'Analysis of micro-scale EDM process', *The International Journal of Advanced Manufacturing Technology*, vol. 25, pp. 923–928.

Kondratenko, VS 1997, 'Method of splitting non-metallic materials', U.S. Patent 5609284.

König, W, Dauw, DF, Levy, G & Panten, U 1988, 'EDM-future steps towards the machining of ceramics', *CIRP Annals*, vol. 37, no. 2, pp. 623–631.doi:10.1016/s0007-8506(07)60759-8.

Kutz, M 2002, 'Handbook of materials selection – part 2. Major materials, Section 14', in R.N. Katz (ed.), *Overview of ceramics materials, designs, and application*, John Wiley & Sons, Inc, New York.

Lee, TC & Lau, WS 1991, 'Some characteristics of electrical discharge machining of conductive ceramics', *Materials and manufacturing processing*, vol. 6, no. 4, pp. 635–648.

Li, K & Sheng, P 1995, 'Plane stress model for fracture of ceramics during laser cutting', *International Journal of Machine Tools and Manufacture*, vol. 35, no. 11, pp. 1493–1506.

Lim, H, Wong, Y, Masaki, T, Asad, A, & Rahman, M 2007, 'Integrated hybrid micro/nano-machining', in *Proceedings of the ASME international manufacturing science and engineering conference*, Los Angeles, CA, USA, 4–8 June 2007, pp. 197–209.

Lu, G, Siores, E & Wangm, B 1999, 'An empirical equation for crack formation in the laser cutting of ceramic plates', *Journal of Materials Processing Technology*, vol. 88, pp. 154–158.

Lumley, RM 1969, 'Controlled separation of brittle materials using a laser', *American Ceramic Society Bulletin*, vol. 48, pp. 850–854.

Makino, E, Sato, T & Yamada, Y 1987, 'Photoresist for photochemical machining alumina ceramic', *Precision Engineering*, vol. 9, no. 3, pp. 153–157.

Masuzawa, T 2001, 'Micro EDM', in *Proceedings of the ISEM XIII 2001*, Bilbao, Spain, pp. 3–19.

New Structural materials technologies: opportunities for the use of advanced ceramics and composites, report. Crediting UNT Libraries Government Documents Department; Gainesville, FL, USA, 1986. University of North Texas Libraries, Digital Library, digital.library.unt.edu.

Pandey, PC, & Shan, HS 1980, *Modern machining processes*, McGraw Hill, New Delhi.

Park, JM, Jeong, SC, Lee, HW, Jeong, HD & Lee, E 2002, 'A study on the chemical mechanical micro-machining (C3M) process and its application', *Journal of Materials Processing Technology*, vol. 130–131, pp. 390–395.

Prabhakar, D 1995, *Machining advanced ceramic materials using rotary ultrasonic machining process*, M.S. Thesis, University of Illinois at Urbana-Champaign.

Präzisions Glas&Optik GmbH.

Samant, AN & Dahotre, NB 2009, 'Laser machining of structural ceramics—a review', *Journal of the European Ceramic Society*, vol. 29, no. 6, pp. 969–993. Doi:10.1016/j.jeurceramsoc.2008.11.010.

Savrun, E & Taya, M 1988, 'Surface characterization of SiC whisker/2124 aluminum of Al_2O_3 composites machined by abrasive water jet', *Journal of Materials Science*, vol. 23, pp. 1453–1458.

Schubert, A, Zeidler, H, Kühn, R & Hackert-Oschätzchen, M 2015, 'Microelectrical discharge machining: a suitable process for machining ceramics', *Journal of Ceramic*, vol. 2015 pp. 1–9.

Shin, T, Mohri, N, Yamada, H, Kosuge, M, Furutani, K, Fukuzawa, Y & Tani, T 1998, 'Machining phenomena in EDM of insulating ceramics-effect of condenser electrical discharges', *VDI Berichte*, vol. 1405, pp. 437–444.

Sheikh Ahmed Jamal Y. 2009. *Machining of Polymer Composites*. Springer,

Sprow, EE 1987, 'Cutting composites: three choices for any budget', *Tooling and Production*, vol. 43, no 12, pp. 46–50.

Steen, WM 2003, 'Laser material processing – an overview', *Journal of Optics A – Pure and Applied Optics*, vol. 5, pp. s3–s7.

SubsTech (Substances & Technologies) knowledge source in Materials Engineering.

Tibbles, C 2005, *Analysis of micro-scale EDM process*, Springer, Berlin/Heidelberg, Germany.

Tsai, CH & Liou, CS 2003, 'Fracture mechanism of laser cutting with controlled fracture', *Journal of Manufacturing Science and Engineering*, vol. 125, no. 3, p. 519.doi:10.1115/1.1559163.

Tsai, CH & Chen, HW 2003a, 'Laser cutting of thick ceramic substrates by controlled fracture technique', *Journal of Materials Processing Technology*, vol. 136, no. 1–3, pp. 166–173. Doi:10.1016/s0924-0136(03)00134-1.

Tsai, C-H & Chen, CJ 2003b, 'Formation of the breaking surface of alumina in laser cutting with a controlled fracture technique', *Proceedings of the Institution of Mechanical Engineers, Part B: Journal of Engineering Manufacture*, vol. 247, no. 4, pp. 489–497.

Tuersley, IP, Jawaid, A & Pashby, IR 1994, Review: various methods of machining advanced ceramic materials. *Journal of Materials Processing Technology*, vol. 42, no. 4, pp. 377–390. Doi:10.1016/0924-0136(94)90144-9.

Uhlmann, E, Piltz, S & Doll, U 2005, 'Machining of micro/miniature dies and moulds by electrical discharge machining—Recent development', *Journal of Materials Processing Technology*, vol. 167, no. 2–3, pp. 488–493.doi:10.1016/j.jmatprotec.2005.06.013.

Unger, U & Wittenbecher, W 1998, 'The cutting edge of laser technology', *Glass 75*, pp. 101–102.

Vikulin, V, Kelina, I, Shatalin, A, & Rusanova, L 2004, 'Advanced ceramic structural materials', *Refractories and Industrial Ceramics*, vol. 45, pp. 383–386.

Yoo, H-K, Ko, J-H, Lim, K-Y, Kwon, WT & Kim, YW 2015, 'Micro-electrical discharge machining characteristics of newly developed conductive SiC ceramic', *Ceramics International*, vol. 41, pp. 3490–3496.

Zheng, ZP, Su, HC, & Huang, FY 2007, 'The tool geometrical shape and pulse-off time of pulse voltage effects in a pyrex glass electrochemical discharge micro drilling process', *Journal of Micromechanics and Microengineering*, vol. 17, pp. 265–272.

7 Assisted Machining Technologies

7.1 INTRODUCTION

The performance of machining processes can be enhanced in terms of material removal rate, machinability, product accuracy, and surface characteristics by introducing thermal, vibration, or magnetic assistance. As an example, machining by cutting is assisted using an external thermal effect which heats the workpiece material, in front of the cutting tool, using either a laser beam as in the case of laser-assisted turning (LAT) and laser-assisted milling (LAM) or plasma beam heating source as in the case of plasma-assisted turning (PAT) and plasma-assisted milling (PAM).

Other metal cutting and processes are also assisted using mechanical vibrations which improve the process performance compared to cutting without vibrations. Examples of such assistance include vibration-assisted turning (VAT), drilling (VAD), milling (VAM), and vibration-assisted grinding (VAG).

Some advanced abrasion processes combine a magnetic field together with the abrasion action as in the case of magnetic abrasive machining (MAF), magnetic float polishing (MFP), magnetorheological finishing (MRF), and magnetorheological abrasive flow finishing (MRAFF). Non-conventional machining is also assisted by mechanical vibrations at low or ultrasonic frequency and small amplitudes in VA-ECM, VA-EDM, and VA-AWJM in order to improve their performance. Figure 7.1 shows the different assisted machining processes by thermal, vibration, and magnetic field.

7.2 THERMAL-ASSISTED MACHINING

The use of high-strength materials, such as nickel-based super alloys and titanium alloys, is becoming increasingly common in aerospace, automotive, energy, medical, and mining applications. However, these materials are traditionally considered to be difficult-to-machine. Conventional machining processes for these materials are notoriously affected by slow machining speeds and/or frequent tool changes due to short tool lives.

The term thermal-assisted machining refers to a conventional cutting process in which an external energy source is used to enhance the chip-generation mechanism. Heat-assisted machining (hot machining) is an alternative machining method used for improving the machinability of difficult-to-cut metals and alloys. The principle behind hot machining is to increase the difference in hardness of the cutting tool and workpiece, leading to reduced component forces, cutting power, improved surface finish, and longer tool life. Depending on the heat source, heat-assisted machining is as follows.

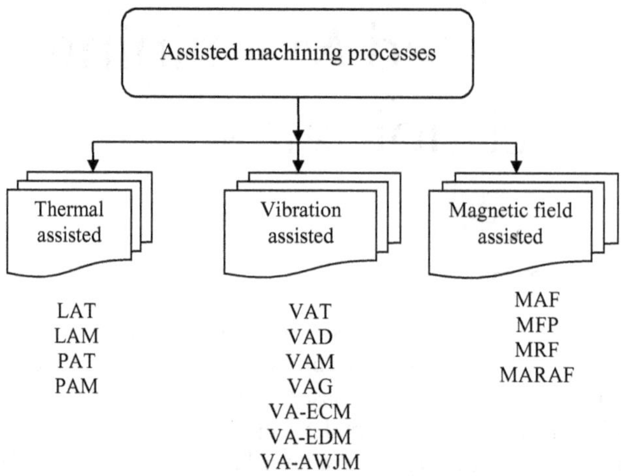

FIGURE 7.1 Assisted machining processes.

7.2.1 Laser-Assisted Machining

The laser-assisted machining (LAM) method became a viable industrial option for cutting difficult-to-machine materials. In LAM, a laser provides intense and localized heating to the workpiece ahead of the cutting tool, which in turn, lowers the material strength in the cutting zone at a certain elevated temperature depending on the workpiece material. LAM achieves lower cutting force/power, less tool wear, higher material removal rate, and better surface quality at low cost. Typical examples include laser-assisted turning (LAT), Figure 7.2, where dual laser beams are used: one is used for heating the work material while the other is focused at the cut

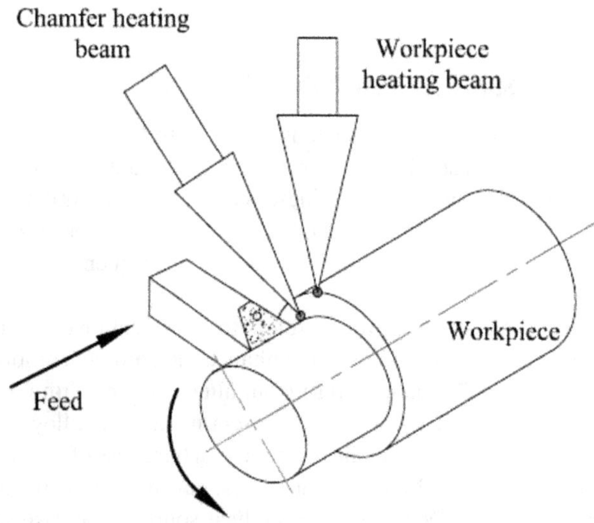

FIGURE 7.2 Laser-assisted turning (LAT).

chamfer. In laser-assisted milling (LAM), Figure 7.3, the beam is heating an area in front of the milling cutter (Nurul and Ginta, 2014).

7.2.2 Plasma-Assisted Machining

Plasma-assisted machining has been applied to the machining of three very low machinability materials: A Ni-base alloy (Inconel 718), a Co-base alloy (Haynes 25) (both belonging to the group of the heat-resistant alloys), and the Ti-base alloy Ti6Al4V. In plasma-assisted turning (PAT), Figure 7.4, it is possible to turn difficult-to-cut materials by conventional methods at a cutting speed of 2 m/min and feed rate 5 mm/rev to produce a surface finish of 0.5 mm R_t. The plasma beam is directed toward the workpiece chamfer at certain angles in front of the cutting tool to control the depth of cut during the machining operation. In plasma-assisted milling (PAM), Figure 7.5, the plasma beam is heating an area in front of the milling cutter. Process parameters include nozzle angle, stand-off distance, and distance from the milling cutter.

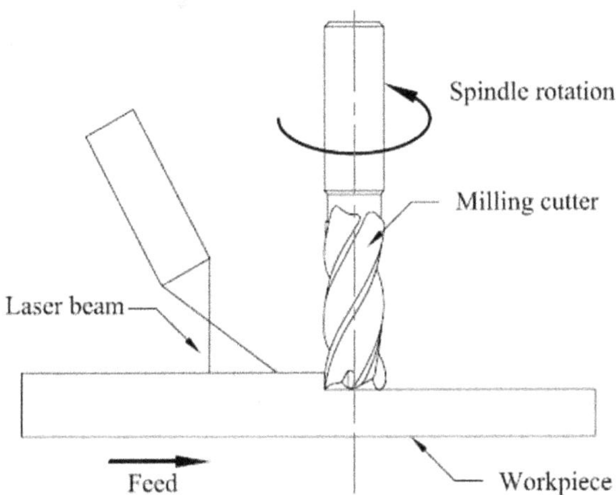

FIGURE 7.3 Laser-assisted milling (LAM).

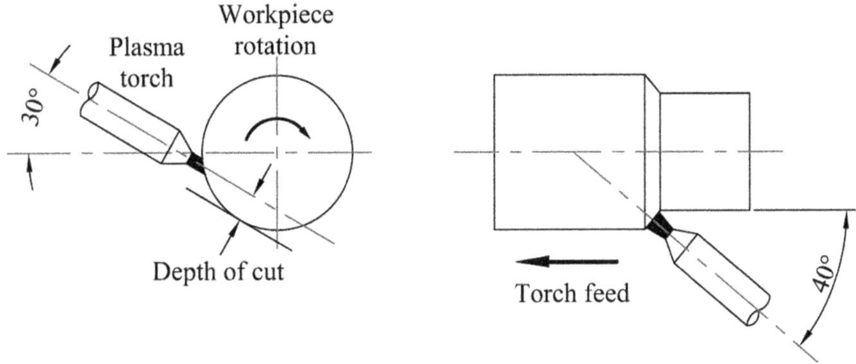

FIGURE 7.4 Plasma-assisted turning (PAT).

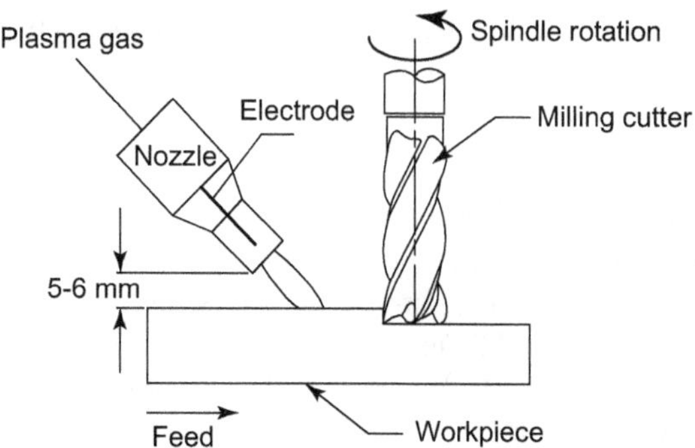

FIGURE 7.5 Plasma-assisted milling (PAM) (López de Lacalle et al., 2004).

7.3 VIBRATION-ASSISTED MACHINING (VAM)

7.3.1 Principles and Aims of VAM

There is a high demand for precision machining of components made from hard and brittle materials such as glasses, lens, ceramics, and high-strength alloys. Such a requirement cannot be achieved by conventional and non-conventional machining methods. To achieve good surface finish, high accuracy, and high precision at low cost, vibration is introduced to any of the workpiece, tool, and in some processes to the working medium during vibration-assisted machining (VAM). A VAM system can be a resonant one that operates at frequencies greater than 20 kHz, and amplitudes less than 6 μm. The non-resonant system operates at frequencies up to 40 kHz and amplitudes ten times greater than the resonant system.

VAM can be performed as one-dimensional (1-D) or two-dimensional (2-D) arrangements as shown in Figure 7.6. The one-dimensional VAM operates in a plane

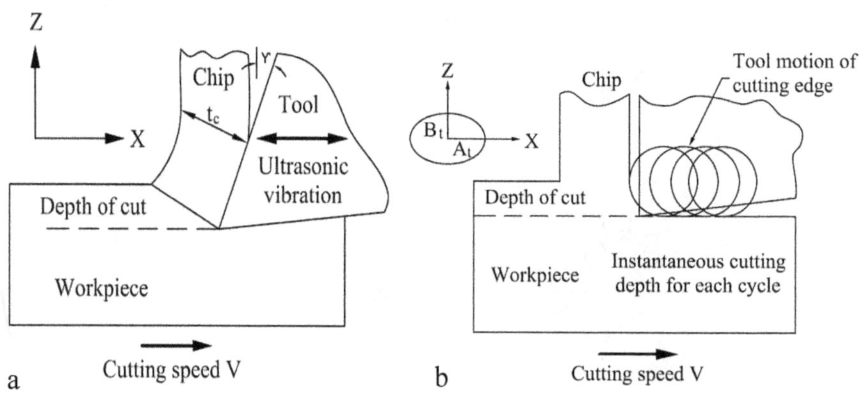

FIGURE 7.6 (a) 1-D and (b) 2-D VAM.

parallel to the workpiece surface, in line with the main cutting force, Figure 7.6a. The 2-D VAM, Figure 7.6b, produces an elliptical tool motion, where the major axis (A_t) of the ellipse is in line with the main cutting force and the minor axis (B_t) is in line with the thrust force. The amplitude of vibration in each axis may or may not be the same and described as $A_t \times B_t$. The resonant 1-D VAM system is the most common type used in applications with small amplitudes and frequency of 20 kHz, respectively. The resonant 2D system operates at frequencies of 20–40 kHz with tool typical path 3 µm × 3 µm to 8 µm × 4 µm. The main categories of VAM are shown in Table 7.1. These include the elliptical cutting which is a 2-D cutting process, and the 1-D VAM when the tool vibrates in the cutting direction or perpendicular to the cutting direction. In continuous cutting, the tool is fed without vibration.

Figure 7.7 shows a 1-D VAM arrangement where the tool is driven harmonically in a linear path, which is superimposed on the feed motion of the workpiece. For VAM frequency f, vibration amplitude A_t, and feed velocity V, the position $x(t)$ and velocity $\dot{x}(t)$ of the tool relative to the workpiece are given by,

$$x(t) = Vt + A_t \cos(2\pi ft)$$

$$\dot{x}(t) = -2\pi f A_t \sin(2\pi ft) + V$$

$$z(t) = 0$$

TABLE 7.1
Cutting Modes

Process Type	Cutting Type	Vibration of the Tool Tip	Tool Tip Trajectory	Tooling Displacement
Continuous cutting	Conventional cutting			$x(t) = Vt$ $z(t) = 0$
1-D VAM	Tool vibration in the cutting direction			$x(t) = Vt + A_t\cos(2\pi ft)$ $z(t) = 0$
	Tool vibration normal in the cutting direction			$x(t) = Vt$ $z(t) = B_t \sin(2\pi ft)$
2-D VAM	Elliptical tool vibration		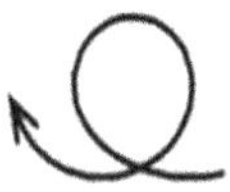	$x(t) = Vt + A_t \cos(2\pi ft)$ $z(t) = B_t \sin(2\pi ft)$

Source: (Xu and Zhang, 2015).

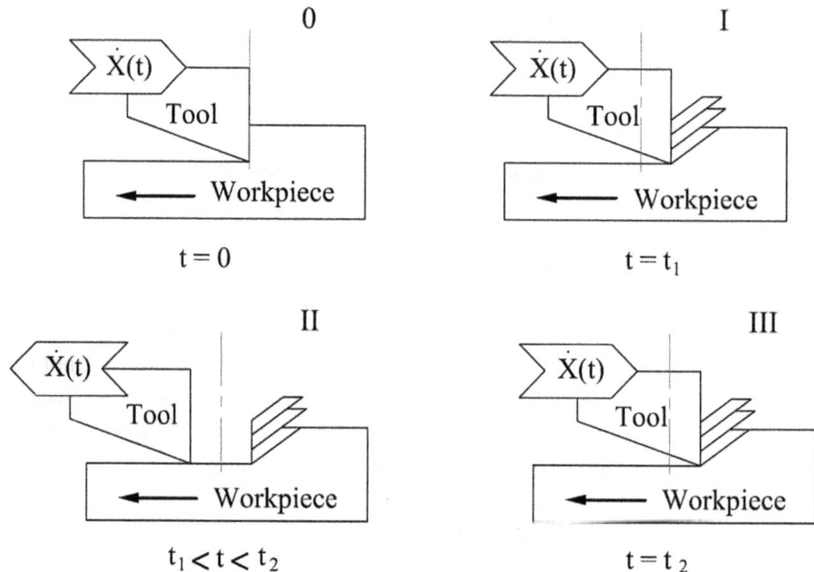

FIGURE 7.7 Principles of machining in 1-D vibration-assisted machining (Maroju et al., 2014).

Figure 7.7 provides a snapshot pictorial of the critical points in the 1D VAM cutting cycle. The tool workpiece contact ratio plays a key role in the process where an increase in the tool vibration parameters and decrease in cutting speed reduce both cutting force and tool wear, improve surface quality, and prolong tool life.

Figure 7.8 shows the 2-D VAM which adds vertical harmonic motion to the horizontal motion of 1D VAM. This causes the tool tip to move in a circle or ellipse, which is superimposed on the feed motion. As can be seen in Table 7.1, the tool moves at frequency f in an ellipse with horizontal amplitude A_t and vertical amplitude B_t. The horizontal position $x(t)$ and velocity relative to the work $\dot{x}(t)$ are again as described above. The vertical position $z(t)$ and vertical velocity $\dot{z}(t)$ are given by,

$$z(t) = B_t \sin(2\pi f t)$$

Figure 7.9 shows the overall advantages of vibration-assisted machining, which include the following:

- VAM mechanisms produce thinner and shorter chips which in turn reduce the average cutting force and induce less stresses to the machined surface.
- Periodic tool-work separation provides a gap to dissipate heat between tool and workpiece. Additionally, better lubrication and cooling effects are achieved compared to conventional machining.
- The reduced forces, cutting temperature, and the periodic separation help the tool to dissipate heat which results in extending the tool life, especially in the case of difficult-to-cut materials.

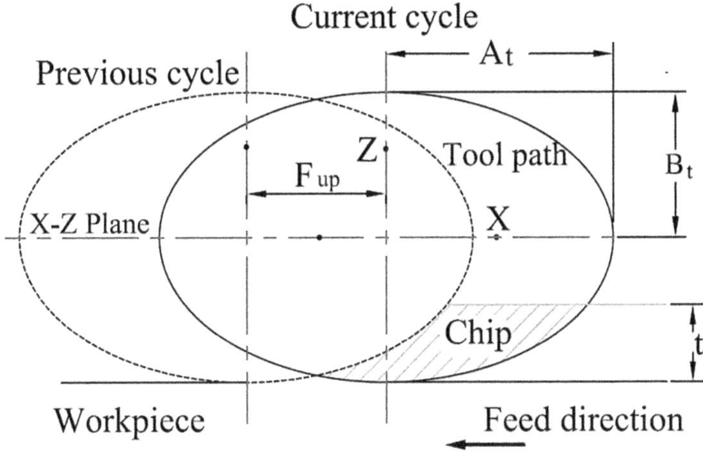

FIGURE 7.8 Tool path in 2-D VAM (Maroju et al., 2014).

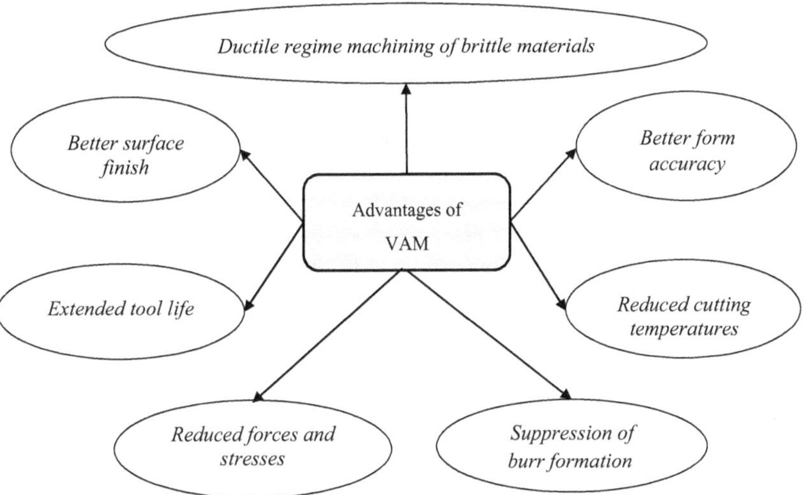

FIGURE 7.9 Advantages of VAM.

- The lower thrust forces decrease the amplitude of tool vibration relative to the workpiece which improves the surface quality.
- The lower friction forces induce lower temperatures which in turn reduce the possible subsurface cracking.
- During VAM, the depth of cut is small, and many brittle materials behave like ductile ones and produce chips by means of plastic flow with minimal subsurface cracking.
- The reduced instantaneous compressive and bending stresses during avoid burr formation.

7.3.2 Vibration-Assisted Traditional Machining Processes

7.3.2.1 Vibration-Assisted Turning

Vibration-assisted turning (VAT) applies ultrasonic frequency of vibration to the cutting tool or the workpiece in addition to the relative main cutting and feed motions to achieve better machining performance. As shown in Figure 7.10, there are three independent principal directions in which ultrasonic vibration can be applied. The process is influenced by the depth of cut, cutting speed, feed rate in addition to the vibration frequency and vibration amplitude. Feed rate has the most significant effect on surface roughness while the depth of cut and cutting speed have a less significant effect on surface roughness. During VAT, the tool face is separated from the workpiece repeatedly leading to an alternating gap between the tool and workpiece. Increasing the vibration amplitude means allowing more cutting fluid to extract the heat generated during the cutting process. This process increases the tool life, improves surface finish, reduces the cutting forces, and decreases the probability of build-up-edge formation. Chipping of the cutting edge can also be prevented which leads to good surface finish. Vibrating the tool reduces the friction force due to the effect of dynamic fluidization of the dry friction. Compared to continuous cutting (CT) without vibrations, VAT decreases the cutting force by 25.0–35.0% while it improves the surface roughness by 12.0–40.0% (Behera, 2011).

7.3.2.2 Vibration-Assisted Drilling

The hard-to-cut materials such as ceramics, glass, and carbon fiber reinforced polymer (CFRP) have high hardness and strength at high temperatures, high wear resistance, and low fracture toughness. They possess poor machinability during conventional drilling (CD) or even vibration-assisted drilling (VAD) using traditional twist drills, Figure 7.11a. Vibration-assisted drilling (VAD) using an abrasive tool, shown in Figure 7.11b, is used to cut hard and brittle materials having poor machinability. The process is mainly applied to machining high precision parts for optics, dental parts, and electronics (Abdelkawy et al., 2018).

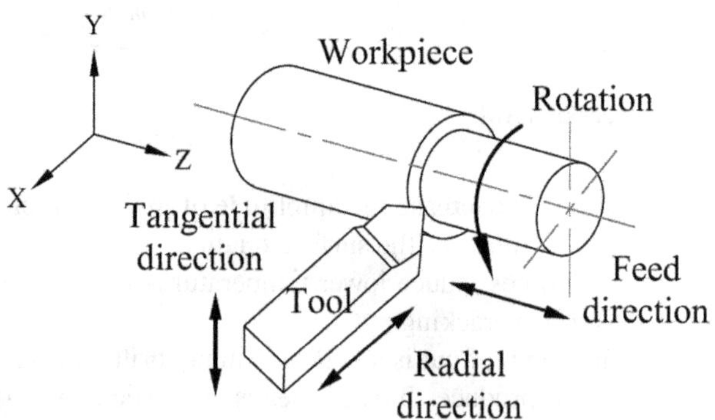

FIGURE 7.10 Principle vibration directions during VAT.

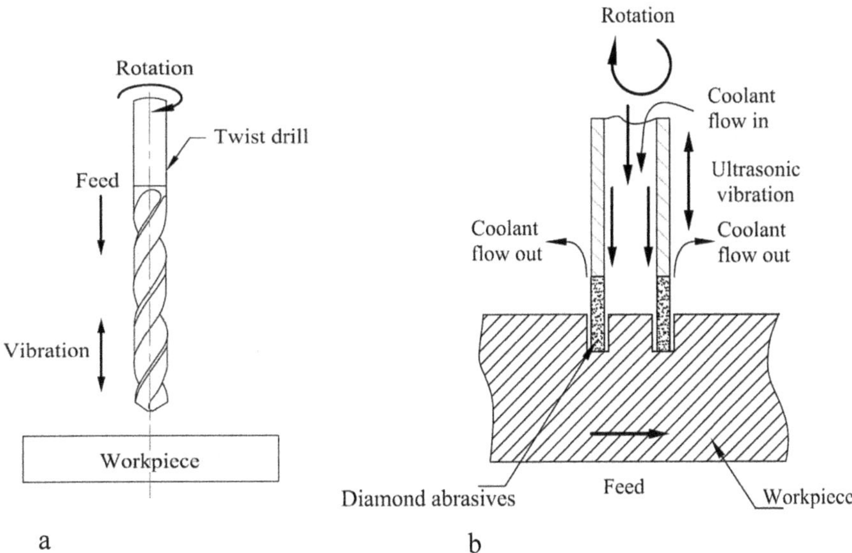

FIGURE 7.11 Vibration-assisted drilling using: (a) Twist drills and (b) VAD using abrasive tool.

In VAD, the cutting force is directly affected by the cutting conditions such as the ultrasonic frequency and amplitude. Moreover, the properties of the part material such as the hardness and fracture toughness affect the process behavior. The cutting forces are less than that in CD due to the reduced interaction time between the tool and workpiece in VAD. VAD reduces the axial cutting force by 40%, the cutting force by 10%, and the torque by 25% compared to CD. The rate of metal removal (MRR) was higher for VAD; however, the surface roughness for both processes was nearly the same (Maroju et al., 2014).

7.3.2.3 Vibration-Assisted Milling

Vibration-assisted milling (VAM) is a conventional milling process with tool or workpiece vibration as shown in Figure 7.12a. The workpiece vibration can be given in the feed or perpendicular to the feed direction while the tool can be vibrated in the axial direction. On the other hand, vibration-assisted milling combines the material removal mechanism of grinding and the milling kinematics with ultrasonic assistance. The process is suitable for machining hard-to-cut brittle materials used in many industrial applications. During VAM, shown in Figure 7.12b, the abrasive tool rotates at a certain speed, vibrates along its axis at high frequency, and the workpiece is fed horizontally to achieve surface or slot milling (El-Taybany and El-Hofy, 2019). Important VAM control variables are those related to the machine (vibration frequency and amplitude, rotational speed, feed rate, cutting depth, and cutting fluid), the cutting tool (abrasive grain size and type, boning type, dimensions and thermal properties), and the workpiece material (hardness, toughness, ultimate tensile strength, thermal properties). The performance obtained from VAM is nearly 6–10 times higher than that of the conventional milling (Marcel et al., 2014).

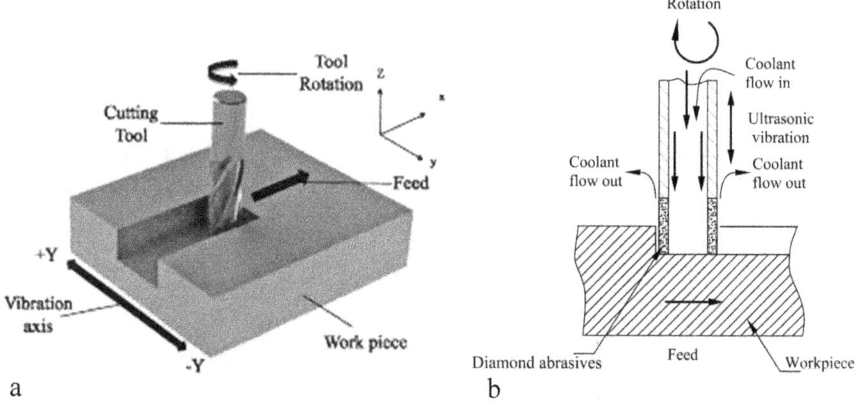

FIGURE 7.12 (a) Vibration-assisted milling and (b) vibration-assisted milling using abrasive tool.

7.3.2.4 Vibration-Assisted Grinding

Conventional grinding (CG) is a finishing process for hard materials due to its multiple abrasive cutting edges. The grinding wheel condition changes as grits wear and chips accumulate (loading) in the spaces between fine grits which deteriorates the surface finish and increases grinding forces and temperature. Consequently, with a higher temperature, the rate of abrasive wear increases. Methods for reducing wheel loading include frequent wheel dressing, application of cryogenic cooling, and optimum position of coolant. As an alternative to these techniques, vibration-assisted grinding (VAG) is used.

Vibration-assisted grinding (VAG) is a novel grinding technology which is characterized by the superposition of a conventional grinding (CG) operation with an ultrasonic vibration. Industrial application of VAG include surface grinding and the drilling of through-holes or grooves as shown in Figures 7.11b and 7.12b, where the abrasive tool is excited at frequency above 20 kHz and amplitude of 10 μm. Wheel loading decreases to almost 80% with increased vibration frequency and amplitude. The temperature reduction observed in vibration-assisted grinding is considered the primary reason for loading reduction in VAG. The change in chip size distribution in vibration-assisted grinding is a secondary reason for loading reduction. There is no deterioration in surface quality or subsurface integrity in comparison to CG.

7.3.3 Vibration-Assisted Non-Traditional Machining Processes

7.3.3.1 Vibration-Assisted Electrochemical Machining (VAECM)

Electrochemical machining (ECM) is used for machining difficult-to-cut materials without tool wear and residual stress. ECM techniques allow the ability to accomplish some difficult machining operations without direct contact between the tool and the workpiece, and with high stock removal rates, regardless of the mechanical properties of the workpiece. One of the methods to improve machining accuracy is to apply electrochemical machining with harmonic tool electrode oscillations and pulse current synchronized with them. The use of pulsed current improves the dimensional accuracy by reducing the electrolyte temperature. It enables a better opportunity to flush the machining products away from the interelectrode gap.

ECM is unavoidably associated with passivation that reduces the dissolution action in the interelectrode gap. Forced depassivation is therefore essential in industrial applications using various methods that activate the gap between the tool and the workpiece. In order to remove the passive layer and other compounds from the anode surface and improve ECM productivity, ECM is assisted by the ultrasonic depassivation process which results in a greater working current, and thus a higher material removal rate. The distributions of gas, sludge, and temperature affect the electrolyte conductivity and determine the machining accuracy in ECM. The use of low-frequency vibration of the workpiece or the cathodic tool enhances electrolyte flushing while US vibration is an effective method to break the passive layer and therefore is adopted to enhance the flushing of by-products in ECM. In principle, two basic techniques of ultrasonic assistance of ECM processes include the direct vibration of the tool or workpiece and the indirect method by introducing US vibration to the electrolyte medium (El-Hofy, 2018, 2019).

A. **Direct Method**

1. Applying US vibration to the cathodic tool improved the replicating accuracy and processing speed. Complex vibration, Figure 7.13, had more of an effect on both the accuracy and processing speed compared to the individual lateral vibration. The use of higher feed rate improved both the processing speed and the replicating accuracy. Figure 7.14a shows that the machining products stagnate in the inter-electrode gap (IEG) in normal ECM. The effect of workpiece vibrations in evacuating such accumulated products is clear when using the US vibrations in Figure 7.14b (El-Hofy, 2019).

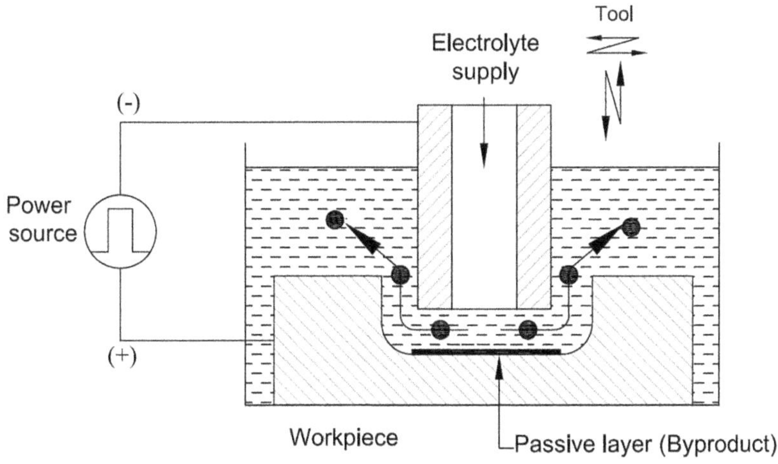

FIGURE 7.13 Complex tool vibration (Natsu et al., 2012).

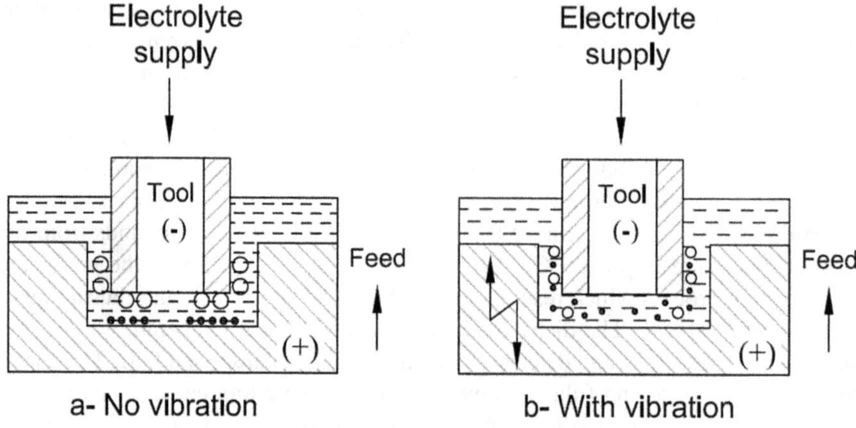

FIGURE 7.14 (a) ECM without vibration and (b) ECM with US vibration (Liu et al., 2013).

2. When using low-frequency vibration VA-WECM, Figure 7.15, bubbles were driven out of the machining gap. The large amplitudes and high speeds of wire vibration and large amplitude and high frequency of anode reduced edge radius and improved homogeneity of the slit width.

B. **Indirect Method**

In the indirect method, the electrolyte is activated by vibration as in the case of the following ECM applications.

1. Embedding US wave in fast-flowing electrolyte is more practical than vibrating a large tool or workpiece electrode as shown in Figure 7.16. Using US wave at 40 kHz significantly enhanced part quality by reducing surface roughness from 2.5 μm to 1 μm. the combined effect

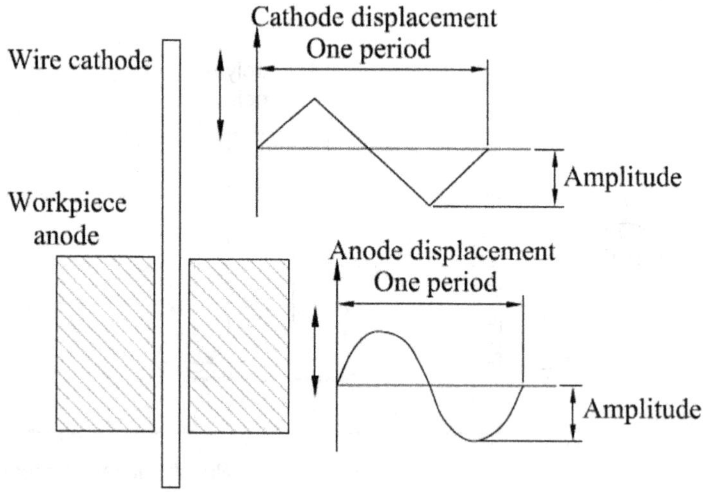

FIGURE 7.15 VA-WECM using wire and anode vibrations (Xu, Zeng, Li, and Zhu, 2017).

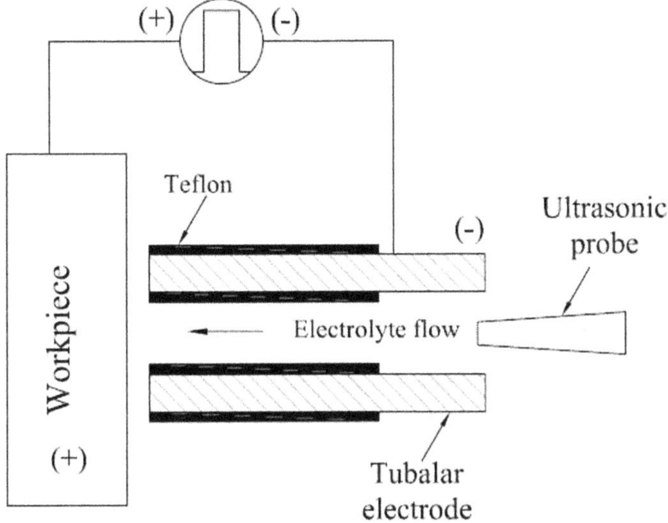

FIGURE 7.16 VA-ECM using US waves to the flowing electrolyte (Patel, Feng, Villanueva, and Hung, 2017).

of pulsed current and US wave reduced the taper angle of the drilled holes from 11° to 1° at the expense of MRR.

2. In order to improve the material removal rate and surface quality in ECM, a pulsating flow at 10 Hz frequency and pressure of 0.2 MPa is delivered through the electrolytic supply nozzle as shown in Figure 7.17.
3. The ultrasonic-assisted jet electrochemical micro-drilling process (Jet-ECMD) is a variation of that combines ultrasonic vibrations to the electrolyte jet as shown in Figure 7.18. It resulted in the improved MRR, reduced the hole taper, and enhanced the aspect ratio of the grooves due to passivation layer breakdown.

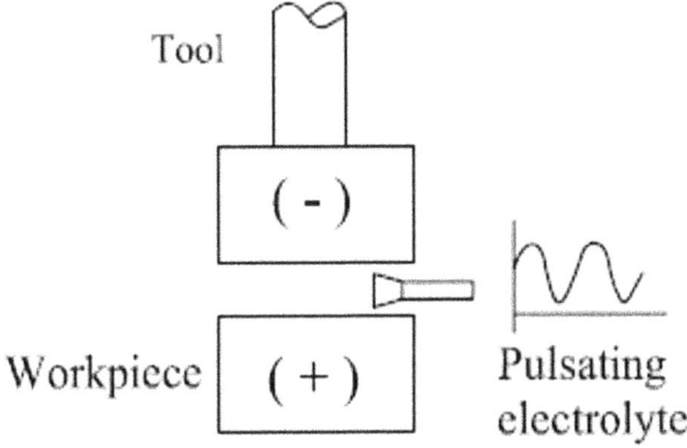

FIGURE 7.17 Pulsating electrolyte configuration (Fang, Qu, Zhang, Xu, and Zhu, 2014).

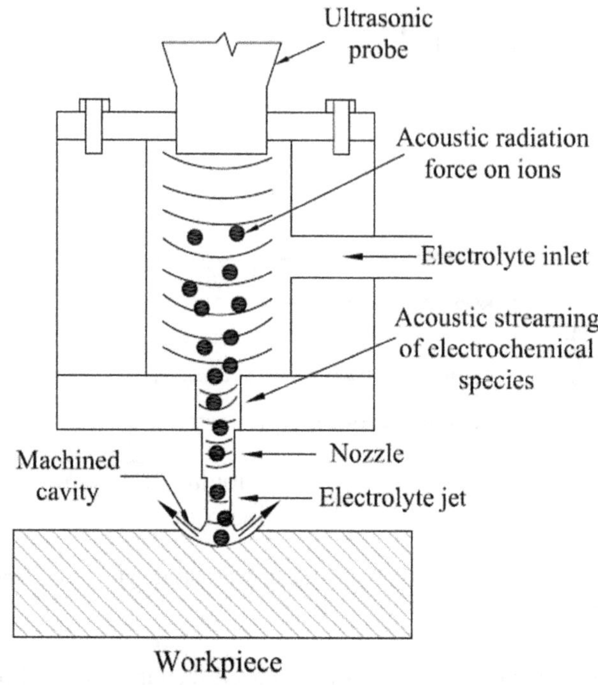

FIGURE 7.18 Vibration-assisted jet electrochemical micro-drilling (Goel and Pandey, 2017).

4. Vibration-assisted electrochemical polishing (VECP), Figure 7.19, increased the MRR and the productivity and is used for extremely smooth surface generation (Kim and Park, 2013).

7.3.3.2 Vibration-Assisted Electrodischarge Machining

In vibration-assisted EDM, two methods of vibration assistance are adopted.

A. **Direct Method**

The workpiece that vibrates at *low frequency* has a relatively significant effect on the metal removal rate when compared to normal EDM. The direct vibration of workpiece or the tool, shown in Figure 7.20a, enhances the flushing effect, and creates better dielectric circulation between the tool electrode and workpiece. During the up movement of the workpiece, the size of the sparking gap increases and allows the entrance of fresh dielectric fluid. When the workpiece is moved down, it pumps the dielectric fluid which is contaminated with debris away from the interelectrode gap. The material removal rate is increased by around 23% at 600 Hz and amplitude of 0.75 μm. The surface roughness and tool wear rate measured from EDM without vibration are higher than in EDM with low-frequency vibration. Direct ultrasonic vibration of the tool or WP improves flushing and stabilizes μEDM of high aspect ratio structures. The retracting movement avoids

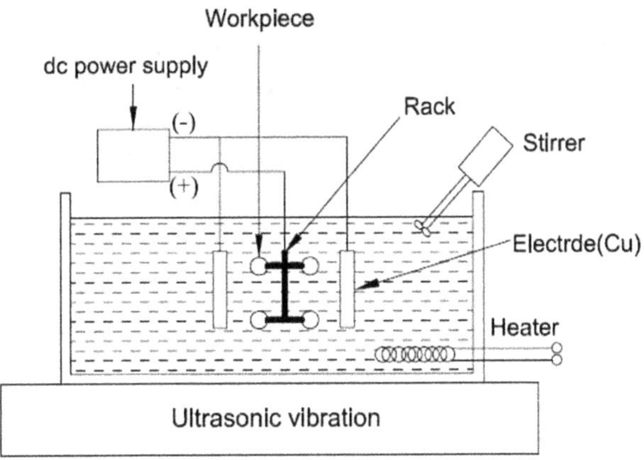

FIGURE 7.19 Vibration-assisted electropolishing (VECP) schematic.

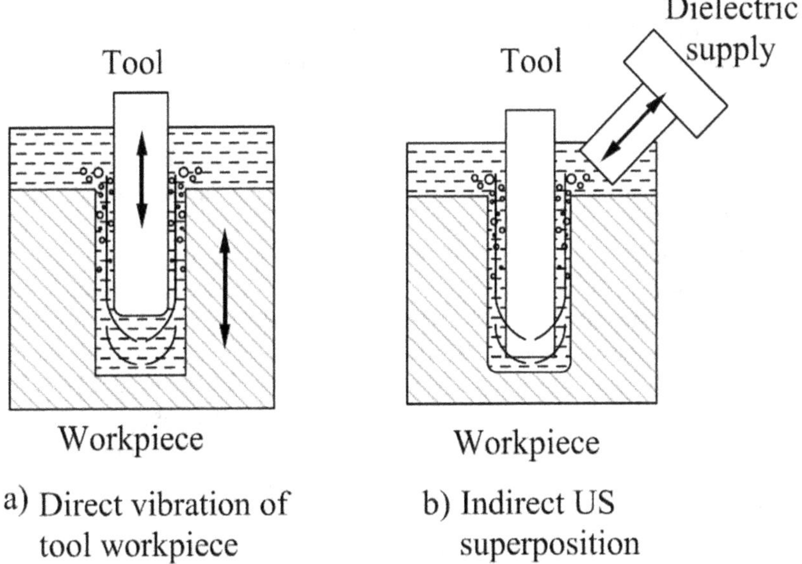

FIGURE 7.20 Vibration-assisted μEDM: (a) Direct vibration of tool or workpiece and (b) indirect US superposition.

arc discharges, the process speed is significantly enhanced, and more complex structures can be machined. The high-frequency vibrations superimposed onto the continuous movement of the tool lead to μEDM stabilization with superior surface finish in comparison to conventional machining (Schubert et al., 2013).

Using the direct *ultrasonically* aided μEDM achieves the following benefits:

- Increasing the machining speed by up to 40%
- Increasing the efficiency to eight times greater than μEDM

- Eliminating arc discharges which reduce the geometric deviations
- Machining complex shapes at high aspect ratio
- Machining bores <90 μm in diameter with aspect ratios >40
- Decreasing the electrode tool wear rate
- Improving surface quality and dimensional accuracy of the microholes

B. **Indirect Method**
The application of indirect ultrasonic vibration is achieved through the application of pressure waves within the dielectric at an angle of 60°, Figure 7.20b, enhances the process speed by 5% compared to normal EDM (EDM with direct vibration of tool or WP).

7.3.3.3 Vibration-Assisted Laser Beam Machining

A vibration-assisted laser takes one of the following options:

1 Vibrating the optical objective lens with a *frequency of 500 Hz* and various displacements (0–16.5 mm) to improve laser machining quality of metals during a femtosecond laser machining process. It is found that both the wall surface finish of the machined structures and the aspect ratio obtained using the low-frequency vibration-assisted laser machining are improved, compared to those derived via laser machining without vibration assistance Park et al., 2012).
2 *US-assisted* laser is used to improve laser hole drilling quality by exciting the work material with ultrasonic frequency during a femtosecond laser drilling process. Both the aspect ratio (depth over diameter) and the wall surface finish of the microholes fabricated using the ultrasonic vibration (US)-assisted laser drilling are improved, compared to those lasers machined without US assistance. This is because the introduction of US into the femtosecond laser drilling process reduced the re-solidified and re-deposited particles on the wall surfaces (Zheng and Huang, 2007).
3 The *ultrasonic-assisted* underwater laser ablation has been used to overcome the limitations of the normal laser machining process. It uses ultrasound to vibrate water while a workpiece is being ablated by a laser beam in water. The ultrasonic-assisted underwater laser micromachining technique could be an alternative micromachining process to gain a higher material removal rate and a better cut surface quality than other methods Charee et al., 2016).

7.3.3.4 Vibration-Assisted Abrasive Water Jet Machining

Vibration-Assisted Abrasive Water Jet Machining (VA-AWJM) is a machining technique used for cutting hard-to-cut materials such as glass, composites, ceramics, and graphite. The workpiece is vibrated in ultrasonic frequency as shown in Figure 7.21. The main material removal mechanism is by mechanical abrasion by impinging ultra-high velocity entrained abrasive particles. The ultrasonic vibration is used to enhance the material removal rate. The process finds applications in precision machining in the photo-electronics, semiconductors, and aerospace industries. During VA-AWJM of glass, the depth of erosion and amount of material volume removal rate increase when the ultrasonic vibrations are induced (Hou et al., 2018).

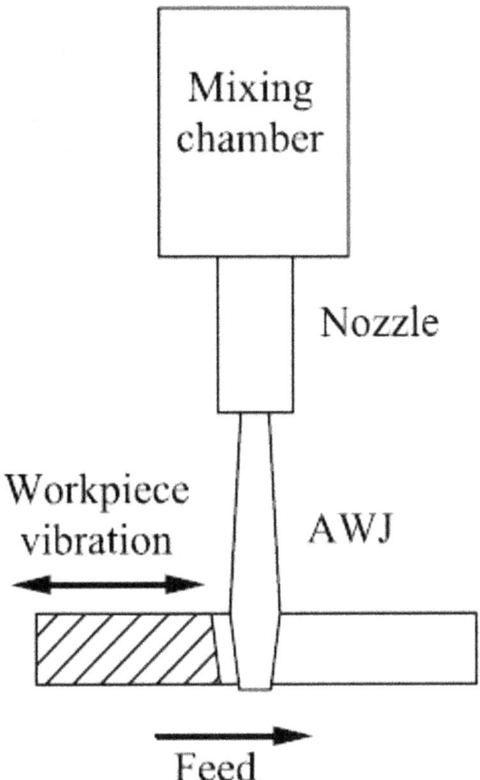

FIGURE 7.21 Vibration-assisted abrasive water jet machining system.

7.4 MAGNETIC FIELD-ASSISTED PROCESSES

Surface finish has a vital influence on important functional properties such as wear resistance and power losses due to friction on most of the engineering components. Poor surface finish will lead to the rupture of oil films on the peaks of the micro irregularities, which lead to a state approaching dry friction, and results in excessive wear of the contacting surfaces. Therefore, fine finishing processes are employed in machining the surface of many critical machined components to obtain a very high surface finish apart from high dimensional accuracies. Such processes include grinding, lapping, and superfinishing among the traditional methods. These traditional methods could not achieve the requirements of producing precise parts of high surface finish, and free of surface defects and scratches while finishing materials. These cracks can significantly reduce the strength and reliability of the components in working. Although grinding is more efficient for removing material than other finishing methods, it is still difficult to achieve a mirror-like finish. In lapping, it is essential that the abrasive grains be fine and of uniform size. Suitable lapping pressures have to be selected to avoid microcracks on the polished surface. In superfinishing, there are certain limitations especially when applied to complex surface finishing (Jayakumar, 2011).

To realize the requirements of high surface quality and accuracy, magnetic field-assisted processes have been recently used. Magnetic field-assisted processes are non-traditional processes in which the machining forces are controlled by a magnetic field. Accordingly, finish polishing is achieved without the need for expensive, rigid, ultra-precision, vibration- and error-free machine tools by incorporating the magnetic polishing elements necessary into the existing machine tools. There are two types of magnetic field-assisted processes: Magnetic abrasive finishing (MAF), which uses a brush of magnetic abrasives for finish machining, and magnetic float polishing (MFP), which uses magnetic fluid that is a colloidal dispersion of sub-domain magnetic particles in a liquid carrier with abrasives. Although MAF originated in the United States during the 1940s, it was in the Soviet Union and Bulgaria that much of the development took place in the late 1950s and 1960s. During the 1980s, the Japanese followed the work and conducted research for various polishing applications.

Table 7.2 illustrates the characteristic features of various abrasive polishing processes indicating the surface finish achievable and the amount of material that can be removed with these processes.

7.4.1 Magnetic Abrasive Finishing (MAF)

The magnetic abrasive finishing (MAF) process, which was introduced during the late 1940s, has emerged as an important non-traditional metal finishing process. The process has found applications in a wide range of fields such as nuclear energy, aerospace, medical, electronics, precision dies and molds, and other advanced industries, which need production of parts with highly smooth and defect-free surfaces. Magnetic abrasive grit size, magnetic field intensity, working gap, relative speed between workpiece and magnet, and vibration of workpiece or magnet are the principal parameters that control the surface characteristics. MAF is the process that is being developed for efficient and precision finishing up to nano-level of cylindrical or flat workpieces made from hard-to-machine materials.

TABLE 7.2
Comparison of Characteristic Features of Some Abrasive Machining Processes

Process	Surf. Roughness Achieved R_a, (μm)	Material Removed, (μm)
Lapping	0.1–0.2	12–15
Superfinishing	0.2–0.5	15–20
MAF	0.01–0.04	1–5
MFP	0.02	0.5

Source: (Jayakumar, 2001).

7.4.1.1 Finishing of Outer Cylindrical Surfaces and Typical Machining Conditions of MAF

Figure 7.22 shows a schematic diagram and a photo of MAF apparatus. A cylindrical workpiece is clamped into the chuck of the spindle that provides the rotating motion. The workpiece can be a magnetic (steel) or a non-magnetic (ceramic) material; the magnetic field lines go through the workpiece. Axial vibratory motion is introduced in the magnetic field by the oscillating motion of the magnetic poles relative to the workpiece. A mixture of fine abrasives held in a ferromagnetic material (magnetic abrasive conglomerate, Figure 7.23 is introduced between the workpiece and the magnetic heads where the finishing process is exerted by the magnetic field. This conglomerate is magnetized by external field and used as the tool for polishing pre-machined surfaces. Typically the sizes of the ferromagnetic conglomerates D are 50 to 400 microns and the abrasives d (A12O3 or SiC) are in the 1 to 10 micron range.

The surface of the conglomerates is armored and bonded with the abrasive grains forming bonded abrading mixture. Some abrasives may be free, forming with the armored conglomerates unbonded abrading mixture. Fox et al. (1994) have investigated machining of stainless steel with MAF, using both bonded and unbonded abrading mixtures. They have depicted that the bonded conglomerates improve the surface quality, while the unbonded realize higher material removal rate. After five minutes, the initial surface roughness R_a decreased from 0.22 μm to 7.8 nm when

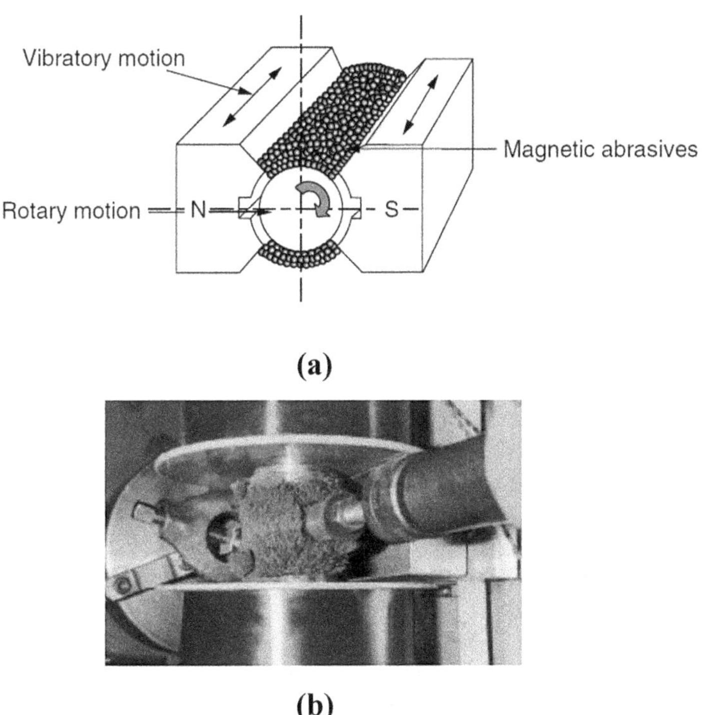

FIGURE 7.22 MAF apparatus: (a) Schematic, (b) photo.

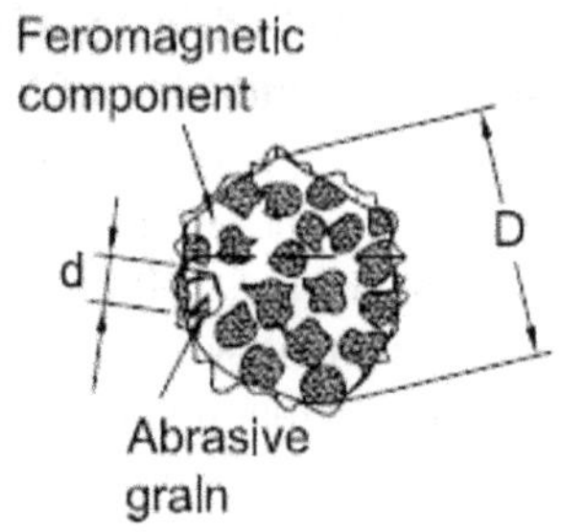

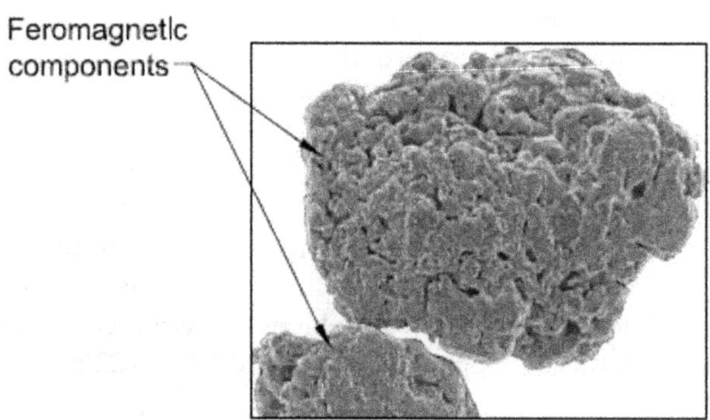

FIGURE 7.23 Typical magnetic abrasive conglomerate.

using the bonded conglomerates. It is remarkable that the roughness obtained when using the unbonded was 15–20 fold that obtained by the bonded. Therefore, it is concluded that the free abrasives in the unbounded penetrate deeper in the machined surface than the bonded conglomerates do. Accordingly, for economical machining, MAF with unbonded conglomerates should be used for surfaces having high initial roughness (pre-machined by turning) as a first stage, which is followed by an MAF operation with bonded conglomerates. On the other hand, if the surfaces are of low initial roughness (pre-machined by grinding), then bonded conglomerates are to be used (Fox et al., 1994).

With non-magnetic work materials, the magnetic abrasives are linked to each other magnetically between the magnetic N and S poles along the lines of the magnetic forces, forming flexible magnetic abrasive brushes.

MAF operates with magneto abrasive brushes where the abrasive grains arrange themselves with their carrying iron particles to flexibly comply with the contour of the work surface. The abrasive particles are held firmly against the work surface, while short stroke oscillatory motion is carried out in the axial workpiece direction. MAF brushes contact and act upon the surface protruding elements that form the surface irregularities.

While surface defects such as scratches, hard spots, lay lines, and tool marks are removed, form errors such as taper, looping, and chatter marks can be corrected with a limited depth of 20 microns. The material removal rate and surface finish depend on the workpiece circumferential speed, magnetic flux density, working clearance, workpiece material, size of magnetic abrasive conglomerates including the type of abrasives used, and its grain size and volume fraction in the conglomerate. Fox et al. (1994) concluded that the average surface finish R_a of a ground rod can be finished to about 10 nm. Increasing the magnetic flux density raises the rate of finishing. High removal rates and the best finish were obtained with an increase in the axial vibration amplitude and frequency. The axial vibration and rotational speed has to be taken into consideration for obtaining the best cross pattern that would give the best finish and high removal rate. Singh et al. (2004) recommended a high voltage level (11.5 V), low working gap (1.25 mm), high rotational speed (180 rpm), and large mesh number for improving the surface quality.

Typical Machining Conditions of MAF

Rotational speed: 500–2000 rpm, depending on the part diameter

Magnetic head axial oscillation:

Amplitude $\eta = 2$ mm

Oscillation freq. $f = 5$–25 Hz

Working gap: About 1 mm, depending on the size of the conglomerate

Magnetic field strength: $H = 0.17$–0.35 Tesla for non-magnetic materials such as ceramics and austenitic stainless steels, and 0.7–1.4 Tesla for magnetic materials such as steels.

For better performance, mineral oils or solid lubricant (Zn-stearates) can be used.

Figure 7.24 illustrates the two-dimensional magnetic force in the x direction and the perpendicular equipotential lines in the y direction when machining non-magnetic materials such as ceramics and austenitic stainless steels.

Equation (7.1) represents the two magnetic components Fx (radial) and Fy (tangential) at the point A in the machining zone.

$$F_x = V \cdot c \cdot H(\partial H/\partial x), F_y = V \cdot c \cdot H(\partial H/\partial y) \tag{7.1}$$

Where

V = volume of spherules (conglomerates)

χ = susceptibility of conglomerates

H = magnetic field strength at point A (x, y)

$\frac{\partial H}{\partial x}, \frac{\partial H}{\partial y}$ = gradients of magnetic field in directions x, y

The component Fy drives the conglomerates radially toward the workpiece to perform a smoothing action, while the component Fx acts tangentially to perform a shearing action on the machined surface. The gradients of magnetic field $\frac{\partial H}{\partial x}, \frac{\partial H}{\partial y}$ in directions x, y maximize the values of the force components Fx and Fy to enforce the conglomerates to move toward the machining zone.

Figure 7.25 also illustrates the two-dimensional magnetic and the equipotential lines, however when machining a magnetic material such as steel. In this case, the magnetic lines penetrate the workpiece. Accordingly, the intensity of the magnetic

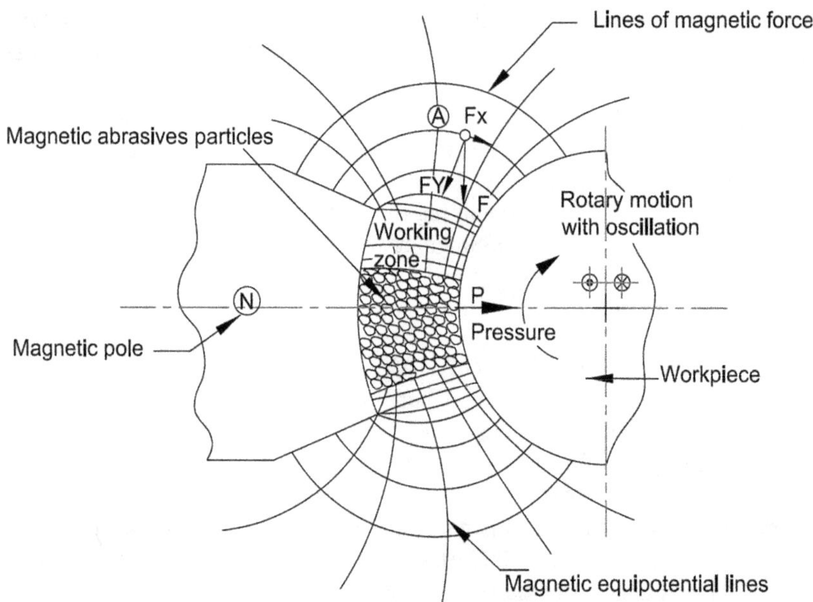

FIGURE 7.24 Magnetic force lines and equipotentials, in the case of a non-magnetic workpiece (Shinmura et al., 1990).

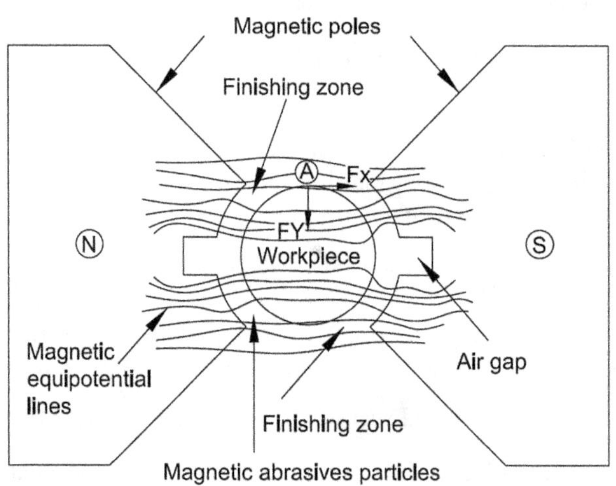

FIGURE 7.25 Magnetic force lines penetrate the magnetic workpiece (Fox et al., 1994).

field in the machining zone decreases, calling for an increase in the field intensity H to four-fold that which is needed for non-magnetic materials.

Effects of machining conditions on removal rate and surface quality in MAF:

- Machining time
- Linear speed of the workpiece
- Vibration frequency
- Intensity of magnetic field
- Working clearance between the workpiece and the magnetic head
- Work material
- Volume of the conglomerates, and the volume and type of abrasives embedded
- Type of lubricant used

The removal rate and the surface roughness rapidly decrease within few minutes with increasing machining time. Figure 7.26a illustrates the effect of the magnetic field intensity H on the surface roughness of non-magnetic material in MAF using bonded conglomerates. The figure shows also that the surface roughness decreases with increasing the magnetic field intensity H. The machining conditions are given in the figure.

Figure 7.26b illustrates the effect of lubricant on the surface roughness. A minimum surface roughness of 30 nm is realized using 5% wt of Zn-stearate, for the same workpiece rotating at the same surface speed, and a flux density H = 0.37 T.

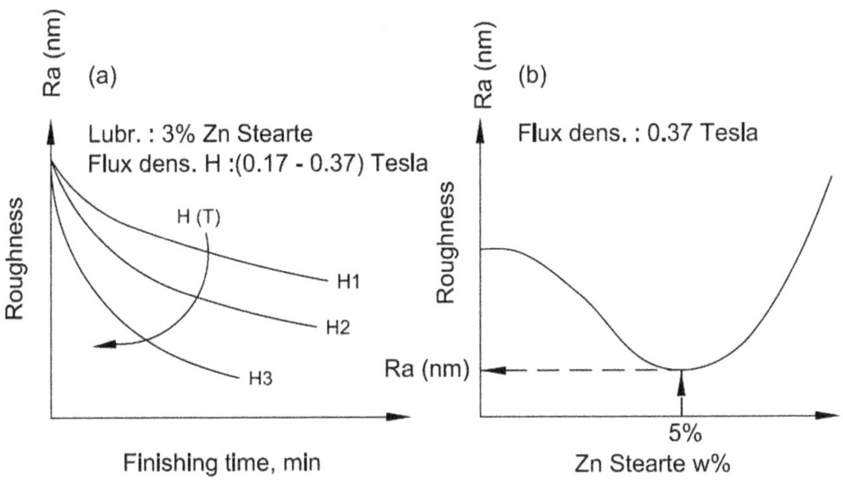

FIGURE 7.26 Effect of finishing time (a) and lubricant (b) on the surface roughness in MAF (Fox et al., 1994).

To realize optimum removal rate and surface quality in MAF, the frequency of the vibratory movement of the magnetic head, along with the rotational movement of the workpiece, should be adjusted to attain half-included angle ranging from 15–35° (Fox et al., 1994).

Finally, the size of the bonded conglomerates D and the embedded grains of size d have a decisive effect on both the removal rate and surface quality of the product. Shinmura et al. (1990) reported that while the size of the abrasive d has no significant impact on the removal rate, it has been found that it affects the surface roughness which increases with increasing abrasive size. On the other hand, they claimed that both the removal rate and surface roughness increased with increasing the size of conglomerates D. From that, it can be concluded that to achieve optimum machining condition in MAF, D must be chosen first to realize the optimum removal rate, then d is selected to realize the required surface quality.

7.4.1.2 MAF Finishing of Inner Cylindrical Surfaces

Clean gas and liquid piping systems need to have highly finished inner surfaces that prevent contaminant from accumulating. Figure 7.27 shows the two-dimensional schematic view of the internal finishing of a non-ferromagnetic tube using MAF. The magnetic abrasives, inside the tubes, are converged toward the finishing zone by the magnetic field, generating the magnetic force needed for finishing. By rotating the tube at a higher speed, the magnetic abrasives make the inner surface smoother. Figure 7.28 shows the case of ferromagnetic tube finishing where the magnetic fluxes mostly flow into the tube (instead of through the inside of the tube) due to their high magnetic permeability. Under such conditions, the abrasives hardly remain in the finishing zone when the tube is rotated. Shinmura et al. (1995) achieved mirror finishing and removed burrs without lowering the form accuracy.

7.4.1.3 Semi-Magnetic Abrasive Finishing (SMAF)

SMAF is another approach of magnetic abrasive finishing (MAF), in which specially made semi-magnetic abrasive grains such as Al_2O_3 or SiC possessing certain

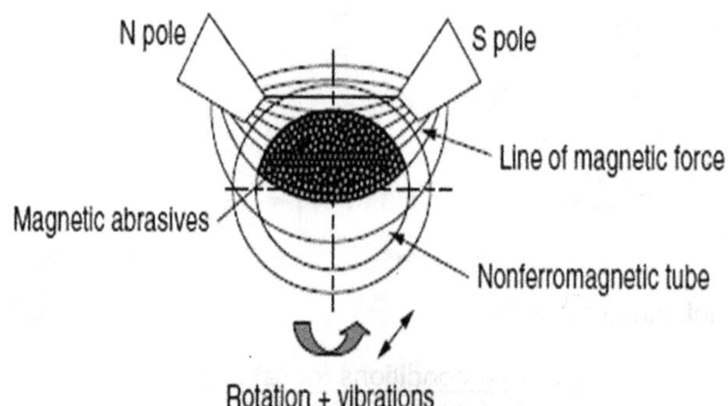

H = 0.17 – 0.35 Tesla

FIGURE 7.27 Internal magnetic finishing of non-magnetic tubes.

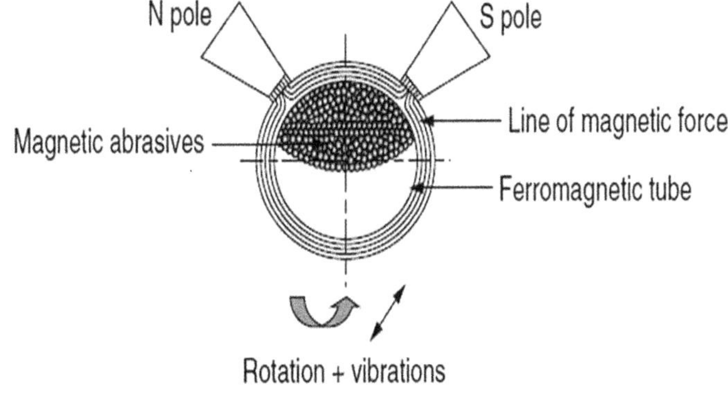

FIGURE 7.28 Internal magnetic finishing of magnetic tubes.

magnetic properties are directly magnetized by external magnetic field and are used as polishing tools. These abrasive particles are coated with iron particles when they are milled and hence exhibit magnetic properties.

7.4.1.4 Other MAF Applications

The process can be applied in many other fields such as:

1. Polishing of fine components such as printed circuit boards
2. The removal of oxide layers and protective coatings
3. Chamfering and deburring of gears and cams
4. Automatic polishing of complicated shapes
5. Polishing of flat surfaces

7.4.2 Magnetic Float Polishing (MFP) or Magnetic Fluid Grinding (MFG)

Advanced ceramics, such as aluminum oxide, zirconia, silicon nitride, and silicon carbide are difficult materials to finish by conventional grinding and polishing due to their high hardness and low fracture toughness. Unfortunately, the failure of these materials is generally initiated from the surface and the defects on the surface play a significant role on the life and reliability of parts made from these materials. However, the extent of damage due to brittle fracture can be minimized significantly by "gentle" grinding and polishing using submicron abrasives where the material removal is limited to low depths of cut (a few nanometers or less) and consequently low levels of force (on the order of 1 N or less per ball).

Conventional finishing of ceramic balls, for bearing applications, uses low polishing speeds and diamond abrasives as a polishing medium. The long processing time and the use of expensive diamond abrasives result in high processing costs, Figure 7.29. Diamond abrasives at high loads can result in deep pits, scratches, and

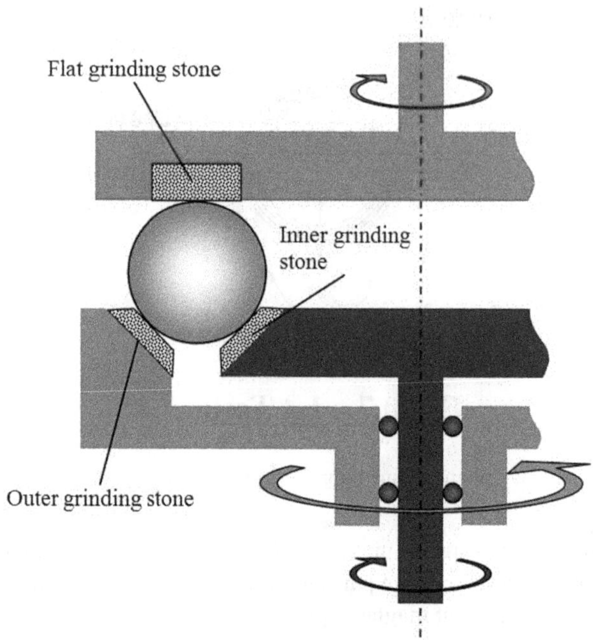

FIGURE 7.29 Conventional grinding of balls, spin angle controlled grinding (Kurobe et al., 2004).

microcracks. Consequently, the high processing cost and the lack of the machining system reliability form possible limitations in conventional grinding.

To minimize the surface damage, gentle polishing conditions are required, namely, low levels of controlled force and abrasives not much harder than the work material. A recent development involves the use of a magnetic field to support abrasive slurries in polishing ceramic balls and bearing rollers. A magnetic float polishing technique for finishing ceramic balls was initiated in Japan chiefly by Umehara and Kato (1996). They termed it magnetic fluid grinding (MFG). However, the term magnetic float polishing (MFP) is also used, as this is more a polishing (3-body) process rather than a grinding (2-body, fixed abrasive) operation. In literature, both terms are found.

Figure 7.30 is a schematic diagram of the magnetic float polishing apparatus showing permanent magnets located at the base of the apparatus. The magnets are located with alternate N and S poles underneath the float chamber. A guide ring is mounted on top of the float vessel to contain the magnetic fluid. Magnetic fluid containing fine abrasive particles is filled into the chamber. The silicon nitride ball blanks are held in 3-point contact between the float at the bottom, chamber wall on the side, and a shaft at the top.

The shaft is connected to the spindle of the Bridgeport CNC milling machine which is capable of operating in the speed range of 60–6000 rpm. The balls and float are pushed upwards against the shaft by the magnetic buoyancy force. This force increases at an exponential rate as the float moves closer to the magnets

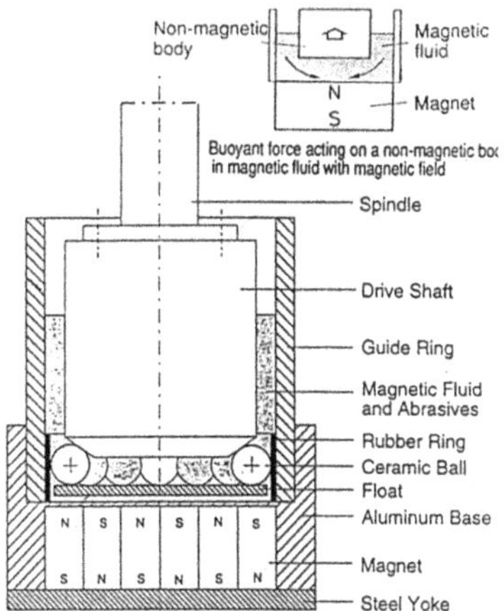

Work material HIP'ed Silicon nitride balls of 10.2 mm diam.

Chemical Composition Si_3N_4 :94 wt.%; Al2O3: 4 wt. %; and Y_2O_3: 2 wt %

Density, g/cm^3 3.228

Bending Strength, MPa 980

Hardness, kg/mm^2 1600-1800

Ferro-fluid water based (W-40) Density: 1.4 kg/cm^3 Viscosity: 2.25 Pa.s. (25°C); Magnetization: 400G at 8KOe

Abrasive: Types Cr203 and Al2O3

Size, |μm 1-5

Concentration 10% by volume

Spindle speed, rpm 1000, 2000, 4000, and 6000

Working load/ball, 1N

FIGURE 7.30 MFP schematic of ceramic balls (Umehara and Kalpakjian, 1994).

Fox et al., 1994).When a magnetic field is applied, the ceramic balls, abrasive grains, and the float of non-magnetic material all float and are pushed upwards by the magnetic fluid. Silicon nitride balls are pressed against the drive shaft and are finished by the rotation of the drive shaft. The drive shaft is made from austenitic stainless steel (non-magnetic) with a 30° chamfer at the edge. The chamfer angle is measured from a plane perpendicular to the axis of the shaft. The chamber wall was covered with urethane rubber and the float material used was acrylic resin.

The magnetic fluid is a colloidal dispersion of extremely fine (100 to 150 A) sub-domain ferromagnetic particles, usually magnetite (Fe_3O4), in various carrier fluids, such as water or kerosene. The ferro-fluids are made stable against particle agglomeration by the addition of surfactants.

When a magnetic fluid is placed in a magnetic field gradient, it is attracted toward the higher magnetic field side. If a non-magnetic substance (e.g. abrasives in this case) is mixed in the magnetic fluid, it is discharged toward the lower side. When the field gradient is set in the gravitational direction, the non-magnetic material is made to float on the fluid surface by the action of the magnetic levitation force. The polishing operation in this process occurs due to the magnetic levitation force. The process works on the basis of the magneto-hydrodynamic behavior of a magnetic fluid that can float non-magnetic abrasives suspended in it. The process is considered highly effective for finish polishing because a levitation force is applied to the abrasives in a controlled manner. The forces applied by the abrasives to the part in this process are extremely small (about 1 N/ball or less). This technique can be extremely cost-effective and viable for the manufacture of ceramic balls, and the surfaces produced have little or no defects.

Table 7.3 shows the form accuracy (sphericity) along with the maximum roughness R_t of Si_3N_4 spheres machined by MFP, after 3 hours machining time.

Figure 7.31 illustrates a practical application for MFP machining of tapered ceramic rollers. Below the figure, the profiles of the rollers are given as the machining proceeds. Also, the operating conditions are given below the same figure.

7.4.3 Advantages of MFAP

As previously mentioned, MFAP comprises two main techniques: MAF and MFP. Both will be compared and the advantages of each will be highlighted:

- MAF is characterized by its higher removal rate, whereas MFP realizes the highest surface quality.
- MAF could be operated using a conventional lathe, which leads to economical merits.
- Two different technologies are applicable with MAF. For economical machining, MAF with unbounded conglomerates for surfaces having high initial roughness, which is followed by MAF operation with bonded conglomerates for initial smooth surfaces. The two technologies differ as regards removal rate and surface finish. The suitable process should be selected according to the particular case.
- In MAF, a short polishing time (2–5 min) is needed, while MFP requires a high polishing time, which may extend to many hours, depending on the accuracy and the surface quality needed.

7.4.4 Magnetorheological Finishing (MRF)

MRF relies on a smart fluid known as magnetorheological (MR) fluid which exhibits dynamic field strength of 50–100 kPa for applied magnetic field of

TABLE 7.3
Machining Accuracy and Roughness of Si_3N_4 Spheres Polished By MFP

Si_3N_4 spheres	As sintered	MFP-machined
Sphere diameter	7.7 mm	7.1 mm
Roughness R_t	10 μm	0.1 μm
Sphericity	500 μm	0.14 μm

Source: (Umehara and Kalpakjian, 1994).

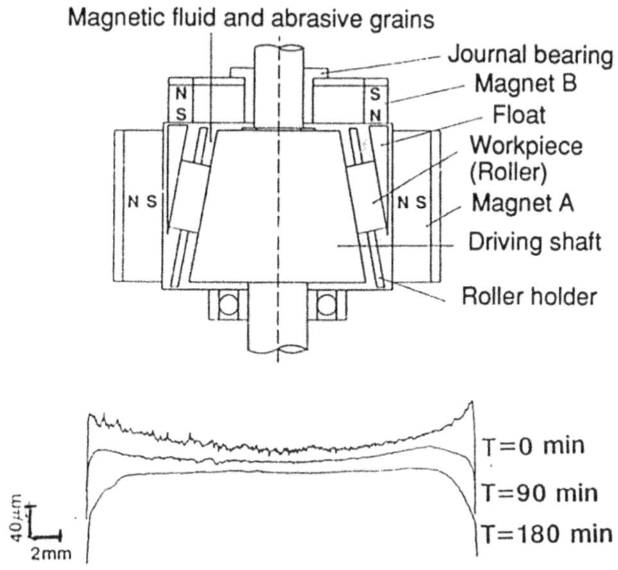

FIGURE 7.31 MFP of tapered rollers (Umehara and Kalpakjian, 1994).

150–250 kA/m. In MRF, a convex, flat, or concave workpiece is positioned above a reference surface and the MR fluid ribbon is formed on the rotating wheel rim. By applying magnetic field in the gap, the stiffened region forms a finishing spot. Surface smoothing, removal of subsurface damage, and figure correction are accomplished by moving the workpiece surface at a constant speed while sweeping the workpiece about its radius of curvature through the stiffened finishing zone. The MR polishing fluid lap carries heat and debris away from the polishing zone and does not load like grinding wheels. It does not lose its shape and is self-deformable.

7.4.5 Magnetorheological Abrasive Flow Finishing (MRAFF)

The MRAFF process combines the advantages of both the AFM and MRF processes. The process maintains the versatility of AFM and the controllability of the rheological properties of the polishing medium in MRF. For its performance, the MRAFF process relies on smart MR polishing fluids whose rheological behavior is controllable by means of external magnetic field (Figure 7.32). Abrasion occurs only where the magnetic field is applied, keeping other areas unaffected. In MRAFF, the viscosity of the abrasive medium can be manipulated and controlled by a magnetic field. The use of machining setup similar to AFM will remove shape limitations on the workpiece surface to be finished (Das, 2011).

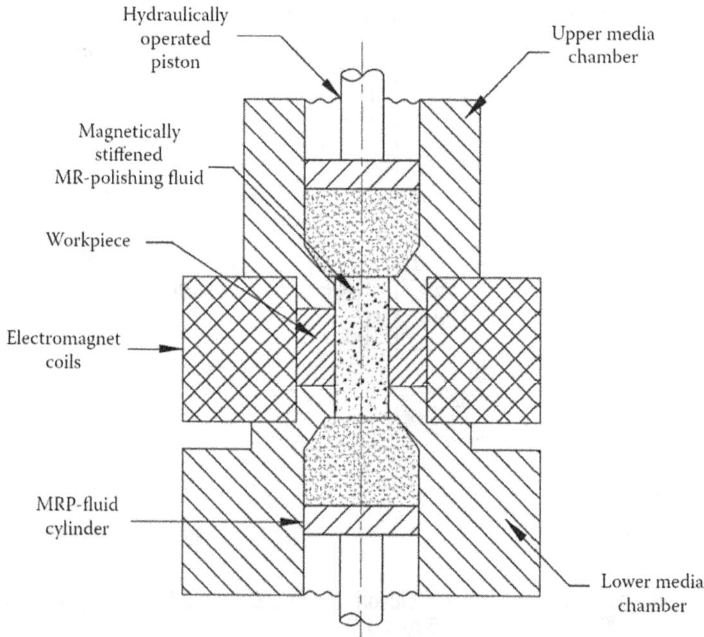

FIGURE 7.32 Schematic of the MRAFF process (Das,2011).

7.5 REVIEW QUESTIONS

7.5.1 State the advantages of using vibration-assisted machining (VAM).
7.5.2 Use a line sketch to show the different types of VAM systems.
7.5.3 Draw the different steps of chip removal during 1-D VAM.
7.5.4 Use a line sketch to show the tool path during 2-D VAM.
7.5.5 Show the kinematics of the VAT, VAD, VAM, and VAG processes.
7.5.6 Show how vibration is induced to the VA-ECM and VA-ED processes.
7.5.7 Explain the necessity for magnetic-assisted machining to replace conventional methods such as grinding, superfinishing, and lapping in producing highly finished surfaces.
7.5.8 Define MAF and describe the cutting action. What are parameters that influence the removal rate of MAF?
7.5.9 Compare the advantages and disadvantages of MAF and MFP.
7.5.10 Explain the principles of MRF and MFP.
7.5.11 Show how MRAFF is developed from AFM and MRF.
7.5.12 Show the schematic diagram of the MRAFF process.

REFERENCES

Abdelkawy, A, Hossam, M & El-Hofy, H 2018, 'Mathematical model of thrust force for rotary ultrasonic drilling of brittle materials based on the ductile-to-brittle transition phenomenon', *The International Journal of Advanced Manufacturing Technology*. doi:10.1007/s00170-018-2943-4.

Behera, BC 2011, *Development and experimental study of machining parameters in ultrasonic vibration-assisted turning.* Master of Technology Thesis, Rourkela-769008, India, National Institute of Technology.

Charee, W, Tangwarodomnukun, V & Dumkum, C 2016, 'Ultrasonic-assisted underwater laser micromachining of silicon', *Journal of Materials Processing Technology*, vol. 231, pp. 209–220 doi:10.1016/j.jmatprotec.2015.12.031.

Das, M 2011, *Experimental investigation of rotational-magnetorheological abrasive Flow Finishing (R-MARAFF) process and a CFD based numerical study of MRAFF process*, PhD Thesis, Kanpur, India, Indian Institute of Technology.

El-Hofy, H 2018, *Fundamentals of machining processes: conventional and nonconventional processes*, 3rd edn, Taylor and Francis Ltd.

El-Hofy, H 2019, 'Vibration assisted electrochemical machining: a review', *International Journal of Advanced Manufacturing Technology*. doi:10.1007/s00170-019-04209-9.

El-Taybany, Y & El-Hofy, H 2019, 'Mathematical model for cutting force in ultrasonic-assisted milling of soda-lime glass', *The International Journal of Advanced Manufacturing Technology*, 103, pp. 3953–3968. doi:10.1007/s00170-019-03399-6.

Fang, X, Qu, N, Zhang, Y, Xu, Z & Zhu, D 2014, 'Effects of pulsating electrolyte flow in electrochemical machining', *Journal of Materials Processing Technology*, vol. 214, no. 1, pp. 36–43. doi:10.1016/j.jmatprotec.2013.07.012.

Fox, M, Agrawal, K, Shinmura, T & Komanduri, R 1994, 'Magnetic abrasive finishing of rollers', *CIRP Annals*, vol. 43, no. 1, pp. 181–184. doi:10.1016/s0007-8506(07)62191-x.

Geskin, ES, Tismentsky, L, Bokhroi, E & Li, F 1995, *Investigation of Ice Jet Machining*, ISEM XI, Lausanne, Switzerland, pp. 883–890.

Goel, H & Pandey, PM 2017, 'Experimental investigations into the ultrasonic assisted jet electrochemical micro-drilling process' *Materials and Manufacturing Processes*, vol. 32, no. 13, pp. 1547–1556. doi:10.1080/10426914.2017.1279294.

Hou, R, Wang, T, Lv, Z & Liu, Y 2018, 'Experimental study of the ultrasonic vibration-assisted abrasive waterjet micromachining the quartz glass', *Advances in Materials Science and Engineering*, 9 p, doi:10.1155/2018/8904234.

Jayakumar, P 2011, 'Semi magnetic abrasive machining', *4th International Conference on Mechanical Engineering*, Dhaka, Bangladesh, pp. V 81–85 Section V: Applied Mechanics.

Kim, US & Park, JW 2013, 'Vibration-assisted electrochemical polishing for extremely smooth surface generation', *Advanced Materials Research*, vol. 813, pp. 475–478.

Kurobe, T, Morita, T & Tsuchihashi, N 2004, 'Super fine finishing of Si3N4 ceramic ball using spin angle controlled machining method', *Journal of the Japan Society for Precision Engineering, Contributed Papers*, vol. 70, no. 11, pp. 1392–1396.doi:10.2493/jspe.70.1392.

Liu, Z, Zhang, H, Chen, H & Zeng, Y 2013, 'Investigation of material removal rate in micro electrochemical machining with lower frequency vibration on workpiece', *International Journal of Machining and Machinability of Materials*, vol. 14, no. 1, p. 91. doi:10.1504/ijmmm.2013.055131.

López de Lacalle, LN, Sánchez, JA, Lamikiz, A & Celaya, A 2004, 'Plasma assisted milling of heat-resistant superalloys', *Journal of Manufacturing Science and Engineering*, vol. 126, no. 2, pp. 274. doi:10.1115/1.1644548.

Marcel, K, Marek, Z & Jozef, P 2014, 'Investigation of ultrasonic assisted milling of aluminum alloy AlMg4.5Mn', *Procedia Engineering*, vol. 69, pp. 1048–1053. doi:10.1016/j.proeng.2014.03.089.

MarojuNaresh Kumar, KanmaniSubbu S, Vamsi Krishna, P & Venugopal, A 2014, 'Vibration assisted conventional and advanced machining: a review', *Procedia Engineering*, vol. 97, pp. 1577–1586.

Natsu, W, Nakayama, H & Yu, Z 2012, 'Improvement of ECM characteristics by applying ultrasonic vibration', *International Journal of Precision Engineering and Manufacturing*, vol. 13, no. 7, pp. 1131–1136. doi:10.1007/s12541-012-0149-5.

Nurul Amin, AKM & Ginta, TL 2014, 'Heat-assisted machining', *Comprehensive Materials Processing*, pp. 297–331. doi:10.1016/b978-0-08-096532-1.01118-3.

Park, J-K, Yoon, J-W & Cho, S-H 2012, 'Vibration assisted femtosecond laser machining on metal', *Optics and Lasers in Engineering*, vol. 50, no. 6, pp. 833–837.doi:10.1016/j.optlaseng.2012.01.017.

Patel, JB, Feng, Z, Villanueva, PP & Hung, WNP 2017, 'Quality enhancement with ultrasonic wave and pulsed current in electrochemical machining', *Procedia Manufacturing*, vol. 10, pp. 662–673. doi:10.1016/j.promfg.2017.07.013.

Schubert, Andreas, Zeidler, Henning, Hackert-Oschätzchen, Matthias, Schneider, Jörg & Hahn, Martin 2013, Enhancing micro-EDM using ultrasonic vibration and approaches for machining of nonconducting ceramics. *Journal of Mechanical Engineering*, vol. 59, no. 3, pp. 156–164.

Shinmura, T, Takazawa, K, Hatano, E, Matsunaga, M & Matsuo, T 1990, 'study on magnetic abrasive finishing', *CIRP Annals*, vol. 39, no. 1, pp. 325–328. doi:10.1016/s0007-8506(07)61064-6.

Singh, DK, Jain, VK & Raghuram, V 2004, 'Parametric study of magnetic abrasive finishing process', *Journal of Materials Processing Technology*, vol. 149, no. 1–3, pp. 22–29. doi:10.1016/j.jmatprotec.2003.10.030

Umehara, N & Kalpakjian, S 1994, 'Magnetic fluid grinding – a new technique for finishing advanced ceramics', *CIRP Annals*, vol. 43, no. 1, pp. 185–188. doi:10.1016/s0007-8506(07)62192-1.

Umehara, N & Kato, K 1996, 'Magnetic fluid grinding of advanced ceramic balls', *Wear*, vol. 200, no. 1–2, pp. 148–153. doi:10.1016/s0043-1648(96)07297-3.

Xu, K, Zeng, Y, Li, P & Zhu, D 2017, 'Vibration assisted wire electrochemical micro machining of array micro tools', *Precision Engineering*, vol. 47, pp. 487–497. doi:10.1016/j.precisioneng.2016.10.004.

Xu, W-X & Zhang, L-C 2015, 'Ultrasonic vibration-assisted machining: principle, design and application', *Advances in Manufacturing*, vol. 3, no. 3, pp. 173–192.doi:10.1007/s40436-015-0115-4

Zheng, HY & Huang, H 2007, 'Ultrasonic vibration-assisted femtosecond laser machining of microholes', *Journal of Micromechanics and Microengineering*, vol. 17, pp. N58–N61. doi:10.1088/0960-1317/17/8/N03.

8 Design for Machining

8.1 INTRODUCTION

Manufacturing cost is the key factor to the economic success of a product. Economic success depends on the profit margin earned on each sale of the product and how many units the firm can sell. The number of units sold and the sales price depends on the product quality. Successful design therefore is ensured by maintaining high product quality while minimizing the manufacturing cost. Design for manufacturing (DFM) is one method of achieving this goal. Effective DFM practices lead to low manufacturing costs without sacrificing product quality. The following principles aid designers in specifying components and products that can be produced at minimum cost:

1. *Simplicity of the product.* This means the minimum number of parts, the least intricate shape, the fewest precision adjustments, and the shortest production sequence.
2. *Standard material and components.* This enables benefits of mass production and simplifies inventory management, avoids tooling and equipment investment, and speeds up the manufacturing cycle.
3. *Standard design of the product.* When several similar products are to be produced, specify the same materials, part, and subassemblies for each as much as possible.
4. *Specify liberal tolerances.* The higher costs of tight tolerance arise due to:
    i. Extra machining operations, such as grinding, honing, or lapping after primary machining operations
    ii. Higher tooling cost
    iii. Longer operating cycles
    iv. Higher scrap and rework costs
    v. The need for more skilled and highly trained workers
    vi. Higher materials cost
    vii. High investment for precision equipment

Table 8.1 shows the approximate relative cost for achieving certain tolerances and surface finishes. Accordingly, it is recommended to consider the following:

1. Use the most machinable materials available.
2. Avoid secondary operations such as deburring, inspection, plating, painting, and heat treatment.
3. Design should be suitable for the production method that is economical for the quantity required.

TABLE 8.1
Approximate Relative Cost for Machining Tolerances and Surface Finishes

	Tolerance		Roughness (R_a)	
Machining Process	± (mm)	Relative Cost	μm	Relative Cost
Rough machining	0.770	100	6.25	100
Standard machining	0.130	190	3.12	200
Fine machining (rough grinding)	0.030	320	1.56	440
Very fine machining (ordinary grinding)	0.010	600	0.80	720
Fine grinding, shaving, honing	0.005	1100	0.40	1400
Very fine grinding, shaving, honing, lapping	0.003	1900	0.20	2400
Lapping, burnishing, superhoning, polishing	0.001	3500	0.18	4500

4. Use special process capabilities to eliminate many operations and the need for separate costly components.
5. Avoid process restrictiveness and allowing manufacturing engineers to choose a process that produces the required dimensions, surface finish, and other characteristics.

8.1.1 General Design Rules

The general design rules for economic production are as follows:

1. Simplify the design by reducing the number of parts required.
2. Design for low labor cost operations wherever possible.
3. Avoid generalized statements on drawings that may be difficult for the production personnel to interpret.
4. Dimensions should be made from specific points or surfaces on the part itself.
5. Once the functional requirements are met, designers should strive for minimum weight.
6. Dimensions should be made from one datum line rather than from a variety of points to simplify tooling and gauging, and to avoid overlap of tolerances.
7. Design to use general purpose tooling rather than special ones.
8. Avoid sharp corners for ease of production and avoid stress concentration on the part.
9. Design a part so that many operations can be performed.
10. Space holes in machined parts so that they can be made in one operation without tooling weakness.
11. Whenever possible, cast, molded, or powder-metal parts should be designed without stepped parting line and with uniform wall thickness.

8.2 GENERAL DESIGN RECOMMENDATIONS

Because of the highly competitive nature of machining processes, the question of finding ways to reduce cost is ever-present. A good starting point for cost reduction is in the design of the product. The design engineer should always keep in mind the possible alternatives available in making their design. Unfortunately, designers often consider that their job is to design the product for performance, appearance, and reliability and think that it is the manufacturing engineer's job to produce whatever has been designed. Of course, there is often a natural reluctance to change a proven design for the sake of a reduction in the machining cost. As a subject, design for machining hardly exists as compared with design for strength.

For obvious reasons, machining is considered a wasteful process and many engineers will feel that the main concern should be to design components that do not require machining. As 80–90% of manufacturing machines are designed to machine metal, the view that machining should be avoided must be considered rather impracticable for the immediate future. However, the trend toward the use of processes that conserve material is clearly increasing, and when large-volume product is involved, this approach should be foremost in the designer's mind. In this chapter, certain design principles that can help to simplify the machining of components and reduce costs are introduced. Machined parts are used in applications for which precision is required. Machining is also involved if surface finish, flatness, roundness, circularity, parallelism, or close fit is involved. Additionally, if the part is in motion, or fits precisely with another part, machining operations will be employed. Machined parts can be as small as miniature screws, shafts, or gears. They can be as large as huge turbines, turbine housings, and valves found in hydroelectric power stations. Machined components are made from ferrous and non-ferrous materials. However, plastics, rubber, carbon, graphite, and ceramics are also employed.

The following are some important regulations that should be followed by the part designers.

1. Avoid machining operations if the surface or the feature required can be produced by casting or forming.
2. Specify the most liberal surface finish and dimensional tolerances consistent with the function of the surface to avoid costly grinding, lapping, and other finishing operations (Figure 8.1).
3. Design the part for ease of fixation and secured clamping during the machining operation.
4. Avoid sharp corners and sharp points in cutting tools to avoid their breakage.
5. Use stock dimensions whenever possible (Figure 8.2).
6. Avoid interrupted cuts during single-point machining operations.
7. Design parts that are rigid enough to withstand clamping and cutting forces.
8. Avoid tapers and contours that simplify tooling and setups.
9. Reduce the number and the size of shoulders, as they require extra materials and operations.
10. Avoid undercuts because they involve more operations and special ground tools.
11. Substitute a stamping operation for the machined component (Figure 8.3).
12. Avoid the use of hardened or difficult-to-machine materials unless their functional properties are required.
13. For thin and flat parts that require machining, allow sufficient stock for rough and finish operations.

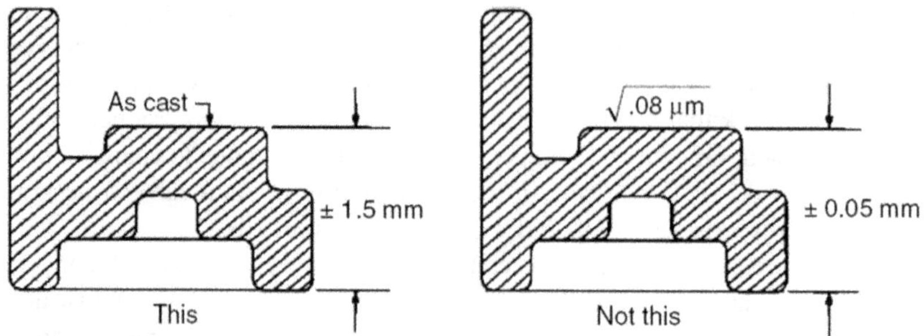

FIGURE 8.1 Avoid tolerances that necessitate machining if as-cast, as-forged, or as-formed dimensions and surface finishes are satisfactory for the parts function.

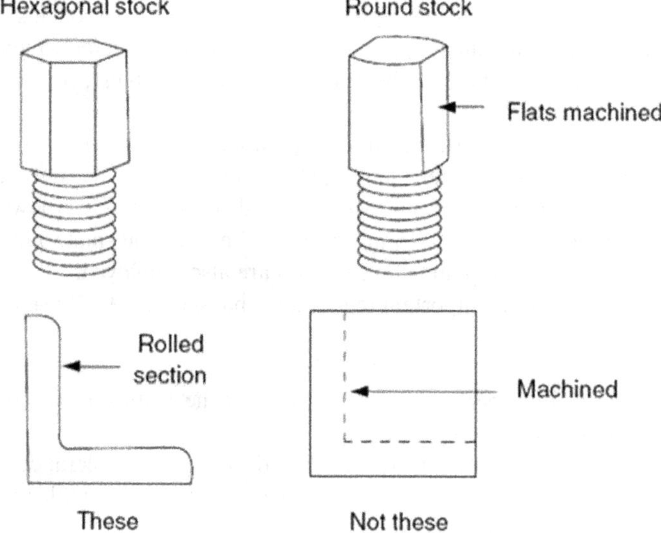

FIGURE 8.2 Use stock dimensions and minimize the machining allowance.

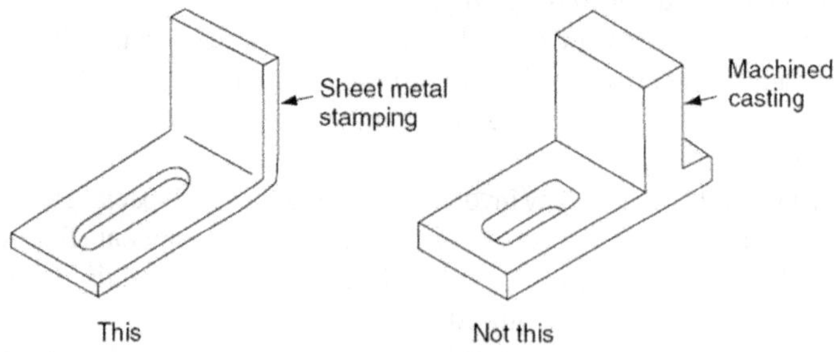

FIGURE 8.3 Metal-formed parts are better than machined castings.

14. Put the machined surfaces in one plane.
15. Provide access room for cutters, bushings, and fixture elements.
16. Design parts so that standard cutters can be used (Figure 8.4).
17. Avoid the use of parting lines or draft surfaces for clamping and locating.
18. Avoid projections and shoulders which interfere with the cutter movement.
19. Provide relief space for burr formation and furnish means for easy burr removal.

8.3 DESIGN FOR MACHINING BY CUTTING

8.3.1 Turning

Turning is a conventional material removal process that produces surfaces of rotation on the WP (Figure 8.5). Turning operations are used for production quantities ranging from one piece to many millions. Depending on the production quantity and part specifications, turning operation ranges from manual, numerical, computer, or completely automatic mechanical control. Tooling costs for parts machined on engine lathes are very low. Turret lathes are used for 10–25 or more parts.

In general, turning machine operations are more frequently used for lower ranges of production quantities. The most economic range for each machine tool is as follows:

- Engine lathes: Very low to low quantity
- Turret lathes: Low to medium quantity
- Tracer lathes: Low to medium quantity
- NC and CNC lathes: Low to medium quantity
- Single spindle (chucking type): Medium to high quantity
- Multi-spindle chucker: High to very high quantities

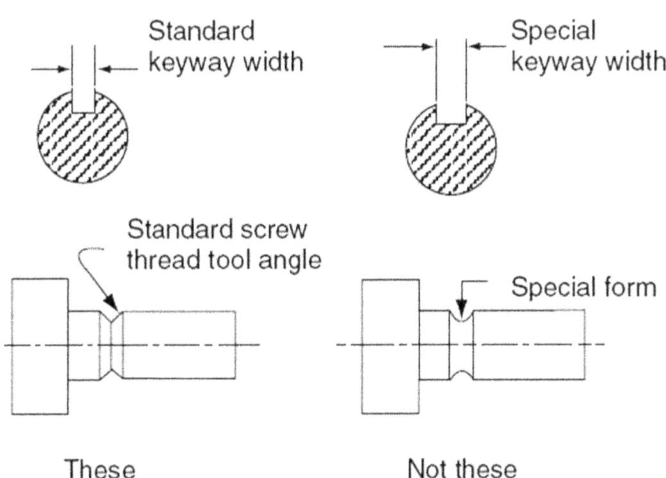

FIGURE 8.4 Design parts to be machined by standard tools.

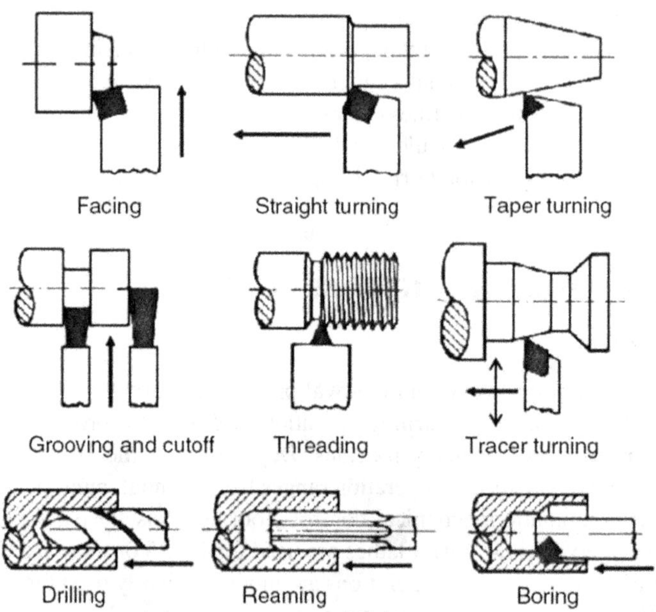

FIGURE 8.5 Basic turning operations.

Automatic screw machined parts are generally cylindrical in shape and may have several outside diameters as well as a hexagonal or square-surfaced portion. They may include threads on one or both ends and may have an internal axial hole with more than one diameter. The hole may be chamfered or tapered. Threads may be different in size and pitch and may be both external and internal. The diameter ranges from the smallest watch parts to 200 mm. Thread length can be as short as 1 mm and as long as 1 m. Such parts can be produced using one of the following machines:

- Swiss-type (shafts, pinions, contacts for electrical devices, pins, and valves)
- Turret-type (rivets, nuts, bolts, shafts, spacers, washers, pulleys, valve stems, spools gear blanks, rollers, and push rods)
- Multi-spindle automatic machine

8.3.1.1 Economic Production Quantities

The economic production quantity depends on the machine tool used. In this regard, the following considerations should be made:

1. The output rate ranges from a few seconds to about 5 min per piece for single-spindle machines.
2. Higher rates are possible using multi-spindle automatic machines.
3. One operator can tend four or more single-spindle machines and may be assigned to ten or more.
4. Two or more multi-spindle machines are assigned per operator.

5. Setup times are from 1 to 8 hours.
6. Multi-spindle machines require more tooling than single-spindle machines.

8.3.1.2 Design Recommendations for Turning

Designers should follow these recommendations:

A. *Stock Size and Shape*
   1. The largest diameter of the component should be taken as the diameter of the bar stock in order to conserve material and save machining time.
   2. Standard sizes and shapes of bar stock should be used in preference to special diameters and shapes.

B. *Basic Part Shape Complexity*
   1. Keep the design of parts as simple as possible to reduce the number of tool stations and gauging processes required.
   2. Use standard tools as much as possible by specifying standard, common sizes of holes, screw threads, knurls, slots, and so on.

C. *Avoiding Secondary Operations*
   1. The part should be complete when cut off from the bar material.
   2. Secondary operations such as slots and flats should be small and performed when the part is held in the pickoff attachment.
   3. Internal surfaces and screw threads should be located at one end so that they can be performed before cutoff and without the need for rechucking (Figure 8.6).

D. *External Forms*
   1. The length of the formed area should not exceed two and a half times the minimum WP diameter (Figure 8.7).
   2. Sidewalls of grooves and other surfaces that are perpendicular to the axis of the WP should have a slight draft (Figure 8.8) of 1/2° or more to prevent tool marks when the tool is withdrawn.

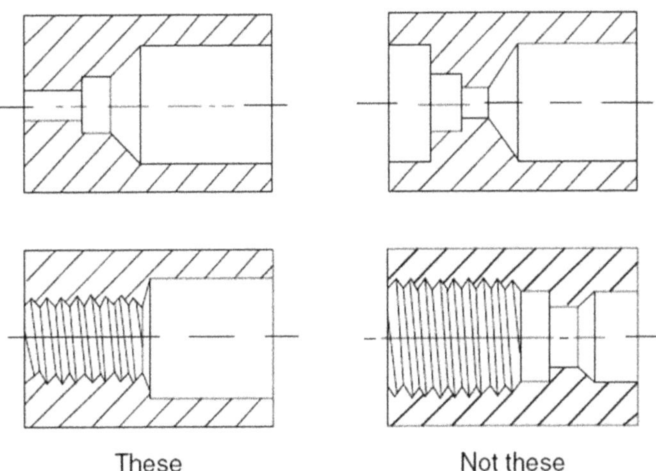

FIGURE 8.6 Operations should be finished without the need for rechucking.

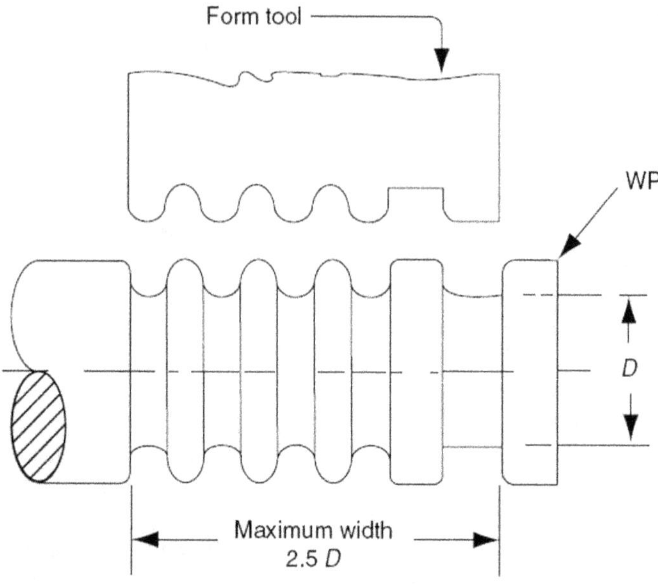

FIGURE 8.7 Form tool width limitations.

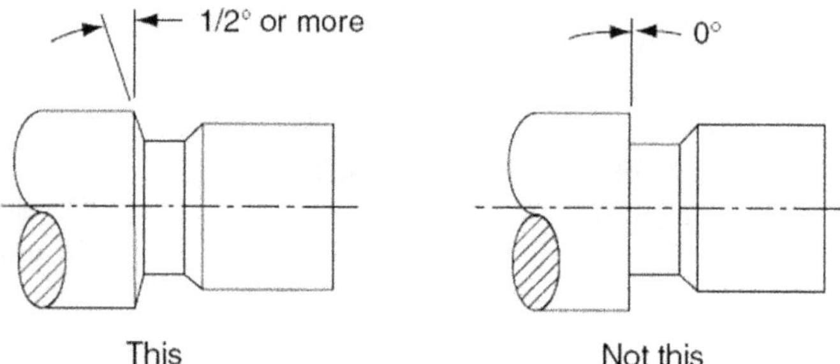

FIGURE 8.8 Provide a slight angle to sidewalls and faces to prevent tool marks when the tool is withdrawn.

3. When turning from square or hexagonal stock, the turned diameter is the distance between two opposite flats of the stock. It is advisable to design turned parts to be about 0.25 mm or smaller than the bar stock size.
4. Avoid deep narrow grooves and sharp corners.

E. *Undercuts*
1. Avoid angular undercuts and use undercuts obtainable with traverse or axial tool movements.
2. External grooves are machined more economically than internal recesses.

F. *Holes*
1. The bottom shape of blind holes should be that made by a standard drill point (Figure 8.9).

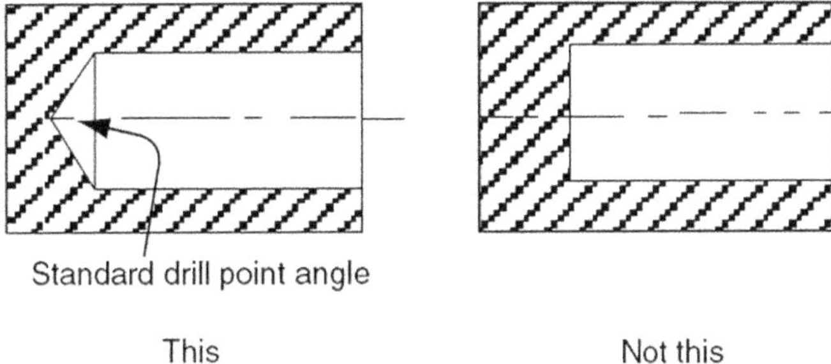

FIGURE 8.9 The bottom of holes should allow the use of a standard drill point angle.

G. *Screw Threads*
  1. Avoid the formations of burrs in threaded parts (Figure 8.10).

H. *Knurls*
  1. *Knurled width should be narrow (≤WP diameter).*
  2. *Specify the approximate number of teeth per inch, type of knurl, general size, and use of knurl.*

I. *Sharp corners*
  1. Avoid sharp corners (external and internal) as they cause weakness or more costly fabrication of form tools.
  2. Provide a commercial corner break of 0.4 mm by 45°.
  3. An internal sharp corner can be made by providing an undercut at the corner (Figure 8.11).

J. *Spherical Ends*
  1. Design the radius of the spherical end to be larger than the radius of the adjoining cylindrical surface (Figure 8.12).

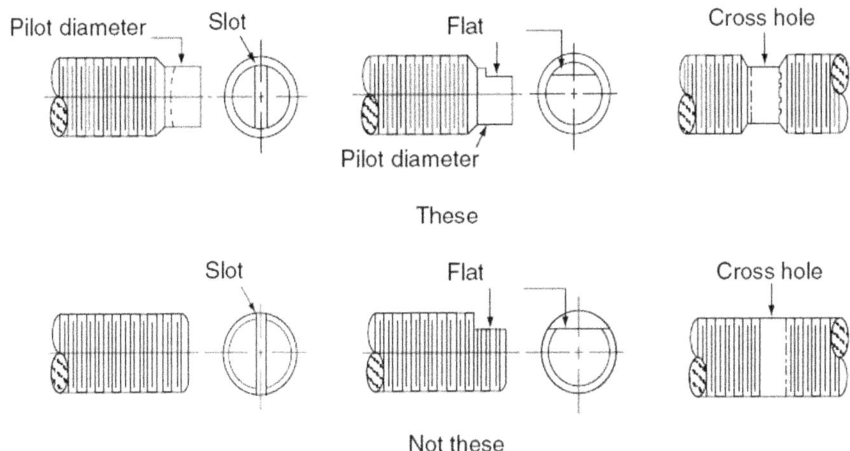

FIGURE 8.10 Avoid burrs at the thread starts.

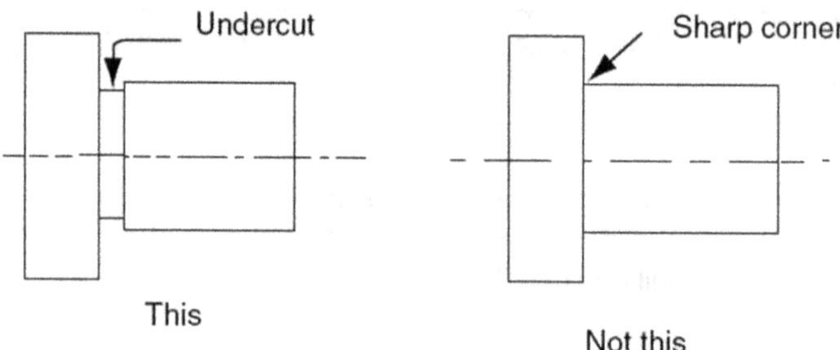

FIGURE 8.11 Undercuts avoid the problems of sharp corners.

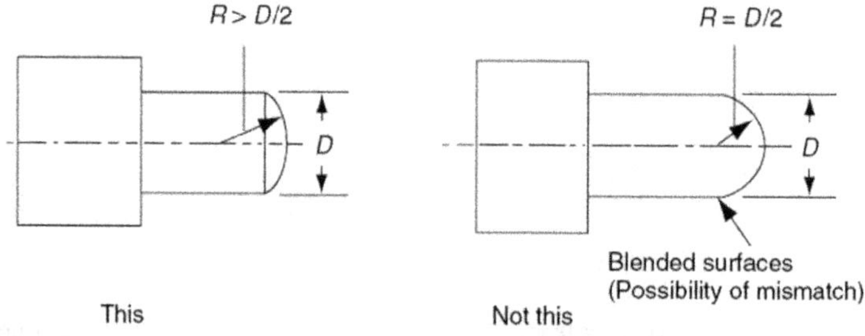

FIGURE 8.12 Avoid blended surfaces formed by a separate cutter.

K. *Slots and Flats*
   1. Slots are produced with a concave surface at the bottom or end (milling cutter radius) (Figure 8.13).

L. *Marking*
   1. Position impression marking so that roller marking tools can be used (Figure 8.14).

Designers should also follow these additional recommendations:

1. Incorporate standard tool geometry at diameter transitions, exterior shoulders, grooves, and chamfer areas.
2. Minimize unsupported, delicate, small-diameter work whenever possible to reduce work deflection. In this regard, short, stubby parts are easier to machine (Figure 8.15).
3. Avoid interrupted cutting actions that may be caused by hole intersections, slant surface drilling, and hole or slotting operations before turning.
4. Castings and forgings with large shoulders or other areas to be faced should have 2–3° from the plane normal to the axis of the part to provide edge relief for the cutting tools (Figure 8.16).
5. Radii, unless critical for the part function, should be large and conform to standard tool-nose radius specifications (Figure 8.17).
6. Specify a break of sharp corners where sharpness or burrs may be hazardous or disadvantageous to the function of the part (Figure 8.18).

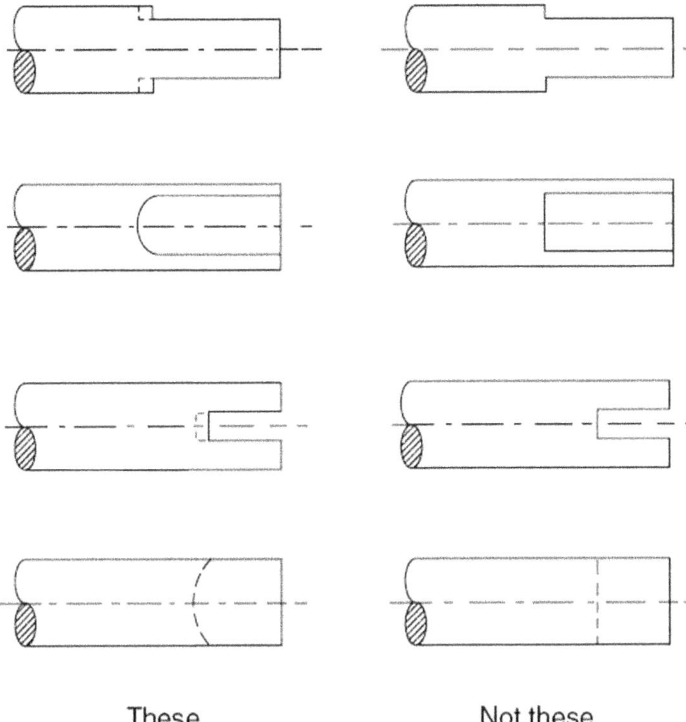

FIGURE 8.13 Permit curved bottoms of slots and flats if possible.

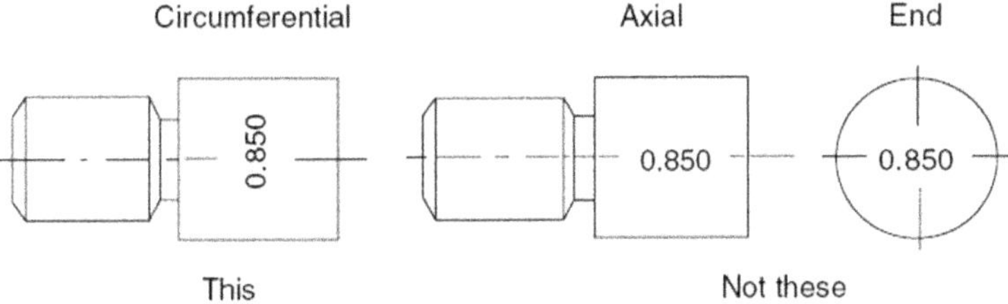

FIGURE 8.14 Marking should allow the use of roller tools.

7. Avoid clamping or locating the part using parting lines, draft angles, and forging flash (Figure 8.19).
8. For tracer-controlled parts, easy tracing is necessary with a minimum number of changes of the stylus and cutting tool. Grooves with parallel or steep sidewalls are not possible and undercuts should be avoided.

8.3.1.3 Dimensional Control

The dimensional tolerances are inversely proportional to the part size. In this regard, the produced tolerances depend on:

- Machine construction

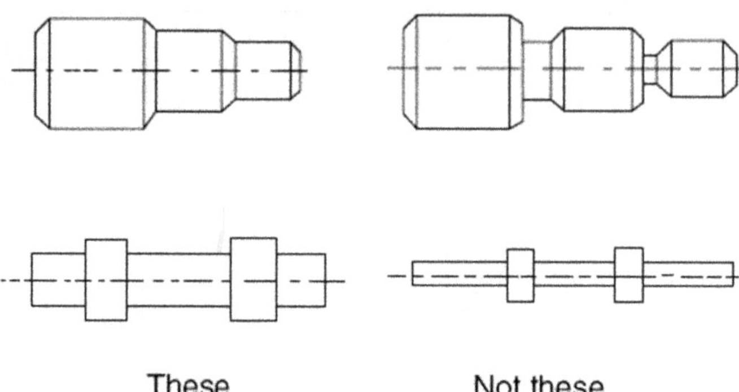

FIGURE 8.15 Avoid slender and long parts to avoid deflection.

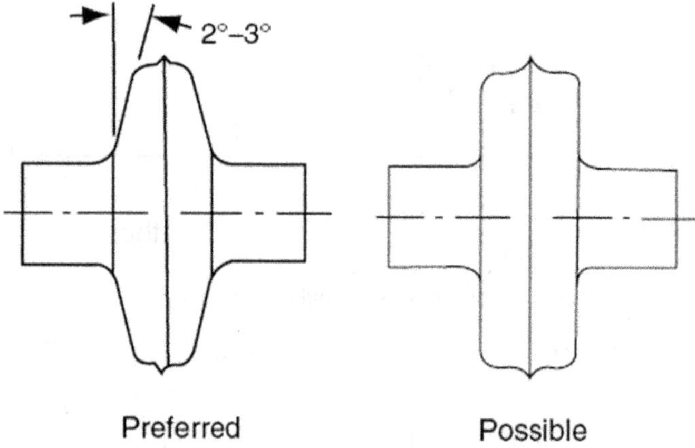

FIGURE 8.16 Allowing a relief on cast or forged parts to be faced provides tool clearance.

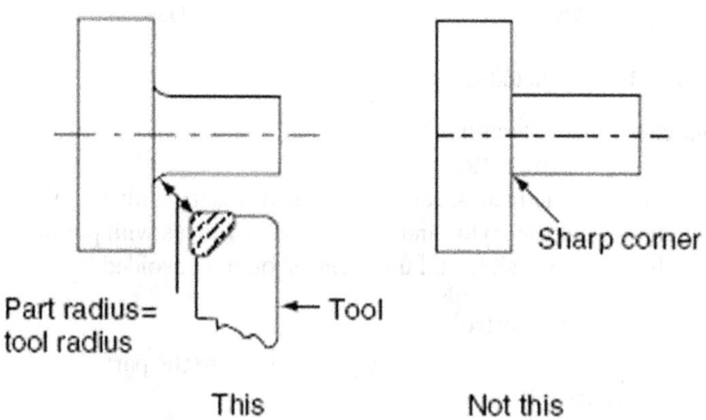

FIGURE 8.17 Avoid sharp corners.

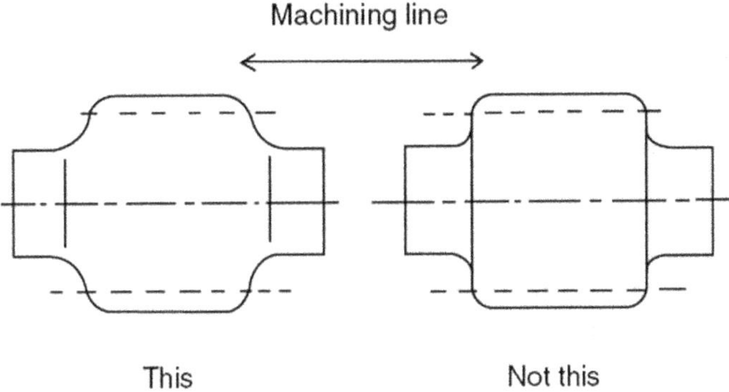

FIGURE 8.18 Minimize sharp corners and burrs by providing chamfers or curved surface to the part before machining.

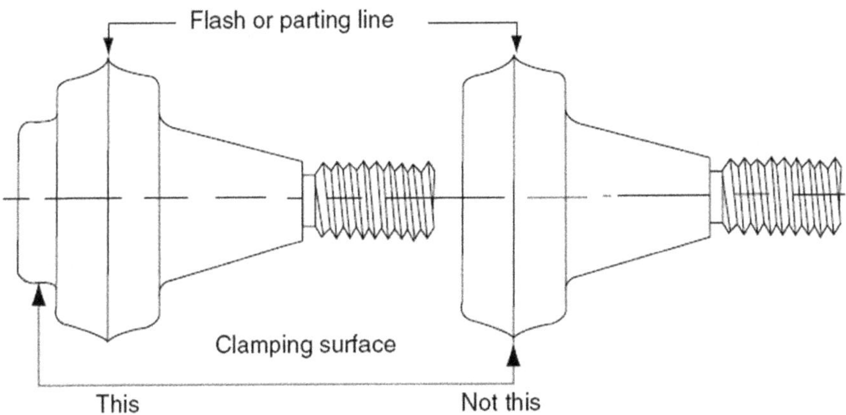

FIGURE 8.19 Avoid designs that require clamping on parting lines or flash areas.

- Control of operational disturbances such as vibrations, deflection, thermal distortion, and wear of operating parts
- Part deflection, tool wear, measuring tool accuracy, and operator skill

Surface finish is also related to the aforementioned factors. It is directly related to the tool feed rate, tool sharpness, tool and WP geometry, and tool materials. Typical machining tolerances are specified by Bralla (1999) for well-maintained turning machines.

8.3.2 Drilling and Allied Operations

Hole-making operations include drilling, boring, trepanning, and gun drilling. Counterboring and countersinking are secondary operations for existing holes. Drilling and reaming can produce holes in the following range:

- Usual diameter range: 1.5–38 mm
- Minimum diameter range (spade-type microdrills): 0.025 mm
- Minimum diameter (reaming): 0.3 mm
- Maximum diameter: 80–90 mm
- Usual maximum depth: Eight times diameter
- Hardness of drilled material: Usually less than HRC 30
- Maximum hardness: HRC 50 rarely to HRC 60

Boring is used when a particular accuracy of diameter, location, straightness, or direction is required. The normal dimensional limits for bored holes are:

- Minimum diameter: 2.5 mm, with fishtail-type solid cutting tool
- Maximum diameter: 1.2 m
- Maximum hardness of material: 60 HRC
- Maximum length of conventional boring bars: Five times diameter
- Maximum length of solid boring bars: Eight times diameter

8.3.2.1 Economic Production Quantities

- NC drilling: One to moderate size
- Normal drilling: <100 piece
- Drilling using drilling jig: >100 piece
- Multi-spindle drill press: >10,000 unit
- Precision jig boring machine: 1 unit

8.3.2.2 Design Recommendations for Drilling and Allied Operations

The following are recommended design practices for:

A. *Drilling*
   1. The drill entry surface should be perpendicular to the drill bit to avoid starting problems and to ensure proper location (Figure 8.20).
   2. The exit surface of the drill should be perpendicular to the axis of the drill to avoid drill breakage when leaving the hole (Figure 8.20).
   3. For straightness requirements, avoid interrupted cuts to avoid drill deflection and breakage (Figure 8.21).

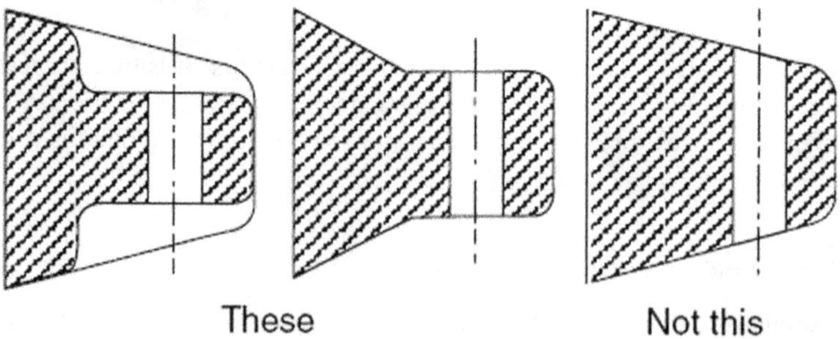

FIGURE 8.20 Entrance and exit surfaces should be perpendicular to the drill axis.

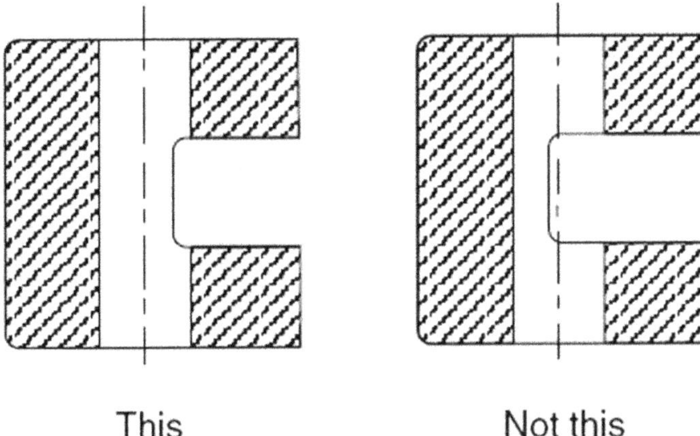

FIGURE 8.21 Keep the center of the drill point in the work throughout the drilling operation.

4. Use standard drill sizes whenever possible.
5. Through holes are preferable to blind holes, as they provide easier clearance for tools and chips.
6. Blind holes should not have flat bottoms because they require a secondary machining operation and cause problems during reaming.
7. Avoid deep holes (over three times diameter) because of chip clearance problems and the possibility of straightness errors (Figure 8.22).
8. Avoid designing parts with very small holes if they are not truly necessary (3 mm is the desirable minimum diameter).
9. If large holes are required, it is desirable to have cored holes (casting) in the WP before drilling.
10. If the part requires several drilled holes, dimension them from the same surface to simplify fixturing (Figure 8.23).
11. Rectangular rather than angular coordinates should be used to designate hole locations (Figure 8.24).
12. Design parts so that all can be drilled from one side or from the fewest number of sides.

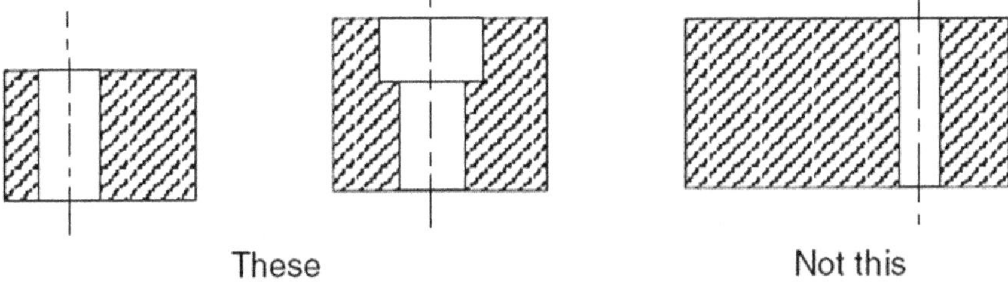

FIGURE 8.22 Avoid deep, narrow holes (depth <3 diameter or consider stepped diameter).

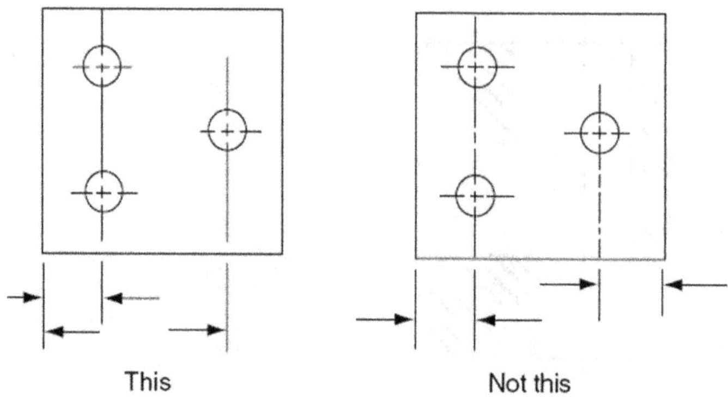

FIGURE 8.23 Locate holes from one datum surface.

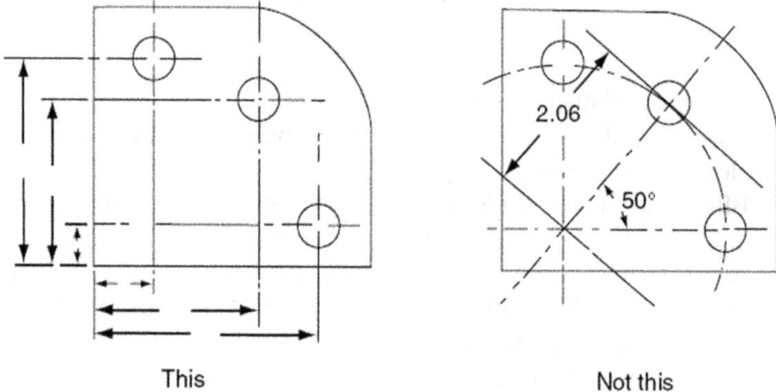

FIGURE 8.24 Rectangular coordinates are preferable to angular coordinates for describing the point location of holes.

13. Design parts so that there is room for the drill bushing near the surface where the drilled hole is to be started (Figure 8.25).
14. Standardize the size of holes, fasteners, and screw threads as much as possible.
15. For multiple-drilling operations, the designer should bear in mind that there are limitations as to how closely two simultaneously drilled holes can be spaced (for 6 mm diameter or less, spacing should not be less than 19 mm center to center).

B. *Reaming*

1. Even when using guide bushing, do not depend on reaming to correct location or alignment discrepancies unless the discrepancies are very small.
2. Avoid intersecting drilled and reamed holes to prevent tool breakage and burr removal problems (Figure 8.26).
3. If blind holes require reaming, increase the drilled depth to provide room for chips (Figure 8.27).

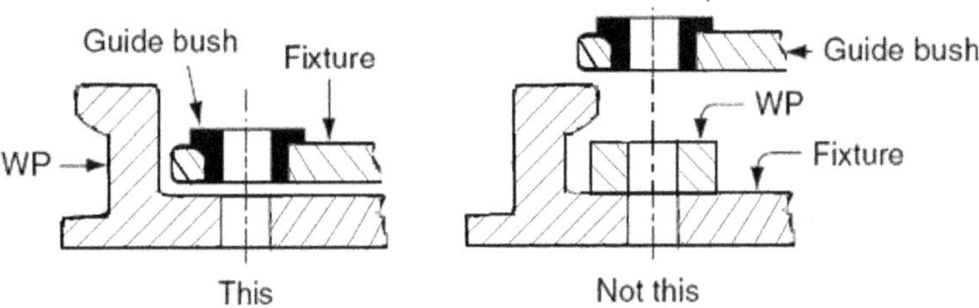

FIGURE 8.25 Allow room for drill bushes close to the drilled surface.

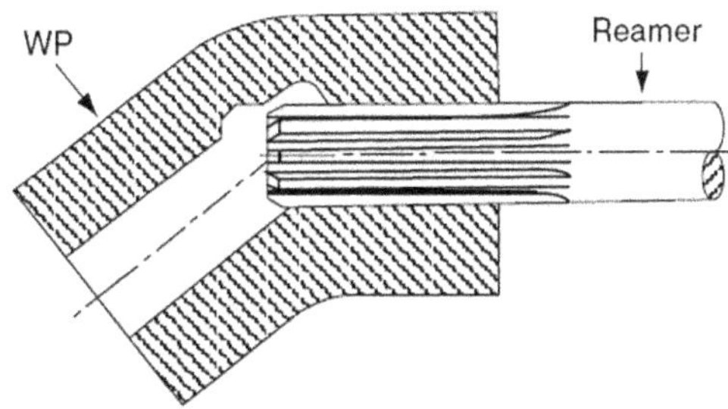

FIGURE 8.26 Avoid intersecting drilled and reamed holes.

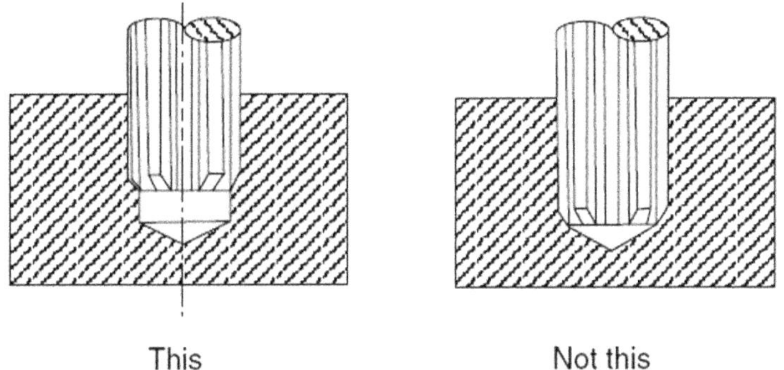

FIGURE 8.27 Provide extra hole depth when reaming blind holes.

C. *Boring*

1. During boring, avoid designing holes with interrupted surfaces, as they cause out-of-roundness errors and tool wear.
2. Avoid designing holes with a depth-to-diameter ratio of over 4:1 or 5:1 to avoid inaccuracies caused by boring-bar deflection. This ratio becomes 8:1 for carbide boring bars.
3. For larger depth-to-diameter ratios, consider the use of stepped diameters to limit the depth of a bored surface (Figure 8.22).
4. Use through holes whenever possible.

5. If the hole must be blind, allow the rough hole to be deeper than the bored hole by 1/4 hole diameter.
6. Use boring only when the accuracy requirements are essential.
7. Do not specify bored-hole tolerances unless necessary.
8. The bored part must be rigid so that deflection or vibrations caused by the cutting forces are reduced.

8.3.2.3 Dimensional Control

The following factors affect the dimensional tolerances of drilling, reaming, and boring operations:

- The drill sharpness affects the accuracy of the diameter and straightness of drilled holes.
- The play and lack of rigidity in the drill spindle.
- Drill bushing reduces the bell mouthing of holes as well as the possible hole oversize.
- Thermal expansions of the material to be drilled.
- The location and direction of reamed holes are affected by the previous drilling operation even when using guide pushing.
- Temperature changes of the WP caused by heat generated by cutting or from other causes affect the accuracy of bored holes.
- WP distortion from clamping.
- Machine condition and rigidity and boring-bar rigidity are essentially important in jig boring operations.

Bralla (1999) provided recommended dimensional tolerances for drilled, reamed, and bored holes.

8.3.3 Milling

8.3.3.1 Design Recommendations

The following general product design rules apply to other machining and milling operations:

1. Sharp inside and outside corners should be avoided.
2. The part should be easily clamped.
3. Machined surfaces should be accessible.
4. Easily machined material should be specified.
5. Design should be as simple as possible.

Additional recommendations particularly applicable to milling are as follows:

1. The product design should permit the use of standard cutter shapes and sizes rather than special ones (Figure 8.28).
2. The product design should permit manufacturing preference as much as possible to determine the radius where two milled surfaces intersect or where profile milling is involved (Figure 8.29).

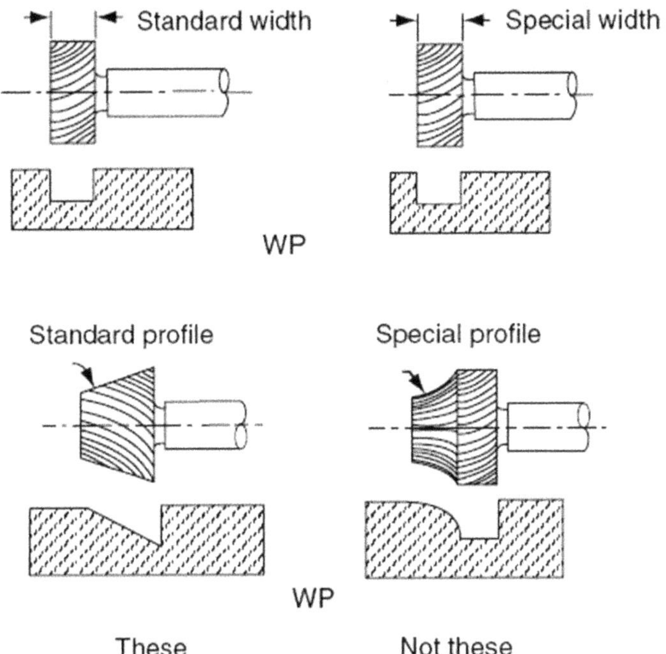

FIGURE 8.28 Allow the use of standard cutter shapes and sizes rather than special ones.

3. When small flat surface is required, the product design should permit the use of spot facing, which is quicker than face milling (Figure 8.30).
4. When spot faces are specified for casting, provide a low boss for the surface to be machined (Figure 8.31).
5. When the outside surfaces intersect and a sharp corner is not desirable, the product design should allow a bevel or chamfer rather than rounding (Figure 8.32).
6. When form-milling or machining rails, do not blend the formed surface to an existing milled surface (Figure 8.33).
7. Keyway design should permit the keyway cutter to travel parallel to the center axis of the shaft and form its own radius at the end (Figure 8.34).
8. A design that requires the milling of surfaces adjacent to a shoulder should provide clearance to the cutter path (Figure 8.35).
9. A product design that avoids the necessity of milling at parting lines, flash areas, and weldments will generally extend the cutter life.
10. The most economical designs are those that require the minimum number of operations.
11. For more economical machining, the product design should allow staking so that a milled surface can be incorporated into a number of parts in one gang milling operation (Figure 8.36).
12. The product design should provide clearance to allow the use of larger size cutters rather than small-size ones to permit high removal rates.
13. In end-milling slots, the depth should not exceed the diameter of the cutter (Figure 8.37).

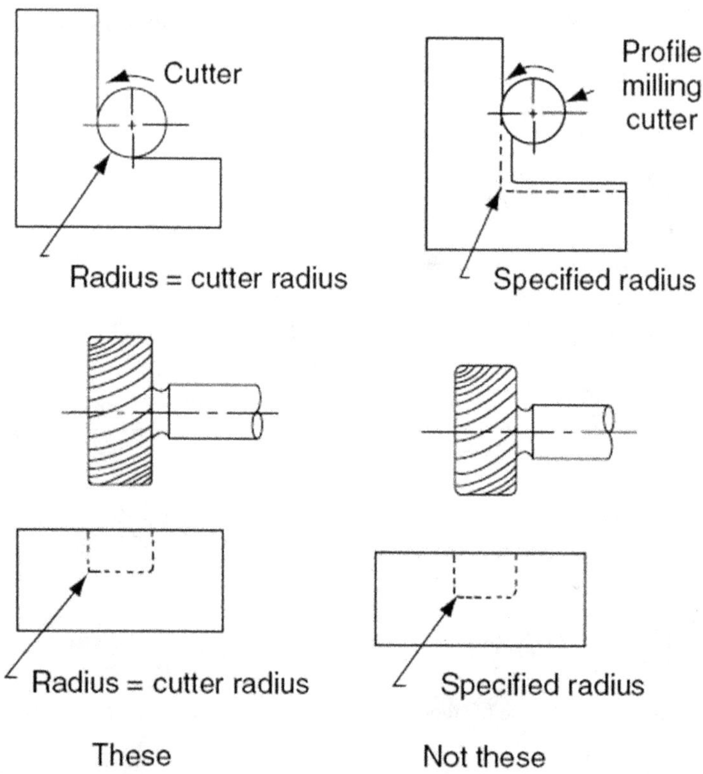

FIGURE 8.29 Allow the use of radii generated by the milling cutter.

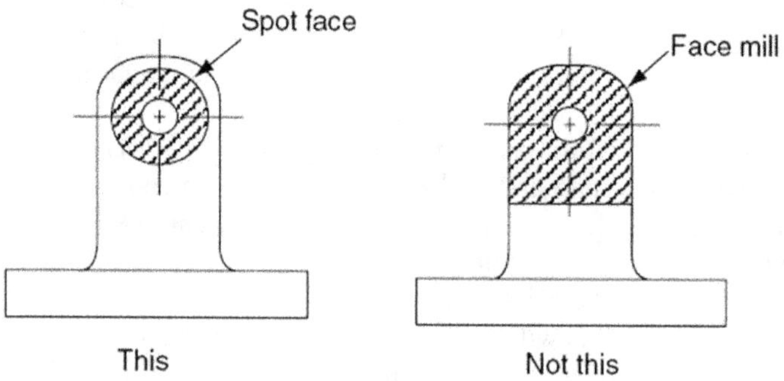

FIGURE 8.30 Spot facing of small surfaces is preferred over face milling.

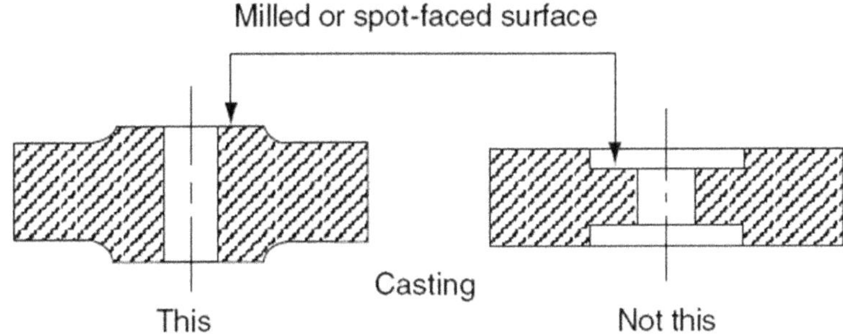

FIGURE 8.31 A low boss simplifies the machining of a flat surface.

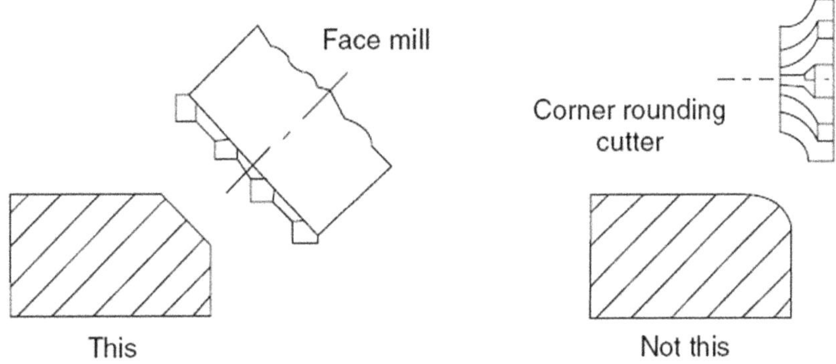

FIGURE 8.32 Allow beveled rather than rounded corners for economical milling.

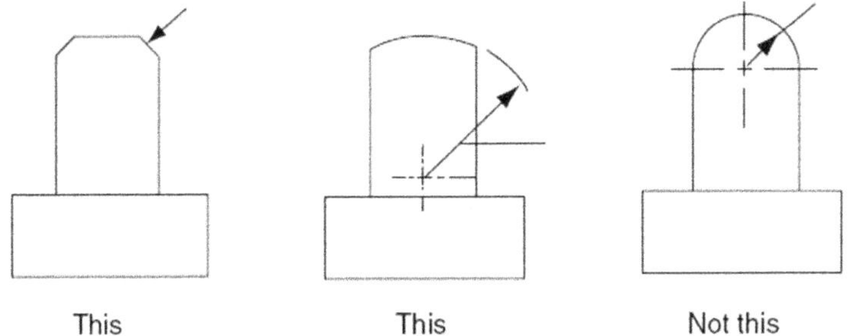

FIGURE 8.33 Do not specify a blended radius on machined parts.

8.3.3.2 Dimensional Factors and Tolerances

The tolerance-holding capabilities of the operation depend on the cutter, machine, and work-holding devices. Additionally, operational disturbances such as tool wear, machine wear, defection, vibration, and rigidity and stability of the WP itself affect the produced tolerances. WP materials of good machinability, fine grain structures,

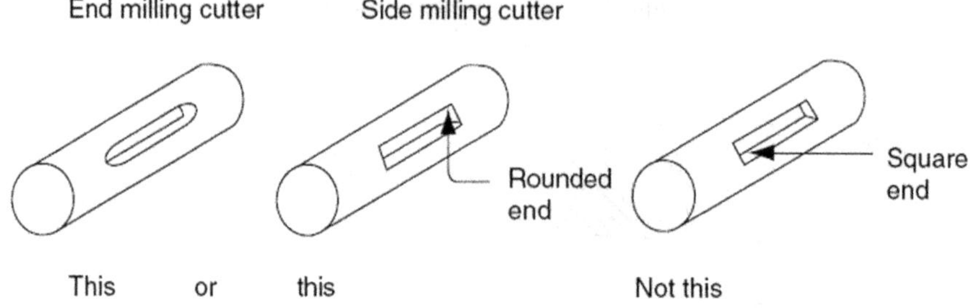

FIGURE 8.34 Design keyways so that a standard milling cutter finishes its sides and ends in one operation.

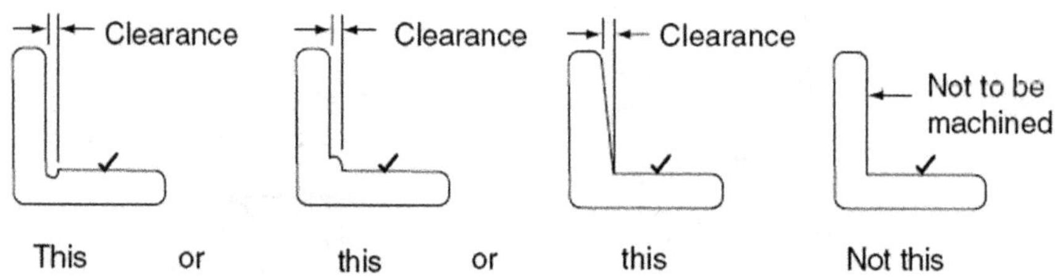

FIGURE 8.35 Provide clearances for milling cutters.

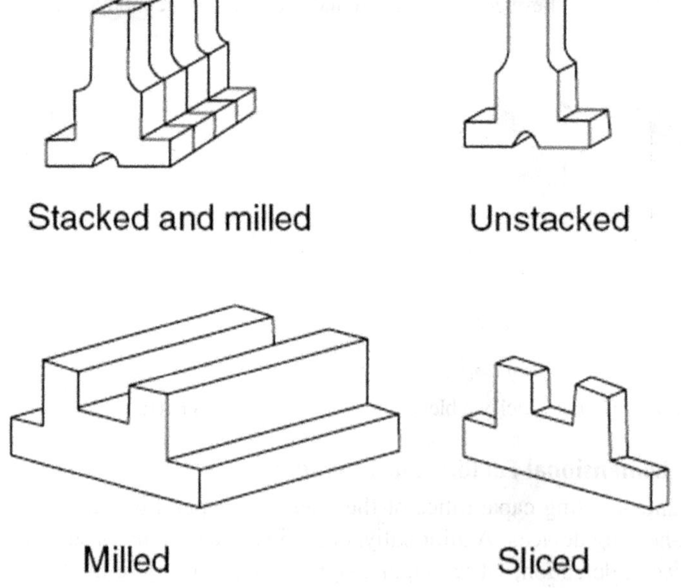

FIGURE 8.36 Designs that permit stacking or slicing are more economical.

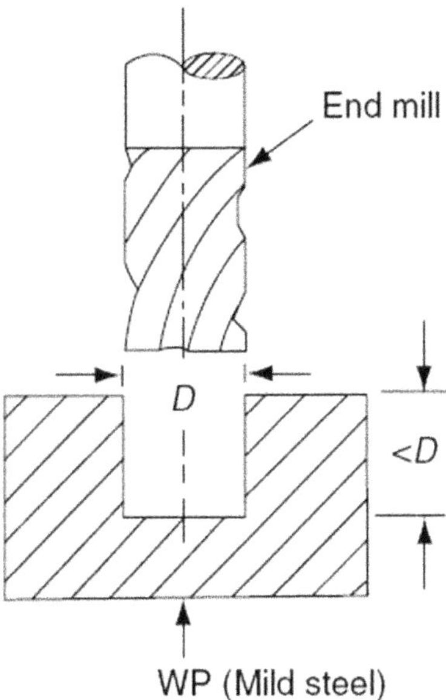

FIGURE 8.37 The depth of end-milled slots should not exceed the cutter diameter.

and reasonable hardness machine more precisely than very hard or very soft large-grained materials. Regarding the surface finish, low feed per tooth, cutting fluid, high cutting speed, and machinable materials produce the desired surface smoothness. Surface finish ranges from 1.5 to 3.8 μm (R_a value) for milling free-machining steels and non-ferrous materials.

8.3.4 Shaping, Planing, and Slotting

8.3.4.1 Design Recommendations

The following are rules that should be adhered to either for economy of operation or for dimensional control:

1. Design parts so that they can be easily clamped to the worktable and are rigid enough to withstand deflection during machining (Figure 8.38).
2. It is preferable to put machined surfaces in the same plane to reduce the number of operations required.
3. Avoid multiple surfaces that are not parallel to the direction of tool reciprocation, which would need additional setups.
4. Avoid contoured surfaces unless a tracer attachment is available and then specify gentle contours and generous radii as much as possible.
5. With shapers and slotters, it is possible to cut to within 6 mm of an obstruction or the end of a blind hole (Figure 8.39). If possible, allow a relived portion at the end of the machined surface.

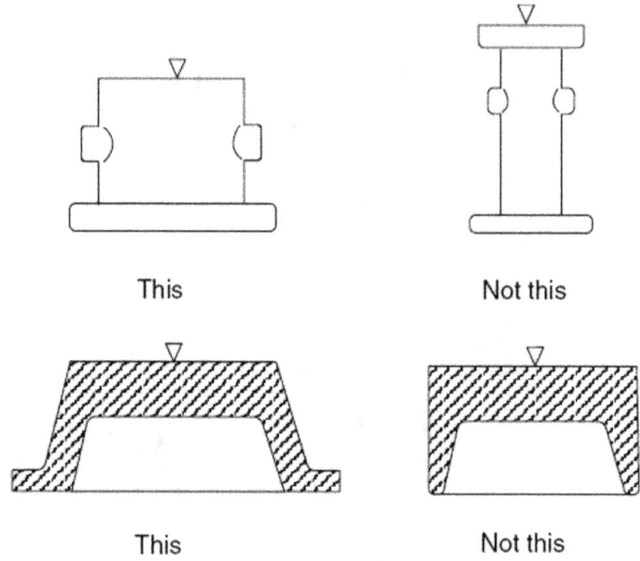

FIGURE 8.38 Shaped and planed parts should withstand cutting forces and provide solid clamping.

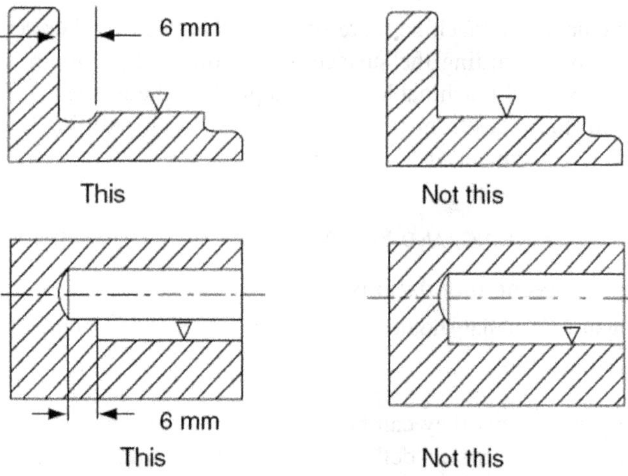

FIGURE 8.39 Avoid machining close to an obstruction at the end of a stroke.

6. For thin, flat WPs that require surface machining, allow sufficient stock for a stress-relieving operation between rough and finish machining or, if possible, rough machine equal amounts from both sides to allow 0.4 mm for finish machining on both sides.
7. The minimum size of hole in which a keyway or a slot can be machined with a slotter or a shaper is about 25.54 mm (Figure 8.40).
8. Because of the lack of rigidity of long cutting tool extensions, it is not feasible to machine a slot longer than four times the hole diameter (Figure 8.40).

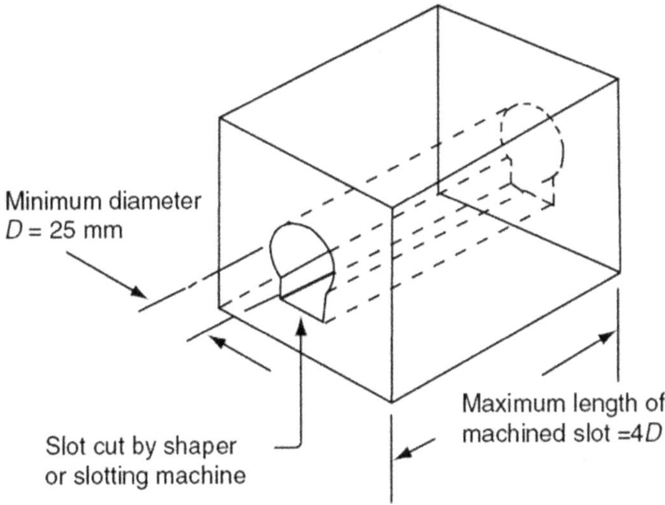

FIGURE 8.40 The minimum size of a hole for machining a slot is D and the slot length should be <4D.

8.3.4.2 Dimensional Control

Dimensional variations occur from human factors, the design and condition of the part itself, and the clamping method. The squareness and flatness of the clamping surface is affected by spring back after machining. Additionally, the dimensional variations are affected by

- Deflection of the part by the cutting forces
- WP warping as a result of the release of the internal stresses in the material during machining
- Tool rigidity, especially in slotting operation or shaping of internal surfaces

Generally, slower cutting speeds, lighter cuts with finer feeds, and the use of lubricants improve the product accuracy. Similarly, sharp tools, correctly ground, and fine feeds facilitate smooth surface finishes. Bralla (1999) presented the recommended tolerances for dimensions and surfaces produced by shapers, planers, and slotters.

8.3.5 Broaching

Broaching usually requires high-volume production that justifies the initial cost of the broaching tools and the need for a special machine. However, it can be applied when there is no machining alternative or when standard broaching tools are available. Production rate ranges from 15 to more than 100 times higher than with alternative machining methods. Tooling a machine for more than one part or for a group of similar parts with the same machined surface makes broaching an economical choice for small-lot quantities.

8.3.5.1 Design Recommendations

Designers should follow specific recommendations regarding the following issues (Bralla, 1999):

A. *Entrance and Exit Surfaces*
   1. The product design should allow easy location and holding of the part.
   2. Surface configuration to the machined area should be square and relatively flat.
   3. Avoid location of parting lines and gates to prevent poor support during machining.
   4. The designer should visualize how the part is supported and retained and avoid the possibility of uneven or inconsistent surfaces in these areas (uneven or inclined surfaces cause side forces that affect the accuracy of finished holes in internal broaching).

B. *Stock Allowance*
   1. Forgings should be held to as close dimension as possible, allowing the minimum stock for finishing to avoid overloading of broach tools during machining (Figure 8.41).
   2. Castings and cold-punched (pierced) holes require greater stock allowance to make sure that clean surfaces are produced.

C. *Wall Sections*
   1. It is advisable to avoid thin wall sections and to maintain a uniform thickness for any wall subjected to the machining forces.

D. *Families of Parts*
   1. The designer should attempt to design parts so that a group of parts use the same broaching tool and, if possible, the same holding fixture.

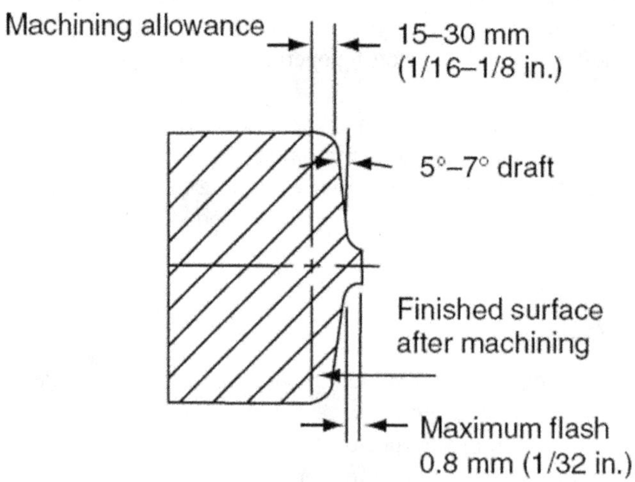

FIGURE 8.41 Stock allowance for broaching a forged part.

E. *Round Holes*

1. Starting holes may be cored, punched, bored, drilled, flame cut, or hot-pierced.
2. For drilled or bored starting holes, consider 0.8 mm stock for 38 mm in diameter and 1.6 mm on larger holes.
3. When cored holes are broached, draft angles, surface texture, and size variation must be taken into consideration in determining the size to assure part cleanup.
4. Long holes should be recessed to improve accuracy and reduce cost (Figure 8.42).

F. *Internal Forms*

1. Symmetrically shaped internal forms are usually broached from round holes.
2. Irregularly shaped internal forms may be started from round holes (Figure 8.43), cored, punched, pierced or machined holes, or machined irregular holes.
3. For casting, stamping, or forging, it is always advisable to leave a minimum amount of stock in addition to the draft, mismatch, and out of roundness errors for complete part cleanup.

G. *Internal Key Ways*

1. Whenever possible, design keyways to the ASA specifications so that standard keyway broaches can be used.
2. Pilot holes for internal keys should be on the same centerline as the finished hole.
3. Balanced designs having more than one key equally spaced are preferred to prevent broach drifting (Figure 8.44).

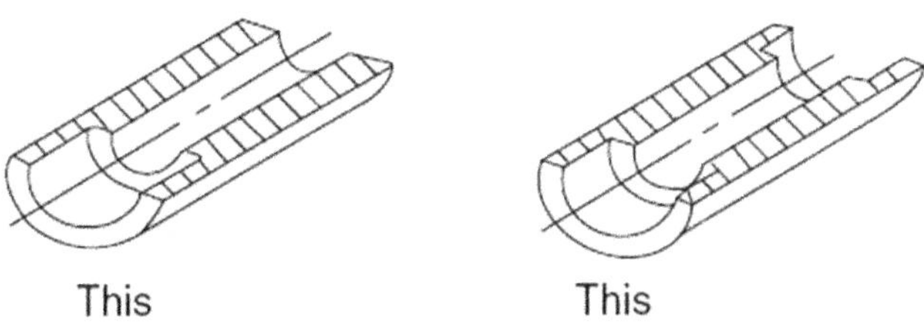

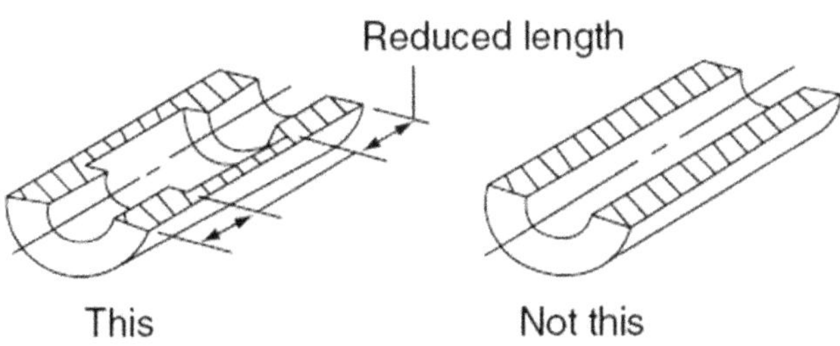

FIGURE 8.42 Long holes should be recessed.

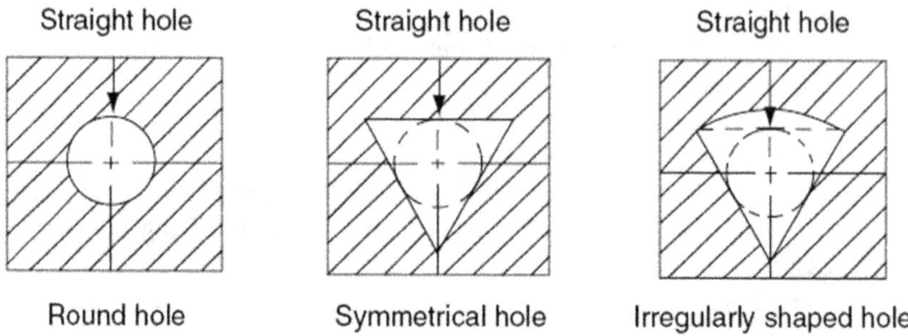

FIGURE 8.43 Irregularly shaped broached holes are started from round holes.

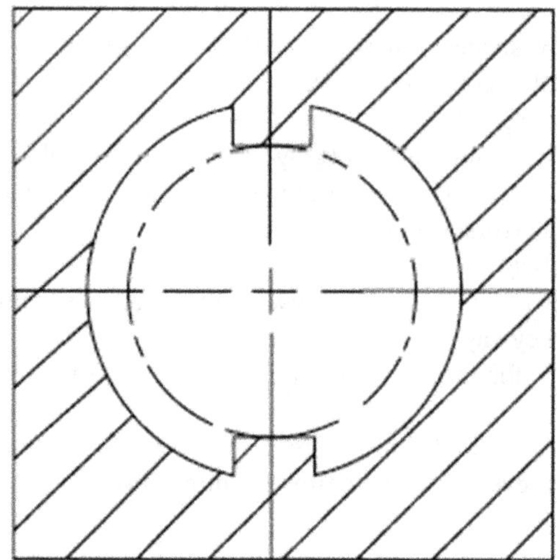

FIGURE 8.44 Balanced WP shapes avoid broach drifting.

H. *Straight-Splined Holes*
   1. Parallel or straight-sided holes should be designed to the SAE standard.
   2. Involute splines should be designed to SAE, Deutsches Institut für Normung (DIN), or the American Gear Manufacturers Association (AGMA) standards.
   3. Fine diametral pitches and stub-tooth forms are advisable to reduce the length of broach required.
   4. Long holes should be recessed or relieved.
   5. The designer is allowed to modify the spline profile to allow room for the upset burr (Figure 8.45).
   6. Dovetail or inverted-angle splines should be avoided whenever possible (Figure 8.46).

I. *Spiral Splines*
   1. In addition to the preceding guidelines, spiral splines with helix angles greater than 40° cannot be broached by conventional methods; use the lowest helix angle possible.

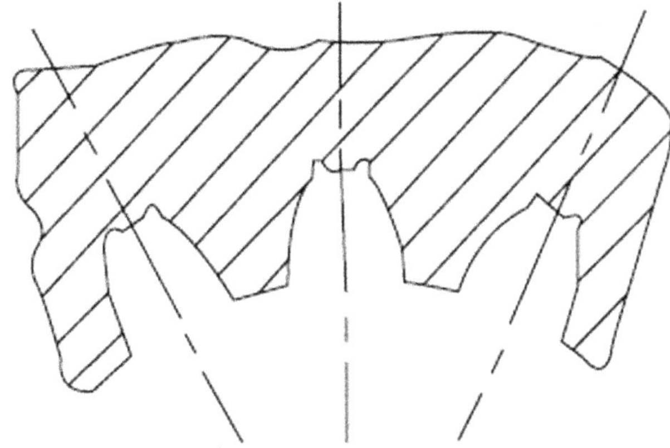

FIGURE 8.45 Allow a room for upset burr.

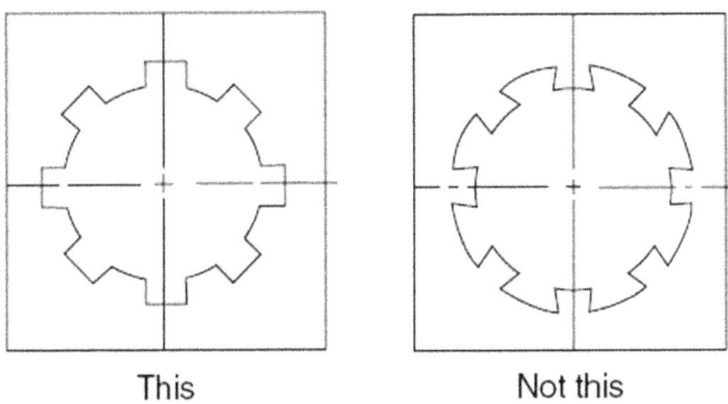

FIGURE 8.46 Avoid dovetail or inverted angle splines.

2. Splines with helix angles >10° require the broach rotation while traveling; the designer should consider preventing the WP rotation.

J. *Tapered Splines*
   1. Tapered splines cannot be produced by broaching (Figure 8.47).

K. *Square and Hexagonal Holes*
   1. It is advantageous to use slightly oversized starting holes, particularly for square holes (Figure 8.48).
   2. Avoid sharp corners at the major diameter to reduce broach costs (Figure 8.49).

L. *Saw Cut or Split Splinded Holes*
   1. When the part will have an interesting cut into the splined hole, the splinded hole should be designed with an omitted space, as shown in Figure 8.50. This allows room for burr produced by the saw cut.

M. *Blind Holes*
   1. Blind holes should be avoided if possible.
   2. Allow a relief at the bottom of the broached area to allow for the chip to break off and be retained in that space (Figure 8.51).

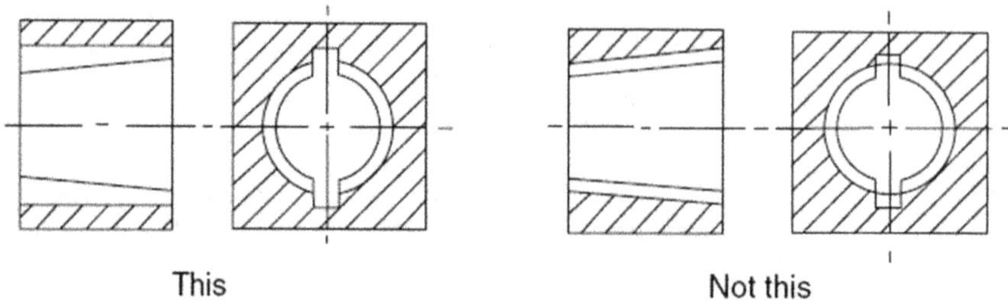

FIGURE 8.47 Avoid tapered splines.

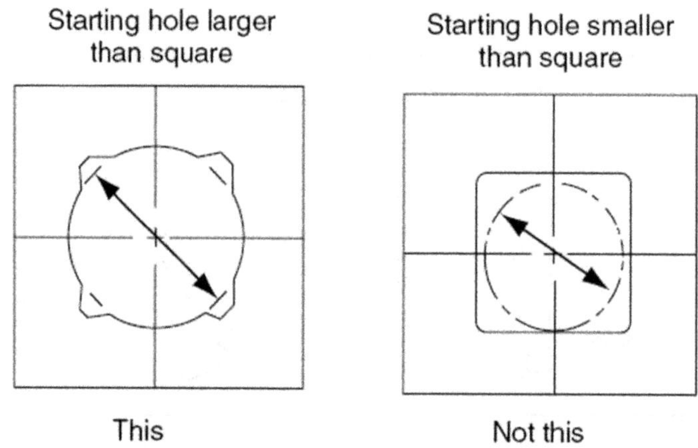

FIGURE 8.48 Use a slightly oversized starting hole.

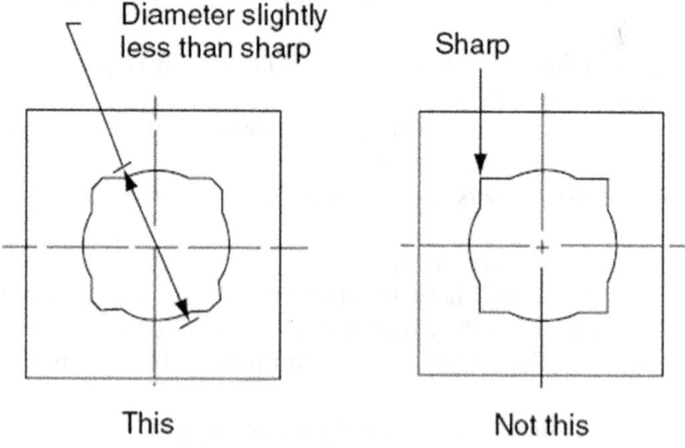

FIGURE 8.49 Avoid sharp corners on major diameters.

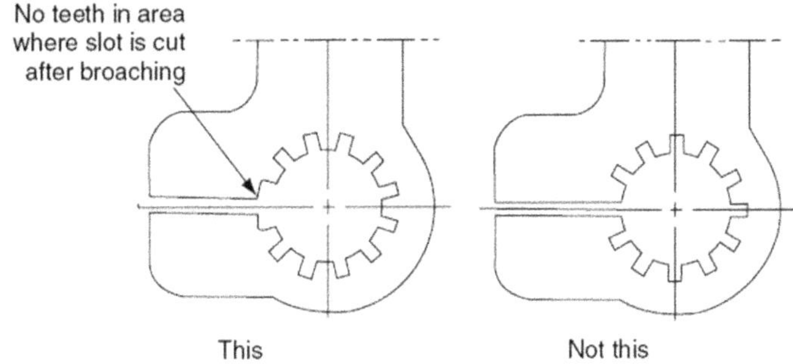

FIGURE 8.50 Allow room for the burr produced by the saw cut.

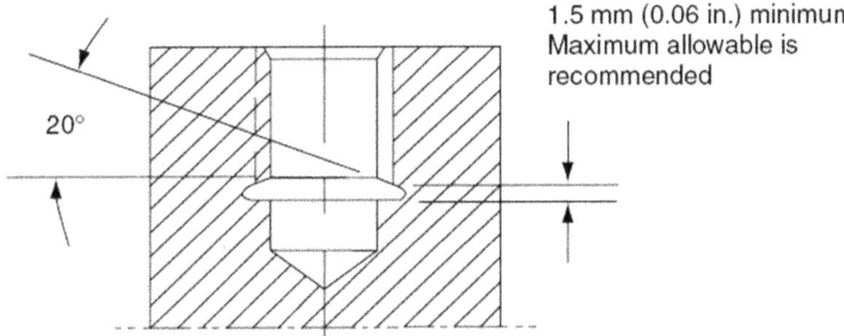

FIGURE 8.51 Allow a relief at the bottom of a blind hole spline.

N. *Gear Teeth*
   1. Gear teeth should be given the same consideration as internal involute splines.

O. *Chamfers and Corner Radii*
   1. In all situations that require breaking the corner, chamfers are preferred over radii.
   2. Sharp internal corners should be avoided to eliminate stress concentration and minimize broach tooth edge wear (Figure 8.52).
   3. Sharp corners or edges of intersecting outer broached surfaces should be avoided whenever possible.
   4. Castings, forgings, and extrusions should be designed with a corner break that does not require machining.
   5. After broaching, outer corners or edges that must be machined should be chamfered rather than rounded (Figure 8.53).

P. *External Surfaces*
   1. External surfaces should be relieved to reduce the area that must be broached.
   2. Reliefs of undercuts in the corners simplify the broaching operation (Figure 8.54).
   3. Large surfaces should be broken into a series of bosses whenever possible (Figure 8.55).

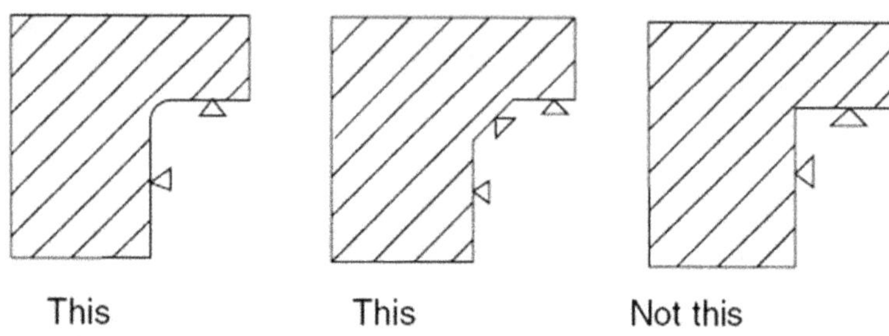

FIGURE 8.52 Internal corner design.

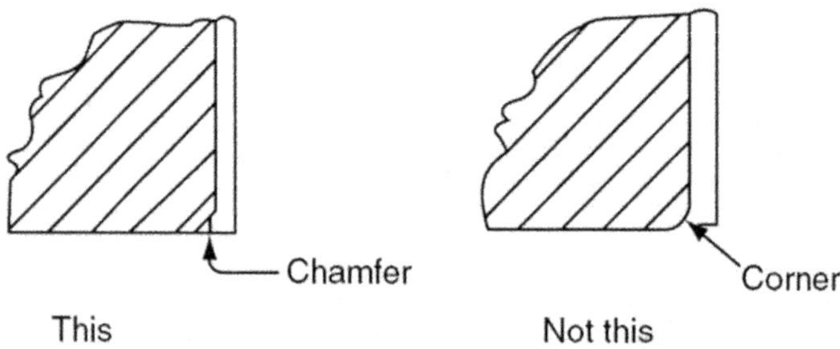

FIGURE 8.53 Chamfer outer corners.

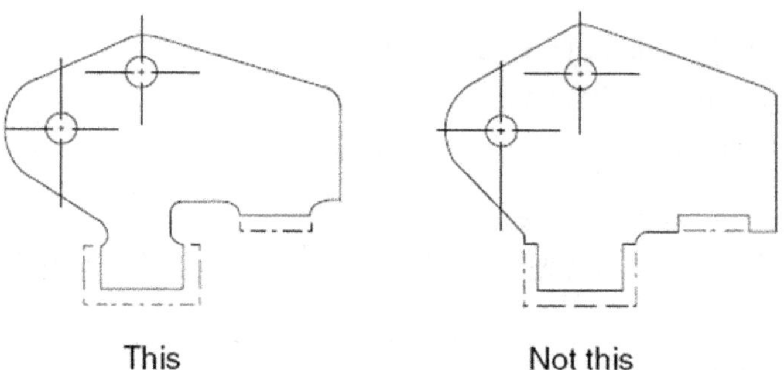

FIGURE 8.54 Reliefs of undercuts in the corners simplify the broaching operation of external surfaces.

Q. *Undercuts*
1. Machined undercuts should be as shallow as possible.
2. Avoid sharp or narrow undercut configurations.

R. *Burrs*
1. Chamfers or reliefs at the exit edge of the surface to be broached are recommended to contain the burr produced and eliminate the need for deburring operation.

S. *Unbalanced Cuts*
1. Unbalanced stock conditions caused by cross holes or other interruptions that cause tool deflection should be avoided.

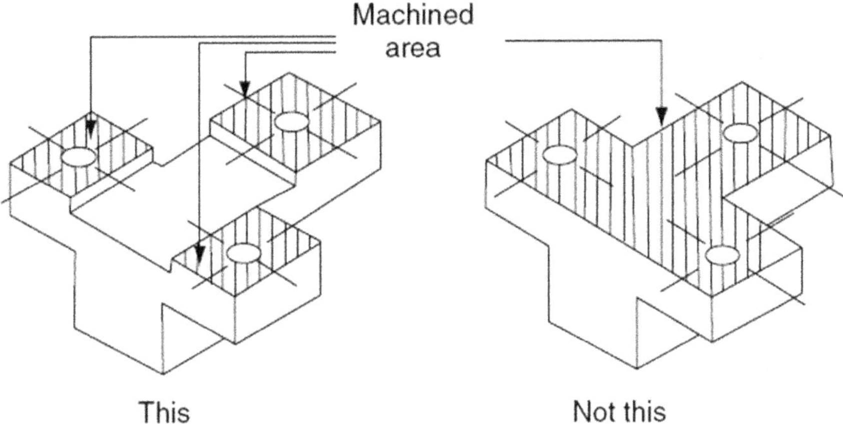

FIGURE 8.55 Break large surfaces into a series of bosses.

8.3.5.2 Dimensional Factors

1. Internal broaching, especially single-pass operations, produces tighter tolerances than external applications.
2. Broaching applications that require tool guiding or multiple passes are subject to residual stresses.
3. Uniformity of material, consistency of datum faces, and strength of the part affect tolerance control.
4. Tool maintenance and resharpening, machinability of material, and proper design of the tool are essential for controlling size and finish required.

8.3.5.3 Recommended Tolerances

According to Bralla (1999), the range of tolerance recommended for broaching operation can be summarized as follows:

Surface finish. Surface finish produced by broaching does not match the grinding finish. However, it is superior to the finish produced by most other machining methods. Burnished finishes can be guaranteed by employing good tool design and proper cooling oils for highly machinable materials.

Flatness. Parts of uniform sections having sufficient strength to withstand cutting forces can be machined within ±0.025 mm total indicator reading (TIR).

Parallelism. Parallelism of machined surfaces machined in the same stroke should be within ±0.025 mm TIR in good-to-fair machinability rated materials.

Squareness. For parts that can be clamped and retained on true surfaces, a squareness of ±0.025 mm TIR is possible and tolerances of ±0.08 mm can be obtained under controlled conditions in good-machinability-rated materials.

Concentricity. The concentricity error caused by broach drift should not exceed ±0.025 to ±0.05 mm for round or similarly shaped holes in good-to-fair-machinability-rated materials.

Chamfers and radii. Tolerances on chamfers and radii should be as liberal as possible. Radii under 0.8 mm should have a minimum tolerance of ±0.13 mm; ±0.025 mm should be allowed for larger sizes. Large tolerances reduce broach manufacturing and maintenance cost.

8.3.6 Thread Cutting

Cutting screw threads in free cutting materials raises the rate of production and thus reduces the machining and tool cost, while non-free-machining metals are difficult to thread. In this regard, brasses and bronzes cut better and at higher speeds than steels. Cast aluminum is quite abrasive and causes excessive tool wear. It is difficult to cut threads in steels having <160 hardness Brinell (HB). Materials of HRC >34 are not suitable for die chasers and taps, which are made from HSS. Carbide tools are used for single-point cutting for materials above HRC 34. The most suitable materials for thread grinding are the hardened steels and any material having HRC >33. Aluminum and comparable soft materials tend to load the GW and cause burning and therefore are difficult for thread grinding.

8.3.6.1 Design Recommendations

Designers should follow these recommendations:

1. Provide a space (1.5–19 mm) for the thread cutting tool (Figure 8.56).
2. Allow chip clearance space when cutting internal threads (through holes are best) (Figure 8.57).
3. Consider the use of a reduced height thread form, which machines more easily (Figure 8.58).
4. Keep the thread as short as possible, which machines quicker and provides longer tool life.
5. Include a chamfer at the top and the end of external threads and a countersink at the top and the end of internal threads.
6. The surface of the starting thread must be flat and perpendicular to the thread's center axis.

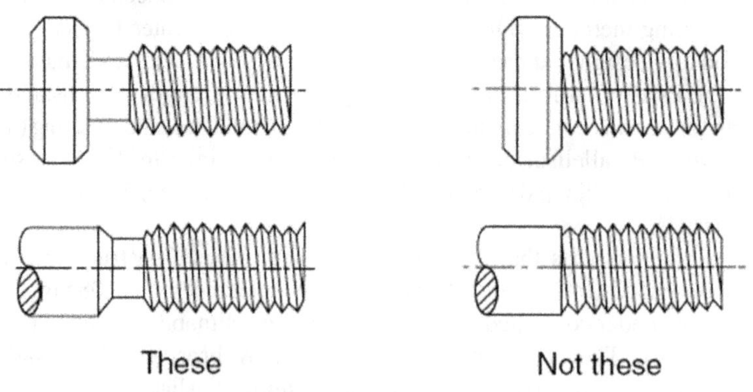

FIGURE 8.56 Allow thread relief at the end of a threaded length.

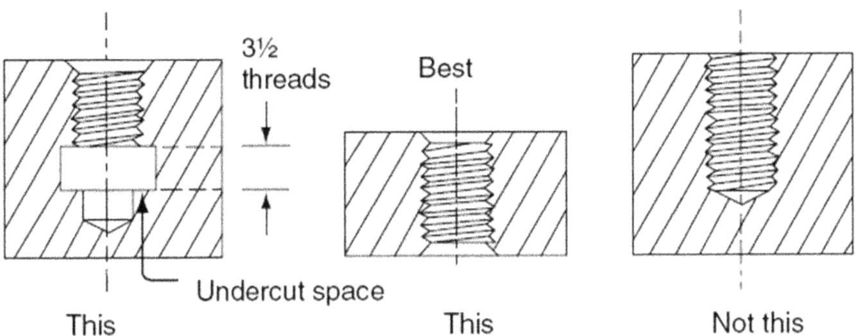

FIGURE 8.57 Allow chip clearance for internal threads.

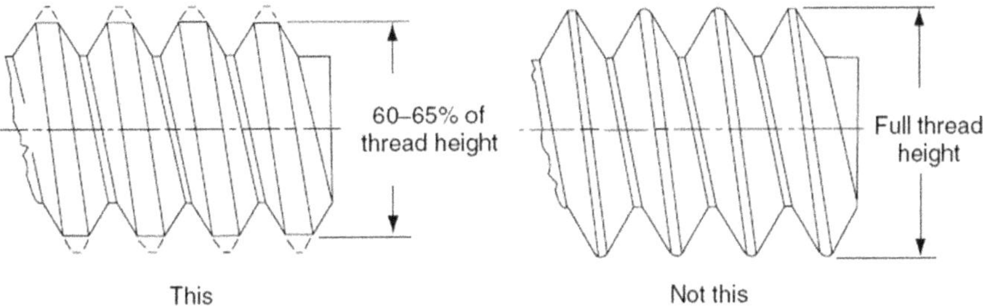

FIGURE 8.58 Reduce thread height for easier machining.

7. Avoid slots, cross holes, and flats that intersect with the cut threads.
8. When cross holes are unavoidable, consider countersinking of such cross holes.
9. Do not specify closer tolerances than required (class 2 is commonly satisfactory).
10. Ground threads should be provided with corners of 0.25 mm at the root.
11. The length of centerless ground threads should be larger than the thread diameter.
12. Coarse threads are more economical to produce and assemble faster than fine threads.
13. Tubular parts must have a wall thickness that withstands the cutting forces.

8.3.6.2 Dimensional Factors and Tolerances

The following factors affect the accuracy of machined thread:

- The accuracy and condition of the tooling and equipment
- The skill of the operator
- The suitability of the WP material
- The feed rate of the threading tool

Threads of classes 4 and 5 can be produced by machining while dies are capable of producing threads of classes 1 and 2 only. Surface finishes smoother than 1.6 μm are not obtained by thread cutting. Using thread milling, dimensional tolerances of ±0.025 mm are obtainable. It is capable of producing accurate classes of thread in materials of poor machinability. Thread grinding achieves threads of classes 4 and 5.

8.3.7 Gear Cutting

Gear machining methods include milling, hobbing, shaping, and broaching. Shaving, grinding, lapping, and burnishing are used to improve accuracy and surface finish. Sufficient strength and machinability are the most important prerequisites of machined gears. The high machinability ratings make it easier to achieve precise machining and smooth surface finishes. Other requirements include corrosion resistance, dimensional stability, wear resistance, natural lubricity, noise damping properties, and low cost. Machined gears are most frequently made from steel, which has a high strength and low cost. Carbon steel, which is low in cost, is satisfactory for machining and case hardening, and is very commonly used for commercial gears. Alloy steels have the advantages of strength, heat treatment, and corrosion and wear resistance. However, they also have poor machinability and are higher in cost compared to plain carbon steels. Leaded and resulfurized (free-machining) steels should be used for machined gear whenever possible; however, they have low impact strength and are less suitable for high-power applications. Stainless steels are only used when corrosion resistance is essential. They are more expensive, difficult to machine, have low wear resistance, and are not heat-treatable. CIs have good machinability, are low in cost, and have vibration damping characteristics. Apart from malleable CI, they have low shock resistance. Cast steels have better physical properties than CI but are more expensive, less machinable, and lack good damping characteristics. Bronze is a superior gear material that has excellent machinability and wear and corrosion resistance. However, its material cost is high. Aluminum is suitable for lightly loaded gears. It is machinable and provides good surface finish and corrosion resistance.

8.3.7.1 Design Recommendations

When designing gears for machining, the designer must consider choices that have a significant effect on the cost and performance of gears. Some points of consideration described by Bralla (1999) include the following:

1. The coarsest pitch that performs the required function is the most economical to cut (Figure 8.59).
2. Helical, spiral, and hypoid gears are difficult to machine than spur gears.
3. Dimensional tolerances and surface finishes should be as liberal as the function of gears permit.
4. Shoulders, flanges, or other portions of the WP larger than the root diameter of the gear should be located away from the gear teeth to allow sufficient clearance for the gear-cutting tool.
5. Herringbone gears should have a groove between halves; internal gears in blind holes require an undercut groove or other recessed space for cutter over-travel (Figure 8.60).

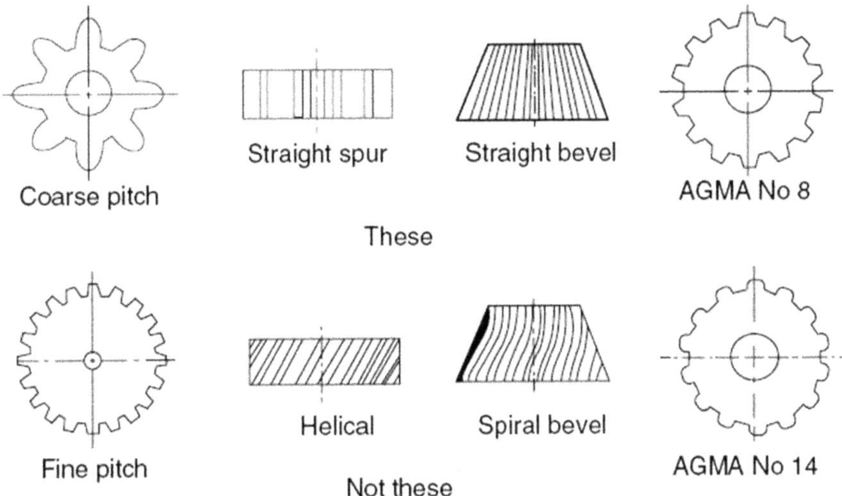

FIGURE 8.59 Use economical gear designs of coarse pitch, straight teeth, and small AGMA number.

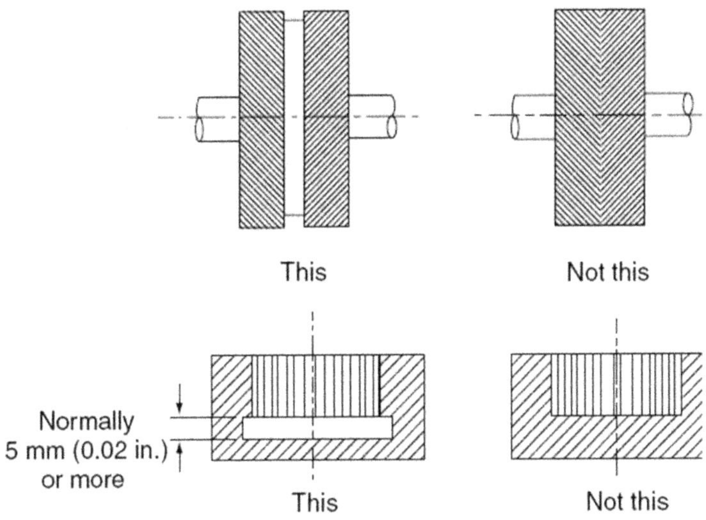

FIGURE 8.60 Allow a groove between halves of herringbone gears and clearance or undercut for internal gears for the cutting tool.

6. Use non-heat-treated gears of larger size instead of heat-treated ones.
7. Heat-treated gears should be of uniform cross section to minimize heat treatment distortion using nitriding.
8. It is advisable to machine gears as separate parts and assemble them afterward onto shafts or other machine components.
9. Use as small helix angles as possible (Figure 8.61).

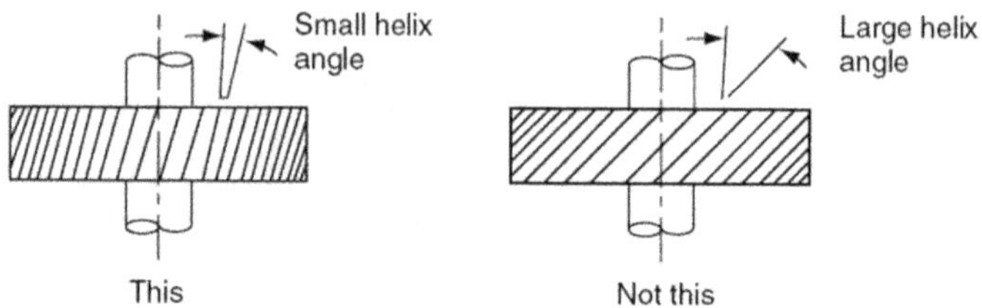

FIGURE 8.61 Avoid large helix angles whenever possible.

10. Avoid wide-faced gears, which are more difficult to machine to the given tolerance (Figure 8.62).
11. Design gear blanks that withstand the clamping and cutting forces without distortion.
12. When gears are press-fitted to a shaft or other components, the fitting surface should not be too close to the teeth.
13. Use standard pitches to minimize tooling costs.
14. When gears are subjected to a finishing process, specify the proper stock allowance that minimizes the production cost.
15. The involute form of the tooth should be the standard specified tooth form for all normal gearing.

8.3.7.2 Dimensional Factors

The dimensional variations in gears result in noise, vibration, operational problems, reduced load-carrying capacity, and reduced life. The AGMA provided dimensional

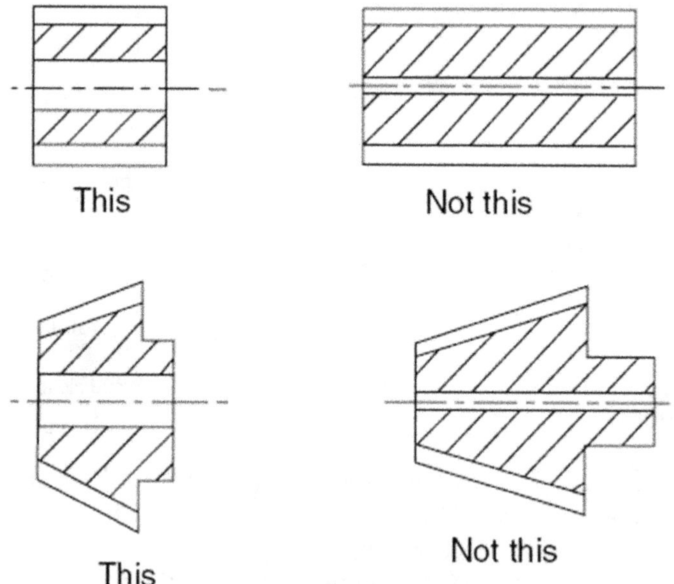

FIGURE 8.62 Avoid wide-face narrow-bore gears.

accuracy levels in terms of quality numbers that apply to gears of varying size, diametral pitch, and type (see AGMA Standard 390.03). Control of accuracy to high AGMA quality levels requires control of the environment and all machining conditions and necessitates secondary machining operations such as grinding, shaving, and lapping.

8.4 DESIGN FOR GRINDING

8.4.1 Surface Grinding

Parts are produced by surface grinding to the required dimensions and surface finish. Horizontal-spindle reciprocating table machines are used for low-quantity production. Small hand-operated tool room surface grinders are ideally suited for single parts such as dies, molds, gauges, and cutting tools. Large-volume production is most effectively carried out on the vertical-spindle rotary-table machines where automatic part loading and unloading is possible. Carbon and alloy steel grades and other high tensile strength metals are ground with aluminum oxide wheels. CI, soft brass, bronze, aluminum, copper, plastics, and other non-metallic materials are ground with silicon carbide wheels. Hardened tool-and-die steels use CBN as an abrasive material.

8.4.1.1 Design Recommendations

Surface grinding is a necessary step in finishing a part and therefore consideration must be given to the design so that the grinding operation can be performed easily:

1. For large-volume production, the part should be designed such that it can be ground on the vertical-spindle surface grinders.
2. The part should not have any surface higher than the surface to be finished (Figure 8.63).
3. The part should be magnetic or easily clamped in an automatic fixture.
4. For large-volume production that requires form grinding or has projections above the surface to be finished, a horizontal-spindle-powered table surface grinder is used.
5. For wider tolerances and rough surface finish of milling or other machining operations, avoid surface grinding of the part.

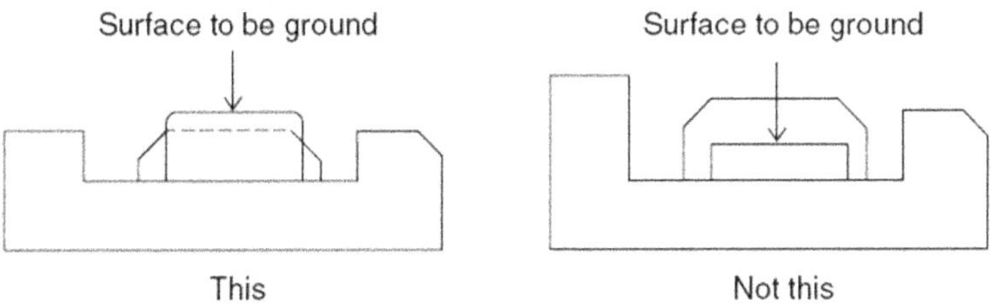

FIGURE 8.63 The ground surface should be higher than other surfaces to avoid wheel obstruction.

6. Design the part to be held by magnetic chucks; large and flat locating surfaces are the best.
7. Do not specify a better ground finish than necessary because this requires longer machining times.
8. Design the part so that surfaces to be ground are all in the same plane (Figure 8.64).
9. Avoid openings in the flat ground surfaces, as the GW cuts deeper at the edges of interrupted surfaces.
10. Avoid unsupported surfaces that may deflect under GW pressure (Figure 8.65).
11. Whenever grooves or other forms are ground on surface grinders, corner relief is more preferable than sharp corners.
12. Avoid blind cuts where the wheel must be stopped during the cut or reversed with too little clearance provided.
13. Design the part for minimum stock removal by grinding, especially if horizontal-spindle machines are used.
14. Avoid extremely thin sections that cause burning or wrapping of the part.
15. When possible, avoid dissimilar materials that cause wheel loading problems.
16. Indicate clearly on drawings the permitted straightness and parallelism required.

8.4.1.2 Dimensional Control

It is possible to achieve a thickness tolerance of ±0.0025 mm (±0.0001 in.) and flatness of less than one light band on surface grinders. Dimensional variations are

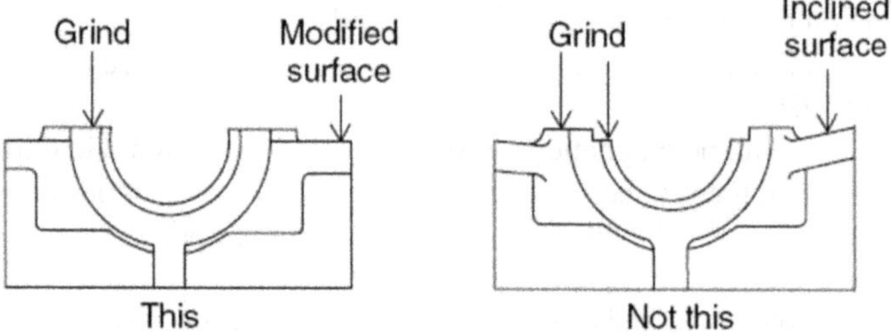

FIGURE 8.64 Design parts so that they can be machined in a single setup without wheel obstruction.

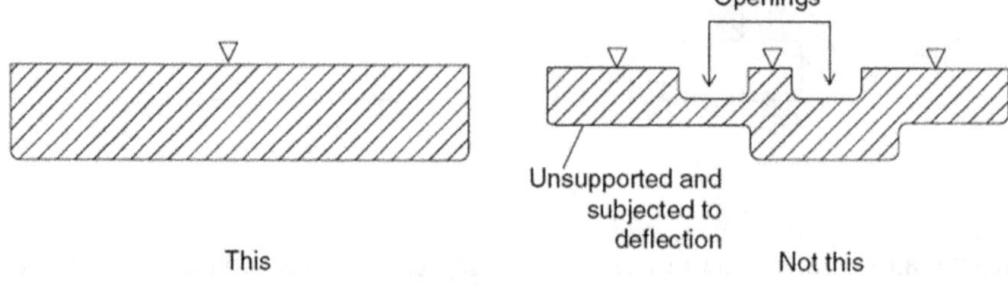

FIGURE 8.65 Avoid openings and unsupported surfaces.

affected by machine condition, accuracy, and cleanliness of the chuck or fixture, suitability of the wheel and coolant, wheel speed, depth of cut, traverse rate, grindability of the WP material, uniformity of temperature, and the freedom of the WP from internal stresses.

Straightness and flatness are affected by the operator technique and surface finish is a function of the process time. Small speeds and slower traverse rates favor improved finishes. Fine grit wheels and the use of grinding fluids, slow traverse, diamond dressing of the wheel, good wheel balancing, and the use of hardened work materials of high grindability enhance the quality of surface finish.

8.4.2 Cylindrical Grinding

The process is capable of producing shafts and pins as well as parts with steps, tapers, and ground forms. Diameters as large as 1.8 m (72 in.) can be ground in these machines. However, the process tackles minimum diameters of 3 mm short cylinders. Typical parts include crankshaft bearings, bearing rings, axles, rolls, and parts with interrupted cylindrical surfaces. Plunge-type grinding is limited to ground surfaces shorter than the GW width. Typical output rates range from 10 to 130 pieces/h, with about 60 pieces/h being a fair average for operations involving a single surface or a single cut.

8.4.2.1 Design Recommendations

When designing components for center-type grinding, the following recommendations are followed:

1. Keep the parts as well balanced as possible for better surface finish and accuracy.
2. Avoid long small-diameter parts to minimize part deflection by the grinding forces (length ≤20 diameter).
3. Keep profile parts for plunge grinding as simple as possible by avoiding tangents to radii, grooves, angular shapes and tapers, and component radii.
4. Avoid interrupted cuts, as the surface adjacent to interruptions ground deeper than the rest of the continuous surface (Figure 8.66).
5. Undercuts on facing surfaces should be avoided (Figure 8.67).
6. If fillets are used, the designer should consider machining or casting a relief on the WP at the junction of two ground surfaces (Figure 8.68).

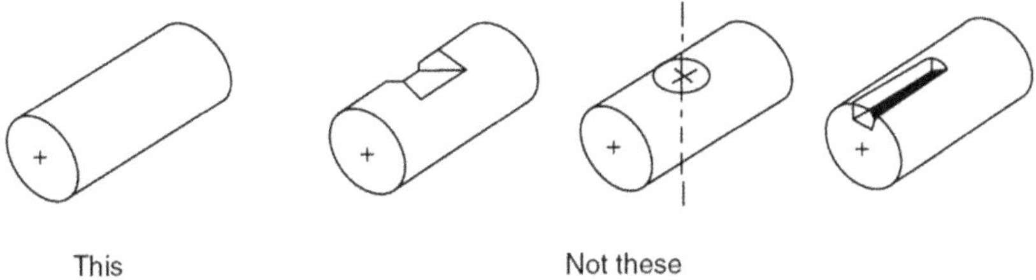

FIGURE 8.66 Avoid interrupted surfaces for better accuracy.

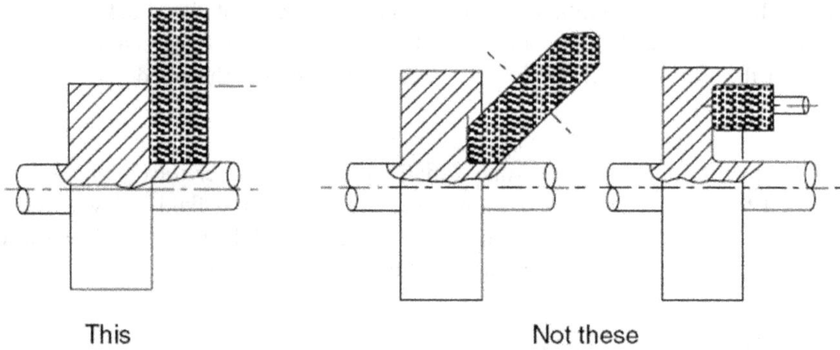

FIGURE 8.67 Avoid undercuts that are costly.

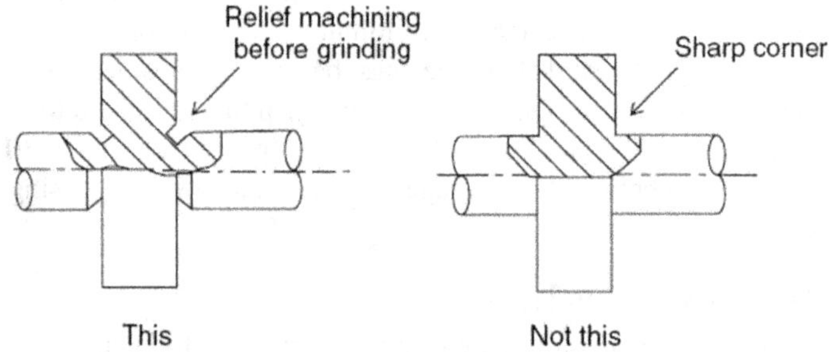

FIGURE 8.68 Machine or cast a relief at the junction of two surfaces before grinding.

7. Center holes on parts held between centers should be made accurately at a 60° angle for accurate cylindrical grinding. They may be lapped in the case of precision grinding.
8. Avoid clamping of thin-walled tubular parts by a three-jaw chuck.
9. Minimize the stock removed by grinding.

8.4.2.2 Dimensional Factors

The accuracy of the final dimensions of machined components reflects the condition of the equipment and the skill of the operator. Worn bearings, centers, machine guide ways, poor coolant action, improper GWs, and the deflection of the WP can adversely affect the finished surface and dimensional accuracy that can be enhanced by:

- The use of correct feed and speed
- The use of steady-rest supports for long parts
- Improving the roundness of center holes in the WP

Recommended dimensional tolerances for cylindrically ground parts under normal production conditions are ±0.0125 mm for diameter, parallelism, and roundness. A surface roughness of 0.2 μm is obtained under normal conditions, which can be improved to 0.05 μm under tight machining conditions.

8.4.3 Centerless Grinding

Centerless grinding is used for solid parts with diameters as small as 0.1 mm and as large as 175 mm. Parts as short as 10 mm and as long as 5 m are also centerless-ground. Pins, shafts, and rings with close tolerances for outside diameters, precise roundness, and smooth finishes are possible. The process is suitable for mass production when short pieces are machined by the through-feed method. Using automatic magazines or hoppers, parts are machined at a rate of 6000 pieces/h. Infeed grinding can be performed at a rate of 30–240 pieces/h. In other words, production rate ranges from 1 to 9 m/min per pass and one to six passes may be used. Brittle, fragile, and easily distorted parts and materials are more suitable for centerless grinding than center-type.

8.4.3.1 Design Recommendations

The following suggestions should be kept in mind by the designer to take the maximum advantage of the process:

1. The ground surface of the WP should be its largest diameter to permit the through-feed operation (Figure 8.69).
2. To avoid the formation of possible tapered or concave- or parallel-shaped surfaces, keep the ground surface sized at least to the diameter of the WP (Figure 8.70).

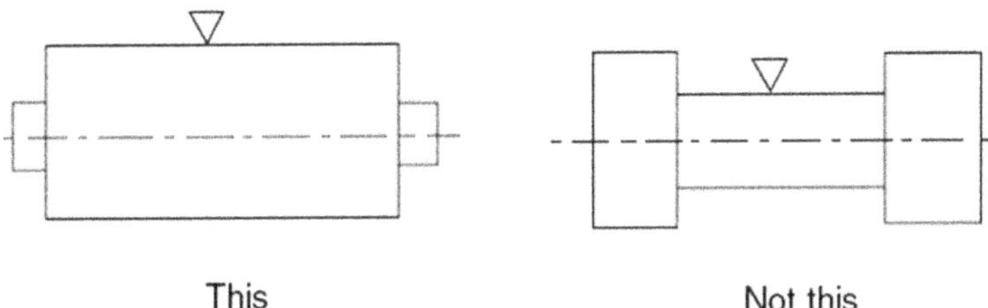

FIGURE 8.69 Avoid through-feed centerless grinding of smaller diameters.

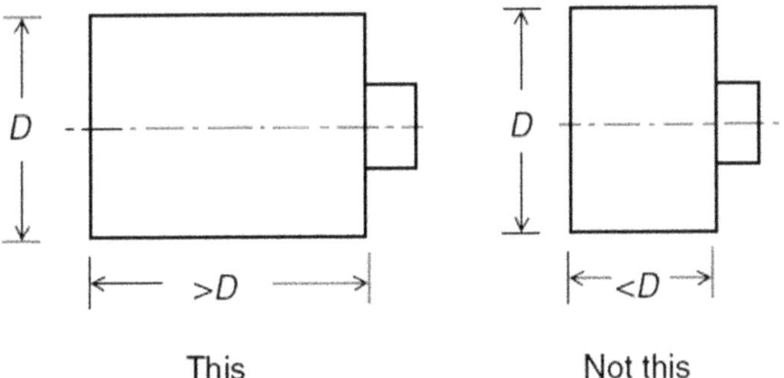

FIGURE 8.70 Avoid shorter WP lengths.

3. Parts of irregular shapes cannot have a ground surface longer than the GW width unless the shape permits a combination of through-feed and infeed grinding (Figure 8.71).
4. Avoid grinding the ends of infeed centerless-ground parts (Figure 8.72).
5. Avoid fillets and radii and instead use undercut or relief surfaces (Figure 8.68).
6. For the form infeed method, keep the form as simple as possible to reduce wheel dressing and other costs.
7. For high-accuracy requirements, avoid keyways, flats, holes, and other interruptions to the surface to be ground or make them as small as possible (Figure 8.66).
8. Avoid unbalanced and end of shaft interruptions (Figure 8.73).

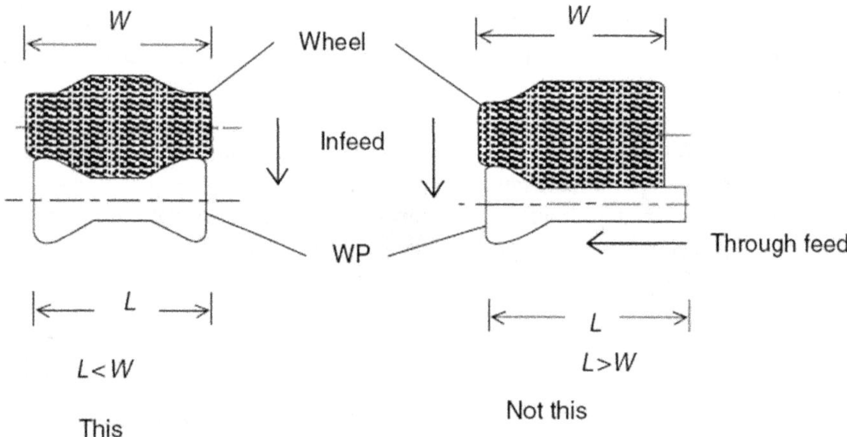

FIGURE 8.71 Avoid long parts of irregular surfaces so that infeed can be applied only to the wheel.

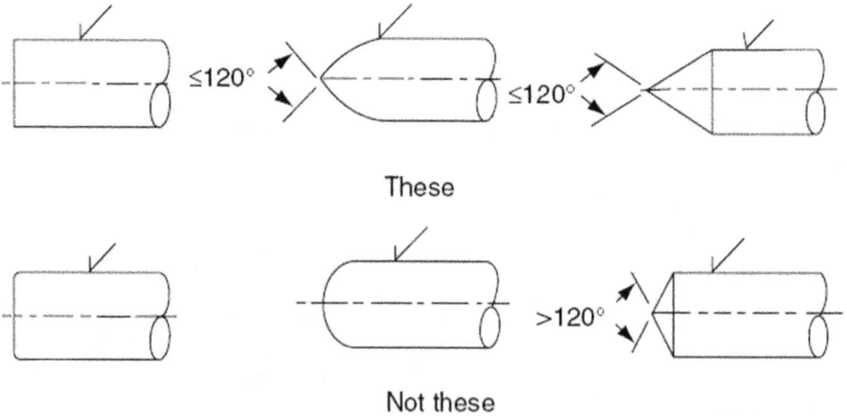

FIGURE 8.72 Avoid grinding the faces of centerless ground parts unless infeed grinding is used and the ends have an included angle of less than 120°.

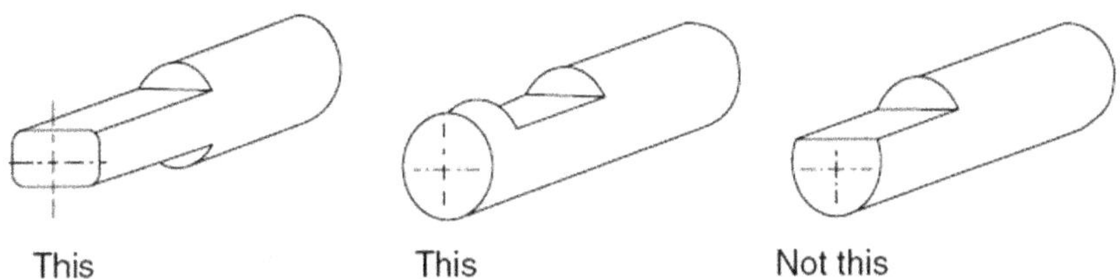

FIGURE 8.73 Avoid unbalanced and end-of-shaft interruptions.

8.4.3.2 Dimensional Control

The condition of the equipment, such as the wheel-spindle bearings, the use of proper wheels and coolant, and the evenness of the temperature of the WP, machine, and coolant, affects the process accuracy. Typical dimensional tolerances are ±0.0125 mm for diameter and parallelism. Surface roughness R_a = 0.2 μm is obtained under normal conditions, which can be improved to 0.05 μm under tight machining conditions.

8.5 DESIGN FOR ABRASIVE FINISHING PROCESSES

8.5.1 Honing

The main purpose of honing is to generate an accurate surface configuration and an improved surface finish. It is used primarily for inside diameters and flat surfaces. The bore diameter ranges from 2.4 mm to 1.2 m, and the length of bore varies between 1.6 mm and 9.1 m. External cylindrical honing can tackle parts of 6.3 mm diameter by 14.7 m long to 450 mm in diameter by 9.1 m long. Flat surfaces of area less than 645 mm^2 and spherical diameters ranging from 3 to 300 mm can also be honed. The nature of a cross-hatched honed surface finish is efficient in moving bearing applications, such as automotive-cylinder bore or a valve body bore. In such applications, the peaks carry the load of the mating parts and valleys provide reservoirs for the lubricating oil. The absence of surface damage by heat is another process advantage. Gear teeth, races of ball and roller bearings, crank-pin bores, drill bushings, gun barrels, piston pins, and hydraulic cylinders are often finished by honing. Honing can be used economically; manual honing suits one-of-a-kind or very limited quantity. When honing is applied in mass production, millions of parts are produced per year using the general-purpose horizontal or vertical honing machines. The most common materials to be honed are steel, CI, and aluminum. Other materials such as bronze, stainless steel, and plastics can be honed at lower speeds.

When designing the WP for honing operation, the following guidelines are recommended:

1. Avoid projections such as shoulders or bosses because honing is applied to the entire surface.
2. Allow a honing stick overrun of 1/2 to 1/4 the length of the honing sticks (Figure 8.74).

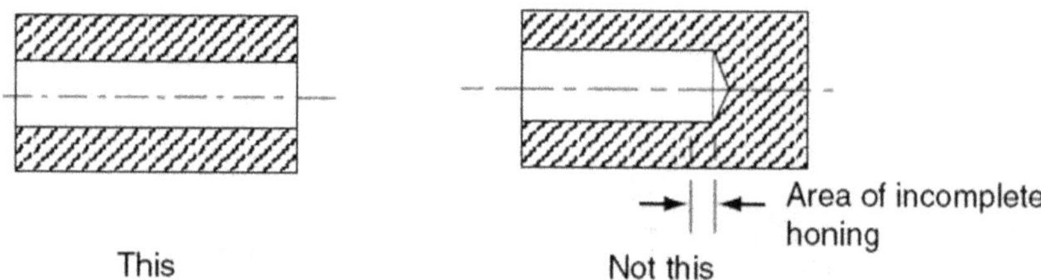

FIGURE 8.74 Design recommendations for honing of internal holes.

3. Avoid/minimize keyways, ports, and undercuts, because they present problems due to the undercut generated by the abrading sticks at this area.
4. Provide easily identifiable locating surfaces.
5. Provide convenient clamping pads that prevent WP distortion during honing.

Bores of automotive engine blocks can be honed to geometric tolerances of roundness, straightness, and size control within 0.008 mm. On hardened steel bores or outside diameter, an accuracy of 0.008 mm can be also produced. Honing produces surface finishes of 0.25–0.4 μm depending on the grit size, coolant used, honing speed, and the generated forces by the honing process.

8.5.2 Lapping

Lapping is normally used to produce one or more of the following:

- An improved surface finish
- Extreme product accuracy
- Correction of minor imperfections in shape
- Precise fit between mating surfaces

Valve spools, fuel injector plungers, seal rings, piston rods, valve stems, cylinder heads, and spherical valve seats are typical applications. Holes or pins as small as 0.8–300 mm in diameter can be lapped. Flat WPs having an area from 6 to 1300 cm^2 can be lapped successfully. For one-of-a-kind fabrication, hand lapping is used. Automatic lapping machines are available for mass production at high rates of 3000 piece/h. Steel and CI are most frequently used by lapping. Glass, aluminum, bronze, magnesium, plastics, and ceramics can also be finished by lapping. Soft materials are less satisfactory for lapping, as the abrasive grits are embedded in the lapped surface.

When designing a part to be lapped, the following factors should be considered:

1. Avoid shoulders, projections, or interruptions.
2. When two opposite sides of a WP must be machined to a highly refined parallelism, the two surfaces should extend beyond the other surface of the WP.

Normal tolerances of lapped surfaces are as follows:

- Diameter or other dimensions: ±0.0006 mm
- Flatness, roundness, or straightness: ±0.0006 mm
- Surface roughness: R_a = 0.1–0.4 μm

8.5.3 Superfinishing

Superfinishing is utilized for surface refining and improving geometric characteristics of the WP. It can be applied to inside and outside diameters, conical surfaces, spherical surfaces, flats, flutes, keyways, and recesses. Typical parts include pistons, piston rods, shaft-sealing surfaces, crank pins, valve seats, bearing recesses, and steel-mill rolls. When using automatic part handling systems, production rate reaches 240 pieces/h. The process tackles any ferrous and non-ferrous metallic parts. It is most frequently used to harden and ground alloy steels. The degree of surface finish depends on the grit size, applied force, relative surface speed, condition of the coolant used, and superfinishing time. Generally, a surface finish of 0.025–0.075 μm is common. Due to the small amount of stock removal capability, refinement of geometric tolerances and control in size is limited by superfinishing.

8.6 DESIGN FOR CHEMICAL AND ELECTROCHEMICAL MACHINING

8.6.1 Chemical Machining

In CHM, parts are machined over their entire surface area. CHM finds applications in the aerospace industry, where large surface areas require a small depth of material removal and other parts require weight reduction. When selective machining in required, maskants are used to protect areas that do not require machining. Scribe-and-peel maskants are used for cuts as deep as 13 mm. With silkscreen masks, the depth is limited to 1.5 mm, but more accurate details are possible. Photoresist provides more accurate details where the depth of etch is limited to 1.3 mm. Photochemical blanking produces very intricate blank shapes from sheet metals by covering the sheet with a precisely shaped mask made using photographic techniques and then removing the unmasked metal by chemical dissolution action. Photochemical milling produces very intricately shaped blanks used for electric motor laminations, shadow masks for color television, fine screens, printed circuit cards, and so on. Sheet thickness ranges from 0.0013 to 3 mm with a common range of 0.0025–0.8 mm. Any material that can be chemically etched or dissolved can be chemically machined. Many ferrous and non-ferrous metals are chemically machined. Material grain size, rolling direction, hardness, freedom from inclusions, and surface quality must be considered.

8.6.1.1 Design Recommendations

When designing a part to be chemically machined, the designer should consider the following points:

1. The smallest hole size or slot width that can be produced by chemical blanking is 1.5 times the metal thickness (Figure 8.75).

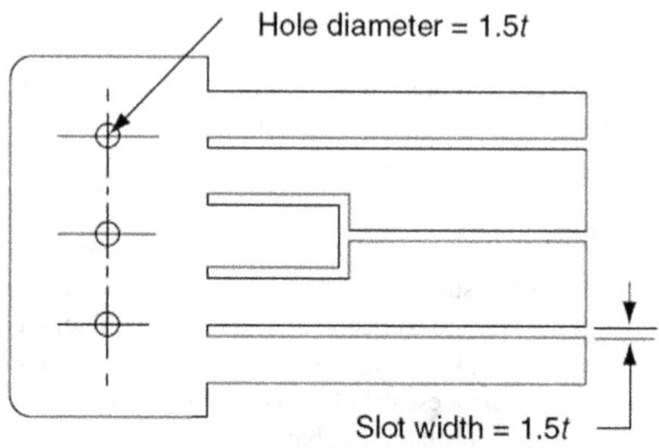

FIGURE 8.75 Maximum hole diameter and slot width by CHM.

2. The minimum land width (Figure 8.76) should be twice the depth of cut but less than 3.18 mm.
3. The radius and the undercut produced by CHM (Figure 8.77) equals the depth of cut.
4. With CHM, sharp corners can be rounded to about 0.7–1.0 mm radius.
5. The normal taper for blanked parts from one side is one-tenth of the blank thickness.
6. The normal taper for parts blanked from two sides is 0.05 of the depth of etch.
7. Avoid part design that has deep, narrow cavities, or folded metal seems to prevent any entrapped chemical solutions.
8. Parts machined by masking techniques should be flat.
9. When machining aluminum, the designer should specify the grain direction and minimize machining across the grains.

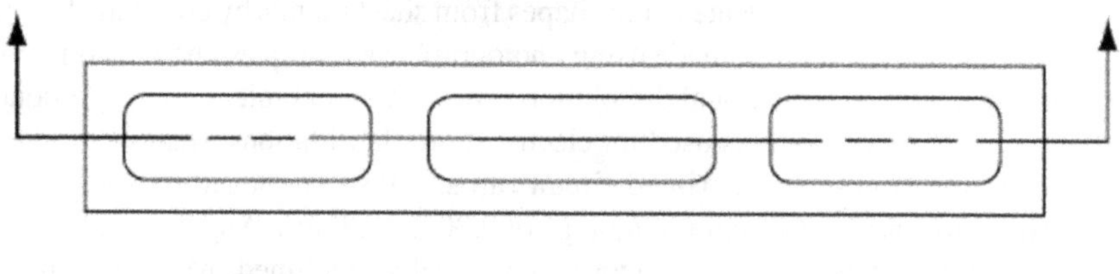

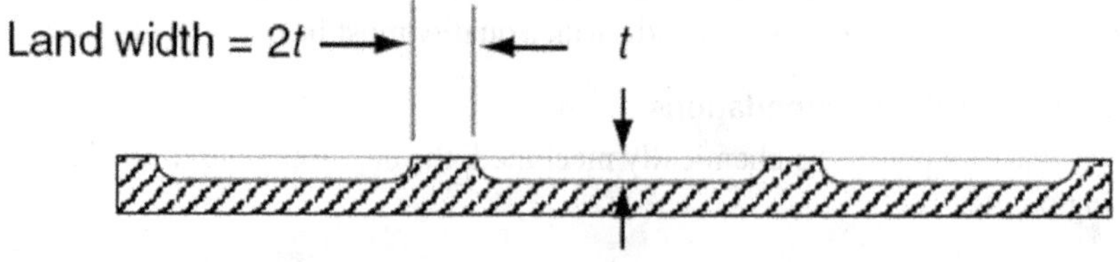

FIGURE 8.76 Minimum land widths.

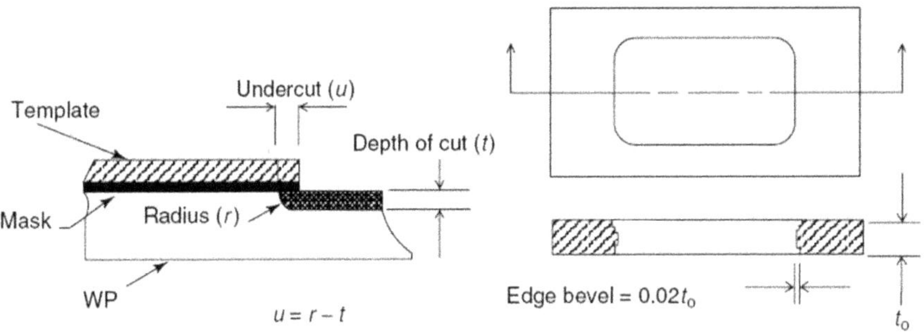

FIGURE 8.77 The radius of undercut and edge bevel.

10. Provide excess material at the periphery of the part for trimming after chemical machining.
11. Sharp internal corners are not produced by CHM, where a radius of 0.5–1 times the depth of etch should be allowed.
12. External corners may have a radius of one-third of the depth of the etch (Figure 8.78).

8.6.1.2 Dimensional Factors and Tolerances

Achieving close tolerances by CHM and PCM depend on the following factors (Bralla, 1999):

- The accuracy of the part's dimensions before CHM and PCM
- The accuracy of the mask
- The undercut allowance
- Uniformity of etching

Tolerance of the length and width of chemically machined parts are as follows:

- With scribe-and-peel maskants and cuts to 1.3 mm deep: ±0.38 mm
- With scribe-and-peel maskants and cuts over 1.3 mm deep: ±0.64 mm

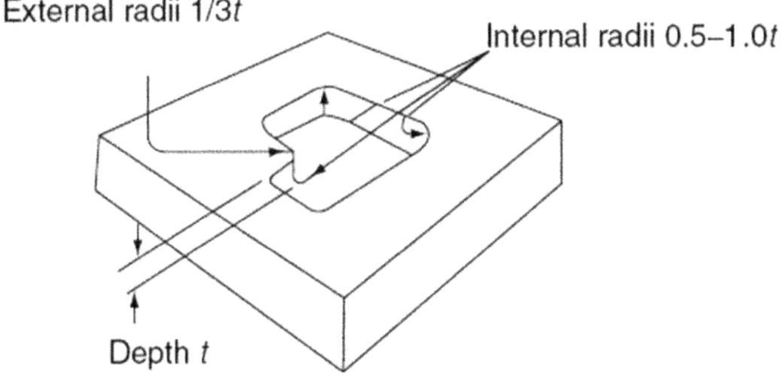

FIGURE 8.78 Minimum radii of CHM corners.

- With silkscreen maskants: ±0.25 mm
- With photoresist mask and a depth 1.0 mm aluminum alloy: ±0.2 mm for other materials and thicknesses

Surface finish produced by chemical machining depends on factors that include:

- The initial surface finishes of the WP
- The etchant used
- WP material
- Depth of etching
- Heat treatment condition of the WP

Normal surface finish R_a is as follows:

- Aluminum (up to 6.3 mm depth of etch): 2.25 μm
- Magnesium: 1.25 μm
- Titanium: 0.63 μm
- Steel: 1.75 μm

8.6.2 Electrochemical Machining

ECM is suitable for materials and shapes that are difficult to machine by conventional methods. ECM finds applications in machining jet engine parts, nozzles, cams, forging dies, and other contoured shapes. ECM is generally used for difficult-to-machine conductive materials such as hardened steel, including alloy and tool steel, nickel alloys, cobalt alloys, tungsten, molybdenum, zirconium, and other refractory metals. The process is suitable for quantity production, as the tool requires special design and testing. However, the tool life is infinite and requires little maintenance. Due to the high machining speeds attainable (25 times as fast as EDM), the process is used at machining high removal rates.

8.6.2.1 Design Recommendations

Designers should follow these recommendations:

1. Avoid the formation of tapers (0.1%) in hole drilling and apply tool insulation that stops stray machining (Figure 8.79).
2. Avoid specifying too-sharp internal corners (<0.4 mm) (Figure 8.80).
3. Avoid specifying too-sharp external radius (0.05 is minimum) (Figure 8.81).
4. When specifying irregular shapes, the designer should allow for ample deviations from the nominal shape to minimize the trial and error development of the tool.
5. Consider an overcut or sidewall tolerance (±0.03 to ±0.76 mm), which can be controlled by adjusting the feed tool rate to within ±0.127 mm.
6. Avoid specifying roundness errors less than 0.013 mm and runout of less than 0.001 mm/mm of hole depth.
7. Hole size of 0.76–100 mm can be achieved.

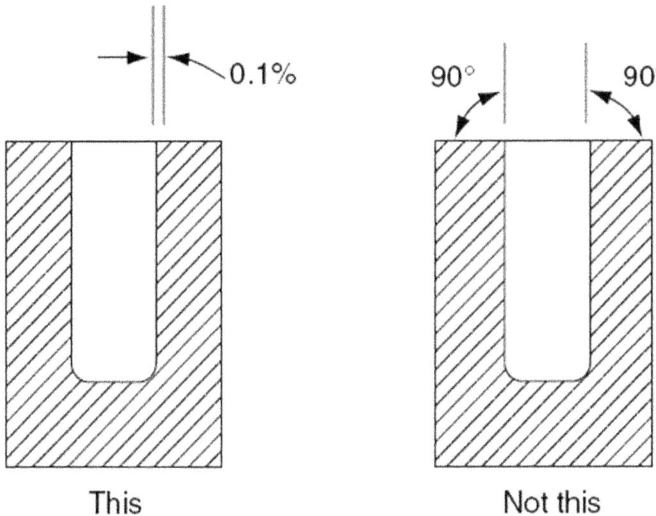

FIGURE 8.79 Allow a taper on the sidewalls of ECM cavities.

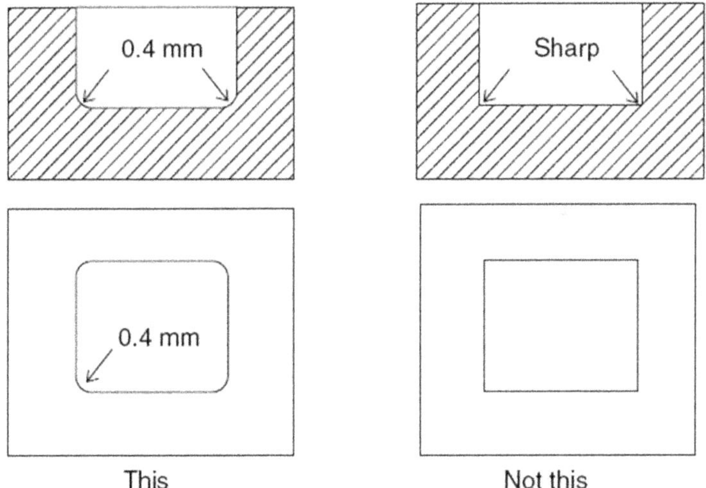

FIGURE 8.80 Minimum corner radii for ECM cavities.

8. Consider the maximum aspect ratio of 180:1.
9. The rough contour to be machined should conform as closely as possible to the profile of the final designed shape.
10. The designer should provide a uniform machining (stock) allowance so that a constant surface finish will be achieved across the entire machined area (Figure 8.82).
11. The tool is not the exact mirror image of the required shape (Figure 8.83).
12. Electrode tools are designed such that when the final depth is reached, the final size and finish of the shape is achieved.

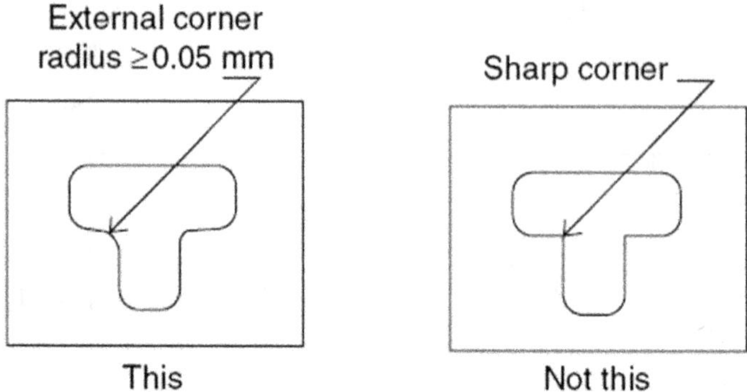

FIGURE 8.81 Allow a radius of 0.05 mm or more for external corners.

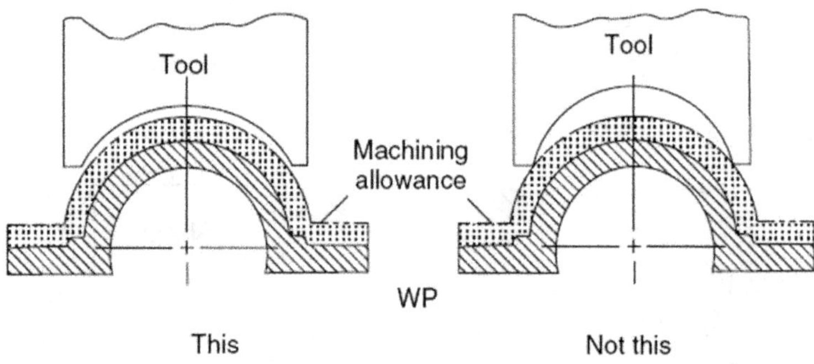

FIGURE 8.82 Allow sufficient machining allowance on castings and forgings (Yankee, 1979).

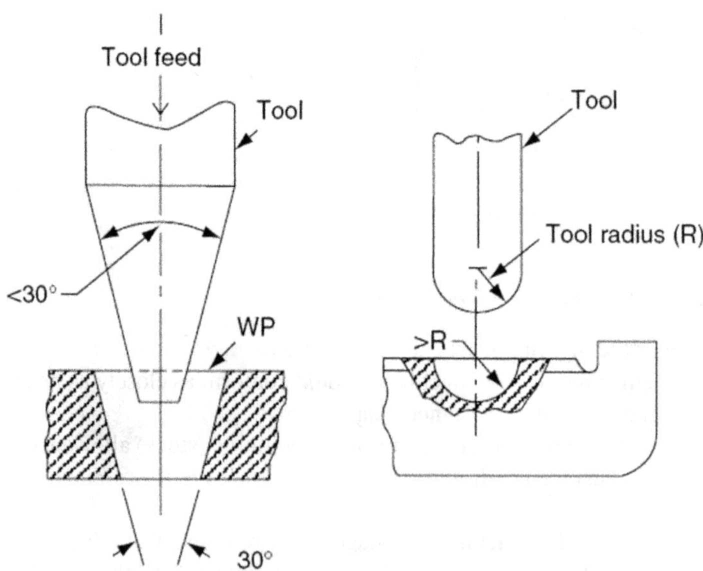

FIGURE 8.83 The electrode tool is less than the true shape of the profile (Yankee, 1979).

13. Sidewalls should be as steep as possible (Figure 8.84).
14. To avoid complex tooling requirements, avoid undercuts into shapes machined by ECM (Figure 8.85).

8.6.2.2 Dimensional Factors

Dimensional variations are mainly affected by the current density, gap volt, electrolyte flow, electrolyte concentration, electrolyte temperature, and electrode feed rate. Other sources of errors include machine deflection (by the electrolyte pressure) and errors in tool manufacturing. Normal dimensional tolerances are within ±0.13 mm, errors in contours ±0.25, frontal surface roughness R_a = 1.6 μm, and surface roughness R_a at the sidewalls in about 3 μm.

8.6.3 Electrochemical Grinding

ECG is used for materials that are difficult to machine conventionally. Fragile materials as well as stress- and burr-free components are good candidates for ECG. Steels (HRC >60), stainless steels, high-nickel alloys, and WC are typical materials. Materials that are fragile or susceptible to heat damage or distortion from normal grinding stresses are tackled by ECG. Examples are aircraft, honeycomb materials, surgical needles, thin-walled tubes, and laminated materials. ECG is used for moderate and high-level production, where the rate of material removal is one to five times the rate attainable by conventional grinding.

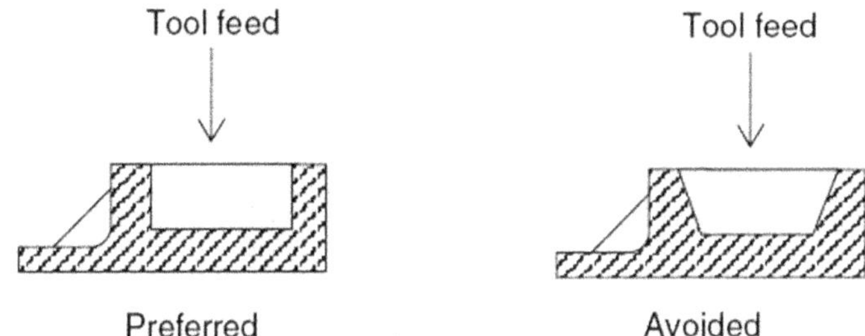

FIGURE 8.84 Keep the sidewalls as steep as possible (use insulated tool).

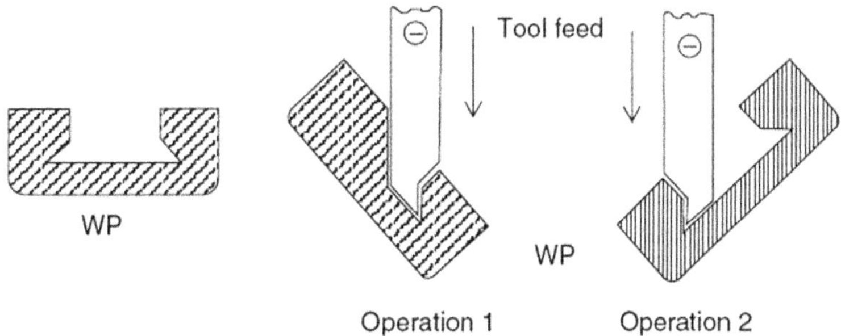

FIGURE 8.85 Avoid undercuts for simplified tooling (Yankee, 1979).

8.6.3.1 Design Recommendations

The product designer should consider the following if the part is to be machined by ECG:

1. Allow 0.75–1.0 mm for inside radii.
2. Specify more liberal tolerances if the groove is deep.
3. Electrochemical action concentrated near the WP corners will round them by 0.13 mm radius.
4. For better accuracy, allow for a final non-electrolytic pass (with the current switched off) over the surface.

8.6.3.2 Dimensional Factors

As the case of ECM, normal tolerances for ECG are as follows:

- Specific dimensions: ±0.025 mm
- Contours: ±0.13 mm
- Surface finish (plunge-grinding carbide): 0.25 μm
- Surface finish (traverse-grinding carbide): 0.40 μm
- Surface finish (steel): 0.75 μm

8.7 DESIGN FOR THERMAL MACHINING

8.7.1 Electrodischarge Machining

EDM is a thermal machining process that removes the WP material by the erosive action of electrical discharges that melt and evaporate the WP material. The process produces intricate shapes in hard materials that are difficult to machine conventionally. EDM is a burr-free process that eliminates the need for secondary operations. Stamping, extruding, wire drawing, die casting, and forging dies and plastic molds are common applications. It is used for one-of-a kind or job-lot quantities. The low cutting rate of EDM (8 cm^3/h) is compensated for by the ease of electrode tool machining from a highly machinable material. Wire EDM cuts at a rate of 130–140 cm^2/h in 5 cm thickness using good flushing conditions. EDM machines any conductive material regardless of its hardness. Hardened steel and carbides are the most commonly machined materials. PCD used in form tools and other cutting tools is currently machined by EDM.

8.7.1.1 Design Recommendations

Designers should adhere to the follow guidelines:

1. It is not economical to use EDM for obtaining an ultrafine finish. This requires many passes, low current, and slower MRRs.
2. Design the part so that the machining allowance is as small as possible.
3. Design the part so that as much of the machining is made by a conventional method or another manufacturing method rather than by the slow EDM process (Figure 8.86).

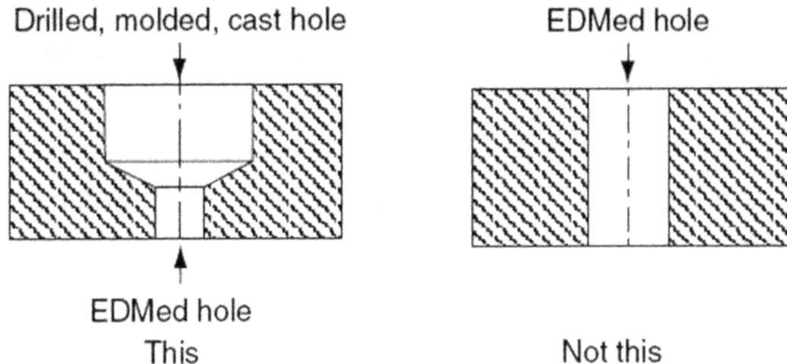

FIGURE 8.86 Perform maximum conventional machining, molding, or casting before the slow EDM process.

4. Design the part so that several parts can be machined simultaneously, or a single part can have several EDM operations simultaneously.
5. Design the part so that the tool electrodes are produced at low cost.
6. Avoid thin and fragile electrode sections, especially in graphite.
7. Complex cavities can be produced by several simple electrode shapes (Figure 8.87).
8. The minimum radius obtainable for internal corners will be equal to the overcut (0.1 mm).
9. It is advisable to specify a cavity tolerance that allows a taper angle of 2–20 min/side.
10. Enlarging or reshaping through holes rather than blind holes permits easier dielectric flow supply.
11. Consider a finishing pass at low removal rate to minimize the adverse effects of the white layer formed on the machined surface.

8.7.1.2 Dimensional Factors

A surface finish of 0.8 μm is obtainable in die sinking by EDM. Better finishes (R_a = 0.4 μm) are possible when using orbiting or rotating tool electrodes. Generally, the surface finish depends on the spark energy. A finish cut may be required at low

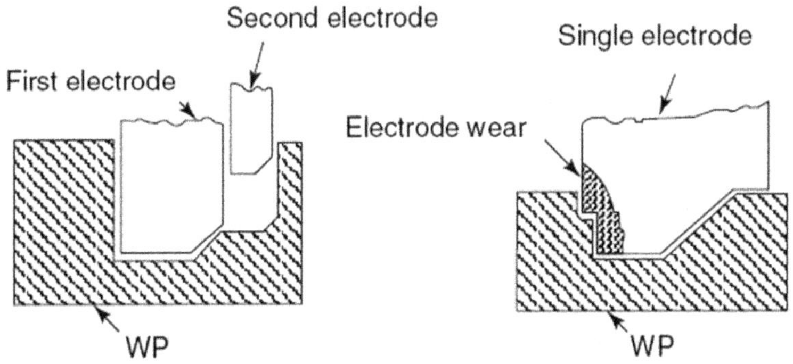

FIGURE 8.87 Complex shapes require special or multiple electrodes (Yankee, 1979).

pulse current, short pulse duration, and a high discharge frequency. Under such conditions, a surface roughness of R_a = 0.4 μm is possible. Dimensional accuracy in EDM die sinking depends on the accuracy of tool electrode manufacture, electrode wear, quality of the power supply, previous stresses existing in the WP material, and the operator skill. Tolerance of ±0.05 to ±0.13 mm is possible in a single cut, and ±0.005 mm with multiple cuts. A dimensional tolerance of ±0.0025 mm is possible with wire EDM.

8.7.2 Electron Beam Machining

EBM utilizes an electron beam that impinges on an area of 0.32–0.46 mm^2 and has a power density of 15 × 10^6 W/mm^2 to melt and evaporate the WP material. It is used for fine cutting and drilling in any material. Holes and slots of few thousands of an inch and very precise contour are possible. EBM is used for drilling accurate holes for diesel fuel injection, gas orifice, wire drawing dies, and sleeve valve holes. Holes of 0.013 and slots having 0.025 mm width can be cut. The depth-to-diameter ratio of drilled holes ranges from 10 to 15 and the maximum depth is 6.4 mm. A taper of 1–2° is expected for WP thicknesses greater than 0.13 mm. Cratering and spattering occur near the hole entrance. The surface of machined holes and slots is non-uniform, with a HAZ of 0.25 mm depth. Cutting rates are rapid for thin materials. However, the volumetric removal rate ranges from 0.8 and 2 mm^3/min. The high cost of equipment, the long time needed to evacuate the machining chamber, and the need for skilled operators are adverse cost factors. Metals, ceramics, plastics, and composites are machined by EBM. Cutting rates are inversely proportional to the melting and evaporating temperatures. Hardened steel, stainless steel, molybdenum, nickel, cobalt, titanium, tungsten, and their alloys are all machined by EBM.

In EBM, designers should follow these recommendations:

1. Keep the part size as small as possible so that many parts can be loaded to the vacuum chamber of the electron beam machine.
2. Avoid specifying internal corners less than 0.25 mm.
3. Do not exceed the maximum cut by 6.3 mm; thinner WP cuts faster with less sidewall taper.
4. Allow for the surface effects, which may be undesirable and may require a secondary operation. A tolerance of ±10% should be allowed on hole diameters and slots. The normal surface roughness R_a ranges between 0.5 and 2.5 μm.
5. The edge of slot walls can be held parallel to a tolerance of 0.05 mm (Figure 8.88).
6. Hole entrance angles can be kept between 20° and 90° to the WP surface.
7. Large diameter holes (>beam diameter) are produced by the trepanning method (Figure 8.89).
8. Blind cuts can be made by switching off the power when reaching the required depth.

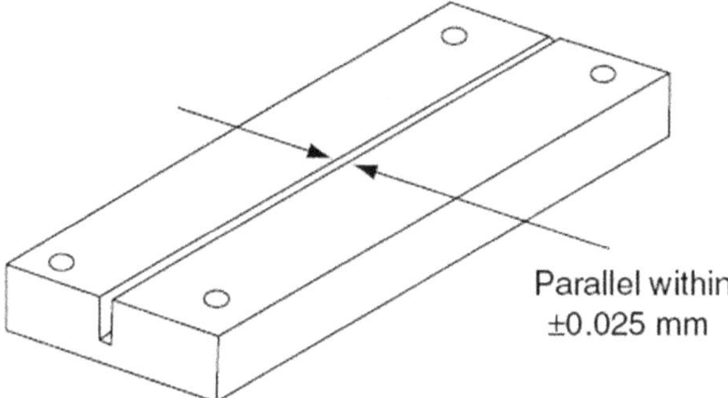

FIGURE 8.88 Parallelism of EBMed parts.

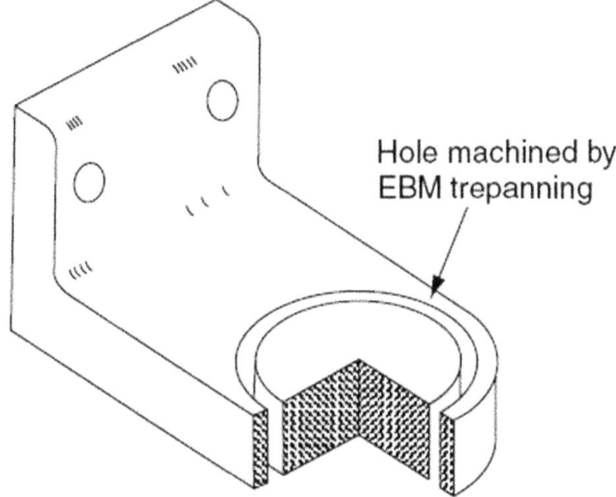

FIGURE 8.89 Trepanning of large-diameter holes (Yankee, 1979).

8.7.3 Laser Beam Machining

LBM is generally used for micromachining of thin parts that are difficult to cut using conventional methods. Holes smaller than 1.3 mm diameter in 5 mm thickness are produced; larger holes can be machined by trepanning. Other laser applications include cutting out blanks from sheet metals or other materials up to 13 mm thickness, slitting, trimming, and perforation. The minimum hole diameter is about 0.005 mm but 0.13 mm is more common. Slits and profile cuts as narrow as 0.4 mm are normally achieved. Laser beam holes deeper than 0.5 mm suffer from a taper, non-uniform diameter, and roundness errors. Cratering at the entrance surface and a narrow damaged layer of 0.13 mm is also common. Due to the high cost of equipment used, higher production quantities and the use of difficult-to-machine material favor the use of the process. The most practical materials for LBM are ceramics, glass, carbides, and some aerospace alloys. Other materials such as copper, aluminum, gold, and silver are not suitable candidates because of their high reflectivity and thermal conductivity. Plastics, rubber, beryllium, zirconium, stainless steel, tungsten, CI, brass, molybdenum, cloth, cardboard, wood, and composites have been machined successfully by LBM.

Designers should follow these recommendations:

1. Surfaces should be dull and unpolished.
2. Through-cut parts should be thin to reduce the time required, taper, and surface irregularities.
3. Allowances should be made for a taper of 3° per side and a heat-affected layer of 0.13 mm depth.
4. A minimum corner radius of 0.01 mm/min is expected.
5. The ideal aspect ratio is 4:1.

Hole-diameter and slot-width tolerances should be ±0.025 mm, which can be increased to ±0.1 mm using assisting gasses. Diameters of blanks cut from sheet metal have a tolerance of ±0.13 mm.

8.8 DESIGN FOR ULTRASONIC MACHINING

USM is used to machine irregular holes in thin sections or shallow, irregular cavities in hard, brittle, and fragile materials. Holes as small as 0.08 mm can be drilled; the maximum possible size is 90 mm. Hole depth ranges between 25 and 50 mm. Larger holes can be machined by RUM using a coring or a trepanning method. Machined surfaces do not suffer from thermal damage and they are burr-free. Cavities machined by USM suffer from an overcut and a sidewall taper. USM is a slow machining process that can be used when other conventional processes are not suitable. Cutting rates range from 0.03 to 4 cm^3/min and tool wear ranges from 1:1 to 1:200 of WP material. USM is most advantageous for hard, brittle, non-conductive materials. Materials having a hardness HRC >64 are best suited for the process and a hardness of HRC <45 is not recommended for USM.

When designing parts to be machined by USM, the designer should follow the below mentioned points:

1. Shallow holes and cavities (depth <2.5 diameter) are more suitable than deep ones.
2. Through holes or holes through passages for abrasive slurry are preferred to blind holes (Figure 8.90).
3. A backup plate should be used when machining through holes/cavities in brittle materials to prevent chipping of edges (Figure 8.91).
4. Allow for a taper of 0.05 mm/mm (depending on the WP material) when machining deep holes (Figure 8.92). This taper can be reduced by the proper tool design or using a finishing pass.
5. Do not specify sharp corners at the bottom of blind holes because tool wear is concentrated at such corners of the tool (Figure 8.93).
6. Allow for an overcut, which equals the tool diameter plus twice the abrasive grain diameter.

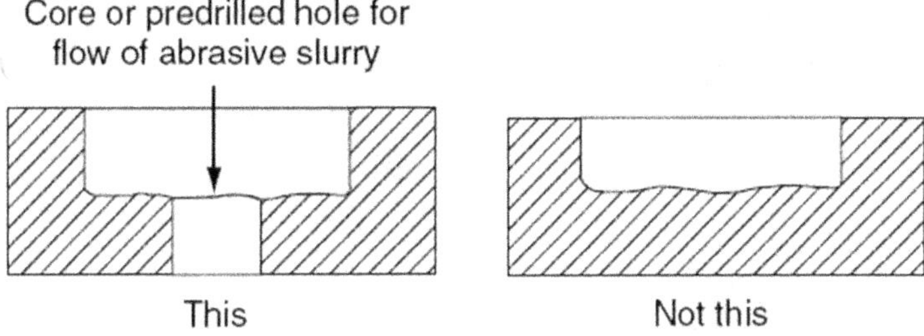

FIGURE 8.90 Provide through passage for the abrasive slurry.

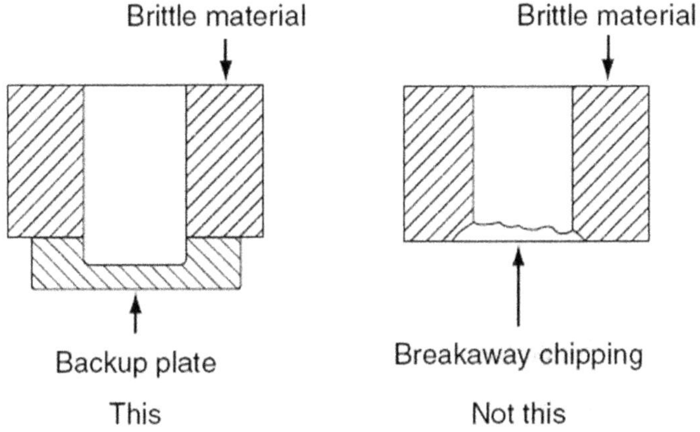

FIGURE 8.91 Avoid breakaway chipping at the exit surface of USM cavities.

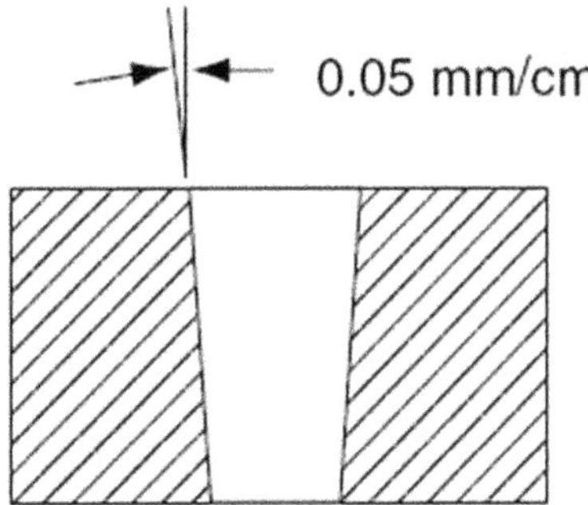

FIGURE 8.92 Allow taper for sidewalls of USM cavities.

The following factors affect the dimensional accuracy of USM machined parts:

- The amount of overcut
- Tool wear
- Machine tool rigidity
- Abrasive grain size

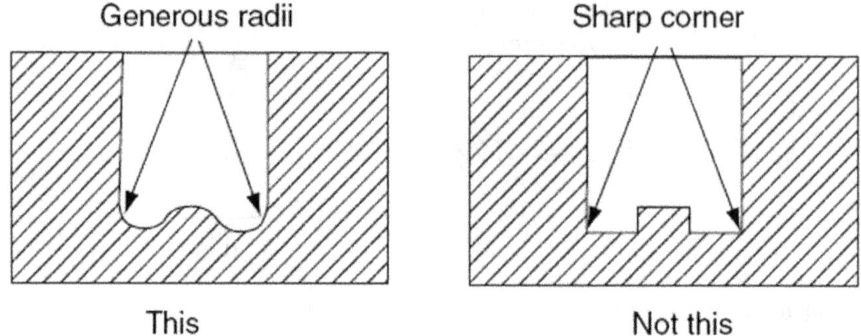

FIGURE 8.93 Allow generous radii at machined corners.

- Abrasive wear
- Tool design
- Slurry conditions

Recommended dimensional tolerances range between ±0.013 and ±0.025 mm. A surface roughness of 0.25–1.0 μm is achievable.

8.9 DESIGN FOR ABRASIVE JET MACHINING

AJM is widely used for cutting, drilling, slotting, trimming, etching, cleaning, deburring, and stripping. The process is also applicable to machining heat-sensitive components. The minimum slot width machinable with AJM is 0.13 mm with sidewall taper that increases with the stand-off distance. The choice of AJM does not depend on the production quantity. Tooling costs are low, which makes the process suitable for small quantities. Removal rates of 0.016 cm^3/min are possible. It is the most suitable process for machining hard, fragile, heat-sensitive materials. Typical materials include ceramic, glass, porcelain, sapphire, quartz, tungsten, chromium/nickel alloys, hardened steels, and semiconductors such as germanium, silicon, and gallium.

Designers intending to use AJM should consider the following allowances in their design (Figure 8.94):

1. The taper of the sidewalls of cuts should be at least 0.05 mm/cm of depth.
2. Allow access room for the jet nozzle.
3. Allow for a kerf, which should be at least 0.13 mm (0.45 mm is preferred).

Provide corners of at least 0.1 mm.

Normal tolerances for dimensions are from ±0.05 to ±0.03 mm; the surface finish can be held between 0.25 and 1.3 μm depending on the machining conditions.

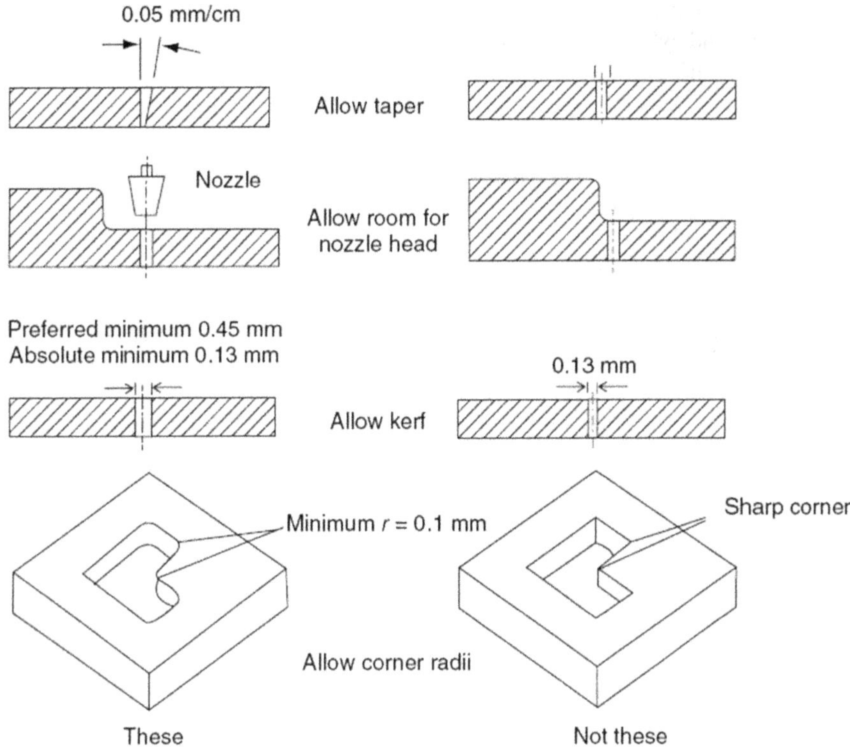

FIGURE 8.94 Design recommendations for AJM.

8.10 REVIEW QUESTIONS

8.10.1 State the general principles of manufacturing components at the minimum cost.

8.10.2 What are the general rules adopted when designing for manufacturing?

8.10.3 List the general guidelines used when designing for economic machining.

8.10.4 State the design recommendations adopted when designing for:

- Turning parts
- Drilling holes
- Reaming and boring

8.10.5 Explain what is meant by dimensional factors and tolerances of milled parts.

8.10.6 Specify the design rules adopted for shaping, planing, and slotting of parts.

8.10.7 Describe the design recommendations that should be followed when broaching splines.

8.10.8 Explain the principles of economic design for thread cutting and gear machining.

8.10.9 Mention the rules followed by the designer when machining WPs on surface and cylindrical grinding machines.

8.10.10 Summarize the design guidelines that should be considered when performing lapping, honing, and superfinishing.

8.10.11 Use line sketches to show the major design factors for ECM and CHM.

8.10.12 Explain what is meant by:

- Design for EDM
- Design for LBM
- Design for USM

REFERENCES

American Gear Manufacturing Association (AGMA) Standard 390.03.

Bralla, J 1999, *Design for manufacturability handbook*, 2nd edn, McGraw Hill, New York.

Yankee, HW 1979, *Manufacturing processes*, Prentice Hall, New York.

9 Accuracy and Surface Integrity Realized by Machining Processes

9.1 INTRODUCTION

The quality of a machined surface is becoming important to satisfy the increasing demands of component performance and reliability. Machined parts used in the military, aerospace, and automotive industries are subjected to high stresses, temperatures, and hostile environments. The dynamic loading and design capabilities of machined components are limited by the fatigue strength of the material, which is commonly linked to the fatigue fractures that always nucleate on or near the surface of the machined components. Stress corrosion resistance is another important material property that can be directly linked to the machined surface characteristics. When machining any component, it is necessary to satisfy the surface technological requirements in terms of high product accuracy, good surface finish, and a minimum of drawbacks that may arise as a result of possible surface alterations by the machining process. The nature of the surface layer has a strong influence on the mechanical properties of the part.

Any machined surface has two main aspects—the first aspect is concerned with the surface texture or the geometric irregularities of the surface, and the second one is concerned with the surface integrity, which includes the metallurgical alterations of the surface and surface layer, as shown in Figure 9.1. Surface texture and surface integrity must be defined, measured, and controlled within specific limits during any machining operation.

9.2 SURFACE TEXTURE

Surface texture is concerned with the geometric irregularities of the surface of a solid material, which is defined in terms of surface roughness, waviness, lay, and flaws, as described in Figure 9.2:

1. *Surface roughness* consists of the fine irregularities of the surface texture, including feed marks generated by the machining process.
2. *Waviness* consists of the more widely spaced components of surface texture that may occur due to the machine or part deflection, vibration, or chatter.
3. *Lay* is the direction of the predominant surface pattern.
4. *Flaws* are surface interruptions such as cracks, scratches, and ridges.

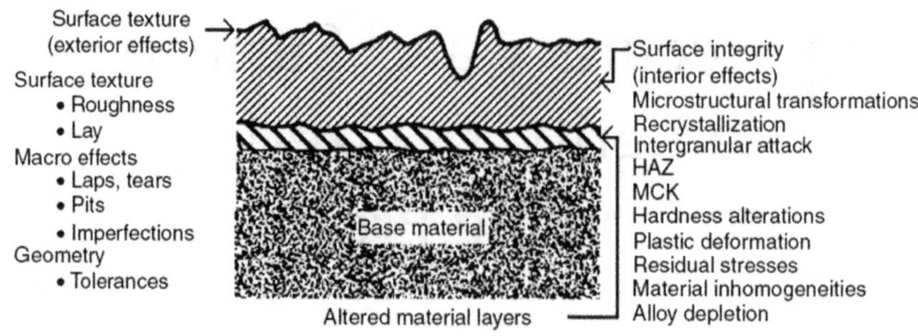

FIGURE 9.1 Surface technology by machining.

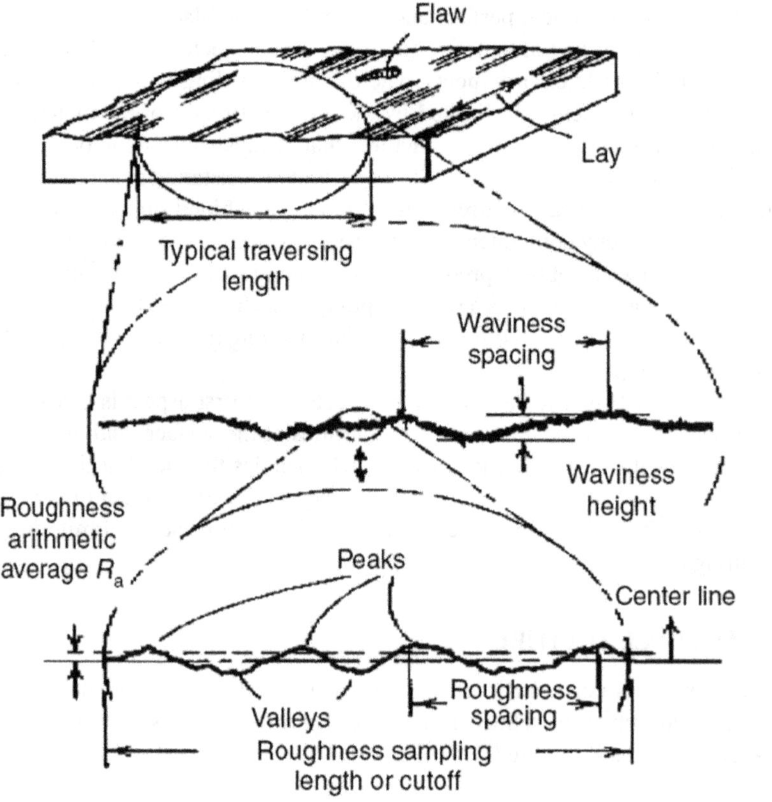

FIGURE 9.2 Surface texture (Surface Texture, 1985).

Stylus contact type instruments are widely used to provide numerical values of surface roughness in terms of the arithmetic average (R_a) or centerline average (CLA), the root mean square (R_q), and the maximum peak-to-valley roughness (R_{max}), as shown in Figure 9.3. Other methods of surface characterization include microphotography and scanning electron microscopy.

The arithmetic average or CLA is determined as follows:

$$R_a = \frac{1}{L}\int_{x=0}^{x=L} |y|\,dx$$

where L is the sampling length and y is ordinate of the profile from the centerline shown in Figure 9.3a.

The root mean square roughness is calculated as follows:

$$R_q = \left(\frac{1}{L}\int_{x=0}^{x=L} y^2\,dx\right)^{1/2}$$

This can be approximated by the following equation (Figure 9.3b):

$$R_q = \sqrt{\frac{(y_1 - Y_M)^2 + (y_2 - Y_M)^2 + \cdots + (y_N - Y_M)^2}{N}}$$

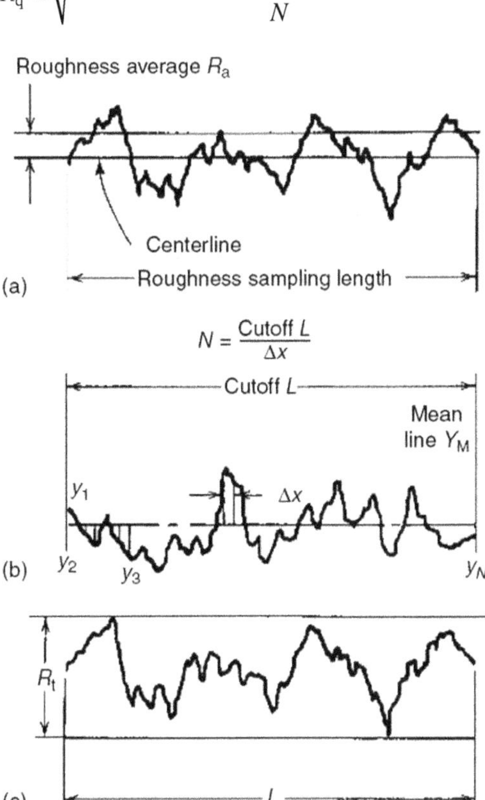

FIGURE 9.3 Commonly used surface roughness symbols: (a) Average roughness (R_a), (b) root mean square roughness (R_q), (c) maximum peak-to-valley roughness height (R_t or R_{max}) (Surface Texture, 1985).

The maximum peak-to-valley roughness (R_t or R_{max}) is the distance between two lines parallel to the mean line that contacts the extreme upper and lower points on the profile within the roughness sampling length (Figure 9.3c).

Enhanced surface texture specifications are essential to improve fatigue strength, corrosion resistance, appearance, and sealing. Typical applications of such surfaces include antifriction and journal bearings, food preparation devices, parts operating in corrosive environments, and sealing surfaces.

9.3 SURFACE QUALITY AND FUNCTIONAL PROPERTIES

The quality of surface finish affects the functional properties of the machined parts as follows:

1. *Wear resistance.* Larger macro irregularities result in non-uniform wear of different sections of the surface where the projected areas of the surface are worn first. With surface waviness, surface crests are worn out first. Similarly, surface ridges and micro irregularities are subjected to elastic deformation and may be crushed or sheared by the forces between the sliding parts.
2. *Fatigue strength.* Metal fatigue takes place in the areas of the deepest scratches and undercuts caused by the machining operation. The valleys between the ridges of the machined surface may become the focus of concentration of internal stresses. Cracks and microcracks (MCK) may also enhance the failure of the machined parts.
3. *Corrosion resistance.* The resistance of the machined surface to the corrosive action of liquid, gas, water, and acid depends on the machined surface finish. The higher the quality of surface finish, the smaller the area of contact with the corrosive medium, and the better the corrosion resistance. The corrosive action acts more intensively on the surface valleys between the ridges of micro irregularities. The deeper the valleys, the more destructive will be the corrosive action that will be directed toward the depth of the metal.
4. *Strength of interference.* The strength of an interference fit between two mating parts depends on the height of micro irregularities left after the machining process.

Figure 9.4 shows the ANSI Y14.36 (1978) standard symbols used for describing part drawings or specifications. These include the maximum and minimum roughness, maximum waviness, and the lay for machined parts. Table 9.1 shows the symbols used to define surface lay and its direction. Accordingly, a variety of lays can be machined, including parallel, perpendicular, angular, circular, multidirectional, and radial ones. The same table also suggests a typical machining process for each produced lay. Figure 9.5 shows the surface roughness produced by common production methods. Several machining processes that employ cutting, abrasion, and erosion actions are also included and compared to some metal forming applications.

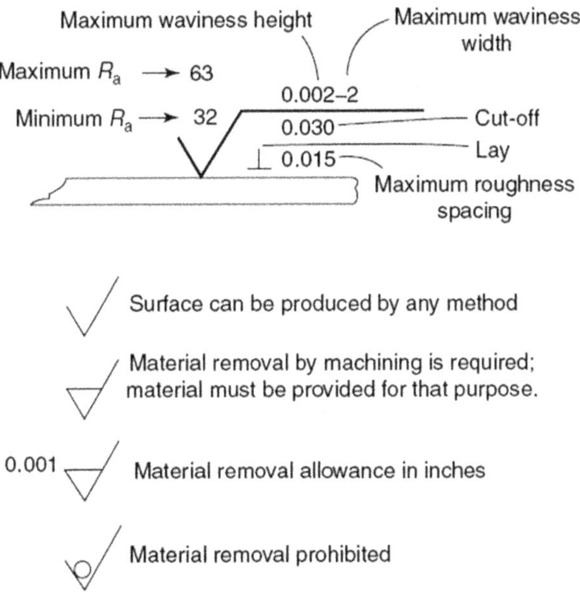

FIGURE 9.4 Surface texture symbols for drawings or specifications (Surface Texture, 1985).

The accuracy of machined parts indicates how a part size is made close to the required dimensions. Produced accuracy is normally expressed in terms of dimensional tolerances. Each machining process has its own accuracy limits that depend on the machine tool used and the machining conditions. Tolerances required for materials that are highly engineered, heavily stressed, or subjected to unusual environments are closely related to surface roughness. In this regard, closer dimensional tolerances require very fine finishes, which may necessitate multiple machining operations that raise the production cost. Figure 9.6 shows typical surface roughness and dimensional tolerances for machining operations.

Theoretically, surface roughness in milling and turning can be calculated as a function of the feed rate, tool nose radius, end cutting edge angle, and the side cutting edge angle. However, the actual surface roughness may be higher, due to the formation of the built-up edge (BUE) and the possible tool wear. Table 9.2 summarizes different factors that affect the surface roughness for different machining operations.

9.4 SURFACE INTEGRITY

Surface integrity is defined as the inherent condition of a surface produced in a machining or other surface generating operation. Surface integrity is concerned primarily with the host of effects that a manufacturing process produces below the visible surface. During machining by conventional methods, the pressure exerted on the metal by cutting and frictional forces, heat generated, and plastic flow change the physical properties of the surface layer from the rest of metal in the part. Similarly, thermal machining by EDM and LBM is accompanied by material

TABLE 9.1
Symbols Used to Define Lay and Its Direction

Symbol	Meaning	Example	Operation
=	Lay approximately parallel to the line representing the surface to which the symbol is applied		Shaping vertical milling
⊥	Lay perpendicular to the line representing the surface to which the symbol is applied		Horizontal milling
X	Lay angular in both directions to the line representing the surface to which the symbol is applied		Honing
M	Lay multidirectional		Grinding
C	Lay approximately circular relative to the center to which the symbol is applied		Face turning
R	Lay approximately radial relative to the center to which the symbol is applied		Lapping
P	Lay particulate, non-directional, or protuberant		ECM, EDM, LBM

melting, evaporation, resolidification, and consequently, the formation of a heat-affected layer. Machining by chemical and EC processes does not impose thermal changes to the WP. However, the surface suffers from several other effects such as pits and intergranular attack (IGA).

As a result of some machining processes, the thickness of the altered layer may reach a considerable value during rough machining operations. The mechanical, thermal, and chemical properties of the WP material determine the extension of surface effects and thickness of the altered layer. Surface alterations have a major influence on the material performance, especially when high stresses or severe environments are used.

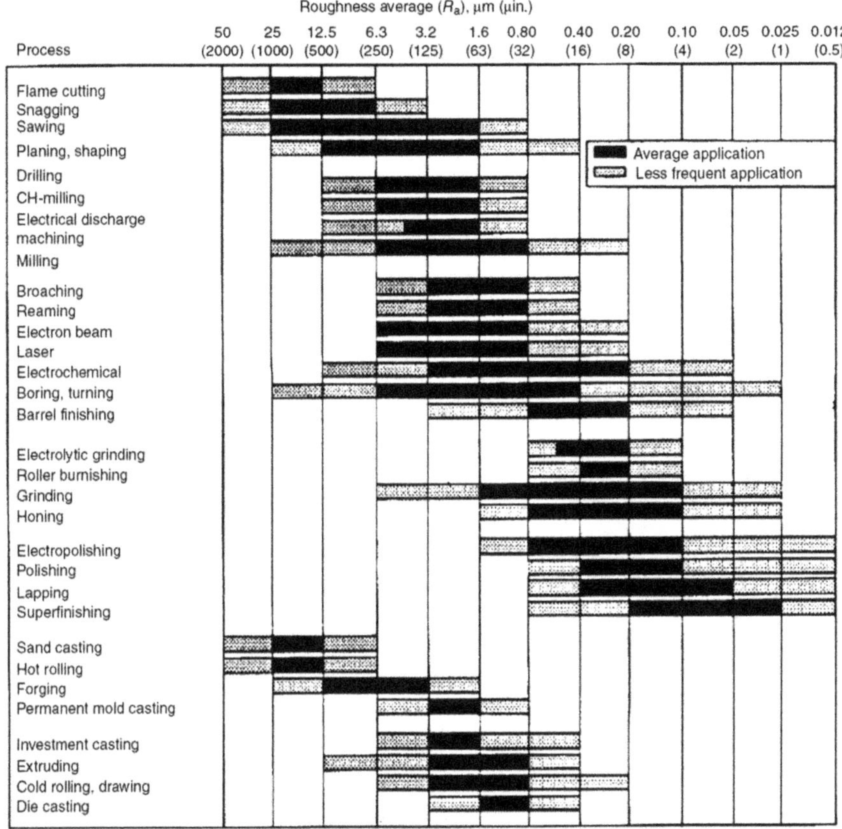

FIGURE 9.5 Surface roughness produced by common production methods (Surface Texture, 1985).

The nature of the surface layer has in many cases a strong influence on the mechanical properties of the machined part. This association is more pronounced in some materials and under certain machining operations. Typical surface integrity problems are as follows:

1. Grinding burns on high-strength steel of landing-gear components
2. Untempered martensite (UTM) in drilled holes
3. Stress corrosion properties of titanium by the cutting fluid
4. Grinding cracks in the root section of cast nickel-based gas turbine buckets
5. Lowering of fatigue strength of parts processed by EDM or ECM
6. Distortion of thin components
7. Residual stress induced in machining and its effect on distortion, fatigue, and stress corrosion

The subsurface characteristics occur in various layers or zones, as shown in Figure 9.1. The subsurface altered material zone (AMZ) can be as simple as a stress

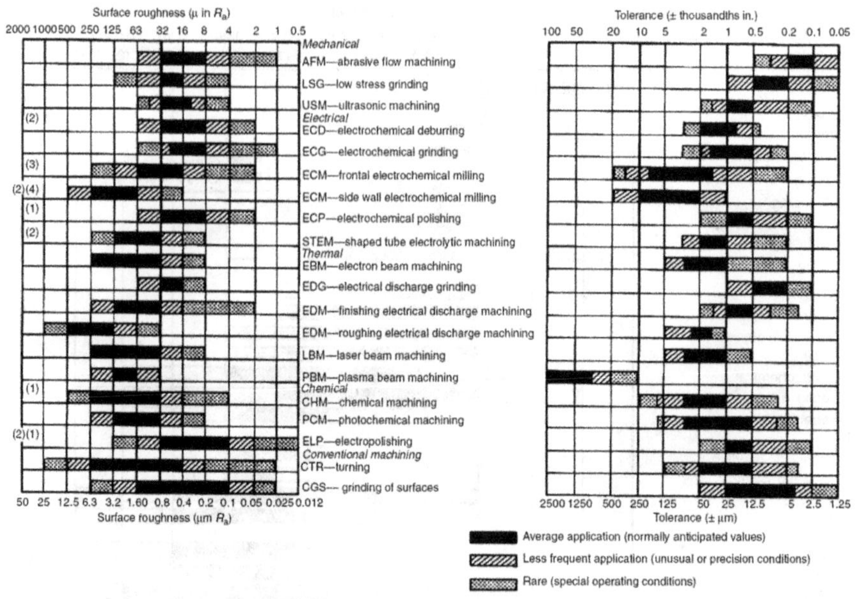

FIGURE 9.6 Different tolerances achieved by machining methods: (1) Depends on state of starting surface, (2) titanium alloys are generally rougher than nickel alloys, (3) high-current density areas, and (4) low-current density areas (*Machining Data Handbook*, 1990).

TABLE 9.2
Factors Affecting Surface Roughness for Various Machining Technologies

Machining Process	Machining Action	Parameters
Chip removal processes Turning, drilling, shaping, milling	Cutting	WP material Tool material and geometry Machining conditions Machine tool Built-up edge Coolant
Abrasive machining Grinding, honing, lapping, superfinishing	Abrasion	Grain-type size Type of bond Machining conditions Machining medium Machine tool
Chemical and EC	Chemical or EC erosion	WP grain size Machining conditions
Thermal machining process EDM, LBM, EBM, PBM	Thermal erosion	WP thermal properties Machining conditions
Mechanical NTM USM, AJM, WJM	Mechanical erosion	WP mechanical properties Machining conditions

condition different from that in the body of the material or as complex as a microstructure change or IGA.

The principal causes of surface alterations produced by the machining processes are as follows:

1. HTs and HT gradients
2. Plastic deformation
3. Chemical reactions and subsequent absorption into the machined surface
4. Excessive machining current densities
5. Excessive energy densities

Surface alterations of machined parts may take one of the forms shown in Table 9.3. This includes mechanical, metallurgical, chemical, thermal, and electrical forms. Table 9.4 shows the altered material zone (AMZ) definitions and symbols used to represent the possible surface alterations. Table 9.5 shows typical surface alterations that may occur by some machining process.

9.5 SURFACE EFFECTS BY TRADITIONAL MACHINING

9.5.1 Chip Removal Processes

The surface layer is the layer from the geometrical surface inward that shows changed physical and sometimes chemical properties, as compared with those with the material before machining. As shown in Figure 9.7, the main parts of such a layer are defined as:

1. Adsorbed and amorphous zone of adsorbed gas, solid, or liquid particles
2. Fibrous zone that occurs by the frictional forces between the tool and WP
3. Compressed layer that occurs due to grain size changes

Surface layer alterations occur when abusive (severe) cutting conditions are used. Under such conditions, high temperature (HT) and excessive plastic deformation is promoted. Figure 9.8 shows the surface alterations produced from drilling with dull tools where cracks, UTM at the surface, and a softer overtempered zone below the untempered surface layer are clear. Figure 9.9 shows the surface produced by face milling of Ti-6Al-4V (aged, 35 HRC). For gentle machining conditions, a slight white layer and no changes in microhardness occur. In abusive machining conditions, a white layer of 0.01 mm and a plastically deformed zone of 0.04 mm are visible. Figure 9.10 shows typical residual stresses in milling operations, which tend to be compressive. Residual stresses and part distortion are closely related, such that the greater the area under the residual stress curve, the greater will be the distortion of the machined WP.

9.5.2 Grinding

Using severe grinding conditions, the process becomes more likely to produce surface damage. In this regard, low-stress grinding (LSG) of AISI 4340 steel

TABLE 9.3
Forms of Surface Alterations by Machining

Form	Description
Mechanical	Plastic deformations (as a result of hot or cold working)
	Tears and laps and crevice-like defects (associated with "built-up edge" produced in machining)
	Hardness alterations
	Cracks (macroscopic and microscopic)
	Residual stress distribution in surface layer
	Process inclusions introduced
	Plastically deformed debris as a result of grinding
	Voids, pits, burrs, or foreign material inclusions in surface
Metallurgical	Transformation of phases
	Grain size and distribution
	Precipitate size and distribution
	Foreign inclusions in material
	Twinning
	Recrystallization
	UTM or OTM
	Resolutioning or austenite reversion
Chemical	IGA
	Intergranular corrosion (IGC)
	Intergranular oxidation (IGO)
	Preferential dissolution of microconstituents
	Contamination
	Embrittlement due to chemical absorption of elements, such as hydrogen, chlorine, and so on
	Pits or selective etch
	Corrosion
	Stress corrosion
Thermal	HAZ
	Recast or redeposited material
	Resolidified material
	Splattered particles or remelted metal deposited on surface
Electrical	Conductivity change
	Magnetic change
	Resistive heating or overheating

produced no visible surface alterations, as shown in Figure 9.11a, compared to abusive grinding, shown in Figure 9.11b. It is therefore clear that abusive grinding produces UTM of 0.03–0.13 mm deep with a hardness of 65 HRC. Below this layer, an overtempered martensite (OTM) zone having a hardness of 46 HRC is clear. The hardness returns back to its normal value at a depth of 0.3 mm below the surface as shown in Figure 9.12a. Abusive grinding produces residual stresses within the altered layer. As shown in Figure 9.12b, LSG produced a surface of low compressive stress compared to the abusive grinding condition. Additionally,

TABLE 9.4
Altered Material Zone (AMZ) Definitions, Symbols, and Examples

AMZ Definition	Symbol	AMZ Definition	Symbol
Cracks. Narrow ruptures or separations with depth-to-width 4:1 that alter the continuity of the surface		***Craters***. Surface depressions with rough edges approximately round or oval and shallow edges	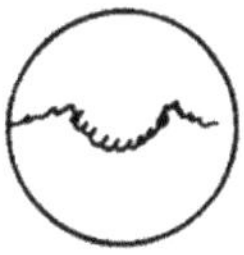
Hardness alterations. Changes in hardness as a result of heat, mechanical deformation, or chemical changes	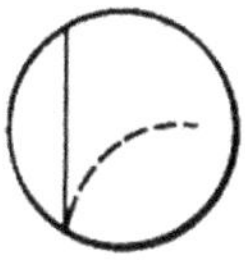	***HAZ***. A layer subjected to sufficient thermal energy that causes microstructure alterations or microhardness alterations	
Inclusions. Small particles in the surface layer in an object, may be either foreign or a part of the normal composition of the material	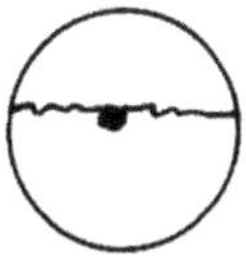	***IGA***. A form of corrosion or attack in which preferential reactions are concentrated at the surface grain boundaries	
Laps, folds, or seams. Defects in the surface from continued plastic working overlapping surface		***Low-stress surface***. A surface containing a residual stress less than 138 MPa or 10% of tensile stress, whichever is greater, at depths below the surface greater than 0.025 mm	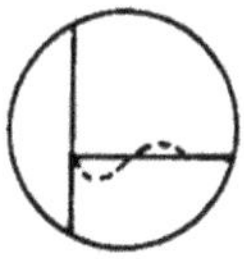
Metallurgical transformation. Micro structural changes resulting from external influences		***Pits***. Shallow depressions such as craters with depth-to-width ratio less than 4:1 or localized selective etching or corrosion that results in holes or pockets left by the machining process	
Plastic deformation. Microstructural changes as a result of exceeding the yield point of the material		***Recrystallization***. The formation of a new strain-free grain or crystal structure from that existing in the material before machining, usually as a result of plastic deformation and subsequent heating	
Recast material. Occurs when some materials become molten and are then resolidified		***Redeposited material***. When some material is removed from the WP in the molten state, and then prior to solidification is attached to the surface (splattered metal)	
Remelted or resolidified material. The portion of the surface that becomes molten, but is not removed from the surface prior to solidification		***Residual stresses***. Those stresses that are present in the material after all external influences are removed	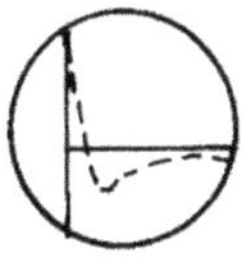
Selective etch. A process or attack in which preferential reactions are concentrated on certain constituents of the base material corrosion			

TABLE 9.5
Examples of Surface Alterations for Some Machining Operations

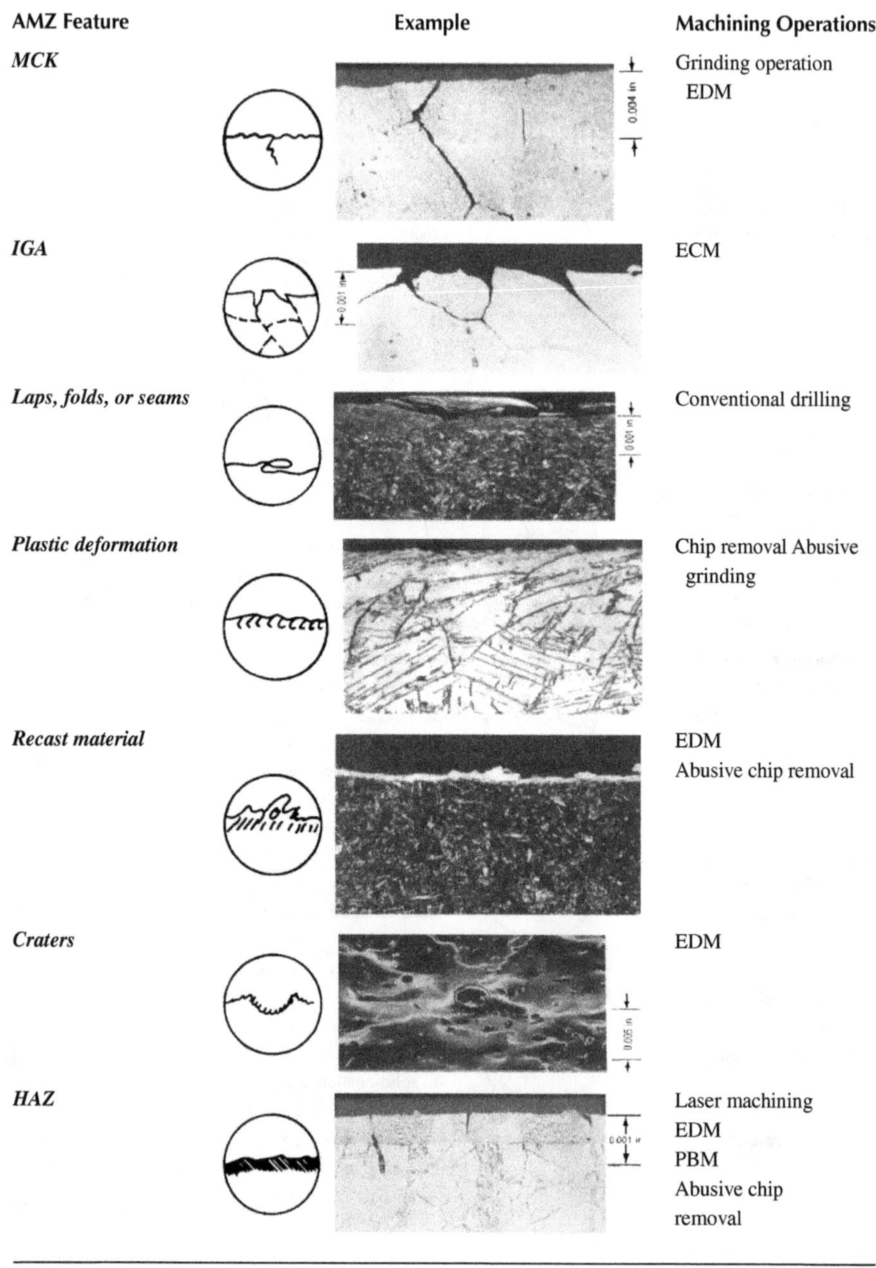

AMZ Feature	Example	Machining Operations
MCK		Grinding operation EDM
IGA		ECM
Laps, folds, or seams		Conventional drilling
Plastic deformation		Chip removal Abusive grinding
Recast material		EDM Abusive chip removal
Craters		EDM
HAZ		Laser machining EDM PBM Abusive chip removal

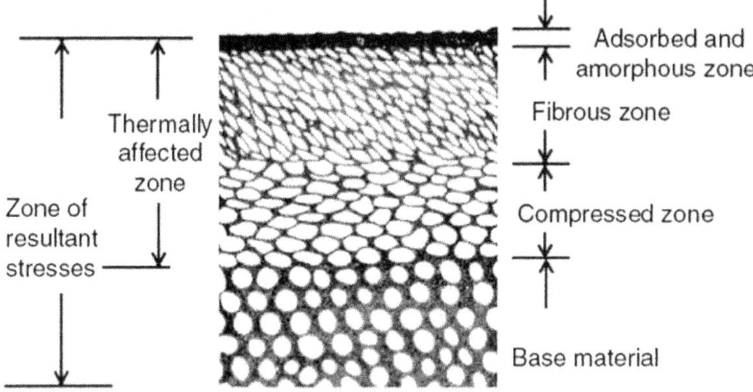

FIGURE 9.7 Surface layer after machining, section perpendicular to the tool (Kaczmarek, 1976).

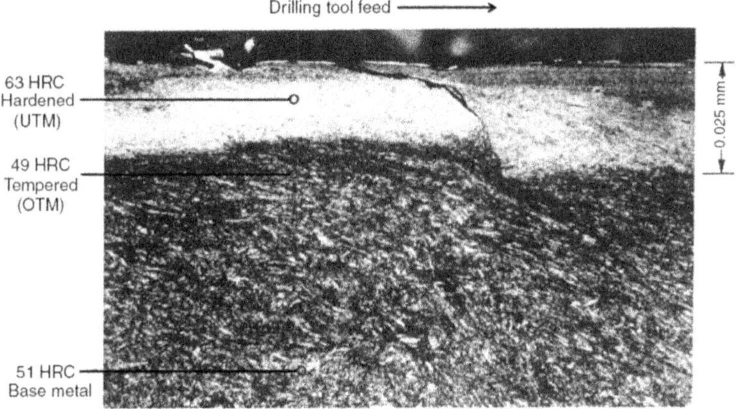

FIGURE 9.8 Surface alterations produced from drilling with dull tools, where a cracked UTM at the surface and softer overtempered zone below the untempered surface layer (Field and Kahles, 1971).

abusive grinding seriously reduces the fatigue strength, as shown in Figure 9.12c. Table 9.6 shows the effect of some machining methods on the fatigue strength together with the percentage change with respect to gentle grinding. Accordingly, the endurance limit of 4340 steel has been decreased by 12% after polishing, compared to gentle grinding. In the case of abusive grinding, there is a tendency to form batches of UTM or OTM on the surface, which is associated with a significant drop in the fatigue strength. Shot peening is used to improve the fatigue strength and stress corrosion properties of most structural alloys that are subjected to high stresses and severe environments. It improves the properties of metals that have been machined and that tends to have degraded fatigue strength and other mechanical properties, as shown in Table 9.7.

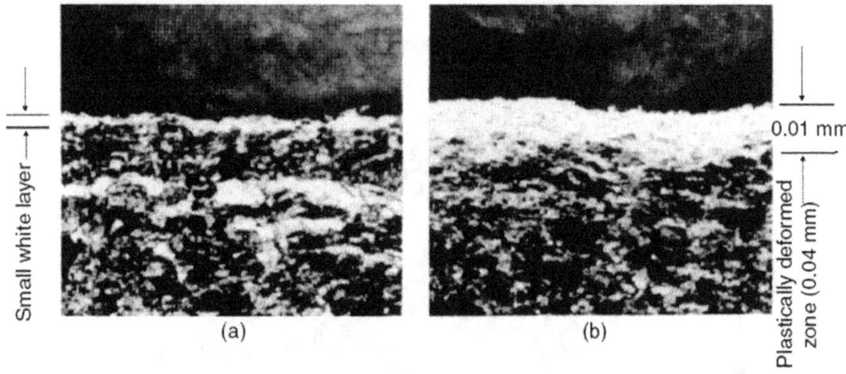

FIGURE 9.9 Surface produced by face milling of Ti-6Al-4V (aged, 35 HRC): (a) Gentle conditions, slight white layer, no changes in hardness and (b) abusive machining conditions, white layer of 0.01 mm and plastically deformed zone of 0.04 mm are visible (Field and Kahles, 1971).

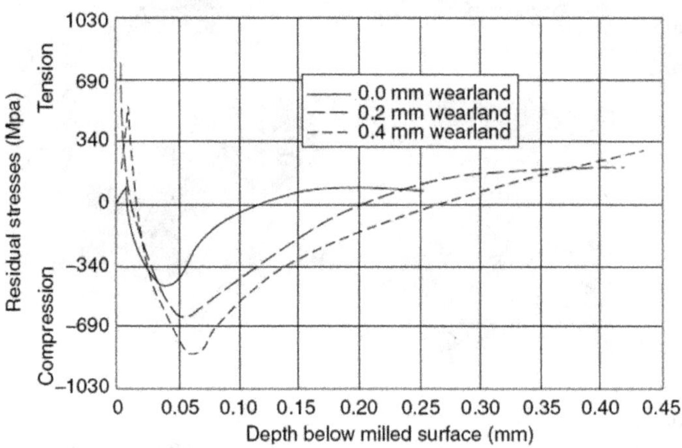

FIGURE 9.10 Residual stresses from surface milling of 4340 steel (quenched and tempered to 52 HRC) (*Machining Data Handbook*, 1990).

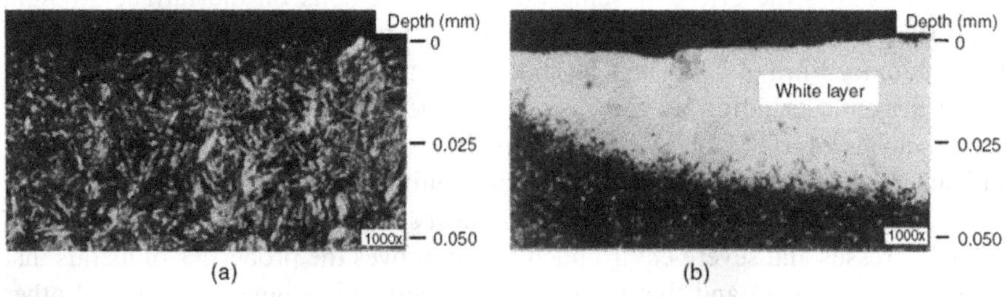

FIGURE 9.11 Surface characteristics produced by surface grinding of AISI 4340 steel: (a) LSG—no visible alterations and (b) abusive grinding (Field and Kahles, 1972).

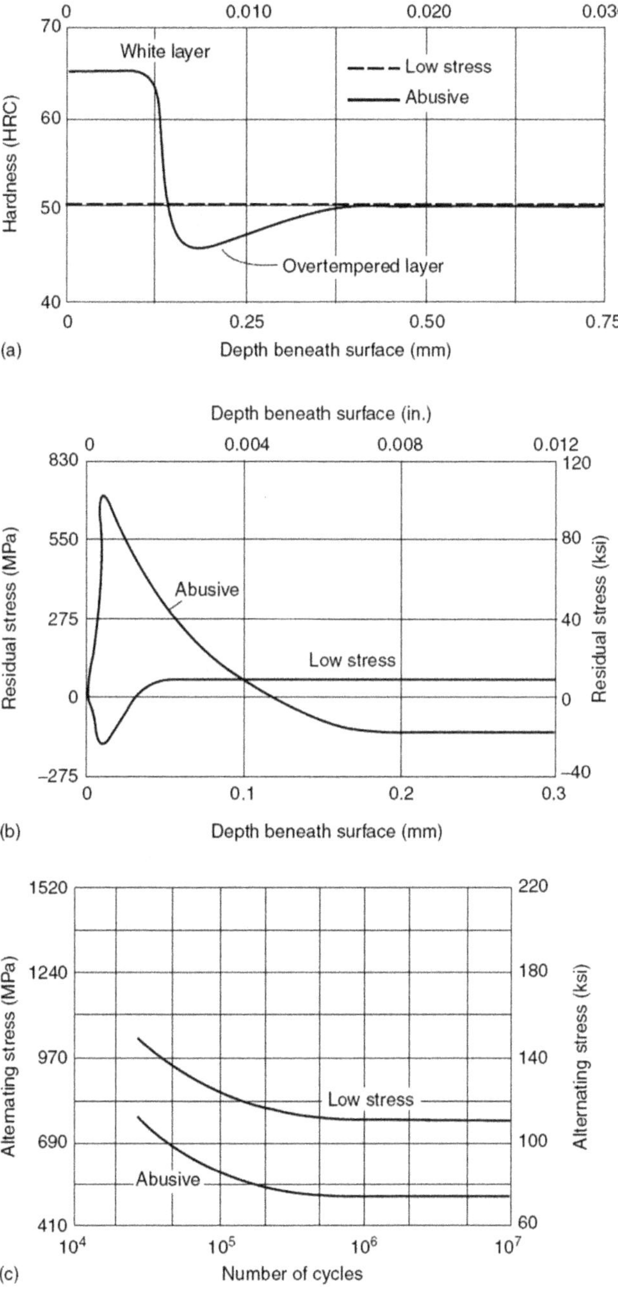

FIGURE 9.12 Surface effects by LSG and abusive grinding of AISI 4340 steel: (a) Hardness, (b) residual stresses, and (c) fatigue strength (Field, Kahles, and Cammett, 1972).

TABLE 9.6
Effect of Machining Method on Fatigue Strength

Alloy	Machining Operation	Endurance Limit in Bending, 10^7 Cycles (MPa)	Change Compared to Gentle Grinding (%)
4340 steel, 50 HRC	Gentle grinding	703	–
	ELP	620	−12
	Abusive grinding	430	−39
Ti-6Al-4V, 32 HRC	Gentle grinding	430	–
	Gentle milling	480	+13
	CH milling	350	−18
	Abusive milling	220	−48
	Abusive grinding	90	−79
Inconel 718, aged 44 HRC	Gentle grinding	410	–
	ECM	270	−35
	Conventional grinding	165	−60
	EDM	150	−63

Source: (Field and Kahles, 1971).

TABLE 9.7
Effect of Shot Peening on Fatigue Strength of Machined Parts

		Endurance Limit in Bending, 10^7 Cycles (MPa)		
Alloy	Machining Operation	Before Shot Peening	After Shot Peening	Percentage Increase
4340 steel, 50 HRC	Gentle surface grinding	703	772	110
	Abusive surface grinding	430	630	146
	Electropolished	620	660	106
Inconel 718, solution treated, and aged 44 HRC	EDM roughing	170	540	317
	EDM finishing	170	480	282
	ECM	285	560	196
	ELP	290	540	186

Source: (Koster, Gatto, and Cammett, 1981).

9.6 SURFACE EFFECTS BY NON-TRADITIONAL MACHINING

NTM methods have been introduced due to components' complex shape and fine finish requirements. It is used whenever the conventional machining processes are not able to cope up satisfactorily with the enhanced properties of difficult-to-machine materials. The principal mechanism of material removal may be chemical during the ECM and CHM processes. The melting and boiling temperatures, heat of fusion, and the specific energy of the machined components play major roles in thermal machining processes. In other cases, the mechanical characteristics are the controlling factors in the removal mechanism associated with mechanical NTM processes.

Some processes combine more than one removal action, such as the double one in ECG and electroerosion dissolution machining (EEDM) and the triplex effect in electrochemical discharge grinding (ECDG). Such processes are generally classified on the basis of the main removal mechanism contributing to the bulk of material removal as a chemical or a thermal machining process (El-Hofy, 2005). Wear resistance, the contact pressure, stress concentration, type of fit, and the corrosion resistance properties determine the performance of any machined component by NTM methods.

9.6.1 Electrochemical and Chemical Machining

In ECM, considerable variations are possible in surface finishes produced due to WP characteristics and machining conditions. Crystallographic irregularities in crystal lattices, such as voids, dissociation and grain boundaries, differing crystal structures and orientations, and locally different alloy composition produce an irregular distribution of current density, thus leaving microscopic peaks and valleys that produce surface roughness (König and Lindenlauf, 1978). Figure 9.13 shows the mechanism of surface roughness generation in an alloy machined by ECM. For non-passivating electrolytes (NaCl), the reduction in electrolyte concentration and the increase of its temperature improves the quality of surfaces. For passivating electrolytes ($NaNO_3$), a low electrolyte concentration and a rise in its temperature increase the formation of a protective layer that causes deterioration to surface quality. Further increase in current density breaks up this layer, so that a smoother surface is produced. Surface roughness depends on the structure of the WP material. In this regard, the more fine-grained and

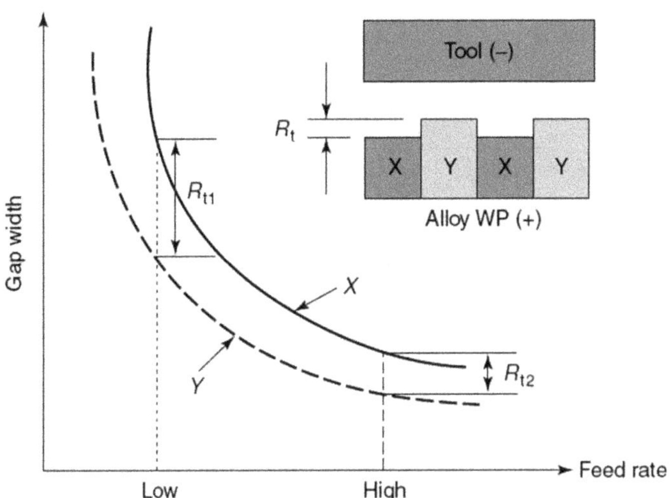

FIGURE 9.13 Surface roughness generation in ECM.

homogeneous the structure, the better the surface quality. The roughness obtained at greater grain sizes was possibly due to the reduced number of grain boundaries present on such surfaces. Surface finish also deteriorates with increasing grain size at lower flow velocities. The different levels of surface roughness R_a and dimensional tolerance that can be obtained by different ECM applications are shown in Figure 9.6.

In CHM, etching rates of 0.025 mm/min with tolerances of ±10% of the cut width (±0.01 to ±0.04 mm) can be achieved, depending on the WP material and depth of etch. The final surface roughness is influenced by the initial WP roughness. It increases as the metal ion concentration of the enchant rises. Table 9.8 shows the dependence of surface roughness on the WP material and production method. In PCM, cutting rates of 0.01–0.2 mm/min are possible. Tolerance of ±15% of the material thickness can be obtained, leaving a surface roughness similar to that of CHM.

Electropolishing (ELP) is a finishing process with metal removal rates between 0.013 and 0.038 mm/min. Because the erosion of the surface asperities takes place at a faster rate than the erosion in the valleys, the surface can be smoothed without considerable material removal. Surface roughness R_a of 0.1–0.8 μm, depending on the initial surface roughness, can be obtained. However, under special machining conditions, a roughness of 0.025–0.05 μm is possible.

ECM, CHM, and ELP produce surfaces that are free from metallurgical alterations. However, under certain conditions, selective etching (SE) or IGA may occur, as shown in Figures 9.14 and 9.15. Surface softening occurs in most materials machined by ECM as well as CHM, as shown in Figures 9.16 and 9.17. Accordingly, the surface is about five points HRC lower in hardness than the interior to approximately

TABLE 9.8
Surface Roughness Achieved in CH Milling after Removing 0.25–0.40 mm from the Surface

Material	Form	Surface Roughness (R_a), μm
Aluminum alloys	Sheet	2.0–3.8
	Casting	3.8–7.6
	Forging	2.5–6.4
Magnesium alloys	Casting	0.75–1.4
Steel alloys	Sheet	0.75–1.5
Nickel alloys	Sheet	0.75–1.0
Titanium alloys	Sheet	0.20–0.8
	Casting	0.75–1.5
	Forging	0.38–1.0
Tungsten	Bar	0.50–1.0
Beryllium	Bar	3.8–6.4
Tantalum	Sheet	0.25–0.5
Columbium	Bar	1.0–1.5
Niobium	Sheet	1.0–1.5

Source: (Wilkinson and Warburton, 1967).

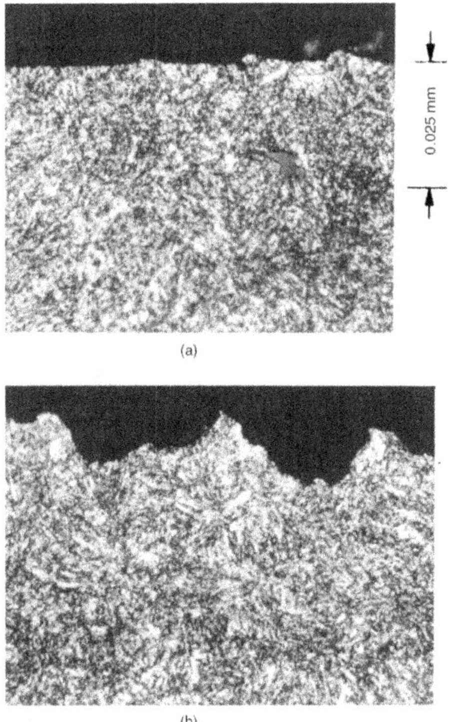

FIGURE 9.14 Surface generation of 4340 steel machined by ECM: (a) Gentle conditions (high current density) and (b) abusive conditions (low current density) (Koster et al., 1970).

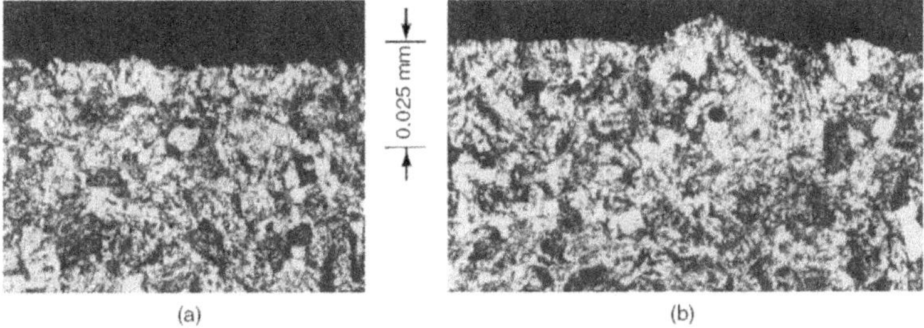

FIGURE 9.15 Surface generation of 4340 steel machined by CHM: (a) Gentle conditions and (b) abusive conditions (Field and Kahles, 1971).

0.05 mm in depth. This softening may be severe enough to affect the fatigue strength and other mechanical properties of metals and may necessitate postprocessing.

As shown in Table 9.6, CHM of Ti-6Al-4V resulted in an 18% drop in fatigue strength, while ECM of Inconel 718 produced a 35% drop in endurance limit compared to gentle grinding. Generally, ECM, CHM, and ELP produce stress-free surfaces. Additionally, the decrease in their fatigue strength resulted from the absence of the compressive stress associated with surface grinding. For CHM of Ti-6Al-4V, the

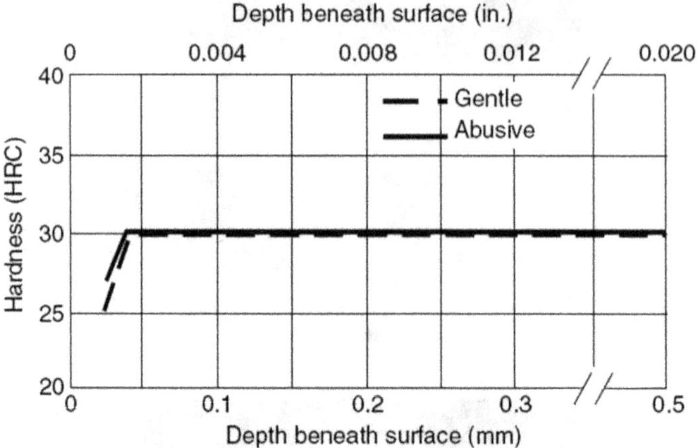

FIGURE 9.16 Hardness of machined surfaces of 4340 steel machined by ECM for gentle and abusive conditions (Koster et al., 1970).

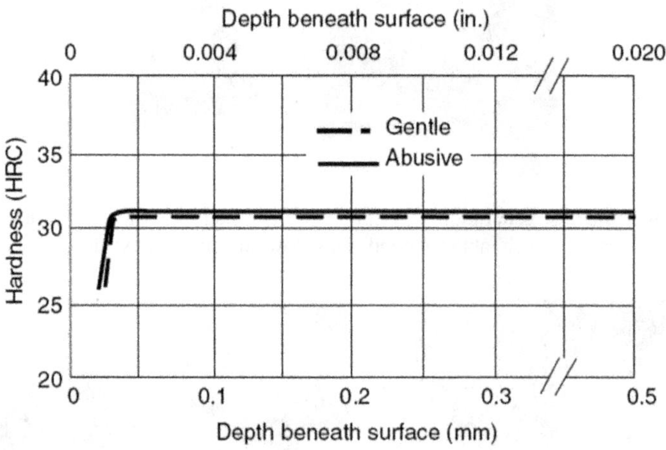

FIGURE 9.17 Hardness of machined surfaces of 4340 steel machined by CHM: (a) Gentle conditions and (b) abusive conditions (Field and Kahles, 1971).

endurance limit is 350 MPa compared to 460 MPa in the case of LSG. Table 9.7 shows how shot peening raises the endurance limit of parts machined by ECM, CHM, and ELP. In ECG, the surface is similar to that obtained using metallographic polishing, free from process-induced residual stresses, and there is no heat-affected layer.

9.6.2 Thermal Non-Traditional Processes

9.6.2.1 Electrodischarge Machining

In this process, the machined surface consists of a multitude of overlapping craters that are formed by the action of microsecond duration spark discharges. These craters

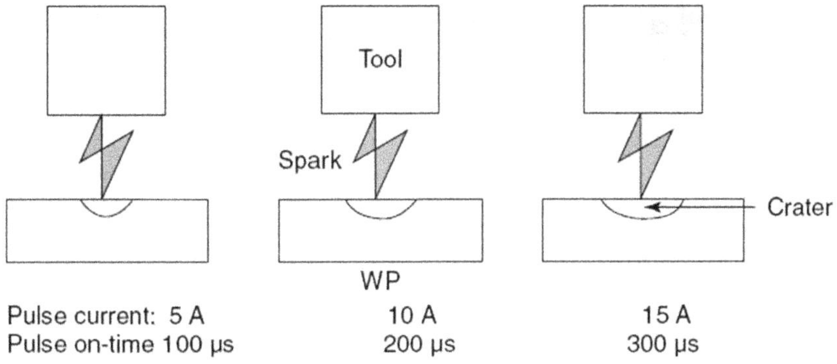

FIGURE 9.18 Effect of pulse characteristics on EDM crater size.

depend on the thermal properties of the material, the composition of the machining medium, the discharge energy, and the pulse on-time, as shown in Figure 9.18. The integral effect of many thousands of discharges per second leads to the formation of the corresponding profile with specified accuracy and surface finish, as shown in Figure 9.6. The peak-to-valley roughness is usually represented by the depth of the resulting crater. The maximum depth of the damaged layer is usually 2.3 times the surface roughness R_a.

According to Rajurkar and Pandit (1984) and Delpretti (1977), the maximum peak-to-valley height was considered to be ten times the average roughness R_a. Surface roughness increases linearly with metal removal rate, and graphite electrodes produce rougher surfaces than the metal ones.

Surfaces produced by EDM have recast splattered metal that is usually hard and cracked, as shown in Figure 9.19. Below such a recast layer, surface alterations occur under roughing conditions where an overtempered zone of 40 HRC to a depth of 100 μm is found (Figure 9.20). As the pulse energy is decreased, surface finish improves and consequently, the depth at which the surface craters disappear (free polishing depth) is reduced. This depth was found to lie between three and six times the root mean square roughness R_q. That depth is important when polishing

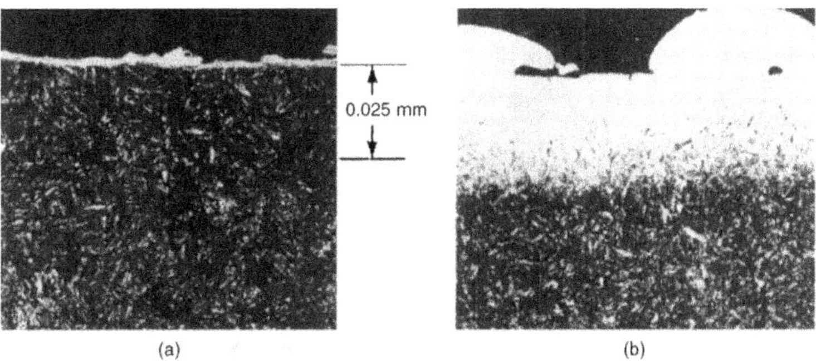

FIGURE 9.19 Surface characteristics of AISI (quenched and tempered 50 HRC) after: (a) Finish EDM and (b) rough EDM (Field and Kahles, 1971).

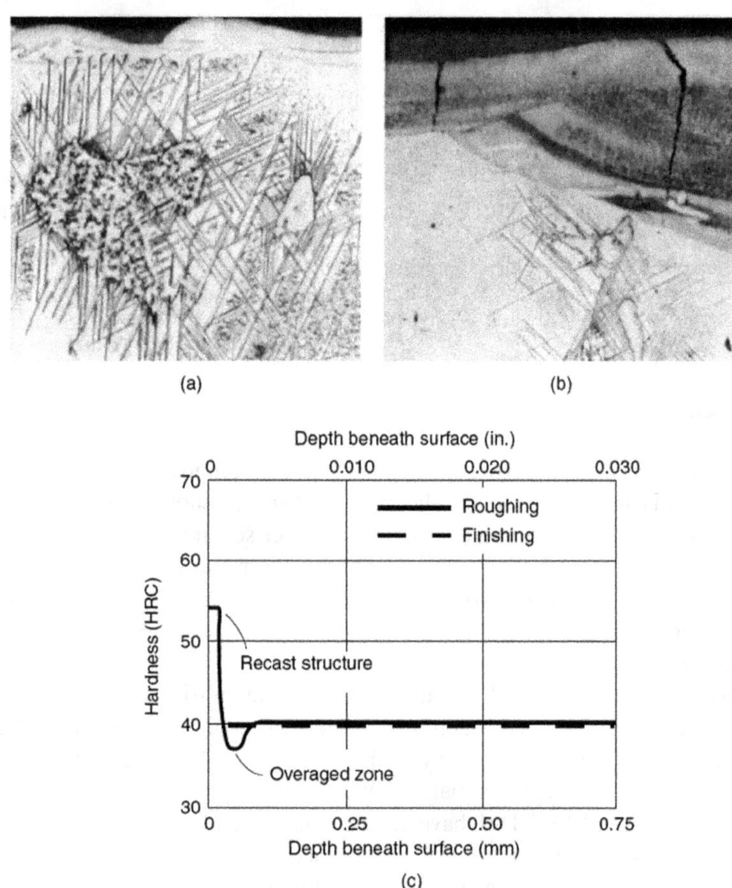

FIGURE 9.20 Surface characteristics of cast Inconel 718 (aged, 40 HRC) by EDM: (a) Finish EDM, (b) rough EDM, and (c) microhardness alterations (Field and Kahles, 1971).

dies and molds and when the residual stresses are to be removed. A quick EC finishing technique, using mate electrode, has been adopted by Masuzawa and Saki (1987). Accordingly, a reduction of surface roughness from 22 to 8 μm R_{max} has been reported together with the removal of the heat-affected layer (Figure 9.21).

Koneda and Furuoya (1991) introduced oxygen gas into the discharge gap which provided extra power by the reaction of oxygen and in turn increased the melting of the WP and created greater expulsive force, thus increasing the metal removal rate and surface roughness. When EDM is used for cusp removal, the silicon powder is suspended in the working fluid, during the stage of finish EDM. Consequently, surface roughness is changed from 45 to 10 μm R_{max} as shown in Figure 9.22. In contrast, the matte appearance of the ED-machined surfaces has been found satisfactory in some applications of texturing described by Aspinwal and Wise (1992) and Amalnik et al. (1997).

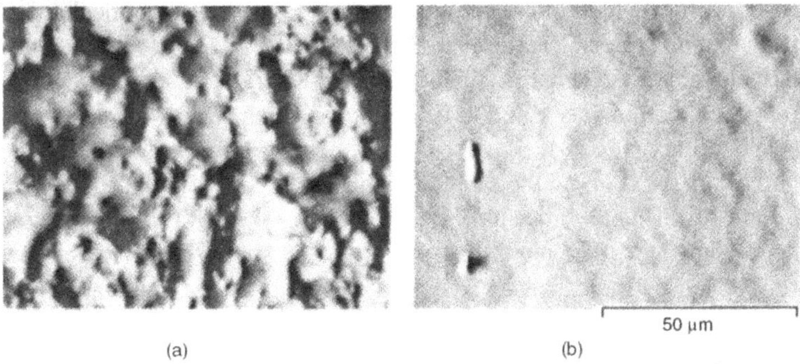

FIGURE 9.21 Use of ECM mate electrode to finish SKS 3 tool steel machined by EDM: (a) After EDM and (b) after ECM finishing (Masuzawa and Saki, 1987).

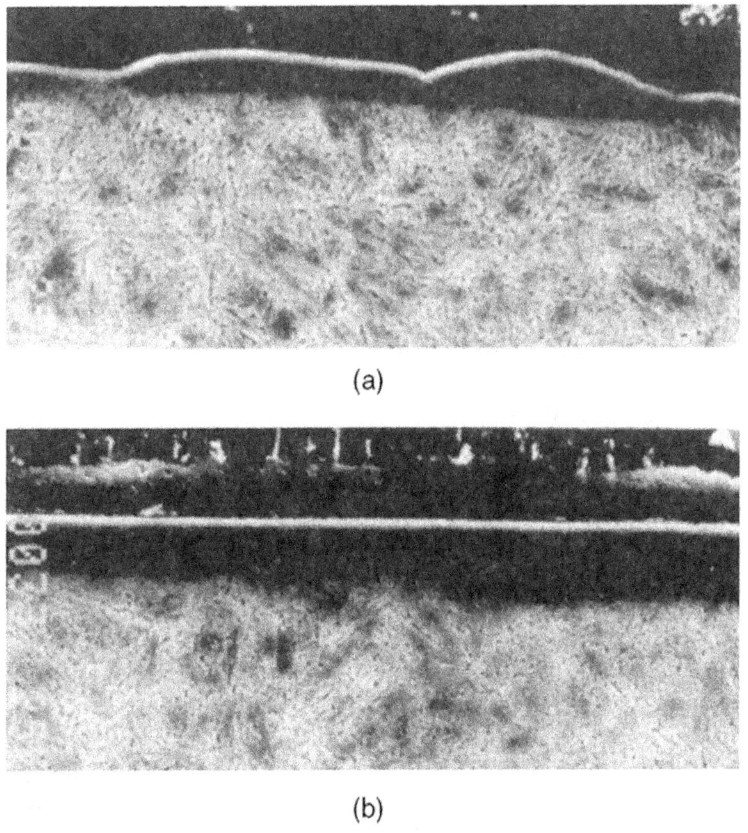

FIGURE 9.22 Surface characteristics after EDM: (a) Conventional EDM and (b) silicon powder mixing EDM (Kobyashi, 1995).

Abusive machining conditions have a minor effect on the static mechanical properties of machined materials. However in the case of EDM followed by stress relief heat treatment, a marked decrease in ductility and tensile stresses have been noticed. The extent of reduction was found to be a function of surface roughness. According to Wells and Willey (1977), the choice of the correct dielectric flow in the gap has a significant effect in reducing the surface roughness by 50%, increasing the machining rate, and lowering the thermal effects in the eroded WP surface. Benhaddad et al. (1991) indicated that for Al/Li alloys, the tensile strength of the machined parts are reduced by increased surface roughness. This reduction was enhanced by increased pulse current. The heat-affected layer reached 200 μm compared to 30 μm for steel due to difference in their thermal conductivities.

In EDG, the range of average surface roughness is 1.6–3.2 μm. However, at low machining rates of 0.03 cm^3/h, a roughness of 0.4 μm can be obtained. A heat-affected layer of 0.013 mm has been reported in the *Machining Data Handbook* (1990) that needs to be removed or modified to ensure the best surface integrity. In electrodischarge sawing (EDS), surface roughness for high feeds is 10–12.5 μm, while at lower ones it is 6.3–10.0 μm. The depth of the recast layer is about 0.025 mm deep. Similarly, tolerances of ±0.076 to ±0.4 μm can be achieved depending on the feed rate. Figure 9.6 depicts surface roughness and tolerances for some thermal machining processes.

Very minor fatigue strength differences exist between finishing and roughing EDM, as shown in Table 9.7. Finishing and roughing EDM are characterized by a wide difference in surface roughness levels (1.25–5 μm R_a) and wide differences in recast layer thicknesses (5–125 μm). High cycle fatigue strength, however, is nearly the same for finishing as for roughing by EDM. The fatigue strength of Inconel 718 alloy machined by EDM reported a reduction by 63% compared with gentle grinding. As shown in Table 9.7, using shot peening has raised the fatigue strength of EDM surfaces by 282% for finish EDM and 317% for rough EDM.

9.6.2.2 Laser Beam Machining

A heat-affected layer ranging from 0.025 to 0.05 mm is created, depending on the machining rate. The surface produced is normally rough and the location accuracy of ±0.025 mm is possible. When a gas stream assists the laser beam, the cutting rate is raised to 7500 mm/min with an accuracy of ±0.1 mm, a heat-affected layer of 0.025–0.25 mm, and a roughness of 3.2–6.3 μm. The heat-affected layer, the surface roughness, and the cutting rate depend on the process parameters, including the gas nozzle diameter and gap.

In LBM of Inconel 718, the heat-affected zone produced a recast layer at the entrance and exit of the hole produced. Such a recast layer was greatly reduced during the experimental work of Yue et al. (1996) and Lau et al. (1990) by inducing ultrasonic vibrations to the WP. It is therefore clear that a much thicker recast layer appears under LBM without vibrations, as shown at the bottom of the hole of Figure 9.23. Additionally, with the help of ultrasonic vibrations, the thickness of the recast layer is reduced by four times. Figure 9.24 shows the boundary of HAZ with and without WP vibration.

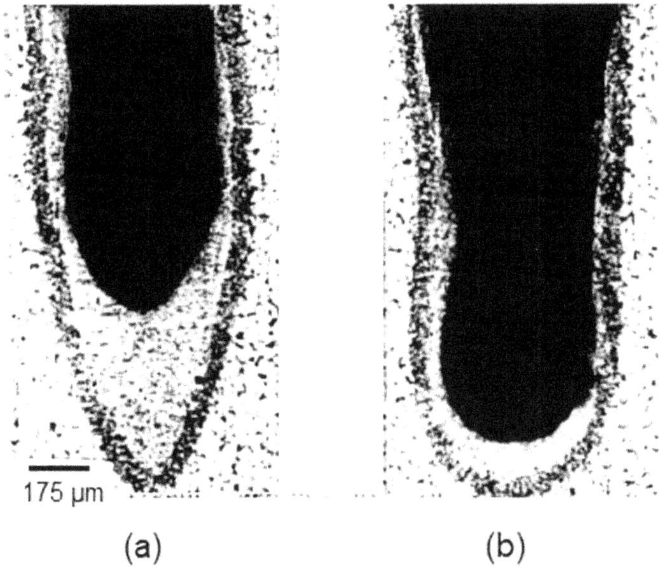

FIGURE 9.23 LBM drilling: (a) Without vibrations and (b) with ultrasonic vibrations (Yue, Chen, Man, and Lau, 1996).

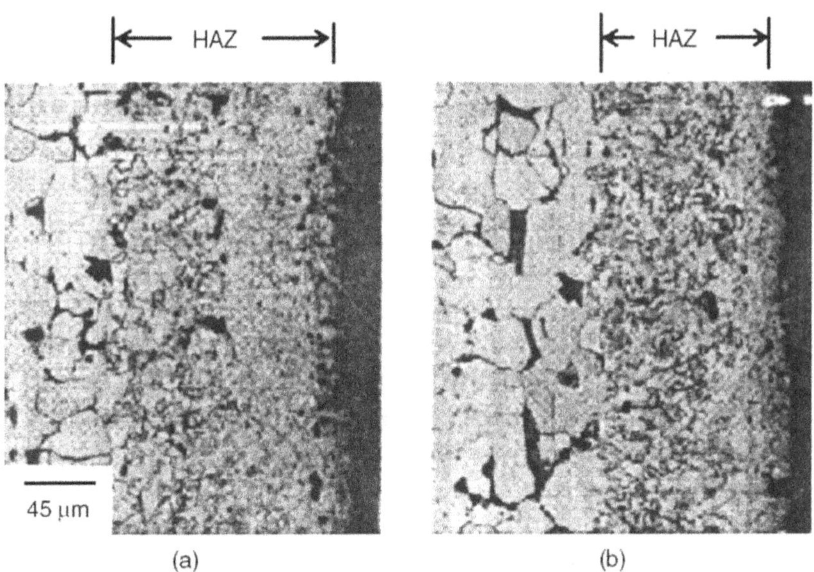

FIGURE 9.24 Laser beam machining: (a) Without vibration and (b) with ultrasonic vibration (Yue, Chen, Man, and Lau, 1996).

9.6.2.3 Electron Beam Machining

In EBM, the quality of the surface produced is influenced by the thermal properties of WP and pulse energy or charge. There is a thin layer of recast or heat-affected material on the cut surface that should be removed or modified. The heat-affected layer increases with pulse energy and therefore the use of short pulses limits the extent of the heat-affected zone. Typical tolerance is about ±10% of the slot width and a heat-affected layer of 0.25 mm is produced.

9.6.2.4 Plasma Beam Machining (PBM)

The depth of fused layer ranges from 0.81 to 4.67 mm and the average roughness R_a is usually between 0.8 and 3.2 μm.

9.6.2.5 Electroerosion Dissolution Machining

Surface generation takes place through electrolytic dissolution of varying intensity depending on the gap size and the consequent crater formation at random locations over the entire machined surface. Saushkin et al. (1982) reported that electrolysis is apparently localized in the proximity of the pits of the craters that are soon made smooth, probably as a result of the HT of the metal and the electrolyte in this zone. The general appearance of the machined surface by EEDM constitutes less turbulence than that of EDM. The crater depth, volume, diameter-to-depth ratio, and surface roughness have been calculated using the roughness profiles. According to El-Hofy (1992, 1996), the combination of ECM and EDM processes markedly reduced the roughness indices together with the absence of a heat-affected layer of machined parts by wire EEDM.

The combination of ECM and EDM in EEDM die sinking produces intrinsic effects on surface roughness, shape geometry, and subsurface layers. The molten crates on the WP surface formed by severe thermal erosive pulses are quenched very rapidly in the surrounding cold electrolyte. The surface and subsurface quickly undergo metallurgical phase transformations. The white damaged layer was clear at the machined die surface. The thickness of such a layer is consistently decreasing in the direction of the electrolyte upstream, owing to the continuous attack by ECD. The thickness of the white layer ranged from 0 to 130 μm. Micro- and macrocracks were also observed in the white layer, occasionally extending to the annealed subsurface layer. Most of the MCK were removed by the electrochemical dissolution action. The pitted surface of the die by selective etching is shown in Figure 9.25a;

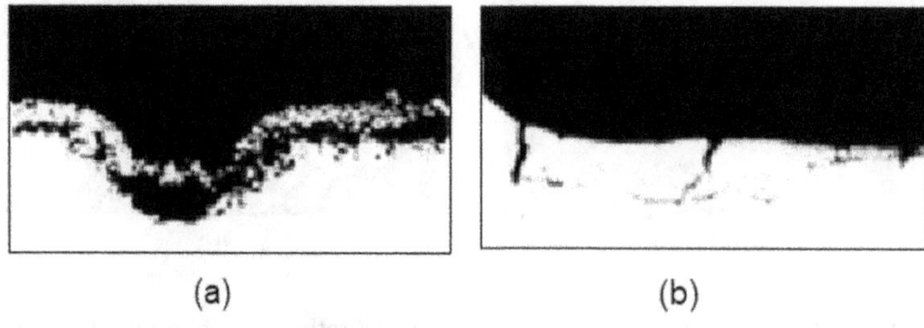

FIGURE 9.25 EEDM surface layer: (a) Pitted surface by selective etching and (b) white layer by EEDM (Khayry, 1990).

Figure 9.25b shows the white layer by the EEDM process at the die bottom section, with clear cracks of micro and macro sizes.

9.6.2.6 Electrochemical Discharge Grinding

The WP, mainly eroded by discharges, is immediately smoothed by mechanical grinding and electrolytic dissolution, thus producing an average roughness of 0.4 μm R_a. The thickness or the altered layer also decreases at lower machining voltages, as well as smaller wheel speeds (Kaczmarek, 1976).

9.6.3 Mechanical Non-Traditional Processes

In USM, the material removal rate and surface finish depend on the feed force, oscillation amplitude, WP material, grain size, and type of abrasives used. Surface roughness increases with oscillation amplitude and grain size. Rougher surfaces are produced when machining glass than when machining hard alloy steels. Generally, the more brittle the material, the greater the resulting surface roughness, as the cavity left by each particular grain is deeper. As the amplitude of vibration is raised, the surface becomes rougher, because the individual grains are pressed further into the surface of the WP. Smoother finish can also be achieved when the viscosity of the liquid carrier is lowered. The roughness in the sides of cut is much higher than that in the bottom. This results from the sidewalls being scratched by grains entering and leaving the machining zone. As a rule, USM produces an average surface roughness R_a of 0.51–0.76 μm, out-of-roundness 10 μm, a taper of 0.005 mm/mm, and production tolerances of ±0.005 to ±0.025 mm. The process leaves no heat-affected zone, and no chemical or electrical alterations. Shallow compressive residual stresses occur at the surface, which promote an increase in the high-cycle fatigue strength of the work material (Rooney, 1957).

In AJM, surface roughness depends on WP material, grit size, and type of abrasives. A material with a high removal rate results in large surface roughness. For this reason, fine grains are used for machining soft metals to obtain the same roughness in hard ones. The decrease of surface roughness with a smaller grain size is related to the reduced depth of cut and the undeformed chip cross section. In addition, the larger the number of grains per unit volume, the larger the number of grains that fall on a unit surface area. Generally, surface roughness of 0.15–1.60 μm R_a with tolerance of ±0.05 mm is possible. In AWJM, a carrier liquid consisting of water with anticorrosive additives has much greater density than air, which contributes to higher acceleration of the grains, with consequent larger grain speed and increased metal removal rate. Moreover, the carrier liquid, when spreading over the surface, fills its cavities and forms a film that impedes the striking action of the grains, and hence bulges and tops of the surface irregularities are the first to be affected and the surface quality improves. Experiments showed that water air jet permits one to obtain, as an average, a roughness number higher by one, as compared with the effect of an air jet. In high-speed WJM of Inconel, Hashish (1992) concluded that the roughness increases at higher feed rates as well as lower slurry flow rates. The surface produced during WJM is generally a polished, scoured clean one of light peening texture. In AFM, stock removal rates and surface roughness depend on grit type and extrusion pressure. The process leaves residual stress of depth 0.025 mm. It can be used for quick polishing where a surface roughness of 0.8–7.6 μm R_a finishes to one-tenth of its original value. When the process is used for deburring of stainless-steel boles, surface roughness changes from 1.6 to 0.4 μm.

9.7 REDUCING DISTORTION AND SURFACE EFFECTS IN MACHINING

Table 9.9 summarizes the possible surface effects produced by different machining processes of some engineering metals and alloys. Reduction of these effects can be achieved by considering the following recommendations:

1. Chip Removal Processes
   - Select machining conditions that lead to long tool life and good surface finish.
   - Use sharp tools.
   - Use rigid and high-quality machine tools.
   - Avoid hand feeding during drilling and reaming.
   - Use deburring to remove sharp corners.
   - Use honing to improve surface quality.
2. Abrasive Machining
   - Use LSG to remove the last 0.25 mm of material.
   - Dress the grinding wheels frequently.
   - Apply cutting fluids.
   - Avoid hand wheel grinding.
   - Allow proper allowance for subsequent cleanup of cutoff parts.
3. Chemical and Electrochemical Machining
   - Consider the prior metallurgical condition of the machined surface.
   - Monitor and control the preselected operating parameters.
   - Control the machining current density.
   - Clean the surface to remove electrolytes or etchants.
   - Adopt post treatment (shot peening) to restore fatigue strength after ECM or CHM.
   - Consider the prior metallurgical condition of the surface.
4. Thermal Machining
   - Consider the prior metallurgical condition of the surface.
   - Monitor and control the preselected operating parameters.
   - Control the magnitude of the energy impinging on that surface.
   - Clean the surface to remove dielectric fluids, beads, and vapor residue.
   - Apply post treatment to restore fatigue strength after EDM of critical or highly stressed surfaces by:
     a. Removal of layers by LSG
     b. Removal of layers by CHM
     c. Addition of a metallurgical-type coating
     d. Reheat treatment
     e. Application of shot peening

TABLE 9.9
Summary of Possible Surface Alterations Resulting from Various Material Removal Processes

	Conventional		Non-Traditional		
Material	**Milling, Drilling, and Turning**	**Grinding**	**EDM**	**ECM**	**CHM**
Non-hardenable 1018 steel	R	R	R	R	R
	PD	PD	MCK	SE	SE
	L and T		RC	IGA	IGA
Hardenable 4340 and D6AC steel	R	R	R	R	R
	PD	PD	MCK	SE	SE
	L and T	MCK	RC	IGA	IGA
	MCK	UTM	UTM		
	UTM	OTM	OTM		
	OTM				
D2 tool steel	R	R	R	R	R
	PD	PD	MCK	SE	SE
	L and T	MCK	RC	IGA	IGA
	MCK	UTM	UTM		
	UTM	OTM	OTM		
	OTM				
Type 410 stainless steel (martensitic)	R	R	R	R	R
	PD	PD	MCK	SE	SE
	L and T	MCK	RC	IGA	IGA
	MCK	UTM	UTM		
	UTM	OTM	OTM		
	OTM				
Type 302 stainless st (austenitic)	R	R	R	R	R
	PD	PD	MCK	SE	SE
	L and T		RC	IGA	IGA
17-4 PH steel	R	R	R	R	R
	PD	PD	MCK	SE	SE
	L and T	OA	RC	IGA	IGA
	OA		OA		
350-grade maraging (18% Ni) steel	R	R	R	R	R
	PD	PD	RC	SE	SE
	L and T	RS	RS	IGA	IGA
	RS	OA	OA		
	OA				
Nickel and cobalt base alloys	HAZ	HAZ			
Inconel alloy 718	R	R	R	R	R
Rene 41	PD	PD	MCK	SE	SE
HS 31	L and T	MCK	RC	IGA	IGA
IN 100	MCK				

(Continued)

TABLE 9.9 (CONTINUED)
Summary of Possible Surface Alterations Resulting from Various Material Removal Processes

	Conventional		Non-Traditional		
Material	**Milling, Drilling, and Turning**	**Grinding**	**EDM**	**ECM**	**CHM**
Ti-6Al-4V	HAZ	HAZ			
	R	R	R	R	R
	PD	PD	MCK	SE	SE
	L and T	MCK	RC	IGA	IGA
Refractory alloy	R	R	R	R	R
molybdenum TZM	L and T	MCK	MCK	SE	SE
	MCK			IGA	IGA
Tungsten (pressed and	R	R	R	R	R
sintered)	L and T	MCK	MCK	SE	SE
	MCK			MCK	MCK
				IGA	IGA

Note: R = roughness of surface, PD = plastic deformation, L and T = laps and tears, MCK = microcracks, HAZ = heat-affected zone, SE = selective etch, IGA = intergranular attack, UTM = untempered martensite, OTM = overtempered martensite, OA= overaging, RS = resolution or austenite reversion, RC = recast, respattered, vapor-deposited metal.

Source: (Field, Kahles, and Cammett, 1972).

9.8 REVIEW QUESTIONS

9.8.1 Differentiate between surface roughness and surface texture.
9.8.2 Describe the theoretical roughness that may be generated during turning.
9.8.3 Sketch the possible lay that may be formed after milling, shaping, grinding, and honing.
9.8.4 State the main reasons behind surface alterations in machining.
9.8.5 Explain how the surface properties in EEDM are better than EDM.
9.8.6 Explain why ECM and CHM produce low fatigue strength compared to LSG.
9.8.7 Differentiate between surface texture and surface integrity.
9.8.8 Explain the effect of shot peening on raising the fatigue strength of machined surfaces.
9.8.9 State the possible forms of surface alteration after ECM, drilling, and EDM.
9.8.10 State the general recommendations that may be considered to reduce the surface alterations in the case of EDM, turning, LBM, and ECM.

REFERENCES

Amalnik, MS, El-Hofy, H & McGeough, J 1997, 'An intelligent knowledge based system for manufacturability evaluation of design for electro discharge texturing', *MATADOR Conference*, Manchester, pp. 418–424.

Aspinwal, DK & Wise, ML 1992, 'Electrical discharge texturing', *International Journal of Machine Tools and Manufacture*, vol. 32, no. 12, pp. 183–193.

Benhaddad, MA, McGeough, JA & Barker, MB 1991, 'Electrodischarge machining of Al-Li alloys and its effect on surface roughness, hardness and tensile strength', *Processing of Advanced Materials*, vol. 6, no. 314, pp. 123–128.

Delpretti, MR 1977, 'Physical and chemical characteristics of superficial layers', *Proceedings of ISEM-5*, Switzerland, Wolfsberg, pp. 209–212.

El-Hofy, H 1992, 'Electroerosion dissolution machining of graphite, Inco 901,2017 Al and Steels', *5th PEDAC Conference*, Alexandria, pp. 489–501.

El-Hofy, H 1996, 'Surface generation in nonconventional machining', *6th MDP Conference*, Cairo University, Egypt, pp. 203–213.

El-Hofy, H 2005, Advanced *machining processes, nontraditional and hybrid machining processes*, McGraw-Hill, New York.

Field, M & Kahles, JF 1971, 'Review of surface integrity of machined components', *Annals of CIRP*, vol. 20, no. 2, pp. 153–163.

Field, M, Kahles, JF & Cammett, JT 1972, 'Review of surface integrity of machined components', *Annals of CIRP*, vol. 21, no. 2, pp. 219–238.

Hashish, M 1992, 'Machining with high velocity water jets', *5th PEDAC Conference*, Alexandria, pp. 461–471.

Kaczmarek, J 1976, Principles *of machining by cutting, abrasion and erosion*, Peter Peregrines Ltd., Stevenage.

Khayry, ABM 1990, 'Die-sinking by electroerosion-dissolution machining', *Annals of CIRP*, vol. 39, no. 1, pp. 191–195.

Kobyashi, K 1995, 'The present and future developments of EDM and ECM', *ISEM-XI conference*, EPFL, Lausanne, pp. 29–47.

Koneda, M & Furuoya, S 1991, 'Improvements of EDM efficiency by supplying oxygen gas into gap', *Annals of CIRP*, vol. 40, no. 1, pp. 215–218.

König, W & Lindenlauf, P 1978, 'Surface generation in electrochemical machining', *Annals of CIRP*, vol. 29, no. 1, pp. 97–100.

Koster, WP, Gatto, LE & Cammett, JT 1981, 'Influence of shot peening on surface integrity of some machined aerospace materials', *Proceedings of 1st International Conference on Shot Peening*, Pergamon Press, France, pp. 287–293.

Koster, WP 1970, *Surface integrity of machined structural components, AFML-TR-70-11*, Metcut Research Associates, Cincinnati, OH, P2.

Lau, WS, Lee, WP & Pans, SQ 1990, 'Pulsed Nd: YAG laser cutting of fiber composite materials', *CIRP*, vol. 39, no. 1, pp. 179–182.

Machinability Data Center, *Machining data handbook* 1990, vol. 2, 3rd edn, Metcut Research Association, Cincinnati, OH.

Masuzawa, T & Saki, S 1987, 'Quick finishing of WEDM products using mate-electrode', *Annals of CIRP*, vol. 36, no. 1, pp. 123–326.

Rajurkar, KP & Pandit, SM 1984, 'Quantitative expressions for some of surface integrity in electro discharge machined components', *ASME Journal of Engineering for Industry*, vol. 106, pp. 171–178.

Rooney, RJ 1957, 'The effects of various machining processes on the reversed-bending fatigue strength of A-110AT titanium alloy sheet, U. S. air force technical report WADC-TR-57-310', Wright Air Development Center, Wright-Patterson Air Force Base, OH.

Saushkin, BP et al. 1982, 'Special features of combined electrochemical and electroerosion machining of elongate machine parts', *Electrochemistry in Industrial Processing and Biology*, vol. I05. no. 3, pp. 8–14.

Surface Texture (Surface Roughness, Waviness, and Lay) ANSI/ASME B 46.1 1985, American Society of Mechanical Engineers, NY.

Surface Texture Symbols, ANSI Y14.36 1978, American Society of Mechanical Engineers, NY.

Wells, PW & Willey, PCT 1977, 'The effect of variation of dielectric flow rate in the gap on wear ratio and surface finish during electrodischarge machining', *Proceedings of ISEM-5*, pp. 110–117.

Wilkinson, BH & Warburton, P 1967, 'Electrochemical machining-machinability', *Iron and Steel Report*, vol. 94, pp. 215–220.

Yue, TM, Chen, TW, Man, HC & Lau, WS 1996, 'Analysis of ultrasonic aided laser drilling using finite element method', *CIRP*, vol. 45, no. 1, pp. 169–172.

10 Environment-Friendly Machine Tools and Operations

10.1 INTRODUCTION

In the early 1970s, public discussion of the consequences and measures necessary to conserve the environment was stimulated by citizens, action groups, and parliamentary movements. In the early 1980s, all parties had integrated environmental protection into their political programs. Ever since, environmental protection has ranked high in public opinion and its importance is growing constantly.

There is an increasing awareness among people in the industry toward the ecological aspects of the machining processes. The increasing sensitivity to environment and health issues is reflected in increasingly stringent legislation and national and international standards. The restrictions resulting from such legislation pose a challenge to scientists and engineers to develop new and alternative manufacturing technologies.

Despite various advantages achieved by the machining processes, they may generate solid, liquid, or gaseous by-products that present hazards for workers, machines, and the environment. The large-scale and long-term environmental threats created by the machining processes lead to direct environmental consequences. It is now appropriate to take direct and immediate action toward understanding the environmental hazards created by the machining processes and analyze their impacts on the environment, with the following suggestions as a guide:

1. Each company's environmental protection policy should keep hazards within acceptable limits through the following steps:
   - Effective management and reduction of emissions in air, water, and land
   - Compliance with relevant legislations covering manufacturing operations
   - Pollution prevention
   - Efficient use of energy
   - Minimization of consumption of natural resources
   - Minimization of waste streams
2. Ensure compliance with the environmental management system standard EN ISO 14001.
3. Set environmental objectives and targets.
4. Monitor and review the environmental performance against the set objectives and targets.

5. Raise staff awareness of the environmental implications of their work.
6. Provide the necessary instructions and resources that help in implementing the environmental policy.

Environmental impacts are measured by the degree of hazard, which is a function of the chemical/physical properties of the substance(s), and the quantities involved; the potential effects are classified under three categories:

- Class A—Major immediate environmental effect
- Class B—Intermediate environmental effect (may be serious but not immediate)
- Class C—Minor environmental effect

In considering a clean machining process, the interaction between economy, ecology, and technology has to be considered, as shown in Figure 10.1. Conflicts may arise among these three factors and a good compromise has to be made. To improve the quality of the machining processes, it is essential to adopt innovative methods that achieve the minimum environmental contamination in addition to their stability, reliability, and acceptable economic conditions. In this regard, the application of near net shape technology to manufacture parts with complex shapes by substituting cutting operations with forming activities provides advantages such as reduced chip volumes, lower cutting forces, reduced volumes of cutting fluids and cutting fluid losses, and simpler machine tools with lower power requirements.

As shown in Figure 10.2, one of the possibilities for minimizing environmental contamination is to modify and develop existing processes and replace conventional processes with alternative ones (rapid prototyping, laser machining, new cutting tool

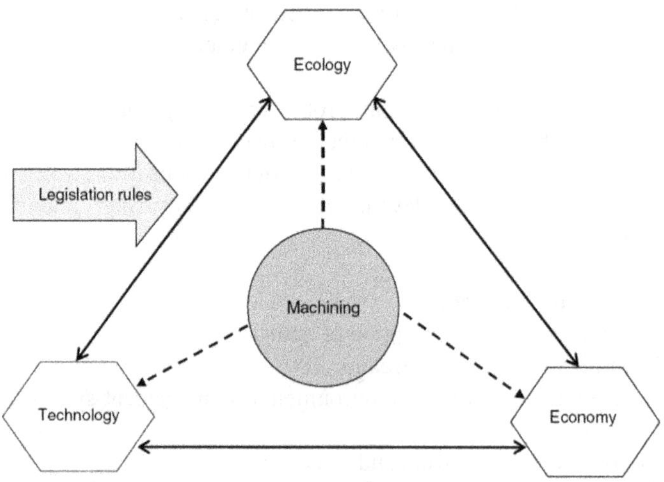

FIGURE 10.1 Interaction between technology, economy, and ecology for machining processes (Byrne and Scholta, 1993).

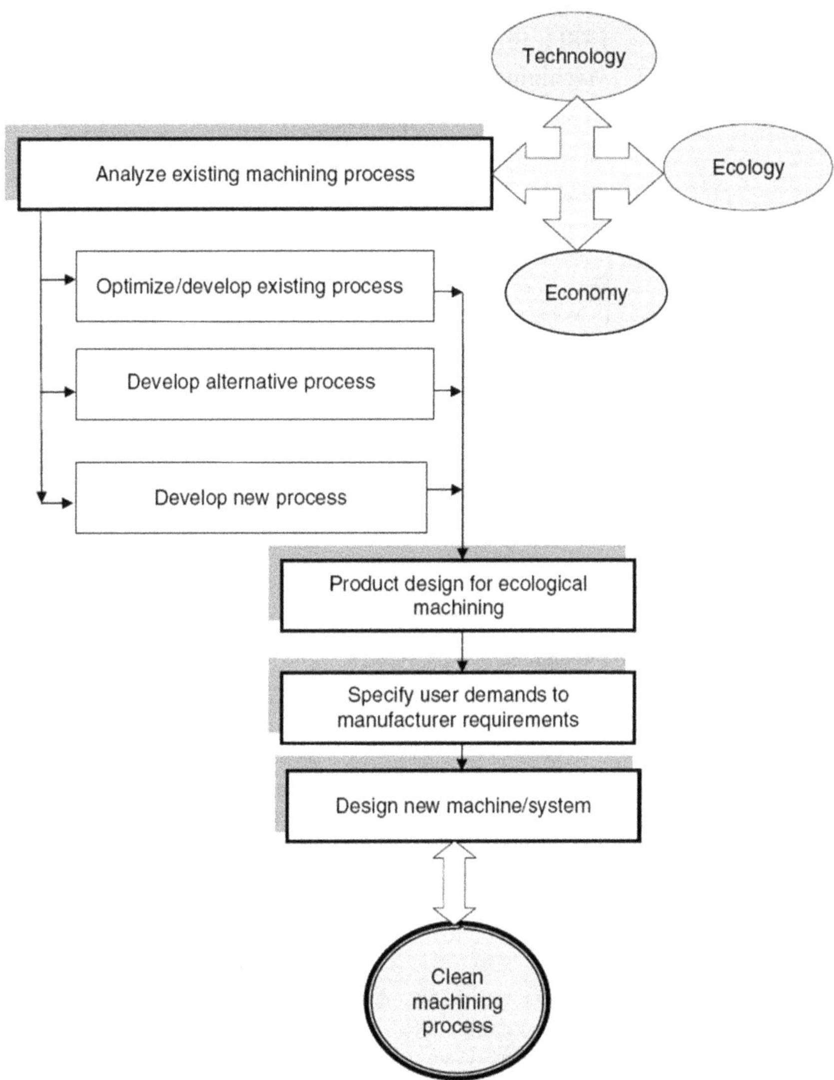

FIGURE 10.2 Achievement of clean machining process.

materials, and so on). It is essential that the new technologies do not lead to environmental hazards of a different nature.

Most machining processes use chemicals and liquids in the form of coolants, lubricants, etchants, electrolytes, abrasive slurry, dielectric liquids, gasses, and anti-corrosive additives as shown in Table 10.1. These chemicals can be transported by a variety of agents and in a variety of forms, which are defined by Hughes and Ferrett (2005) as shown in Table 10.2.

Hazardous substances cause ill health for people at work. These are classified according to their severity and type of hazard that they may present to the workers.

TABLE 10.1
Machining Liquids

Liquid/Medium	Application
Lubricants	Machine tools
Coolants	Cutting operations
Liquid nitrogen	Ecological machining
Etchants	CHM, PCM
Electrolytes	ECM
Dielectric liquids	EDM
Vegetable oil	EDM/lubricants/coolants
Deionized water	EDM wire cutting
Abrasive slurry	USM
Air + abrasives	AJM
Water + abrasives	AWJM

TABLE 10.2
Forms of Chemical Agents and Their Hazard Effect

Form	Description	Hazard Effect
Dust	Solid particles heavier than air suspended in it for some time; 0.4 μm (fine) to 10 μm (coarse).	Fine dust is hazardous because it is respirable, penetrates deep into the lungs and stays there, causing lung disease.
Gas	Substances at temperatures above their boiling point. Steam is a gaseous form of water. Common gasses include carbon monoxide and carbon dioxide.	Absorbed into blood stream.
Vapor	Substances very close to their boiling temperature.	If inhaled, it enters the blood stream causing short-term effects (dizziness) or long-term effects (brain damage).
Liquid	Substances at temperatures between freezing (solid) and boiling temperatures.	Irritation and skin burn.
Mist	Exist near boiling temperature but are closer to the liquid phase.	Produces similar effects to vapors where it penetrates the skin or ingested.
Fume	Collection of very small metallic particles (0.4–1.0 μm) that are respirable.	Lung damage.

Source: (Hughes and Ferrett, 2005).

The most common types are summarized in Table 10.3. The effect of these hazards may be acute or chronic:

Acute effects. These effects are of short duration and appear fairly rapidly, usually during or after a single or short-term exposure to a hazardous substance.

TABLE 10.3
Effects of Hazardous Substances

Effect	Description
Irritant	Non-corrosive substance that causes skin (dermatitis) or lung bronchial inflammation after repeated contact. Many chemicals used in machining processes are irritants.
Corrosive	Substances that attack by burning living tissues.
Harmful	Substances that if swallowed, inhaled, or absorbed by the skin may pose health risks.
Toxic	Substances that impede or prevent the function of one or more organs within the body, such as kidneys, lungs, and the heart. Lead, mercury, pesticides, and carbon monoxide are toxic substances.
Carcinogenic	Substances that promote abnormal development of body cells to become cancers. Asbestos, hard wood dust, creosote, and some mineral oils are carcinogenic.

Chronic effects. These effects develop over a period of time that may extend for many years. Chronic health effects are caused by prolonged or repeated exposures to hazardous substances resulting from the machining processes. Such effects may result in a gradual, latent, and often irreversible illness that may remain undiagnosed for many years. During that period, the individual may experience symptoms. Cancers and mental diseases fall into chronic category.

10.2 TRADITIONAL MACHINING

Machining by cutting is the process of removing the machining allowance from a WP in the form of chips. It requires a tool that is harder than that of the WP, a relative motion between the WP and the cutting tool, and it also requires penetration of the tool in the WP. One of the main machining drawbacks is its impact on environment in terms of noise, leakage, and flying chips. Heat and machining waste cause great hazards and carries a high risk of injuries and disorders. Therefore, safety precautions must be considered to reduce the negative effect of the machining process.

The main hazards created by metal cutting and abrasion processes are shown in Figures 10.3 and 10.4. These include the following:

a. *Noise.* During machining, vibration components of different frequencies that are numerous and not harmonically related to one another are generated and produce noise. Noise levels at 85 dB can damage hearing permanently. That is why the National Health and Medical Research Council has proposed 85 dB as the maximum noise level regarded as safe and tolerable for an eight-hour exposure. When noise levels exceed 90 dB in any work area, ear plugs must be worn. Many machines cause noise due to vibration. Figure 10.5 shows the expected level of noise from some machine tools such as the center lathe, milling, and grinding machines, and Figure 10.6 shows the typical allowable noise levels and their permissible safe exposure times.

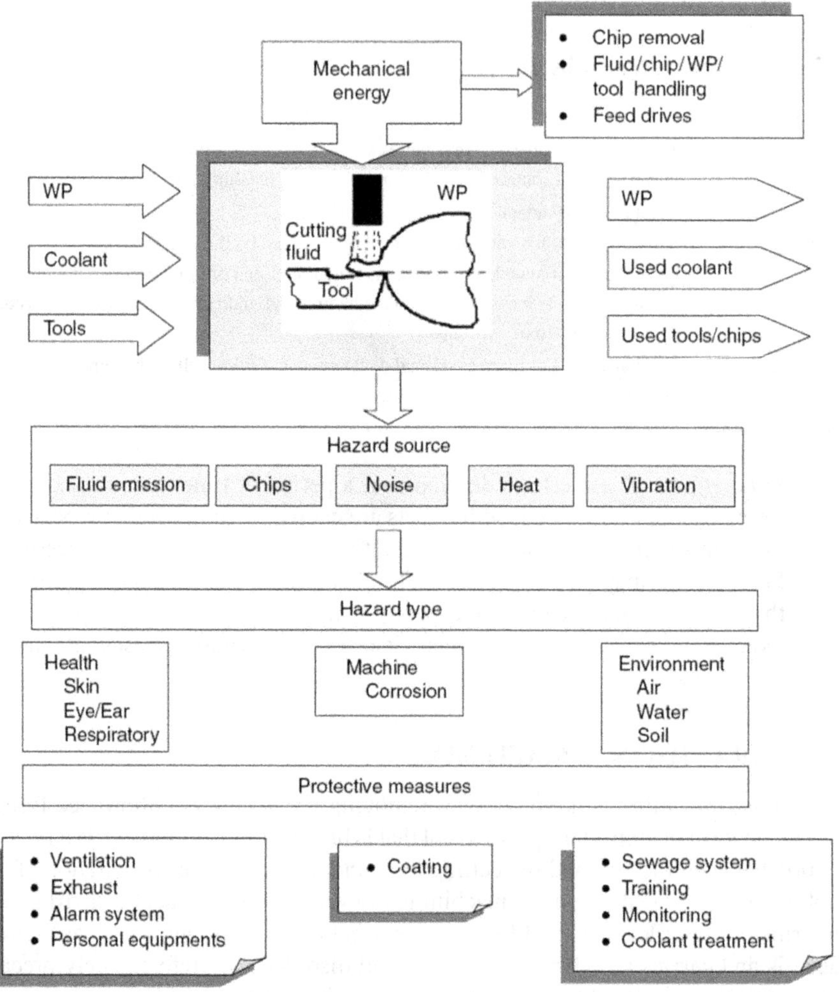

FIGURE 10.3 Environmental impacts of metal cutting processes.

b. *Flying chips.* All metal cutting processes produce chips. Flying chips are considered to be a major environmental impact of machining that causes great hazards. These chips have two main effects on the environment. The first is the danger and the risk they pose to the operator when flying from the machine during the cutting process. The second one is the dust and other flying particles, such as metal shavings, that may result in eye or skin injuries or irritation. Grinding, cutting, and drilling of metal and wood generate airborne particles. Under such circumstances, it is always recommended to wear safety glasses, goggles, or shields.
c. *Cutting tools.* From an environmental perspective, the most significant waste stream is mainly generated from the remaining portion of the tool that is disposed of after its useful life.

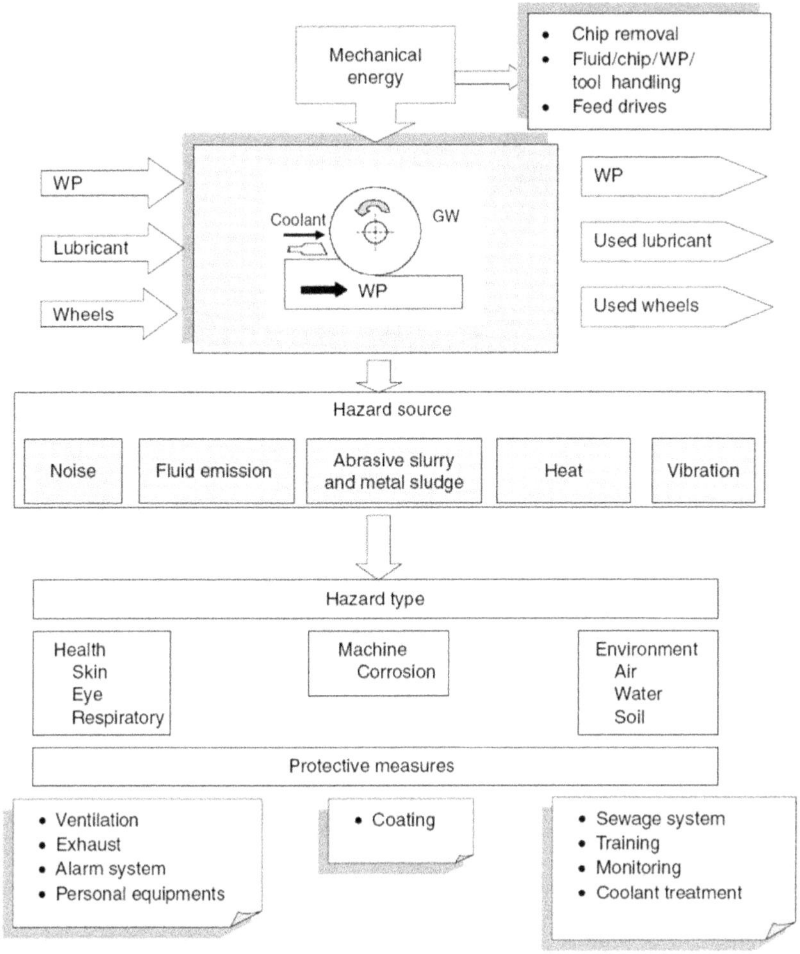

Mechanical energy
Chip removal
Fluid/chip/WP/ tool handling
Feed drives
WP
Lubricant
Wheels
Coolant
GW
WP
WP
Used lubricant
Used wheels
Hazard source
Noise
Fluid emission
Abrasive slurry and metal sludge
Heat
Vibration
Hazard type
Health
Skin
Eye
Respiratory
Machine
Corrosion
Environment
Air
Water
Soil
Protective measures
Ventilation
Exhaust
Alarm system
Personal equipments
Coating
Sewage system
Training
Monitoring
Coolant treatment

FIGURE 10.4 Environmental impacts of abrasive processes.

Noise level (dB)
200
150
100
50
0
62
85
85
100
105
115
140
Conversation
Turning
Milling
USM
Sawing
Grinding
Jet engine
Source of noise

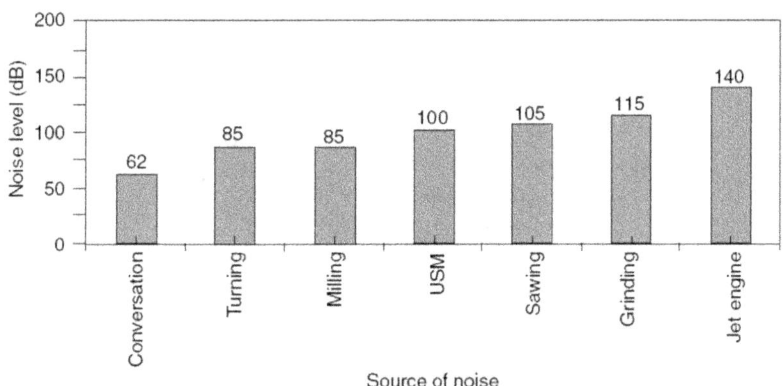

FIGURE 10.5 Noise levels for different machine tools.

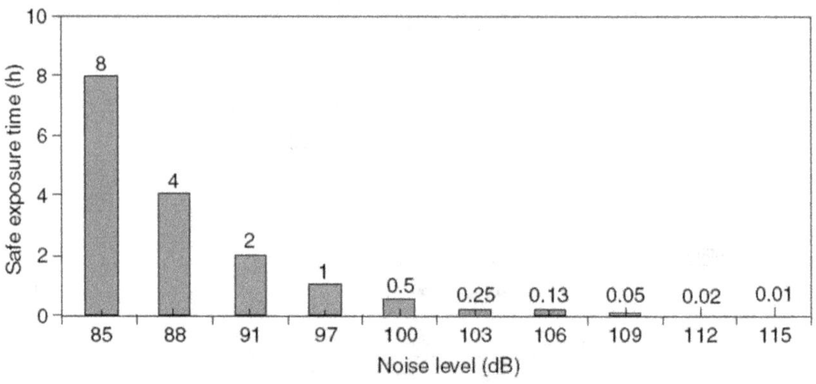

FIGURE 10.6 Safe exposure times for different noise levels.

Environmental impacts of machining by the cutting and abrasion processes can be minimized at the input stage by:

1. Using the minimum material through the near net shape technology
2. Using process parameters for optimum tool life
3. Avoiding dangerous substances and substances hazardous to water
4. Avoiding residuals and emissions (noise, heat, and vibrations)

10.2.1 Cutting Fluids

The main functions of cutting fluids are to:

1. Cool the cutting tool and the WP (cooling effect)
2. Reduce friction between the tool and WP (lubricating effect)
3. Remove the chips (flushing effect)
4. Provide a safe working environment (non-misting, non-toxic, non-flammable, and non-smoking)

10.2.1.1 Classification of Cutting Fluids

Practically, all cutting fluids currently in use fall into one of the following types:

1. *Air.* Compressed air can be used for cooling the machining zone using a pure air jet or air mixed with a fluid.
2. *Water-based cutting fluids.* Water, soluble oil, emulsions, and chemical solutions or synthetic fluids form the cutting fluid.
3. *Neat oils.* Mineral oils, fatty oils, composed oils, extreme pressure (EP) oils, and multipleuse oils.
4. *Liquid nitrogen.* Liquid nitrogen, having a temperature of −196°C, is used as a cutting fluid for difficult-to-cut materials. It is also used to cool WPs of tubular shapes or to cool the tool through internal channels that supply the nitrogen under pressure or by flooding the cutting area.

Cutting fluids contain many additives such as emulsifiers, antioxidants, bactericides, tensides, EP additives, corrosion inhibitors, or agents for preventing foaming. Cutting fluids account for 15% of the shop production cost. The costs of purchase, care, and disposal of the cutting fluids are more than twice as high as tool costs. The tooling cost may be increased by the use of dry cutting or due to the increased tool wear; however, the manufacturing cost may be reduced when compared to the conventional process in which cutting fluids are used.

Despite its benefits, the use of cutting fluids presents potential environmental problems. Cutting fluids are entrained by the chips and WPs and they contaminate machine tools, floor, and workers. Partly the fluid evaporates into the air and partly it also flows into the soil. Cutting fluid has a direct influence on machining economically and ecologically. The intensive contact of the production worker to the cutting fluid can lead to skin and respiratory diseases and there is increased danger of cancer. This is mainly caused by the constituents and additives of the cutting fluids as well as the reaction products and particles generated during the process.

10.2.1.2 Selection of Cutting Fluids

For selecting cutting fluids for environmentally clean manufacturing, the following points must be considered (Byrne and Scholta, 1993):

- The constituents of the cutting fluid must not have negative effects on the health of the machinist or on the environment.
- During use, cutting fluids should not produce contaminants.
- Multifunction oils that can be used for hydraulic systems, for slide way lubrication, and as a coolant and lubrication in machining should have minimum vaporization characteristics.
- Cooling and lubrication should be applied in a manner that minimizes the volume of the fluid used.
- Continuous monitoring of the cutting fluids and machine tool environment with online sensors is desirable.
- Through adequate care and maintenance of cutting fluids, the amount of water required for emulsions can be reduced, thus leading to cost savings.

From the point of view of environmental conservation and protection, complete elimination of the cutting fluids is desirable. The problem of cutting fluid disposal is an important issue in relation to environmental protection. Cleaning components is commonly time-consuming and cleaning agents have to be replaced by environment-friendly agents to reduce possible hazards to the environment and the employees. Through the use of dry machining, there is a complete elimination of hazards due to cutting fluids; the extent of cleaning can be significantly reduced with resulting economic and ecological benefits.

10.2.1.3 Evaluation of Cutting Fluids

The goal of machining is to make an acceptable part as quickly and as cost effectively as possible. Many factors affect the machining cost such as the machine tool, the tooling selected, and the cutting fluid utilized. The issues discussed in the following sections are to be considered in evaluating cutting fluids in machining processes.

1. *Lubricant Stream*

   The cutting fluid is assumed to diverge into four paths during the machining process, as shown in Figure 10.7. Unfortunately, spoiled or contaminated cutting fluids are the most common wastes from the machining process; these are considered hazardous wastes due to their oil content as well as other chemical additives they may contain. Therefore, it is essential to prevent or reduce this waste to avoid the cost of having to frequently replace and dispose of them. Contaminants of these cutting fluids may include the chips, machined parts, dust and other particulates, and moisture in non-aqueous solutions. However, the most troublesome contaminant is the tramp oils that interfere with the cooling fluid, promote bacterial growth, and contribute to unwanted residue on cutting tools. All of these factors contribute to premature fluid degradation and the need for replacement.

   During machining, high cutting speeds are used (>3500 m/min). Such a speed produces HT in the machining zone, which in turn results in high thermal stresses for the cutting tool. To achieve an acceptable tool life, cutting fluids must be used. Vaporization of the fluids and particles that are exposed to elevated temperatures takes place. The emissions entering the atmosphere thus represent a complex mixture of:

   - Vapors due to elements of the WP and cutting tool material
   - Vapors caused by deposits on the surface of the component being machined
   - Vapors resulting from cutting fluids

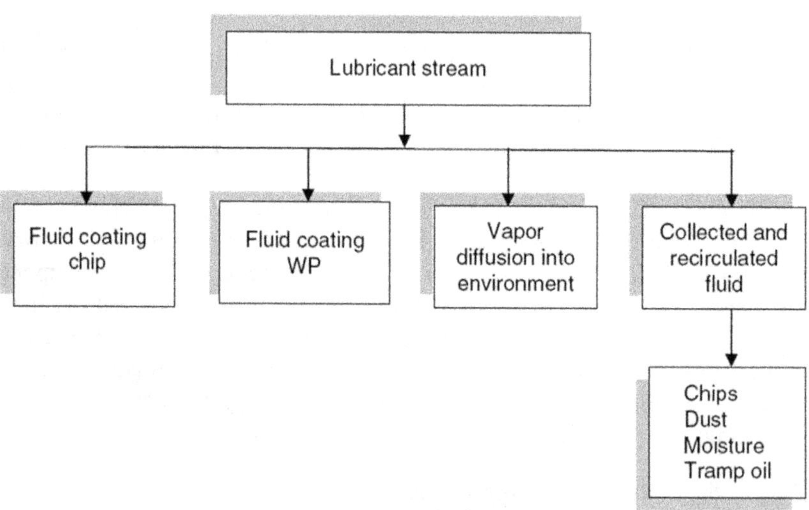

FIGURE 10.7 Lubricant streams.

Regulations under the U.S. Occupational Safety and Health Administration (OSHA), Environmental Protection Agency (EPA), and the Department of Transportation (DOT) are developed to provide the users of metal removal fluids with a user-friendly guide for assessing and minimizing the environmental health and safety (EHS) impacts in the selection, use, and disposal of metal removal fluids. Information on material safety data sheets (MSDS) and labeling of metal removal fluids, a matrix for rating metal removal fluid, and other issues that affect the use of metal removal fluids are also available.

2. *Disposal*
   Disposal is rapidly becoming the second most important factor in selecting new metal removal fluids. This importance is primarily due to the increasing cost and liability associated with these fluids. Whether onsite waste treatment and disposal or a pay-to-dispose method is used, the waste treater should provide details on product components that may adversely affect the waste treatment process, resulting in increased cost to the company.

10.2.2 Hazard Ranking of Cutting Fluids

The following are the main types of hazards that may occur by the cutting fluids:

1. Physical hazards (flammable, explosive, reactive, radioactive)
2. Health hazards (irritant, toxic, carcinogenic)

Two commonly used labeling formats with colors and numbers are supplied by the National Fire Protection Association (NFPA) system and the Hazardous Material Identification System (HMIS). Both the NFPA and HMIS systems use the same colors and numbers to identify the degree of hazard if the chemical is not handled properly. Each color on the label stands for a different type of hazard as follows: Blue = health hazard, red = fire hazard, yellow = reactivity hazard, and white = special hazard (NFPA) or protective equipment recommended (HMIS). The numbers from 0 to 4 rank the degree or severity of hazard if the chemical is not handled correctly: 0 = minimum hazard, 1 = slight hazard, 2 = moderate hazard, 3 = serious hazard, 4 = severe hazard.

10.2.3 Health Hazards of Cutting Fluids

Cutting fluids have negative health effects on the operators which appear as dermatological, respiratory, and pulmonary results. Occupational health hazards and skin problems are the most common for the metal or ceramic industries. Occupational dermatitis takes three forms:

1. Irritation contact dermatitis, which accounts for 50–80% of all cases. It is caused by exposure to fluids that damage the skin.
2. Allergic contact dermatitis, which accounts for 20–50% of all cases. This type of dermatitis is caused due to a worker's allergic intolerance to chemicals and is generally non-curable.

3. Exposure to mists caused by the cutting fluids raises a worker's susceptibility to respiratory problems. This depends on the level of chemicals and particles contained in generated mists.

Factors such as operator's hygiene, plant cleanliness, and air quality contribute to the likelihood of dermatitis among workers. Several factors leading to skin irritation include the following:

- The high pH of the cutting fluids (8.5) compared to 5.5–6 for the human skin
- Certain metals such as nickel, chromium, and cobalt
- Microbial contamination of cutting fluids
- The use of surfactants in cutting fluids

Deterioration of the cutting fluid through use causes adverse effects on process performance:

1. Contamination with small metal particles enhances tool wear and WP surface deterioration.
2. Bacterial growth, additive absorption, demulsification, and oxidation cause fluid failure.
3. Contamination with tramp oil and hydraulic oil provides breeding grounds for microorganisms.
4. Corrosion of the machine components, paint stripping, and foaming caused by improperly formulated cutting fluids.

Several cleaning methods are used to clean the used fluids, including centrifugal and hydrocyclene separators. In this regard, the vibratory-enhanced shear processing (VESP) membrane filter (Figure 10.8) can be used to filter and clean the cutting fluids at a temperature of 80°C, significantly higher than competitive membrane technology. The vibration amplitude and corresponding shear rate can also be varied,

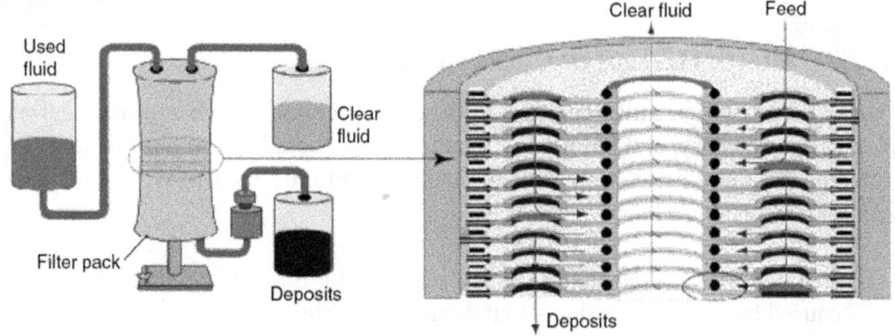

FIGURE 10.8 VESP membrane filter pack (New Logic Research, Inc., Emeryville, CA, www.vsep. com/solutions/technology.html).

which directly affects filtration rates. This system has many applications, such as coolant recycling and metal hydroxide recovery filtration.

10.2.4 Cryogenic Cooling

High-speed machining is characterized by the generation of high cutting temperatures. Such temperature levels adversely affect the tool life, dimensional and geometrical accuracy, and the surface integrity of the machined parts. The HT generated in the cutting zone has been controlled, conventionally by employing flood cooling. In high-speed machining, cutting fluids fail to penetrate the tool-chip interface and thus cannot remove the heat effectively. However, a high-pressure jet of soluble oil reduces the temperature at the tool-chip interface and improves the tool life to some extent.

The application of conventional cutting fluids causes several environmental problems:

a. Environmental pollution due to chemical dissociation of the cutting fluid at the high cutting temperature and thus formation of harmful gases and fumes
b. Biological problems to operators due to physical contact with cutting fluids
c. Water pollution and soil contamination during disposal
d. Requirement of extra floor space and additional systems for pumping, storage, filtration, recycling, chilling, and so on
e. The high cost of disposal of used coolants under tougher environmental laws

Coolant injection at high pressure has reduced the consumption of the cutting fluid by 50%, as well as the cutting temperature and the cutting forces. Cryogenic cooling by a liquid nitrogen jet has been used in the machining and grinding of steels. Under such circumstances, better surface integrity, lower cutting forces, and longer tool life have been achieved. The experimental results of Dhar et al. (2002) indicated the possibility of a substantial reduction in cutting forces by cryogenic cooling, which enabled a reduction in cutting forces and increased tool life due to reduced cutting temperature. Cryogenic cooling is therefore a potential environment-friendly clean technology for control of the cutting temperature.

10.2.5 Ecological Machining

Pending OSHA and EPA regulations in metal cutting fluids have made dry/ecological machining an important issue. Promising alternatives for the commonly used flood cooling (Figure 10.9) include:

1. Minimum quantity lubrication (MQL) in which a very low amount of fluid (<50 mL/h) is pulverized in the flow of compressed air during high-speed machining
2. Dry/ecological machining
3. Liquid gases like N_2

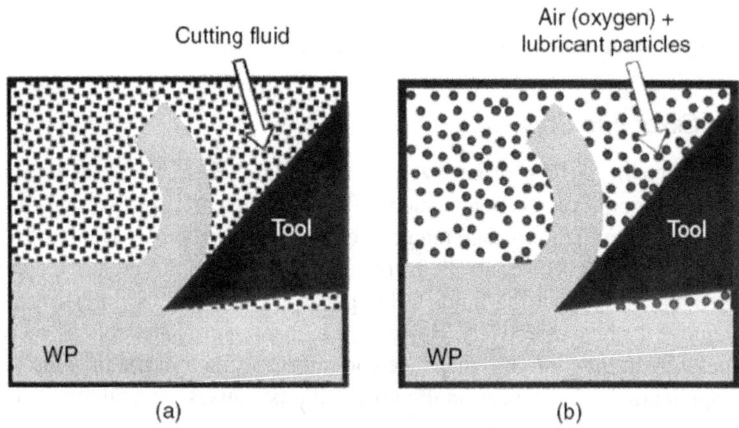

FIGURE 10.9 Conventional and MQL supply of cutting fluids: (a) Conventional and (b) MQL supply (Weinert, Inasaki, Sutherland, and Wakabyashi, 2004).

4. Super-hard tools such as CBN or diamond that do not require any coolant due to their outstanding wear resistance
5. Ionized air

Minimum quantity lubrication (MQL) is an alternative for normal flood cooling application. It uses a mist or a minimum quantity of neat oils or emulsions in the flow of compressed air, thus generating a spray that is directed to the cutting zone to serve as a lubricant and coolant. This technique is suitable for drilling and some grinding operations.

10.2.6 Factors Affecting the Use of MQL

Despite the advantages of using MQL, its wide acceptance in machining technology is affected by the following drawbacks:

1. *Environmental pollution.* Although this technique replaces the flood fluid method, it causes pollution because the pulverization of the oil in the air flow causes the suspension of a lot of oil particles in the air. This problem requires special encapsulation, protection guards, and good exhaust systems with particle control for the machine.
2. *Noise.* A line of compressed air must be used with MQL, which works intermittently during the machining process. These air lines produce a lot of noise, usually higher than the human ear can handle (>80 dB). It also makes communication between persons more difficult, which is also bad for the working environment.

MQL is considered to be an intermediary solution between the conventional use of cutting fluid and dry cutting. Figure 10.10 shows the different factors that affect dry machining. Dry machining benefits are shown in Figure 10.11. To enhance the

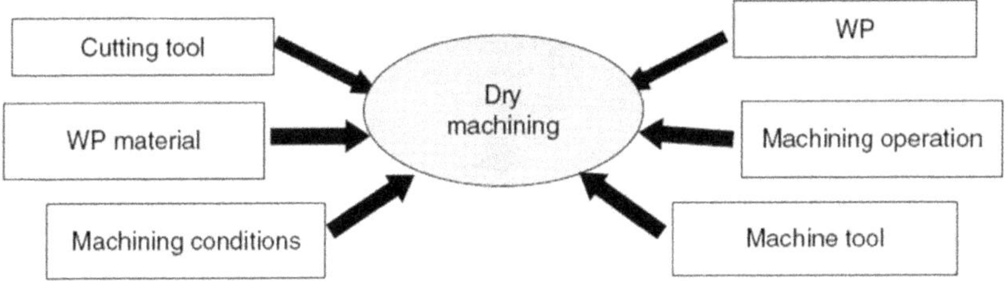

FIGURE 10.10 Factors that affect dry machining.

capabilities of dry cutting, multiple-layer coated-carbides, ceramics, and CBN are recommended. Such tool materials withstand the intensive heat and achieve satisfactory results without the use of cutting fluids. Properly applied coatings offer longer tool life at high cutting speeds and feeds. Recent developments in multilayer coatings on carbide cutting tools are important for dry cutting applications. Figure 10.12 shows the main requirements of cutting tool materials for dry cutting.

10.2.7 Applications of Ecological Machining

Figure 10.13 shows the influence of the machining process on the cooling liquid supply; Table 10.4 shows the different applications of dry and MQL machining. Dry machining applications include the following areas:

Turning. Dry cutting of hardened steel with super-hard cutting tools such as CBN or polycrystalline diamond (PCD) showed good results as an ecological alternative that reduces the operation time and hence improves

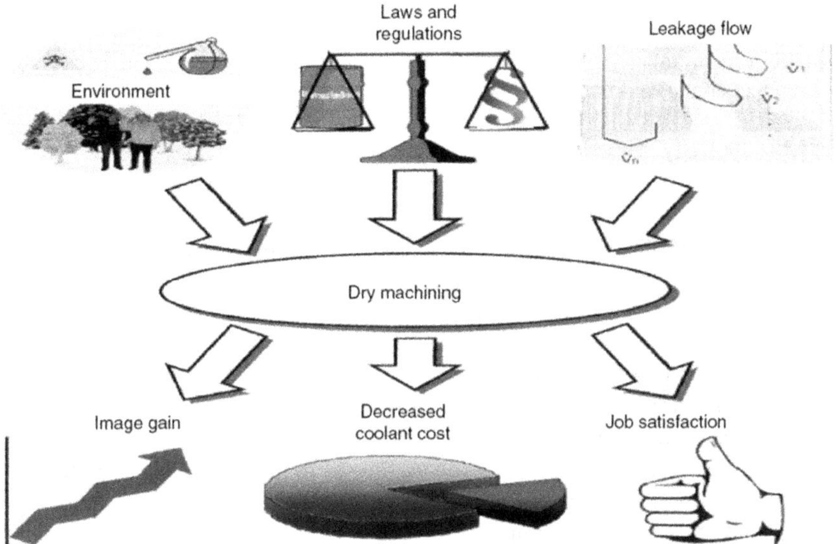

FIGURE 10.11 Benefits of dry machining (Weinert, Inasaki, Sutherland, and Wakabyashi, 2004).

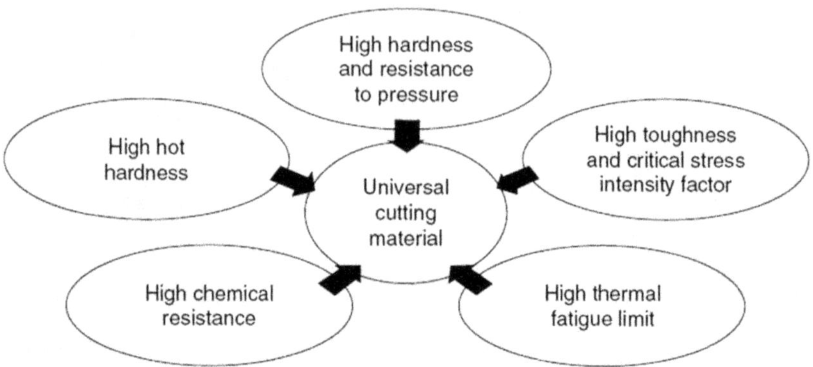

FIGURE 10.12 Optimal cutting tool materials for dry machining (Weinert, Inasaki, Sutherland, and Wakabyashi, 2004).

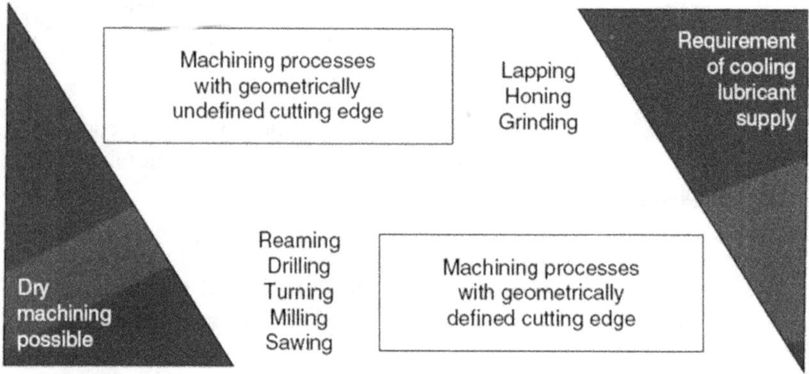

FIGURE 10.13 Influence of machining process on the cooling liquid supply (Weinert, Inasaki, Sutherland, and Wakabyashi, 2004).

productivity. The rationale is related to the tools' lower coefficient of friction, higher abrasion resistance, higher thermal conductivity, and better chemical and thermal stability, which make these materials suitable for high-speed machining applications. Moreover, in case of hardened steels, dry cutting with CBN tools provides a suitable cheaper substitution for grinding.

Drilling. Drilling without coolants or lubricants sticks the chip to the cutting tools and breaks them in a short time. MQL provides good lubrication for drilling using naturally dissolving oils such as vegetable oil or ester oil. It is impossible to perform drilling using dry cutting or using dry compressed air. MQL is normally used with an oil flow of 10 mL/h. However, the roughness, roundness, diameter accuracy, and cylindericity of drilled holes are identical to the flood technique of soluble oil.

Milling. For minimizing the negative effect on the environmental drawbacks when using ample amount of flood coolant, MQL using 8.5 mL/h is used

TABLE 10.4
Application Areas for Dry Machining

	Aluminum		Steel		CI
Material Process	**Cast Alloy**	**Wrought Alloys**	**High-Alloyed Bearing Steel**	**Free Cutting Quench and Tempered Steel**	**GG20 to GGG70**
Drilling	MQL	MQL	MQL	MQL/dry	MQL/dry
Reaming	MQL	MQL	MQL	MQL	MQL
Tapping	MQL	MQL	MQL	MQL	MQL
Thread forming	MQL	MQL	MQL	MQL	MQL
Deep-hole drilling	MQL	MQL		MQL	MQL
Milling	MQL/dry	MQL	Dry	Dry	Dry
Turning	MQL/dry	MQL/dry	Dry	Dry	Dry
Gear milling			Dry	Dry	Dry
Sawing	MQL	MQL	MQL	MQL	MQL
Broaching			MQL	Dry	Dry

Source: (Weinert, Inasaki, Sutherland, and Wakabyashi, 2004).

instead of the 42,000 mL/h used in the normal flood technique. It is clear that MQL achieves a drastic reduction (1/5000 times) in lubricant consumption. There is a considerable reduction in cutting force components for MQL when compared to both dry cutting and flood cooling techniques.

Hobbing. Conventional hobbing requires coolant, which complicates the process, and the disposal of which affects the environment. Dry hobbing is also possible using high-speed steel hobs coated by TiAlN, which makes the hob operate at a higher speed than is normally used in the wet cutting of gears.

Gear shaping. Cutting conditions for gear shaping allows dry machining with TiN, which doubles the tool life.

Grinding. The main functions of grinding fluids are cooling, lubrication, and flushing of chips and abrasives. Currently, soluble coolants containing chemical additives are used. These fluids are environmentally hazardous, and their disposal is expensive. Other methods of cooling in grinding include:

1. Cryogenic grinding using N_2 is uneconomical and is therefore not applicable.
2. Compressed cold air provides significant roughness due to lack of cooling at large depths.
3. Dry grinding induces high thermal wear and grinding forces (and is thus not recommended).
4. Chilled air can be used in precision grinding.
5. Cold air and oil mist reduces the grinding force and surface roughness, and therefore can be used for precision grinding.
6. Solid lubricants such as graphite, calcium fluoride, and barium fluoride are mixed in oil to form a paste that requires a flushing for this process.

Vegetable oils are considered potential candidates to substitute for conventional mineral oil-based lubricating oils and synthetic esters for the following reasons:

- Vegetable oils are non-toxic, have low volatility, pose no workplace hazards, and are biodegradable.
- The polar ester groups that are present in the vegetable-based lubricants are able to adhere to metal surfaces and therefore possess good boundary lubrication properties.
- Vegetable oils have high solubility power for polar contaminants and additive molecules.
- Vegetable oils have poor oxidation stability, so they are highly susceptible to radical attack.
- Vegetable oils maintain their environmental friendliness using additive packages or recently by chemical modification. The use of additives allows an increase in the overall performance of oil and improvement of its physical properties, but they may also raise the cost of lubricants and, in some cases, may even be harmful, as in the case of antioxidant lubricants commonly composed of phenolic and amine molecules.

The use of renewable materials benefits the environment by reducing greenhouse gases, and the use of natural resources improves the economic competitiveness of industry through the development of new markets and produces social benefits by stimulating rural communities.

Vegetable oils have been considered as potential substitutes for lubricant production when used in applications such as hydraulics or metalworking fluids, reporting benefits for the environment. The evaluation has been performed through a life cycle assessment approach of the comparative environmental impact of a traditional mineral dielectric fluid and a renewable resources-based one. From this analysis, a clear benefit has been obtained with regard to the safety of human health by using environment-friendly fluids.

10.3 NON-TRADITIONAL MACHINING PROCESSES

NTM provides alternatives for many machining by cutting operations. It is used for the production of complex profiles in hard-to-cut materials. Most lubricants and machining liquids (Table 10.1) come from petroleum, which is toxic to the environment and difficult to dispose of. That is the reason why the chemical industry is trying to increase the ecological friendliness of their products. NTM processes have many advantages; however, they generate solid, liquid, or gaseous by-products that present hazards for the workers, environment, and the equipment. The increasing awareness of environmental aspects has led the industry to become more sustainable and to adopt the use of environment-friendly products. In this section, the type and quantity of hazardous substances for the NTM methods are discussed.

10.3.1 Chemical Machining

CHM and PCM are important machining processes that have environmental impacts that must be minimized within a manufacturing company. CHM depends on chemical acids that have severe effects on the surrounding environment, difficulties in handling and storage, and damaging effects on different materials. An acid is defined as any substance that when dissolved in water dissociates to yield corrosive hydrogen ions. The acidity of an etchant dissolved in water is commonly measured in terms of pH number (defined as the negative logarithm of the concentration of hydrogen ions). Solutions with pH values of less than 7 are described as being acidic.

Acid deposition influences the environment by attacking structures that are made from steel and fading paint on machine tools. Chemical acids also cause air pollution, which causes significant corrosion of metals. There are many acids involved in CHM, such as:

$FeCl_3$	Ferric chloride
NaOH	Sodium hydroxide
HNO_3	Nitric acid
Hf	Hafnium
$CuCl_2$	Cupric chloride
HCl	Hydrochloric acid

The majority of PCM is carried out with aqueous solutions of $FeCl_3$ used at temperatures over 50°C. $FeCl_3$ is acidic, relatively cheap, and readily available; it is also versatile, as it attacks the majority of commonly used engineering metals and alloys. Environmentally, it is attractive as it is of low toxicity and relatively easy to filter, replenish, and recycle. However, the spent etchant and its rinse water contain heavy metal ions such as nickel and chromium, which are hazardous to the environment. Methods that can be employed to reduce the overall consumption of ferric chloride include prolonging the life of the etchant before disposal, with the drawback of reduced etch rate, and regenerating the spent etchant by *in situ* oxidation, thus maintaining a constant etch rate.

During CHM, exposure to Hf can occur through inhalation, ingestion, and eye or skin contact. Overexposure to Hf and its compounds may cause mild irritation of the eyes, skin, and mucous membranes. Figure 10.14 shows the typical impacts of CHM, which include the following:

1. *Labor.* CHM causes health effects on labor, which include the following points:
   a. Irritation causing inflammation and chemical burns.
   b. Corrosive injuries and burns due to heat as acids and alkaline come into contact with water.
   c. Rapid, severe, and often irreversible damage of the eyes.
   d. Acid fumes may also corrode the teeth.
   e. Direct contact of many organic anhydrides with skin, mucous membranes, eyes, or the respiratory system causes irritation and sensitization.
   f. The exposure may also increase the risk of cancer.
   g. Inhaling mists of inorganic acids containing sulfuric acid involves an elevated risk of larynx and lung cancer of occupational origin.

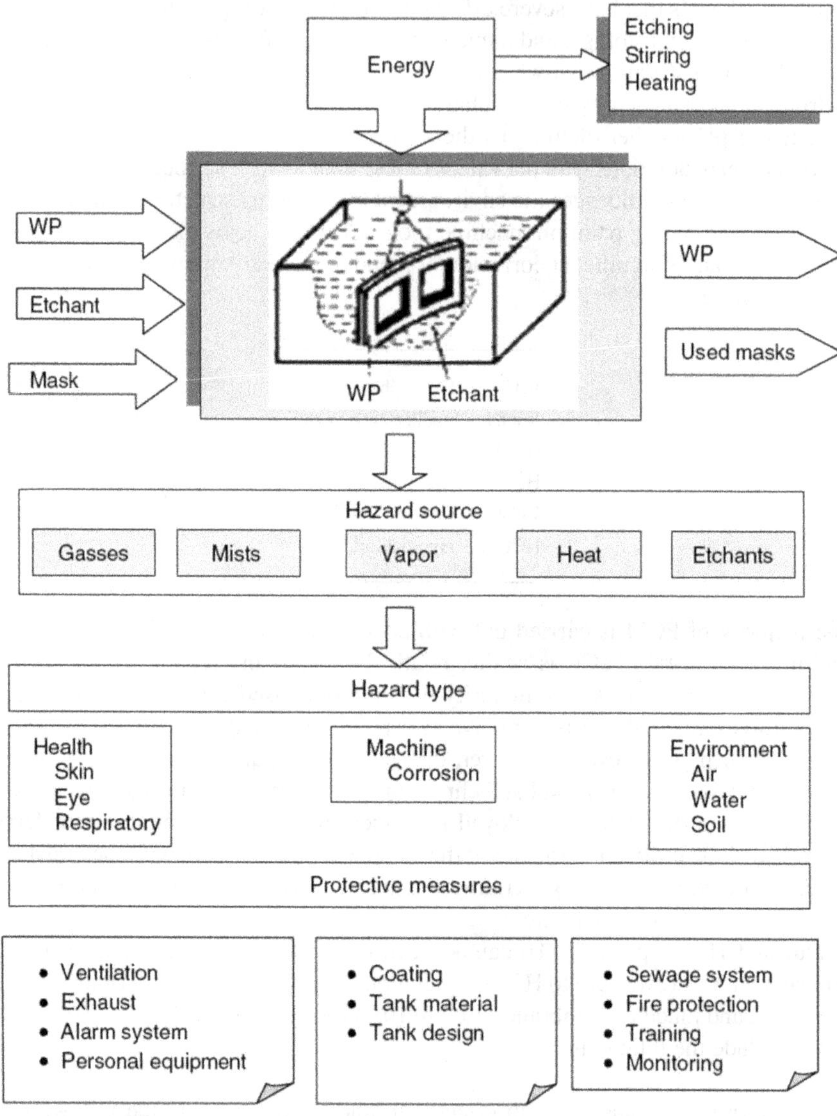

FIGURE 10.14 CHM environmental impact.

The effect of CHM hazards depends on the properties of the substance, the concentration, and time of contact with acids and alkalis. Although a dilute solution may cause irritation, the effect of strong acids and alkalis is experienced within moments of exposure. Depending on the substance and the concentration, the hazard effect may also be delayed.

2. *Machine.* Corrosion is the aging of unprotected materials that are formed from the chemical effects of a gas or liquid. Many metals are often affected by corrosion of many types, which include localized attack and uniform

corrosion. A localized attack affects small areas of metal and creates small holes or cracks. Uniform corrosion affects larger pieces of metal ranging from one foot in diameter to a much larger surface. Rust is probably the most common form of corrosion, which forms mainly on the surface area of steel when steel is exposed to moisture. Rust does not form quickly, and it takes a long time before a sign of rust appears on the metal.

3. Traditionally, covering materials (e.g. paints) have been used to protect iron and steel from rusting. However, the protection provided by covering materials is only as important as the preparation of the materials to be coated. Poor surface preparation is the prime cause of protective coating failure. In this regard, if any rust is left under the coating, it continues to grow both into the metal and eventually through the coating.
4. *Environment.* Improper disposal of CHM etchants changes the level of acidity and alkalinity which affect the flora and fauna in soil and water. The change of pH from 7 (neutral water) has an adverse effect on aquatic life. At pH 6, crustaceans and mollusks start to disappear and moss increases. At pH 5.5, some fish such as salmon, trout, and whitefish start to die, and salamander eggs fail to hatch. Acidity of pH 4 has a lethal effect on crickets and frogs. Some alkalis such as ammonia also have an acute toxic effect on fish.
5. *Soil.* Soil is classified as contaminated when due to acidity, it has a pH value of 4–5 and heavily contaminated when the pH is 2–4. When soil has a pH value of 9–10, it is contaminated due to alkalinity, and at pH 10–12, it may be classified as heavily contaminated.
6. Aerosols of solid (nitrogen and sulfuric oxides) or liquid corrosive substances are air pollutants, and so are corrosive gases. These gases may combine with water to form acids that precipitate with rain. Acid gases and acid fumes damage plants. Effects may be specific; for example, acetic acid fumes harm trees with leaves. Neutralization does not always remove the hazards to the environment, as the salts produced in this reaction may also be harmful.

10.3.2 Electrochemical Machining

ECM is known from the past as an environmental polluting process; however, with the development in the treatment of electrolyte, this process is currently less environmentally polluting. By realizing a closed electrolyte treatment system, the disposals in the sewage system are shut off. The sources of hazards in ECM are shown in Figure 10.15 and include those caused by gases, electrolyte, mist, and vapor as follows:

a. *Hydrogen gas.* There are difficulties in safely removing and disposing of the explosive hydrogen gas generated during the electrolyzing process. At the cathode, the reaction is likely to be the generation of hydrogen gas and the production of hydroxyl ions:

$$H_2O + 2e^- \rightarrow H_2 + 2OH^-$$

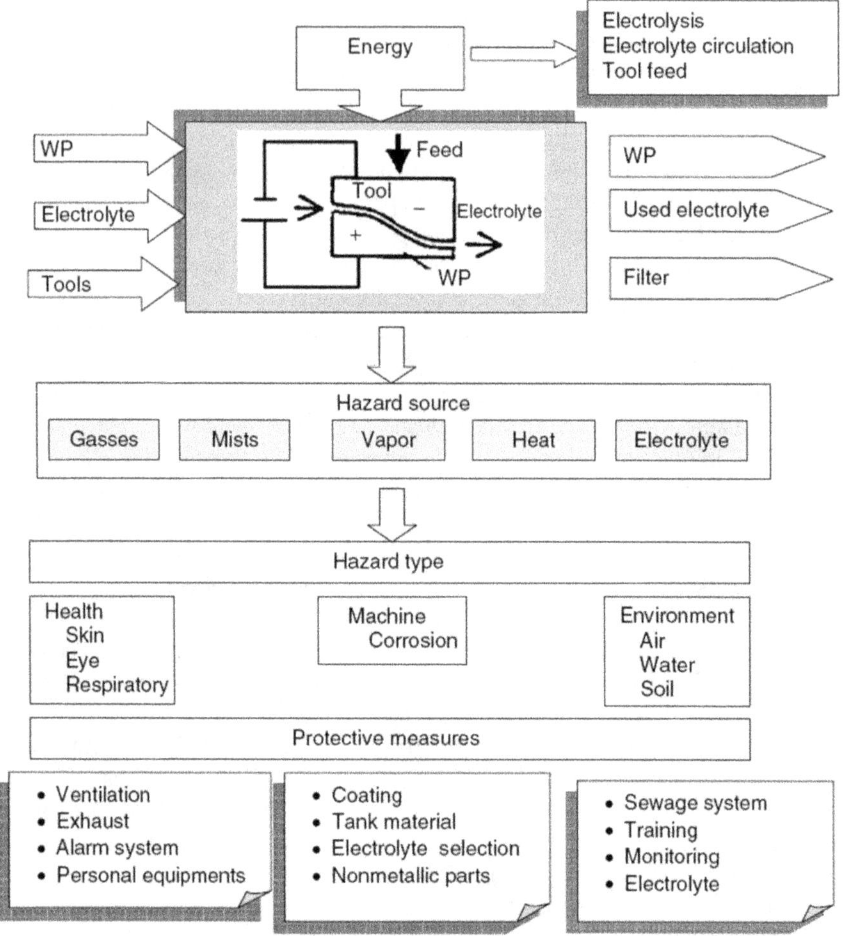

FIGURE 10.15 ECM impacts.

The net reaction is thus

$$Fe + 2H_2O \rightarrow Fe(OH)_2(s) + H_2$$

Local exhaust must be provided to prevent the hydrogen gas from reaching its lower flammability limit and remove the mists from the workers breathing zone.

b. *Electrolyte splash.* Electrolytes are pumped through the gap between the WP and the tool. As the dc flows, metal ions are removed from the WP and swept away by the electrolytes. Electrolytes are aqueous solutions of sodium chloride, sodium nitrate, and other salts that become insoluble hydroxides that deposit out of the solution as a sludge. Skin contact with the electrolyte contents must be controlled by good work practices. The door/cover shall be interlocked with the electrolyte supply system where there is electrolyte splashing. Machining process should be stopped or not started when the interlock of the safeguarding device is open.

c. *Chromate*. Exposure to chromium III compounds occurs through inhalation, ingestion, and eye or skin contact. Chromium III compounds can affect the skin, liver, and kidneys in humans. Dermal contact with trivalent chromium compounds has resulted in contact dermatitis.
d. *Nitrate*. Nitrates and nitrites are known to cause several health effects, including the following:
    - Reactions with hemoglobin in blood, decreasing the oxygen-carrying capacity of the blood (nitrite)
    - Decreased functioning of the thyroid gland (nitrate)
    - Shortages of vitamin A (nitrate)
    - Fashioning of nitrosamines, which are known as one of the most common causes of cancer (nitrates and nitrites)

Methods that are effective in controlling worker exposure to nitrogen and chromium III, depending on the feasibility of implementation, are as follows:

- Process enclosure
- Local exhaust ventilation
- General dilution ventilation
- Personal protective equipment

10.3.3 Electrodischarge Machining

EDM is one of the most important machining processes in the die and mold industry. Materials of high hardness and strength such as hardened steel, WC, or conductive ceramics can be machined. During EDM, the work material is removed by a series of sparks that occur in the dielectric liquid filling the gap between the tool-electrode and WP. The main functions of the dielectric fluids are:

1. Concentrating the discharge channel, which increases the energy intensity
2. Carrying removed material out of the gap
3. Filtering the gasses and liquid phases when expelled from the gap

EDM has several hazard potentials, which are described in Figure 10.16:

- The HT in the working gap generates a hazardous smoke, vapors, and aerosols.
- Decomposition products and heavy metals accumulate in the dielectric and erosion slurry.
- Hydrocarbon dielectrics have a negative effect on the skin.
- Sharp-edged resolidified metallic particles may damage the skin.
- Possible fire hazard.
- Explosions may occur under unfavorable circumstances.
- Electromagnetic radiation causes negative health impacts.

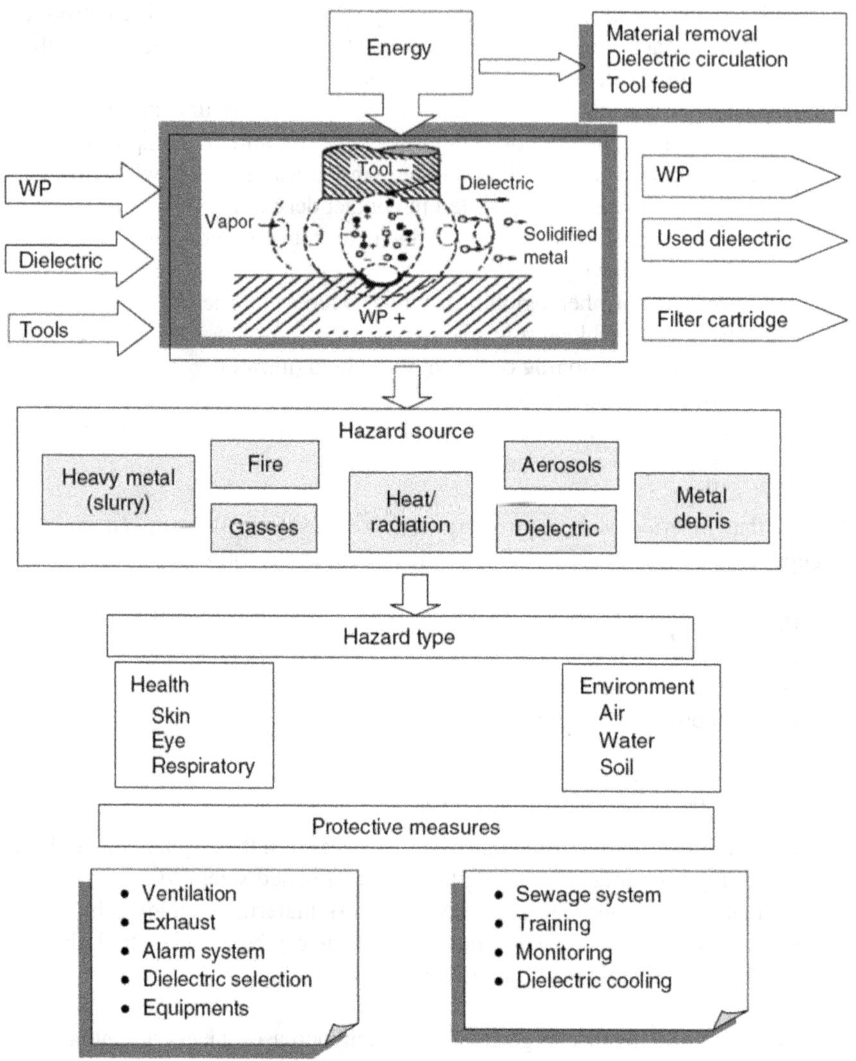

FIGURE 10.16 EDM impacts.

In EDM, the total aerosols and vapor concentrations exceed the limits of 5 mg/m^3 if no protective measures are taken. Fumes, vapors, and aerosols depend on the material removal principle (sinking, wire cutting, roughing, finishing), the dielectric, and the work material. In this regard:

1. Rough machining conditions (die sinking) generate more fumes and aerosols than finish cut by wire EDM.
2. The material composition is of interest when they contain toxic or health attacking substances such as nickel.

3. The type of the dielectric, its composition, and viscosity have high influence on the fumes and vapor developed under the HTs of the plasma channel created in the working gap.
4. Lower viscosity produces less fumes and vapors.
5. The level of the dielectric has to be higher than 40 mm over the erosion spot to condense and absorb a considerable part of the vapor and fumes in the dielectric itself (80 mm is recommended).

When using mineral oils or organic dielectric fluids for EDM die sinking, the following hazardous fumes are generated:

- Polycyclic aromatic hydrocarbons (PAH)
- Benzene
- Vapor of mineral oil
- Mineral aerosols
- Products generated by dissociation of oil and its additives

For hydrocarbon dielectrics, all of these vapors and aerosols appear except for PAH and benzene.

For water-based solvents, normally used in wire EDM, the following hazardous materials are formed:

- Carbon monoxide
- Nitrous oxide
- Ozone
- Harmful aerosols

Improvements in refining processes have led to more acceptable mineral dielectric oils. Research in the field of dielectric fluids includes the use of water-based dielectrics, gas as dielectric, and sometimes solid dielectrics. Although there are synthetic fluids in the market, the highly refined mineral dielectrics are still widely used. That contributes to the fact that 40,000 tons of mineral lubricants are lost every year all around the world, causing important environmental problems and leading to the need for alternative products with minimal environmental impacts.

There is an increasing demand for exploring methods that reduce or eliminate the adverse effect of the working fluid (dielectric in EDM) in the machining industry. One of the greatest challenges for current and future manufacturing industries is to reduce production waste and minimize the related environmental impact to near zero. In die-sinking EDM, electrodischarge grinding (EDG), and abrasive electrodischarge grinding (AEDG), hydrocarbon oils are used as dielectric fluid to constrict the discharge channel, cool the erosion zone, and flush away debris from the interelectrode gap. The dielectric oil has a long operating life, but the debris collected during machining needs an appropriate disposal.

Owing to sparking, metals of the WP and tool electrodes, inorganic substances such as WC, titanium carbide, chromium, nickel, molybdenum, and barium are released and condense in the air. Emissions of organic materials are generated by the vaporization of the dielectrics. Additionally, the rising smoke can carry organic components from substances in the dielectric liquid. The erosion slurry contains eroded WP and tool material and solid decomposition products of the dielectric. This slurry has to be filtered in the dielectric system of the EDM machine.

10.3.3.1 Protective Measures

To reduce the possible hazards that may arise due to machining by EDM, the following measures should be strictly followed:

1. Reduce air pollution to the extent permissible using suitable filters.
2. Incorporate a dielectric cleaning and recirculating system as a part of the EDM machine.
3. Keep the temperature of the media at a constant level of 15°C below flashing point using a proper cooling system.
4. Reduce the emitted electromagnetic radiation by proper shielding of the machine.
5. Reduce the possibility of fire hazard with an automatic fire extinguisher system.
6. Use level sensors for the dielectric level to be above the spark gap by 80 mm.
7. Avoid dielectrics with flashing point of 65°C. A flashing point above 100°C is recommended.
8. Dispose of the waste appropriately.
9. EDM uses very high-voltage electricity. The operator needs to make every effort to be constantly aware of the machine and its surroundings. Coming into contact with any part of the fluid or the electrode can cause severe injury or even death.

10.3.4 Laser Beam Machining

In LBM, a highly collimated, monochromatic, and coherent light beam is generated and focused to a small spot. High power densities (10^6 W/mm^2) are then obtained. The unreflected light is absorbed, thus heating the surface of the specimen. Additionally, heat diffusion into the bulk material causes phase change, melting, and vaporization. Depending on the power density and the time of beam interaction, the mechanism progresses from one of heat absorption and conduction to one of melting and then vaporization. Machining by laser occurs when the power density of the beam is greater than what is lost by the conduction, convection, and the radiation. Moreover, the radiation must penetrate and be absorbed into the WP material. LBM has many impacts related to:

a. *Environment*. Global (greenhouse effect, ozone layer), regional (acidification, entrophication, toxicity), and local impacts (acute toxicity, odor, and noise) occur.
b. *Occupational health*. Chemophysical risks for the workers (process, gas, fumes, laser machine emissions), light emission risks (laser light, secondary light emission), and noise arise due to LBM.

Figure 10.17 shows the possible process impacts on the environment. During machining by lasers, the material is heated, partly vaporized, and chemically transformed.

The hazardous materials have the consistency of gasses or aerosols. Aerosols are solid or liquid substances in the gas with particle sizes in the range of 10–16 nm. They can be characterized by the chemical content of the particles, the size distribution, and the material flow rate. During laser machining of thermoplastics, Tönshoff et al. (1996) reported that 99% of the particles generated have a diameter less than 10 μm and more than 90% are smaller than 1 μm. Most particles are in the range of 0.03–0.5 μm. High alloyed steels emit four to five times more aerosols than carbon steel does. Almost all particles generated by LBM have a diameter in the range 0.042–0.35 μm.

The permissible concentrations are quickly reached, especially when the laser machining process generates toxic or dangerous emissions. Consequently, working

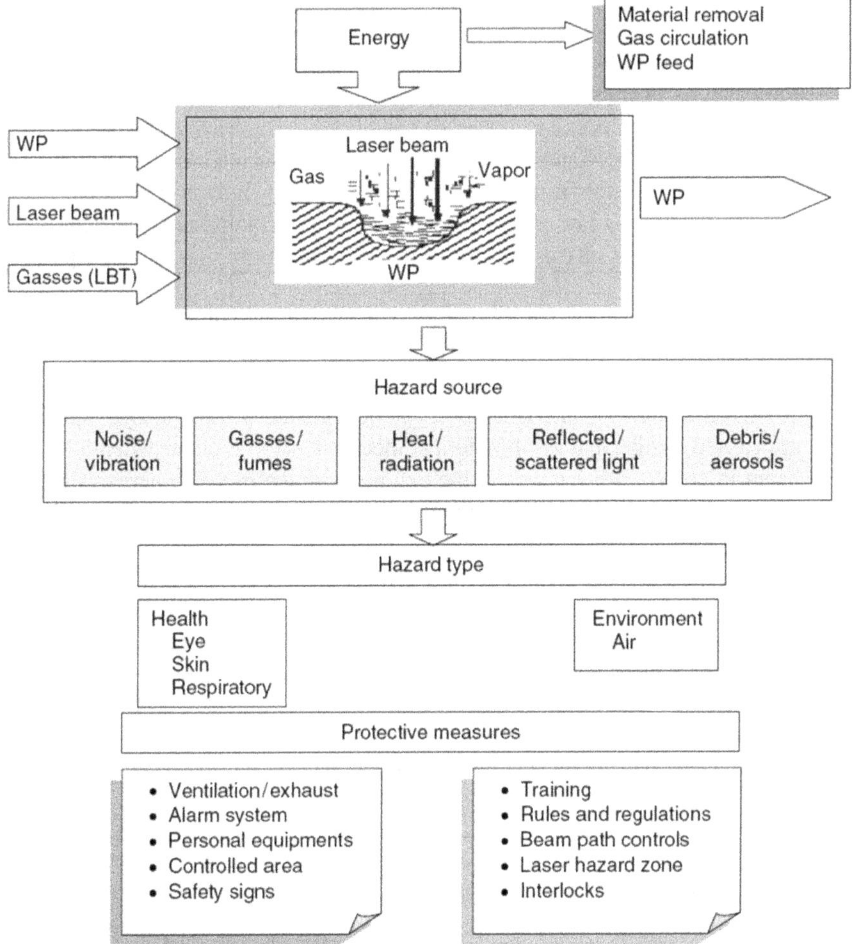

FIGURE 10.17 LBM hazards.

enclosures and exhaust systems have to be applied. A suitable method to extract the hazardous by-products in addition to filtration must be also considered.

Safety aspects are concerned with impacts on the human body; that is, the skin, and especially the eyes. The hazard potential is influenced by power density/intensity, wavelength, exposure time, and whether the beam is visible or invisible.

Laser beam machining causes several hazards that can be summarized as follows:

Beam hazard. The maximum permissible exposure (MPE) to laser radiation ranges between 400 and 1400 nm. Under the worst conditions, the laser radiation can be focused by lens on the retina in a small spot of 10–20 μm. This corresponds to an increase of 0.5×10^6 W/mm^2 in the power density, which forms a serious danger to the retina, especially in the range of wavelength between 400 and 1400 nm. Such a range encloses the wavelength of a Nd:YAG laser (1060 nm), which therefore requires the most attention when protective measures have to be designed. For Nd:YAG lasers, protective measures are compulsory, although their photon is normally lower than that of CO_2 lasers, and the reflection of metals for the wavelength of 1060 nm is normally lower. Nevertheless, because of the properties of the human eye, the MPE is so low that protection is necessary under all conditions. Excimer lasers work in the UV range of 157–353 nm. The biological effect of such a radiation accumulates over the exposure time, as far as UV type B, or C light is concerned.

Skin hazards. Repeated or even a single exposure to certain laser wavelengths can cause damage of varying degrees to the skin more than to other parts of the body.

Eye hazards. Eye hazards are significant when some laser types are used. The first and perhaps the most important factor in determining a laser's eye hazard potential is the wavelength, which determines which part of the eye absorbs the radiation and whether it can be focused by the eye. Eye injuries are caused by thermal or photoletical mechanisms that occur when a laser beam interacts with the eye. If the beam enters the eye, its energy is concentrated by the lens of the eye about 100,000 times at the retina. Therefore, even a small amount of laser light can cause eye damage.

Exposure to a laser beam occurs directly when a person is in the direct path of the laser beam. Indirect exposure occurs when a beam is scattered before it reaches the eye or skin. The material scattering the laser energy may be a rough, non-reflective surface, such as a break wall, or small, airborne particles, such as dust and water vapor. During indirect exposure, the beam energy dissipates rapidly away from the material that caused the beam scatter. LBM protective measures are as follows:

1. *Regulations.* The principal standard for industry is ANSI Z136.1, which provides requirements and recommendations for the safe use of lasers in typical industrial and research environments.
2. *Laser hazard zones.* These are achieved by determining the nominal hazard zone (NHZ), defined as the space within which the level of direct, reflected, or scattered radiation exceeds the level of the applicable MPE.

3. *Beam path controls.* Most industrial lasers fall into the higher classifications of a source of potential eye, skin, and even fire hazards, unless they are totally enclosed, interlocked, and there is no beam access during normal system operation.
4. *Controlled area.* When the beam path is not sufficiently enclosed, the exposure to radiation above the MPE limit occurs. Hence, a laser-controlled area is required. During periods of service, a temporary controlled area may be established.
5. *Engineering controls.* Engineering controls are features designed into the laser machine to minimize the risk of exposure to hazardous beams. The most common engineering controls are:
   a. Protective housings and enclosures that cover the equipment or the beam path
   b. Interlocks are often placed on the protective housings so that if they are removed, the beam is shut off
   c. Beam stops that provide safe termination of the beam path
   d. Labels and signs that give notice of lasers operating in a given area
6. *Protective equipment.* Protective equipment such as barriers or curtains, clothing, or eyewear should be relied upon only if the other control measures do not provide adequate protection.
7. *Administrative and procedural controls.* Administrative and procedural controls consist of a series of rules and regulations that are designed to minimize the risk of laser beam exposure. One of the most effective administrative controls is training.

10.3.5 Ultrasonic Machining

USM is the removal of hard and brittle materials using an axially oscillating tool at ultrasonic frequency (18–20 kHz). During that oscillation, the abrasive slurry of B_4C, Al_2O_3, or SiC (100–800 grit) is continuously fed into the machining zone, between a soft tool (brass or steel) and the WP. The abrasive particles are therefore hammered into the WP surface and cause chipping of fine particles from it. The oscillating tool, at amplitude ranging from 10 to 40 μm, imposes a static pressure on the abrasive grains and feeds down as the material is removed to form the required tool shape (Figure 10.18).

The abrasive slurry is circulated between the oscillating tool and WP through a nozzle close to the tool/WP interface at an approximate rate of 25 L/min. The process finds many industrial applications when machining hard and brittle materials such as ceramics, glass, and carbides. However, it has many environmental and health hazards that include the electromagnetic field (EMF), ultrasonic waves, and abrasives slurry.

10.3.5.1 Electromagnetic Field

In spite of the absence of valid evidence based on solid scientific research, the effects of the EMF on the health of individuals and environments is still of concern to some people. Until research data suggests a need for more extreme action, individuals are advised

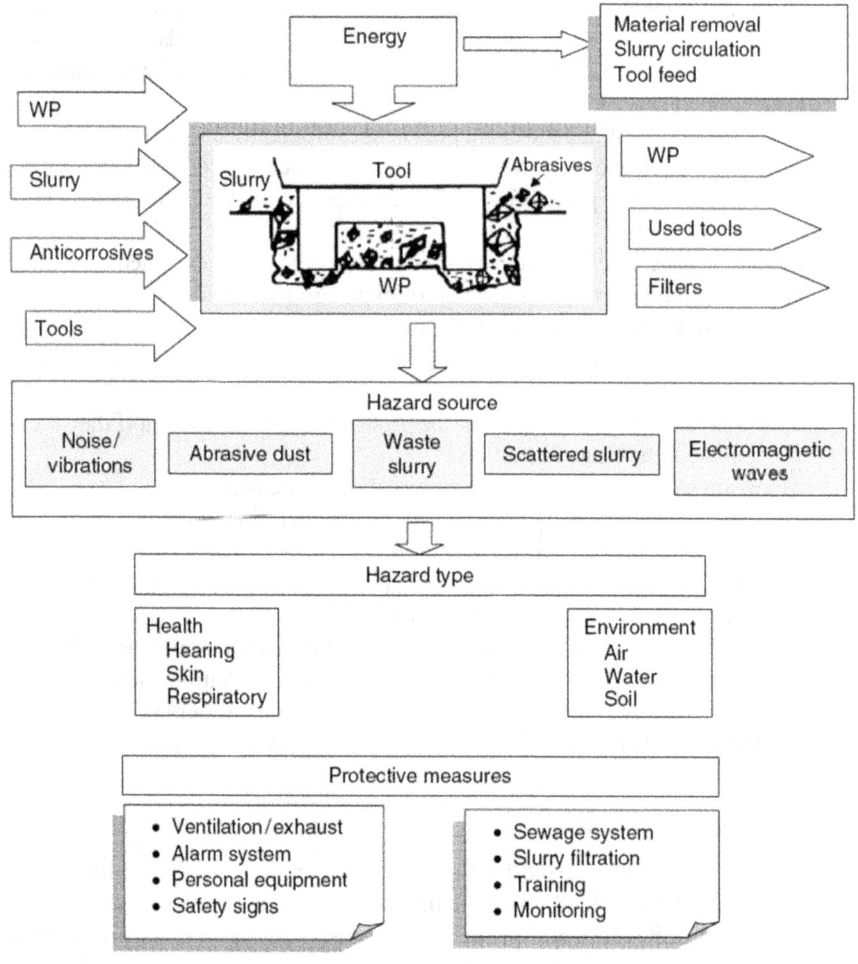

FIGURE 10.18 USM hazards.

to stay away from the EMF sources. The strength of a magnetic field drops quickly just few feet from the source. Additionally, it is not recommended to sleep or sit for long periods of time near electrical devices, especially ultrasonic generators and motors.

10.3.5.2 Ultrasonic Waves

Ultrasonic frequencies have been found to produce sound in the audible range from 96 to 105 dB, although it may not appear noisy to older persons and those with imperfect hearing. The World Health Organization (WHO) has proposed the following maximum exposure levels for unprotected persons: 75 dB for frequencies from 14.1 kHz, and 110 dB for frequencies greater than 22.5 kHz. Excessive noise levels should be reduced usually by enclosures and thin layers of common material that give adequate acoustic insulation at upper audible (and ultrasonic) frequencies. Alternatively, ear protectors can be provided and used. Ultrasound is HF sound that is inaudible, or cannot be heard, by the human ear. However, it may still affect hearing and produce other health effects. Factors to consider regarding ultrasonic waves include:

- The upper frequency of audibility of the human ear is approximately 15–20 kHz.
- Some individuals may have higher or lower (usually lower) limits.
- The frequency limit normally declines with age.

Noise-induced hearing loss is one of the most common occupational illnesses; it is often ignored because there are no visible effects. Hearing loss usually develops over a long period of time, and, except in very rare cases, there is no pain. Actually, a progressive loss of communication, socialization, and responsiveness to the environment occur. In its early stages, it affects the ability to understand or discriminate speech. As it progresses, it begins to affect the ability to hear sounds in general.

The three main types of hearing loss are conductive, sensor neural, or a combination of the two. The effects of noise can be simplified into three general categories:

- Primary effects, which include noise-induced temporary threshold shift, noise-induced permanent threshold shift, acoustic trauma, and tinnitus
- Effects on communication and performance, which may include isolation, annoyance, difficulty in concentrating, absenteeism, and accidents
- Other effects, which may include stress, muscle tension, ulcers, increased blood pressure, and hypertension

10.3.5.3 Abrasives Slurry

Basically, workers are exposed to metal cutting fluids via skin exposure, aerial exposure, and ingestion. Slurry fluids are important causes of occupational contact dermatitis, which may involve either irritant or allergic mechanisms. Water-mixed fluids generally determine irritant contact dermatitis and allergic contact dermatitis when they are in touch with workers' skin. Non-water-miscible fluids usually cause skin disorders such as folliculitis, oil acne, keratoses, and carcinomas.

Mists are aerosols comprised of liquid particles less than 20 μm, which may accommodate abrasive dust. The non-aqueous components of the abrasive slurry, such as the biocide additives, then become a fine aerosol that can enter the workroom air. Mist also may be generated by the spray from the slurry application. Small droplets may be suspended in the air for several hours or several days, possibly in the workers' breathing zones. Inhaled particles (with aerodynamic diameters less than 10 μm) deposit in the various regions of the respiratory system by the complex action of the different deposition mechanisms. Particulates below 2.5 μm aerodynamic diameter deposit primarily in the alveolar regions, which is the most sensitive region of a lung. The size of particulates range from 2.5 to 10 μm deposits, primarily in the airways. The mist droplets can cause throat, pancreas, rectum, and prostate cancers, as well as breathing problems and respiratory illnesses. One acute effect observed is mild and reversible narrowing of airways during exposure to the slurry mist.

Several other epidemiological studies have also suggested that exposure to fluid mist may be associated with increased risk of airway irritation, chronic bronchitis, asthma, and even laryngeal cancer. The OSHA standard for airborne particulate (largely due to fluid mist) is 5 mg/m^3, and the International Union of United Automobile, Aerospace, and Agricultural Implement Workers of America (UAW) has proposed a reduction in the standard to 0.5 mg/m^3.

Anti-misting compounds, such as a polymethacrylate polymer, polyisobutylene, and poly-*N*-butane in concentrations of 0.2% as well as poly (1, 2-butene oxide) have been suggested to be added to cutting fluids. But consideration must be given to the effects of these chemicals upon humans. The most effective way to control mist exposure is to use a mist collector that prevents mist from entering the plant air. Many collectors use several stages of filters in series and other collectors use centrifugal cells or electrostatic precipitators as intermediate stages.

10.3.5.4 Contact Hazards

If fingers or hands are put into an ultrasonic machine, a tickling sensation is instantly experienced on the skin surface followed 2–3 s later by pain in the joints.

10.3.5.5 Other Hazards

Some scientists have asserted that effects such as nausea, dizziness, tiredness, and tinnitus can be caused by exposure to sound in the ultrasonic frequencies and others that the upper audible frequencies are also implicated.

10.3.6 Abrasive Jet Machining

Atmospheric contaminants from abrasives are a major hazard in AJM. The prime hazard in AJM is dust; in particular, silica dust. Prolonged inhalation of crystalline silica dust can cause silicosis. Particles of other materials may also be present in the dust to which the employee is exposed. The nature of these materials depends on the material being machined. Sometimes heavy concentrations of iron oxide are produced during the cleaning of iron castings. Machining toxic metals such as lead, mercury, arsenic, zinc, and cadmium, and toxic dusts may constitute a significant hazard. Employers must ensure that their employees are provided with appropriate respiratory protective equipment to ensure that they are protected against atmospheres containing substances that may be toxic and harmful if breathed.

There are four risk factors with dust:

- Type of particulate involved and its biological effect
- Concentration of airborne particulate in the breathing zone of the work
- Size of the particles present in the breathing zone
- Duration of exposure

Dust generated by AJM falls into three categories:

a. *Inspirable dust* is the dust which a worker can inhale from the air in the work area. It can contain a wide range of particles of different sizes, including particles that are too heavy to be inhaled and captured by the respiratory system to very small particles of respirable size.
b. *Respirable dust* is that portion of inspirable dust consisting of very small particles of dust that can easily penetrate the lung down to the lower

bronchioles and alveolar regions. In this regard, AJM produces high levels of respirable dust.

c. *Silica dust* is a major hazard in AJM, generated when:
    - The abrasive medium contains silica
    - Machining materials contain silica

The major risk from silica dust is silicosis, which is a chronic disease that causes stiffening and scarring of the lungs. Symptoms usually take a number of years to appear. They include shortness of breath, coughing, and chest pain. This leads to degeneration in the individual's health. The risk of disease is directly related to the amount of dust inhaled. There is some evidence that people with silicosis have an increased risk of developing lung cancer.

Measures to control silica dust generated by AJM include using a jet machining medium that does not contain silica, and ensuring the process is isolated or appropriate administrative strategies are implemented. Operators must wear personal protective equipment and wet AJM techniques must be used.

Isolating the process where silica dust is produced may be achieved by:

- Machining in a closed chamber
- Enclosing an area with plastic or other forms of impervious protective sheeting to limit the movement of dust from the source where jet machining in a blast chamber is not practicable
- Setting up designated exclusion zones by:
    1. Installing physical barriers and warning signs to prevent unprotected persons from entering the area
    2. Shrouding the area where jet machining is to occur
    3. Restricting entry of unprotected persons into the AJM area while jet machining is running

If isolation is by means of an exclusion zone, signs must be posted at least 25 m from the perimeter of the exclusion zone of the AJM site to warn workers and others that:

- AJM is in progress
- Access to the work area is restricted to authorized persons
- Respiratory protection must be worn in the work area

Particular care must be taken in the cleanup process. Dust must be cleaned and collected in an appropriate manner to ensure that the level of silica dust does not exceed the exposure standard. Methods of cleaning include the wetting down of dust prior to cleanup. Hazardous impacts associated with AJM include:

1. Worker: The hazardous environment associated with AJM can affect the worker's health and performance through the following:
    - Health hazard
    - Low worker productivity
    - High physical/mental fatigue

TABLE 10.5
Hazardous Materials by Different Machining Processes

		Solid	Chemical										
Process	**Mode**	**Chips**	**Dusts**	**Gasses**	**Vapors**	**Liquids**	**Mists**	**Fumes**	**Slurry**	**Noise**	**Radiation**	**Vibrations**	**Magnetic Field**
Turning	MQL	X		X	X		X	X		X		X	
	Dry	X										X	
Drilling	MQL	X		X			X	X		X		X	
Milling	MQL	X		X	X					X		X	
	Dry	X										X	
Reaming	MQL	X		X	X		X			X		X	
Grinding	MQL		X			X	X	X	X	X		X	
CHM				X		X							
ECM				X	X	X			X				
EDM				X	X	X			X				
LBM				X	X					X	X		X
EBM				X									
PBM					X								
USM			X						X	X		X	
AJM		X	X									X	
WJM		X				X				X		X	

- Low job satisfaction
- High error rates

Worker condition improvement can be achieved by wearing suitable protective clothing and using safety tools and by replacing the worker regularly from time to time to decrease health problems.

2. Product quality is impured as the abrasives may get impeded in the work surface that may be also covered with dust.
3. Workplace is also affected by the mixture of abrasives and WP particles that may lead to the following:
   - Production of dusty air
   - Abrasive particles cause problems to exist in other machines in the same station
   - Other work parts on the other m/cs may be polluted
   - Health problems to other workers in the same station

   Workplace condition improvement can be achieved through good ventilation and implementing an air filtration system. For better machine conditions, isolation of the machined part and the use of a suitable dust collection system are recommended. Table 10.5 summarizes the possible hazards of machining processes.

10.4 REVIEW QUESTIONS

10.4.1 What are the main steps that should be followed by a company's environmental policy?

10.4.2 Use a line sketch to show how a clean machining technology could be achieved.

10.4.3 Illustrate the different machining liquids used for cooling, for lubricating, and as a machining medium.

10.4.4 What are the major chemical agents that may arise during the various machining processes?

10.4.5 Discuss the effect of the various hazardous substances associated with the machining technology.

10.4.6 Using diagrams, explain the hazardous sources and type of hazards for machining by the cutting, abrasion, and erosion processes.

10.4.7 Use a sketch to show the possible streams of cutting liquids.

10.4.8 State the different machining processes that adopt the dry machining method.

10.4.9 What is meant by the hazards ranking system of cutting fluids?

10.4.10 State the possible hazards of cutting fluids.

10.4.11 Explain what is meant by cryogenic cooling and ecological machining.

10.4.12 What are the major health effects of CHM and ECM?

10.4.13 State the hazards potentials and the protective measures for machining by cutting and abrasion.

10.4.14 State the hazards potentials and the protective measures for machining by ECM, CHM, LBM, EDM, and USM.

REFERENCES

Byrne, G & Scholta, E 1993, 'Environmentally clean machining processes—a strategic approach', *CIRP*, vol. 42, no. 1, pp. 471–474.

Dhar, NR, Nada Kishore, SV, Paul, S & Chattopadhay, AB 2002, 'The effect of cryogenic cooling on chips and cutting forces in turning AISI 1040 and AISI 4320 steels, Proceeding of Institution of Mechanical Engineers', *Journal of Engineering Manufacture*, vol. 216, no. Part B, pp. 713–724.

Hughes, P & Ferrett, E 2005, *Introduction to health and safety at work*, 2nd edn, Elsevier, New York.

New Logic Research, Inc., Emeryville, CA, <www.vsep.com/solutions/technology.html>.

Tönshoff, HK, Egger, R & Klocke, F 1996, 'Environmental and safety aspects of electrophysical and electrochemical processes', *CIRP Annals*, vol. 45, no. 2, pp. 553–567.

Weinert, K, Inasaki, I, Sutherland, JW & Wakabyashi, T 2004, 'Dry machining and minimum quantity lubrication', *CIRP Annals*, vol. 53, no. 2, p. 17.

11 Hexapods and Machining Technology

11.1 INTRODUCTION

Most industrial machine tools, such as conventional milling machines, drilling machines, lathes, stacked axis robots, and so on, have a serial or open-loop kinematic architecture, which means that each axis supports the following one, including its actuators and joints (Figure 11.1). These machines are mainly based on the perpendicular composition of three linear axes. Two more rotary axes may be integrated to extend the ease of applying the Cartesian coordinate system to control spatial movements (AKIMA, 1997).

The kinematic analysis of this serial or stacked-axis system is easy. However, it generates cumulative errors, because inexact positioning on one axis dislocates the positioning on the next axis, which multiplies the imprecision with each subsequent station. In addition to this design drawback, the conventional stacked-axis machine, which has a bed, saddle, and so on, requires a massive concrete foundation for stability requirements. This results in a cumbersome unit that takes up considerable floor space and is difficult to transport (Figure 11.1).

Machine tools based on advanced closed-loop parallel kinematics represent a promising new technology of the 21st century that is currently receiving a lot of focus and interest. Several prototypes, developed by famous companies all over the world, have already proved the general feasibility of the idea of the parallel kinematic system (PKS). Such systems are characterized as:

> Mechanisms based on a kinematic structure which allows movements with *N*-Degrees of Freedom, whereby a moving base carrying the cutting tool and a fixed base is connected to each other by *N* independent kinematic chains. Every chain is composed of maximum two segments and the articulation between them has one degree of freedom. The motion of the structure is generated by *N* actuators, one for each chain.
>
> **(AKIMA, 1997)**

Hexapods and tripods (Figure 11.2) are the most famous paradigms of the parallel kinematic mechanism (PKM). In this chapter, the hexapod positional device is introduced. Its historical background, applications in machining technology, design features, elements, and characteristics are highlighted and traced.

11.2 HISTORICAL BACKGROUND

Around 1800, the mathematician Augustin-Louis Cauchy studied the stiffness of the so-called *articulated octahedron*. The earliest known hexapod machine was a car tire tester designed by Gough of Dunlop in 1941. Stewart (1965) described the

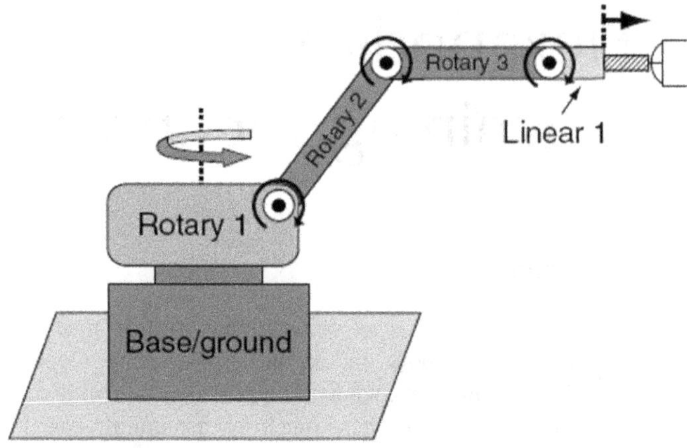

FIGURE 11.1 Conventional architecture (serial mechanism).

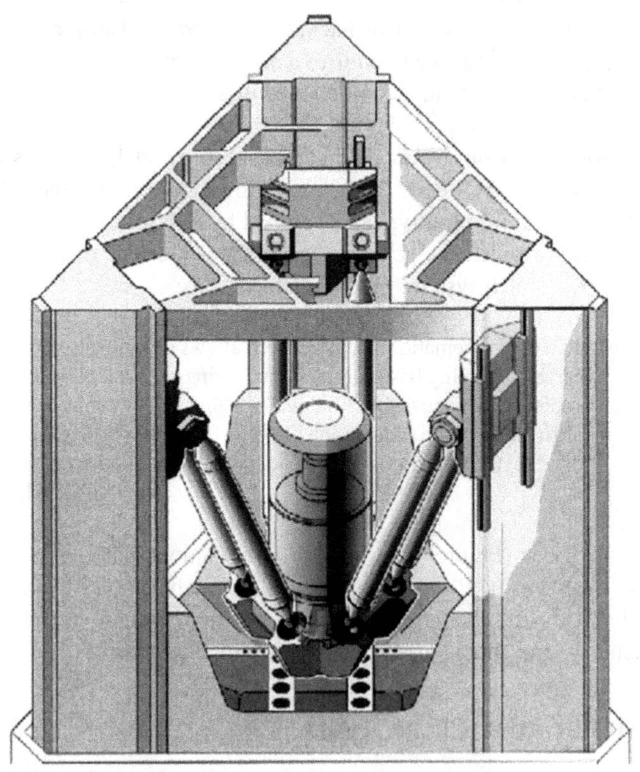

FIGURE 11.2 Tripod parallel mechanism.

attributes of a simple hexapod, thus giving his name to the platform. Figure 11.3 shows a hexapod employed in surgical operations using a flexible movement with a laser beam for operations. This model is used nowadays to simulate the whole operation to train physicians. This platform has been successfully used as a flight simulator to train pilots. Figure 11.4 illustrates nanopods that meet stringent medical safety standards. Figure 11.4 shows a nanopod with three additional legs containing redundant position sensors. A seventh axis is added to increase the linear travel range (Figure 11.4b). In 1995, the Fraunhofer Institute in Stuttgart developed a hexapod that worked successfully as a surgical robot.

Many unsuccessful attempts were made between 1970 and 1990 to make practical hexapod machines. The cost of computing strut lengths of hexapods has until recently been too high for most applications. In the cases where computing costs are justified, some interesting hexapod machines, such as flight simulators, were successfully made.

Recently, the cost of computing has fallen drastically, and therefore many companies are offering hexapod-based machines for various applications that include hexapod machine tools (Geodetic, 1997). In 2006, Hitachi, Seiki, and Toyota joined the U.S. builders Giddings Lewis, Ingersoll, and Britain's Geodetic to develop hexapod technology that started in the early 1990s.

In 1990, Ingersoll announced and exhibited its own prototype of the *octahedral hexapod*. This prototype utilized a 12-node hexapod suspended from an octahedral framework, with the spindle pointing down toward the WP (Ingersoll Co., 2001).

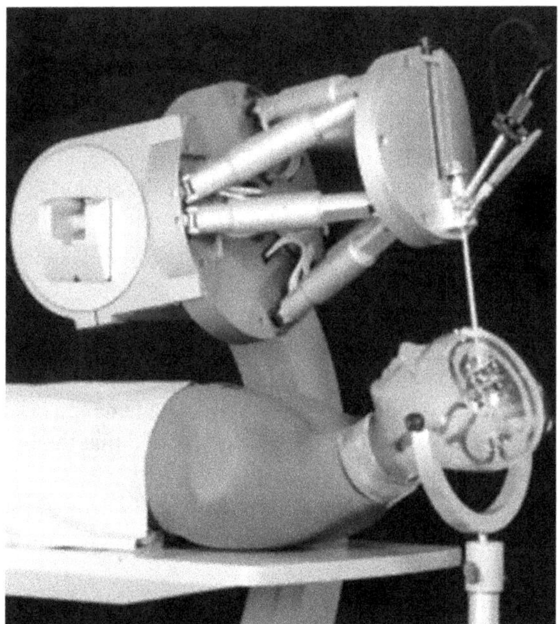

FIGURE 11.3 Hexapod for surgical operations, equipped with laser beam.

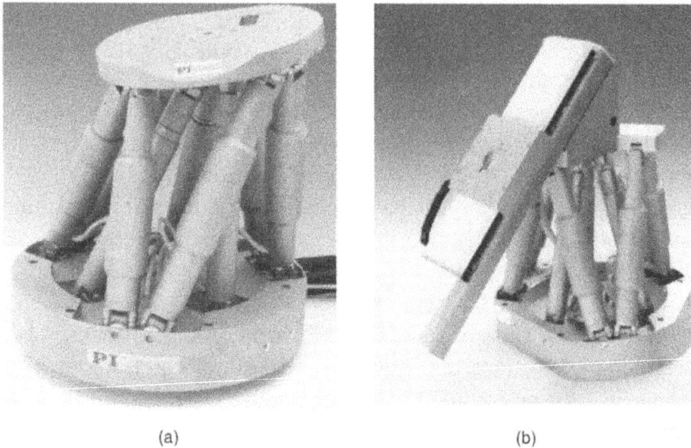

(a) (b)

FIGURE 11.4 Nanopods: (a) Nanopod and (b) nanopod with an additional linear axis for surgical applications.

The Variax Hexacenter milling machine, introduced in 1994 for high-speed milling of aluminum, took its design inspiration from the flight simulator, with the struts crossing over and the spindle pointing downward from the platform toward the working volume enclosed by the machine.

11.3 HEXAPOD MECHANISM AND DESIGN FEATURES

11.3.1 Hexapod Mechanism

The hexapod mechanism consists of a fixed upper dome platform (base) of a hexagonal shape and a moving platform of triangular shape, connected by six struts (telescopic or ball screw) (Figure 11.5). Starting from the six joints on the base, each two struts intersect at three nodes on the moving triangular platform.

The movement of the platform is actuated when the six struts change their lengths in a coordinated manner (Figure 11.6):

a. When the six struts simultaneously expand or contract at the same feed rate, the platform moves downward or upward horizontally in an extending movement.
b. When some struts expand and others contract and change orientation, such that the platform moves horizontally, then it is said that the hexapod exhibits panning movement.
c. When strut orientations and lengths are changed to achieve a specific platform inclination in space with respect to coordinate axes, then the hexapod exhibits rotation.
d. When all struts are of the same length and similarly rotated, such that the platform moves horizontally, then the hexapod is being twisted.

Therefore, through the rotational and axial movements of struts, the moving platform is capable of reaching any point on the WP, and the hexapod becomes an excellent

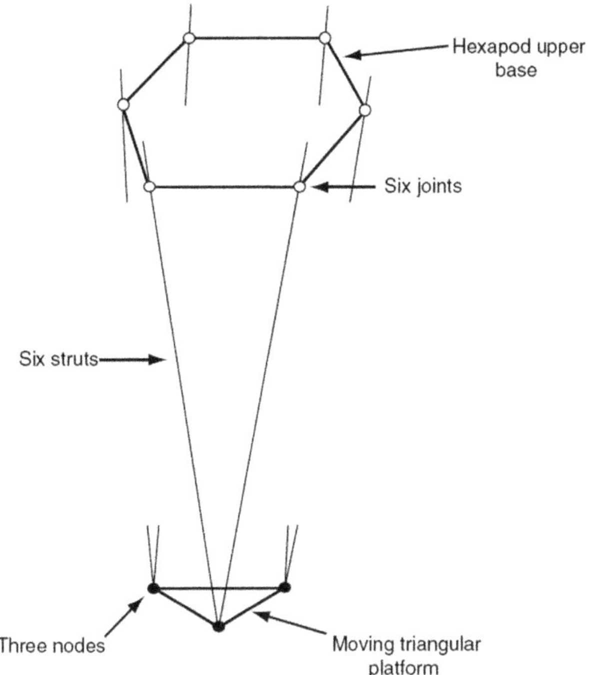

FIGURE 11.5 Hexapod mechanisms.

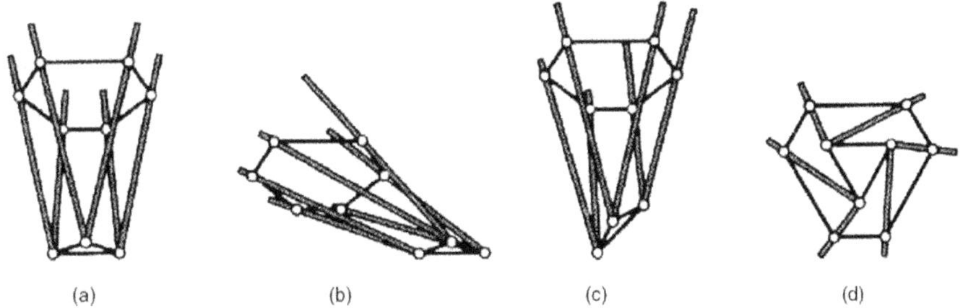

FIGURE 11.6 Models of hexapod movements: (a) Extending, (b) panning, (c) rotating, and (d) twisting (Ingersoll, 1997).

universal positioning apparatus. Of course, provisions should be taken to prevent struts crossing and colliding with each other or with the platforms.

11.3.2 Design Features

Hexapods have two main design features, as described in the following.

11.3.2.1 Hexapods of Telescopic Struts (Ingersoll System)

In this design, the hexapod consists of six hydraulic, telescopic struts. The struts may be of a circular cross section or square and are free to expand or contract between a base and a platform. The platform represents the output element that gets the six degrees of freedom (6 DOF) of the system (Figure 11.7). Both ends of the hydraulic telescopic struts are connected to either the platform or the base by universal joints (Figure 11.8). Such a system was first introduced as a flight simulator positioning system. It is commercially available for a variety of applications that require micron and submicron accuracy.

The relationship between the mobility (M), constraints (c_i), and the total DOF of the system is expressed by the Kutzbach criterion:

$$M = 6(n_e - 1) - \sum_{i=1}^{j} C_i \tag{11.1}$$

where n_e is the total elements of the system.

Calculation of DOF of the telescopic hexapod:

$$\text{total number of parts} = 6 \text{ hydraulic struts, each 2 parts} + 1 \text{ base} + 1 \text{ platform}$$

$$= 2 \times 6 + 1 + 1 = 14 \text{ Then, the total DOF}$$

$$= 6(\mathrm{n_e} - 1) = 6 \times 13 = 78 \text{ DOF}$$

According to Table 11.1, the total constraints = 3 × 6 × 4 = 72 constraints.

Therefore,

$$M = 78 - 72 = 6 \text{ DOF (three translations and three rotations)}$$

By combining an octahedral structural frame (Figure 11.9), with the previously described hexapod actuator, Ingersoll has created the stiffest and the most rigid

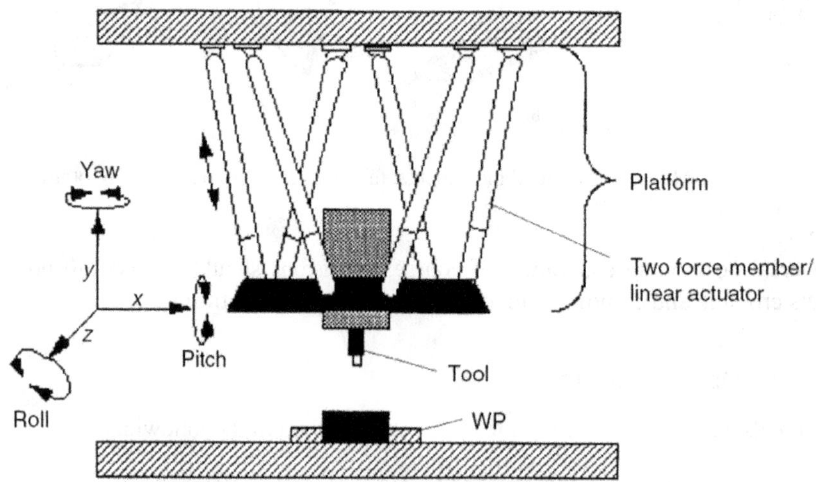

FIGURE 11.7 Hexapod of telescopic struts, Ingersoll system (Ingersoll, 1997).

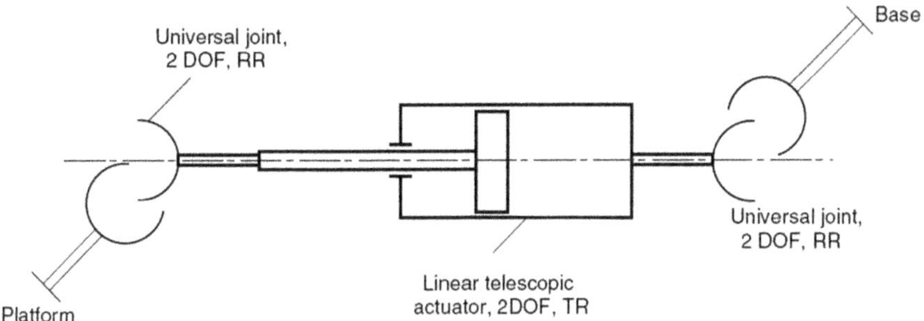

FIGURE 11.8 Telescopic struts with universal joints.

TABLE 11.1
Analysis of Telescopic Strut System

Quantity	Occurrences	Constraints	DOF	Description
6	Base: Yokes 1/2 UJ	4	2	RR
6	Strut Cyl. and rod	4	2	TR
6	Strut rod/platform UJ	4	2	RR

Note: R: Rotational; T: Translational; UJ: Universal joint; Cyl.: Cylinder.

machine tool possible. The stiffness and rigidity have a positive impact on the accuracy, surface quality, and cutting tool durability. The octahedron (the frame structure) consists of 12 beams of similar length that are joined at six junction points (Figure 11.9). This robust structure is self-supporting and needs minimal foundation requirements. The mechanism that guides the spindle, the hexapod with its telescopic struts, is attached to the top of the octahedron at top corners *A*, *B*, *C*. Ingersoll has provided two versions of octahedral hexapods. These are horizontal spindle octahedral hexapod (HOH)-600 and vertical spindle octahedral hexapod (VOH)-1000 hexapods (Figure 11.10).

The technical specifications of the VOH-1000 model are as follows:

Axis travels:	X, Y, Z = 600, 600, 800 mm
Feed rates:	Maximum feed rate (strut axis) = 30 m/min
	Maximum traverse rates (strut axis) = 30 m/min
Acceleration:	0.5 G (4.8 m/s^2) depending on X, Y, Z position
Spindle:	Speed range = 0–20,000 rpm
	Maximum torque = 49 N m
	Maximum power = 37.5 kW
Coolant pressure:	50 bar through the spindle
Tool storage magazine:	40, 80 tools, maximum tool weight = 12 kg
Volumetric accuracy:	20 μm using laser diagonal displacement facility

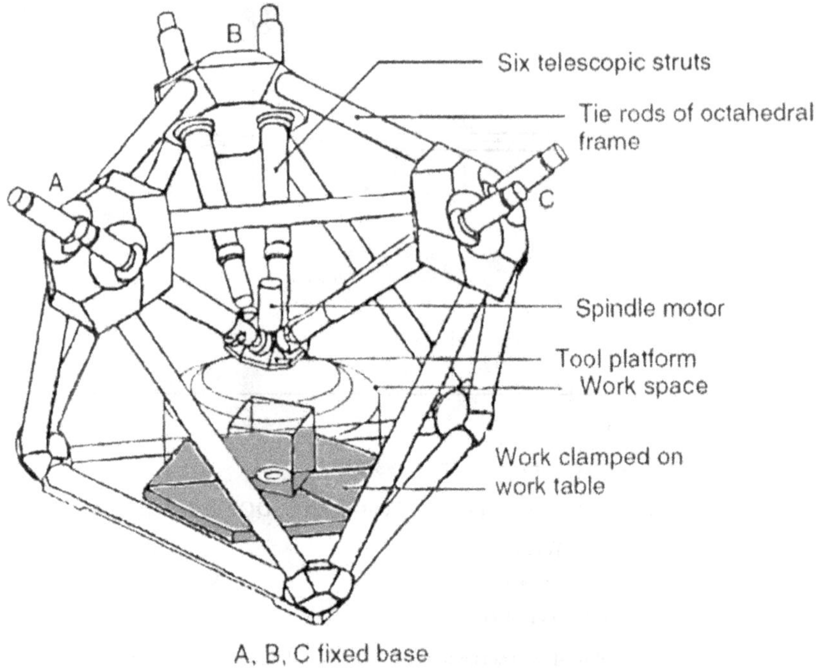

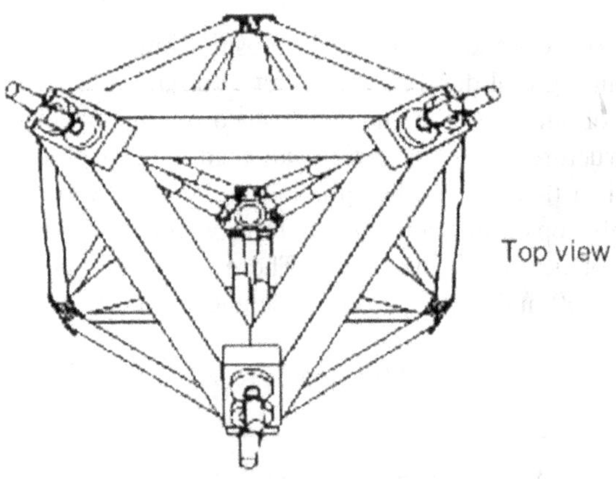

FIGURE 11.9 Hexapod with rigid frame.

11.3.2.2 Hexapods of Ball Screw Struts (Hexel and Geodetic System)

Hexel and Geodetic dramatically simplified their approach by developing a bifurcated ball between pairs of struts meeting at the working platform. This development reduced the number of nodes to nine (six nodes in the work cell and three on the working platform), which improves the stiffness, simplifies the control, and allows for automated calibration. This design approach strives for precision to be derived from software. Such flexibility reduces cost and time, thus leading to a truly soft machine (Ingersoll file).

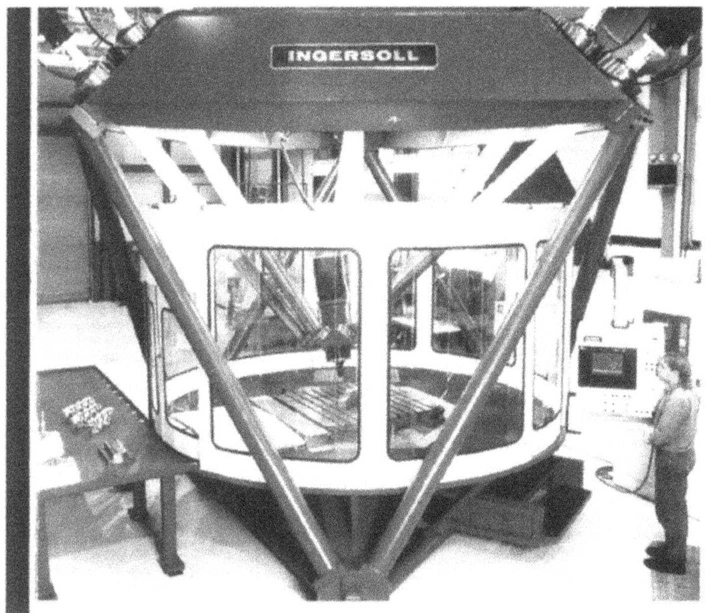

FIGURE 11.10 Vertical spindle VOH-1000 hexapod.

This type of hexapod is extremely stiff and reliable, because it does not require any telescoping mechanism. It is equipped with low-pressure spherical universal joints to facilitate accurate and repeatable calibration while inherently providing excellent strut damping (Ingersoll Co., 2001).

The characteristics and working principle of this type of hexapod are summarized as follows:

1. In this case, the hexapod consists of six precise ball screw struts, each of which extends from the tool platform through an integral spherical servomotor that is mounted on the upper dome platform of the hexapod (Figure 11.11). The strut cross section must be circular to allow rotation.
2. Each of the six ball screw struts is driven by a servomotor (sphere drive) through a ball nut, which is housed in the joint. The servomotor can rotate at 3200 rpm, propelling the 5 mm pitch screw of the ball screw at speeds up to 25 m/min.
3. Through coordinate movement of ball screws, the working platform alters position according to the required contour. The struts join the lower platform at three nodes, with two struts shearing a ball and socket joint known as a bifurcated ball (Figure 11.11).
4. Each individual ball screw strut is independent from the others and possesses a personality file containing information such as:
   - Error mapping (lead pitch variation)
   - Mounting offsets
   - Physical and thermal expansion characteristics

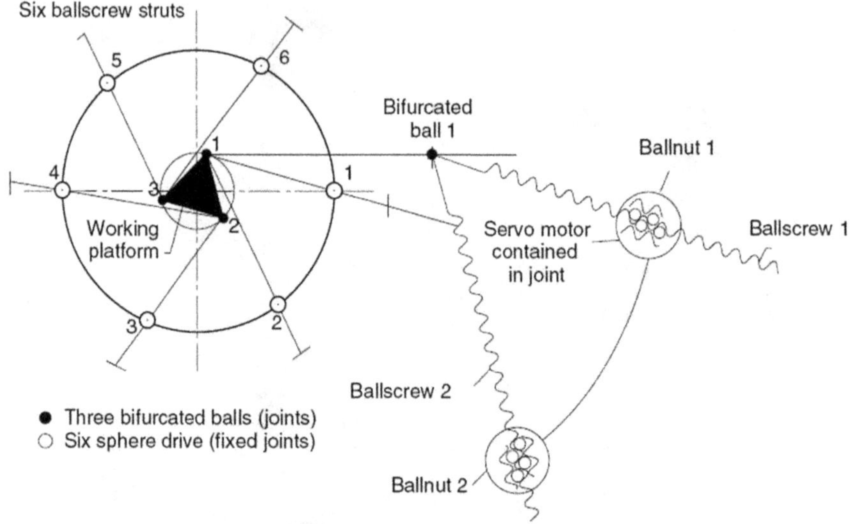

FIGURE 11.11 Ball screw hexapod.

5. The completely interchangeable drives are calibrated offline as discrete units and are automatically compensated when assembled. This concept allows hexapods to be assembled and serviced very quickly without the need for complex alignment or calibration exercises. Replacement of worn or damaged strut drives can be achieved within two hours, thereby minimizing downtime.

Referring again to Equation 11.1 expressing the Kutzbach criterion, the calculation of the DOF of ball screw hexapods is as follows:

$$n_e = \text{Total number of hexapod elements}$$

$$= 6 \text{ struts} + 6 \text{ base struts} + 1 \text{ platform} + 1 \text{ base}$$

$$= 14 \text{ parts}$$

$$\text{Total DOF of the system} = 6\left(n_s - 1\right) = 6 \times 13 = 78 \text{ DOF}$$

According to Table 11.2,

$$\text{Total constraints} = 2 \times 6 \times 3 + 6 \times 5 = 66 \text{ constraints}$$

Therefore,

$$M = 78 - 66 = 12 \text{ DOF} \left(6 \text{ required / basic DOF} + 6 \text{ local motilities}\right)$$

The coordinated motion of struts enables the moving platform (and spindle) to perform 6 DOF: orthogonally (*X*, *Y*, *Z*) and rotary (pitch, yaw, and roll). This makes the spindle very dexterous, easily accessing unusual angles and geometric features.

TABLE 11.2
Analysis of a Ball Screw System

Quantity	Occurrences	Constraints	DOF	Description
6	Base/struts: Ball and socket joint	3	3	RRR
6	Base struts/extensible strut, lower end	5	1	T
6	Strut upper end/platform: Ball and socket joint	3	3	RRR

Note: R: Rotational; T: Transnational.

11.4 HEXAPOD CONSTRUCTIONAL ELEMENTS

In this section, design features of the ball screw hexapod (Hexel and Geodetic system) are mainly considered.

11.4.1 Strut Assembly

The Geodetic ball screw strut assembly is illustrated in Figure 11.12. The ball screw is connected to a movable triangular platform by a bifurcated ball and is then driven through the pivot on the dome by the sphere drive. This strut assembly displays a

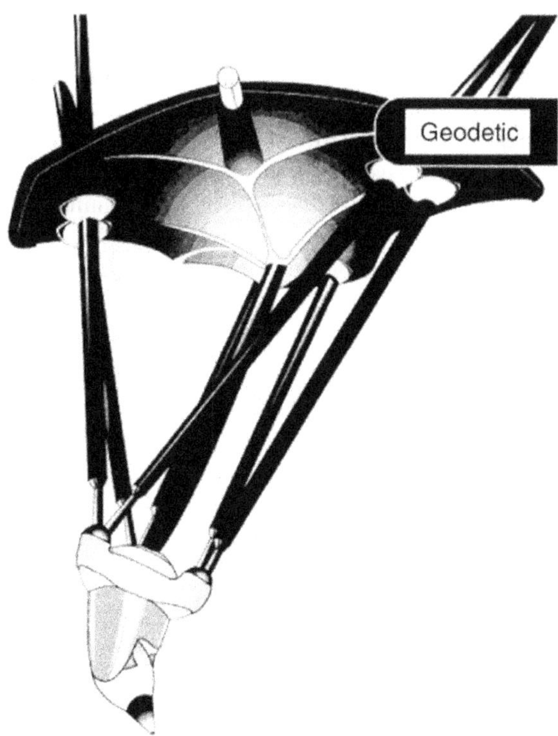

FIGURE 11.12 Geodetic ball screw strut assembly (Geodetic, 1997).

very high extension to contraction ratio, because the ball screw struts can pass into unlimited space behind the pivot. The pivot points (bifurcated nodes and sphere drive joints) are hermetically sealed to retain flexibility and to provide very smooth running conditions, even under shock loads. A variety of lower platforms can be used to accommodate various tool attachments.

11.4.2 Sphere Drive

The sphere drive is a special mechanism that forms the heart of any ball screw strut-type hexapod. It is located in the upper base and provides the accurate positional movements of the hexapod struts. The sphere drive is a hollow ball that accommodates a high-powered, high-specification, frameless, brushless dc motor. A rotor is keyed to a spindle, which runs on two high precision bearings (Figure 11.13). A ball nut is keyed to a flange that forms a part of the internal spindle.

This arrangement drives the ball screw in and out through the hollow-cored servomotor. The unit is water-cooled to maintain reliability and control of thermal inaccuracies.

A high-resolution radial incremental encoder is mounted outside the unit to provide the accurate positional information required by the controller. An integral thermocouple on the motor windings feeds back data to the controller, allowing thermal expansion to be compensated. All critical parts included in the sphere drives are effectively protected from foreign matter. The sphere drive is clamped to the top plate by six bolts; a locating dowel ensures precise and repeatable positioning.

The sphere (Figure 11.14a) containing the drive is retained in an annular ring. Figure 11.14b uses a hydrostatic system to provide the greatest rigidity combined with a low coefficient of friction, while ensuring excellent lubrication. The annular ring suspends the sphere drive within an envelope of high-pressure fluid, which is continuously recirculated. Metal-to-metal contact is minimized, thus providing greater damping and smooth operation. The sphere drive is capable of driving the ball screw at axial feed rates exceeding 40 m/min, with high forces and accelerations.

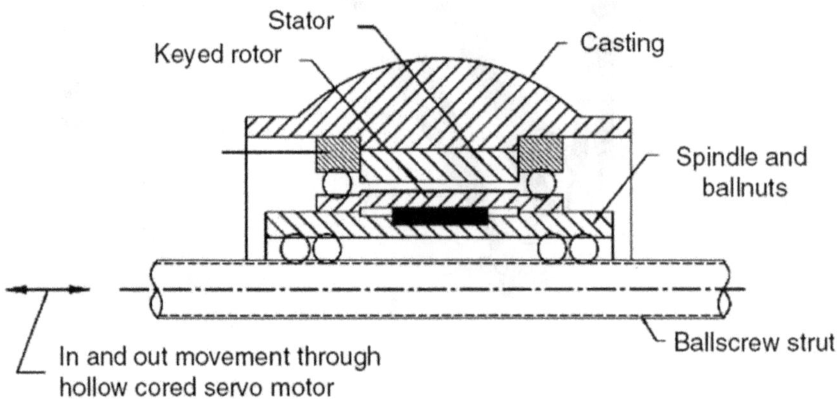

FIGURE 11.13 Sphere drive.

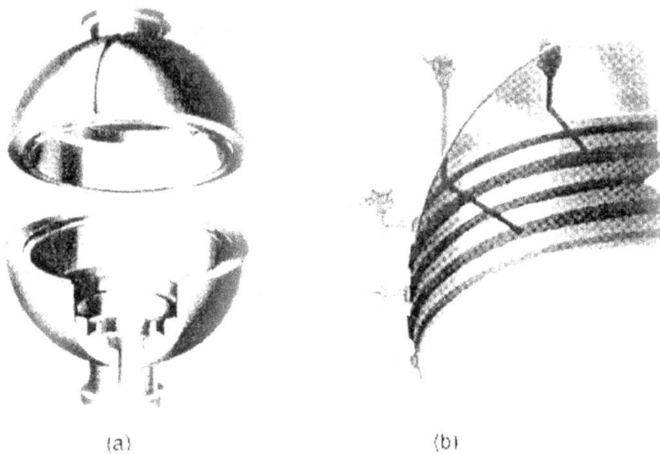

FIGURE 11.14 Casing elements of sphere drives: (a) Sphere and (b) annular rings suspending the sphere drive (Geodetic, 1997).

All sphere drive components such as ball screws, ball nuts, bearings, motors, and encoders are standardized. This makes the sphere drive mechanically simple, cost-effective, and reliable. Moreover, this drive can be easily tailored to meet a wide range of performance requirements for a variety of manufacturing applications.

11.4.3 Bifurcated Balls

This nodal joint allows a pair of struts to terminate at common focal point at the lower platform (Figure 11.15). The design includes a split ball in a variety of socket arrangements (Figure 11.16):

The hydrostatic bifurcated ball joint. This is the latest addition to the family of ball joints. The new design uses a system similar to that of the hydrostatic sphere drives, thus ensuring elimination of clearances. This promotes a smoother, vibration-free, and well-lubricated environment, leading to higher accuracy and better surface quality (Figure 11.17). Figure 11.18 illustrates a layout of a lubricating system of hydrostatic bifurcated balls and sphere drives.

Magnetic bifurcated ball joint. The key feature of this joint is the use of powerful rare-earth neodymium iron boron magnets to retain the split ball in its socket. This permits the ball to be held over less than half of its surface (Figure 11.19), leaving clearance to permit a wide DOF. The magnetic socket has an inherent overload protection feature, allowing the joints to dislocate when loads exceed the holding forces without causing any damage to the mechanism. This joint uses a shear lubricant interface between the ball and the socket to provide critical damping (Geodetic, 1997).

The magnetic joint is frequently used in applications where small forces are exerted.

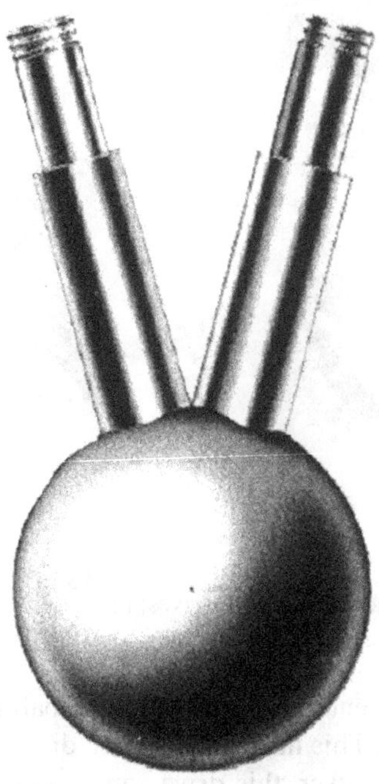

FIGURE 11.15 Bifurcated ball.

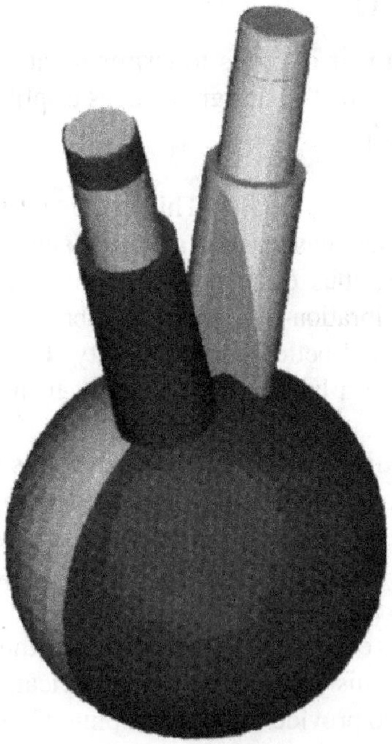

FIGURE 11.16 Split bifurcated ball arrangement (Geodetic, 1997).

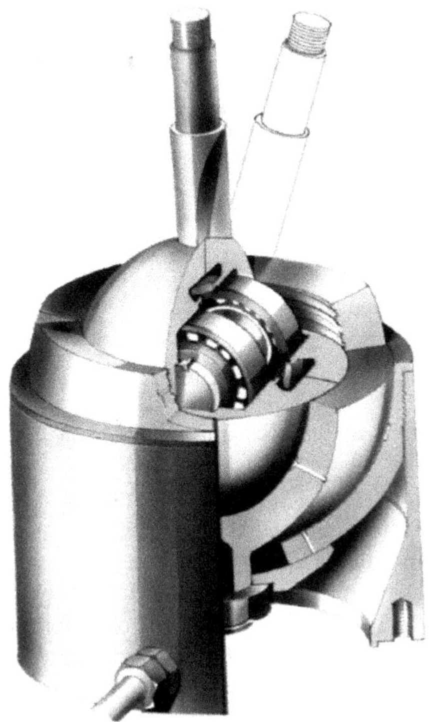

FIGURE 11.17 Hydrostatic bifurcated ball joint (Geodetic, 1997).

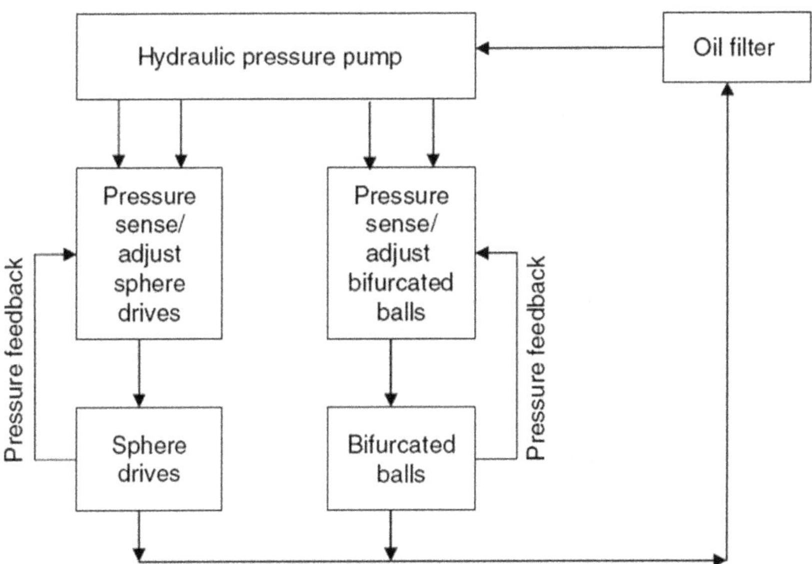

FIGURE 11.18 Layout of the lubricating system of a hydrostatic bifurcated ball joint (Geodetic, 1997).

FIGURE 11.19 Magnetic bifurcated ball joint (Geodetic, 1997).

This is common in laser cutting, WJM, and coordinate measuring machines (CMM).

Mechanical bifurcated ball joints. These have been developed for light machining applications. Similar to magnetic joints, the mechanical joints are made up of a number of stages, each contributing to its freedom of movement. Unlike the magnetic joints, the ball is kept in its cup by a retaining ring (Figure 11.20). This design permits extra forces and eliminates the problem of metal swarf contamination. A limit switch ensures that the joint does not exceed its physical limits.

11.4.4 Spindles

Hexapods are equipped with high-speed spindles that pack enormous power into a very compact unit. Hexapod builders make full use of the conical shape design of their spindles, which reduces interference with the WP (Figure 11.21). Geodetic machining units are equipped with motors of power ratings between 3 and 20 kW at a 20,000 rpm maximum spindle speed. Geodetics also produce a range of air drive spindles for special application. A mechanical drawbar mechanism automatically clamps and releases the tool holder, making use of a high-speed tool changer. Electrically driven spindles are water-cooled. Hexapod spindles provide speed feedback to ensure precise speed control.

To ensure maintenance-free operation at such high speeds, the hexapod spindles should feature self-lubricating precision ceramic roller bearings. The spindles accommodate tools up to 20 mm in diameter and with a length of 200 in. High speed promotes productivity, reduces tool wear, achieves tighter machining tolerances, and provides higher surface quality. Hexapods are designed to perform milling, engraving, and drilling operations (Figure 11.22).

11.4.5 Articulated Head

An articulated head's dexterity is comparable to that of the human hand (Figure 11.23a). The two-axis head incorporates the high-speed spindle, coupled

FIGURE 11.20 Mechanical bifurcated ball joint (Geodetic, 1997).

with 2 DOF, a rotary stage and a tilt stage (Figure 11.23b). This self-contained unit extends articulation to the movements attainable by the human arm (wrist and hand), allowing for machining complex surfaces as well as undercutting.

The two-axis head can tilt by over 90° and rotate over 540° (Figure 11.24). Because of the separation between the center of the platform and the additional pivot axis, the mechanism configures itself in such a way as to bend around obstructions in its working environment. This allows the task to be approached at the optimum angle. The larger the separation between the two axes, the better the reach around obstacles. A powerful positioning water-cooled motor, capable of delivering high torque, delivers precise positioning in excess of 540° of rotation. Tilting is achieved by a geared sector with backlash elimination. Bearings are protected by a powerful air curtain, which prevents contamination.

11.4.6 Upper Platform

Hexapods must be mounted on a stiff base platform that does not deflect significantly under load. The Geodetic upper dome platform is shown in Figure 11.25. The dome is the most logical configuration to achieve the best angular coverage while maintaining maximum rigidity and stiffness.

FIGURE 11.21 A high-speed conical spindle.

FIGURE 11.22 Tools accommodated by the spindle of a hexapod.

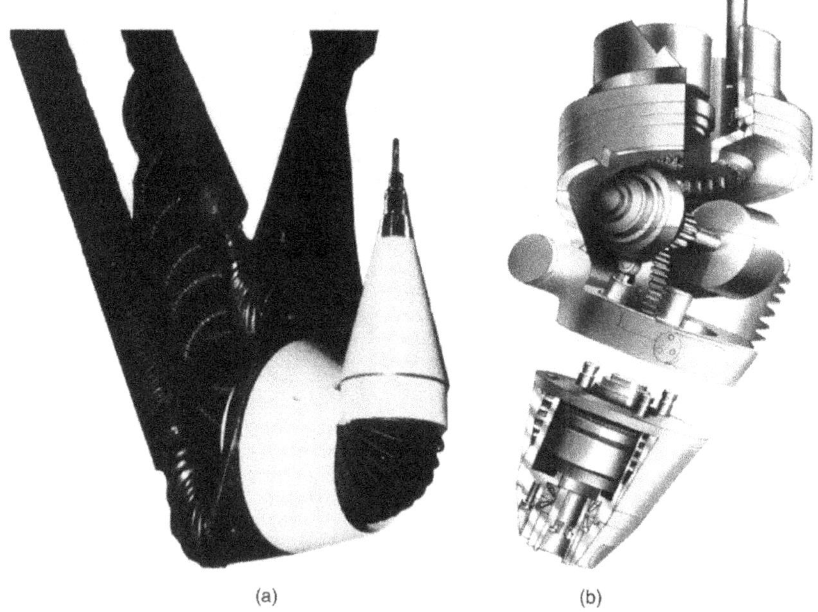

FIGURE 11.23 Articulated heads: (a) Two-axis articulated head unit and (b) rotating and tilting stages of articulated head (Geodetic, 1997).

The circular form ensures that the frame attachment points are as close as possible to stress points; that is, the sphere drive. Integral ribs enclose circuitry and increase stiffness while allowing for sphere drives to be plugged in without the need for complex wiring. The complete unit is embedded into resin, which further dampens vibrations and provides 100% surface contact with the frame. The cast iron dome must be designed with extensive use of finite element analysis (FEA) to achieve a lightweight yet stiff platform to support the hexapod.

11.4.7 Control System

As a machine tool, the hexapod requires sophisticated control versus conventional machine tools. In hexapods, a continuous and exact relationship of strut movements should exist to control the triangular platform movements. The contour to be followed by the cutting tool is controlled by CAD/CAM software, based on Cartesian coordinates *X*, *Y*, and *Z*, and orientation vectors *A*, *B*, and *C* of the six struts. The location coordinates *X*, *Y*, *Z*, *A*, *B*, and *C* of each strut are calculated by the controller online, which necessitates few milliseconds to be performed.

Besides the contour geometry, the calculations of the tool movements require additional data such as contouring speed and acceleration. The contour calculations are then analyzed and tested online to make sure that the dynamic limits of the machine are not exceeded to prevent the damage of the tool on the machine. The hexapod has excessive movement possibilities that possibly lead to collision between

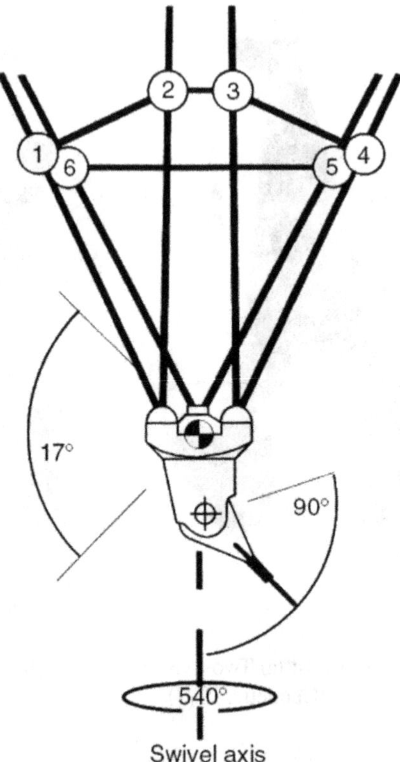

FIGURE 11.24 Tilting angles of the two-axis articulated head (Geodetic, 1997).

its elements. The struts may touch or even cross each other. Such possibilities must be perceived and prevented by the hexapod controller.

Contours must be corrected by the controller taking into consideration the tool diameter and length in the real-time operating system. The controller software must also be compensated for inaccuracies inherent to machine elements. The vibration, the speed profile of struts, and the positioning errors of joints and nodes, due to incorrect calculations, should also be compensated for by the controller. The developed heat due to forces and movements, which leads to complex expansion effects, should also be compensated for and corrected. Kreidler (1997) of Siemens developed a CNC control, SINUMERIK 840D, which permits the integration of different error sources to create an outstanding and efficient controller, which has been used by Geodetic and Ingersoll (Figure 11.26).

Alternatively, Geodetic has used Siemens controller, a real-time, Art-to-Part that is capable of driving an unlimited number of axes simultaneously by using G and M programming codes, CLDATA, and APT.

The new high-performance PC-based controller provides a cost-effective approach to multi-axis machine control. A simple-to-use comprehensive graphical user interface guides the operator through all tedious tasks. Interaction with the machine is handled through a programmable logic controller.

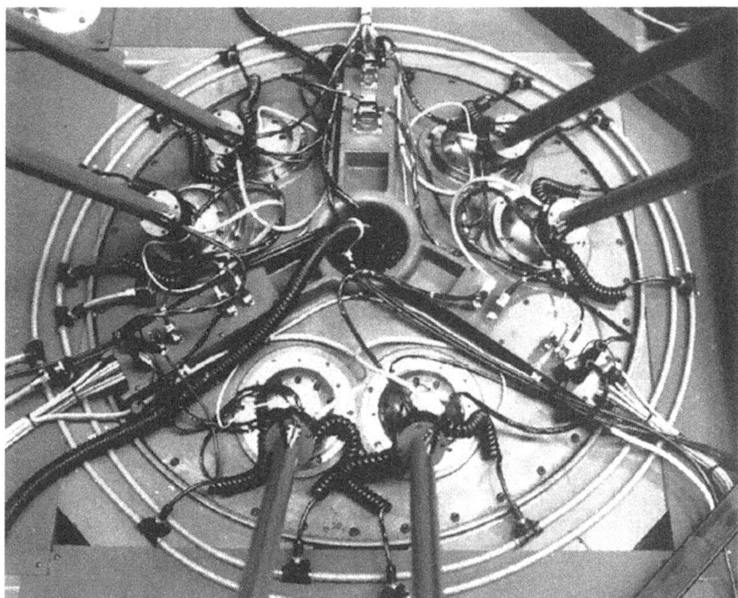

FIGURE 11.25 Geodetic upper dome platform (Geodetic, 1997).

Art-to-Part uses the latest forward and inverse kinematics transform algorithms. These new algorithms have been streamlined and are extremely fast. Art-to-Part is completely hardware-independent. It includes a tool management database, support for automatic tool change, automatic head change, palette changers, probes, and many other features. Written entirely in C++, Art-to-Part can be ported across to any platform with minimal effort.

11.5 HEXAPOD CHARACTERISTICS

The hexapod is a modern technology breakthrough that bridges the gap between robots and machine tools of multi-axis mechanisms that use either orthogonal or rotational movements. The current paradigm in design and manufacturing of hexapods involves integration of numerous hardwares and sophisticated software to create a unique product of extremely high rigidity and accuracy. The objective of this integrated product is to enhance quality and reliability, and to reduce the cost and overall cycle time through the dramatic departure from conventional mechanism design. As development and refinements continue, it is believed that the hexapod will eventually proliferate. A hexapod provides significant benefits to the end-user, since it offers many new attributes for the manufacturing processes.

The merits of hexapod are numerous, and include the following:

1. *Six-degree freedom.* The hexapod, with its six struts, provides the tool platform with 6 DOF. In addition to the extending motions in the orthogonal directions *X*, *Y*, *Z*, the platform is also able to move in other rotational

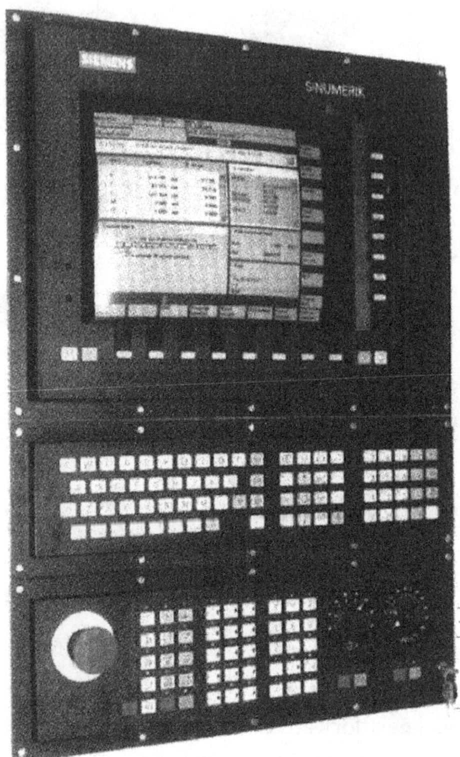

FIGURE 11.26 SINUMERIK 840D control system used by Geodetic, Inc. and Ingersoll, Inc.

compliments (pan, rotate, and twist). This advantage allows the spindle to reach unusual angles and to machine parts of difficult geometrical features such as turbine blades, plastic injection molds, stamping die, and other parts requiring high precision (Ingersoll, 1997).

2. *Flexibility and agility*. Hexapods behave according to flexible (or agile manufacturing scenarios). Flexibility is the ability to react to planned changes and agility is the ability to react to unplanned changes. Either mechanical simplicity plus its foundation independence gives the user the ability to quickly reconfigurate with changes in production lines with the easy option of storing the machines disassembled when they are not needed. The agile strut-supported spindle platform positions the spindle in all 6 DOF.
3. *Productivity*. Hexapods provide higher production rates through:
    - Designing the machine to be above the worktable
    - Continuous processing capability by accommodating a pallet shuttle system that can automatically move WPs in and out of the workspace
    - Making use of a high-speed automatic tool changer
    - Reducing the mass of moving parts to achieve very fast acceleration/deceleration (up to 0.5–1 G) (Figure 11.27a). Many designs of hexapod achieve contouring feed rates up to 30 m/min (Figure 11.27b) while maintaining

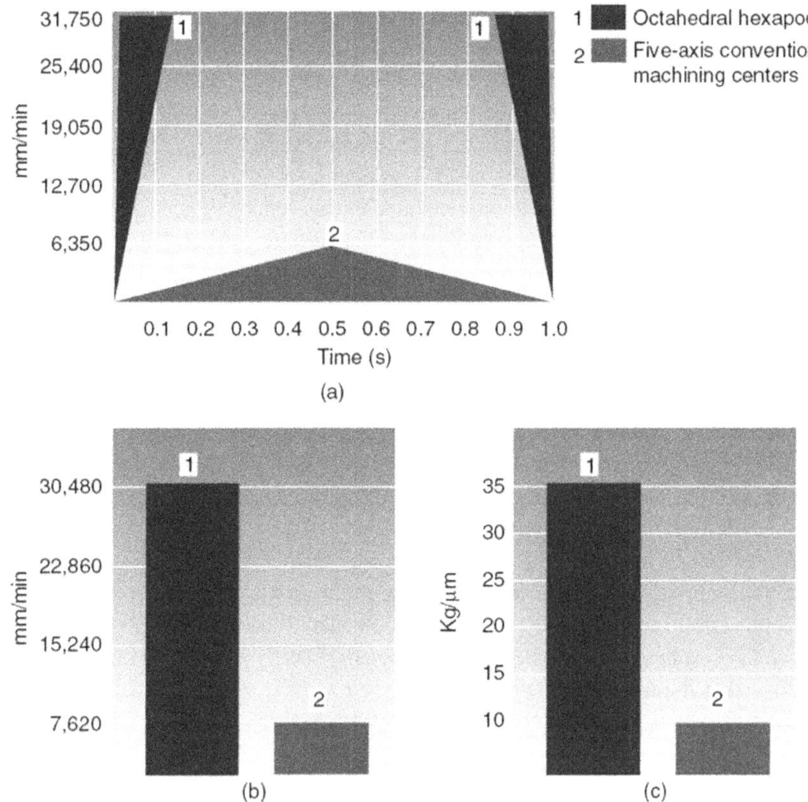

FIGURE 11.27 Hexapod characteristics: (a) Acceleration and deceleration, (b) feed rate, and (c) stiffness (Ingersoll, 1997).

precision. This feed rate is much faster than that of conventional machine tools, which often have to move the WPs as well as the heavy beds

- Using high-speed/high-power precise spindles

These five productivity-enhancing features, together with reduced setup and processing time and consequently reduced overall cycle time, lead to the increased production rate.

4. *Stiffness and rigidity.* A well-constructed hexapod is characterized by its rigid frame, which does not deflect significantly under acting loads. An optimum design approach to a hexapod should check the tendency of struts to buckling. The critical buckling load is proportional to the fourth power of the strut diameter and is inversely proportional to the square of the strut length. Therefore, a small strut diameter is sufficient to make a stiff structure. The high stiffness and rigidity of hexapod elements result in extraordinarily high natural frequencies, which consequently allow high cutting speeds during machining. The stiffness of Ingersoll's octahedral hexapod is about three to four times that of the five-axis conventional machining center of the same rating (Figure 11.27c).

5. *Precision and accuracy.* The accuracy of a hexapod is measured volumetrically. Any loads exerted are transmitted as tension or compression. Therefore, no bending forces occur, which promotes the machining accuracy. The parallel strut arrangement lends itself to error averaging. Hexapods are lighter than conventional machines, and because sliding friction can be virtually eliminated, there is much less backlash on axis reversal, thereby promoting smoother profile movements. In this regard, it is easy to control a strut length to 2.5 μm using sophisticated software control; however, some hexapod models attain submicron accuracy.
6. *Unique installation.* By concentrating all forces of the machining process within the hexapod frame, an important advantage is offered: The lack of a need for a special foundation. Design and installing a foundation for a conventional machine tool represents a substantial cost. The hexapod foundational independence may be demonstrated by using a crane to lift one hexapod corner during machining. A hexapod could function on a ship at sea, laid on its side, or suspended from the ceiling without sacrificing the precision of its performance.
7. *Simplicity.* Another potential of hexapods is the simplicity and ease of manufacturing. The part count in a hexapod is only about 300, compared to about 1000 in conventional machine tools. The other important characteristic is that many are duplicate parts. Assembly is so easy and takes so little time that the hexapod could be sold as a kit.
8. *Portability.* Hexapod is characterized by its high potential for portability. It is a machine that could be taken to jobs, for example, remote oil fields as well as different manufacturing plants.
9. *High load or weight ratio.* The high nominal load (power/weight) is a very important characteristic of hexapods. The cutting force acting on the moving platform is approximately equally distributed on the six parallel struts. It means that each strut suffers only from 1/6 of the total load. Furthermore, the struts are stressed longitudinally either in tension or compression; consequently, there is no need for them to be designed as massive and strongly dimensioned as in conventional machines.
10. *Scalability.* Hexapods are scalable in size, both upward and downward, to accommodate a multitude of applications ranging from micro-assembly and surgery to milling, drilling, turning, welding, painting, inspection, and assembly. Versions of design varying in size from table-type models for the semiconductor industry to units so large that the octahedron frame would form the building structure.
11. *Dexterity.* The hexapods have a complex working volume (a truncated hexa cone) based on the polar sweep of struts between maximum expansion and compression, and the degree of angular freedom. Dexterity extends substantially with the addition of two-axis articulated unit.
12. *Enhanced control systems.* A key step toward hexapod design is the development of a computer control system and software that are capable of processing the complex algorithmic calculations necessary to command the

struts. The processor requires a calculating power equivalent to that of several fast PCs combined. Additionally, the software should be capable to compensate offset data, thermal deviations, and the like.

13. *Cost.* After conventional machine tools comes the realm of hexapods. As more hexapods are built, it is expected in the near future that their prices will be reduced to 20% or less as compared to equivalent CNC machine tools (Figure 11.28). This is so because they are simple and easy to design and assemble. Fast assembly means lower inventory, less space, and lower labor costs. Six identical struts simplify the construction, providing easy and fast assembly and reducing maintenance cost. The replacement of faulty parts subjected to wear is also easy. Control and calibration is facilitated by highly efficient software. Moreover, the power consumption of a hexapod is considerably less than that of conventional machines. They are capable of adapting to a flexible manufacturing system (FMS). Figure 11.29 illustrates comparisons of cost, number of parts, power consumption, and time elements between hexapods and conventional CNC machine tools.

However, as a new design, hexapods still have some problems that need further development and refinements. The main limitations of hexapods include:

1. *Friction.* This issue is a crucial problem for hexapods. Owing to the high coefficient of friction ($\mu = 0.8$), the accuracy and repeatability are negatively affected. In advanced designs, however, where ceramic coating and special lubricants are used, the coefficient of friction may be reduced to 0.2.
2. *Length of struts.* The hexapod accuracy is inversely proportional to the strut length due to the possibility of bending. This problem may be overcome by mapping each strut element before installing in the machine.

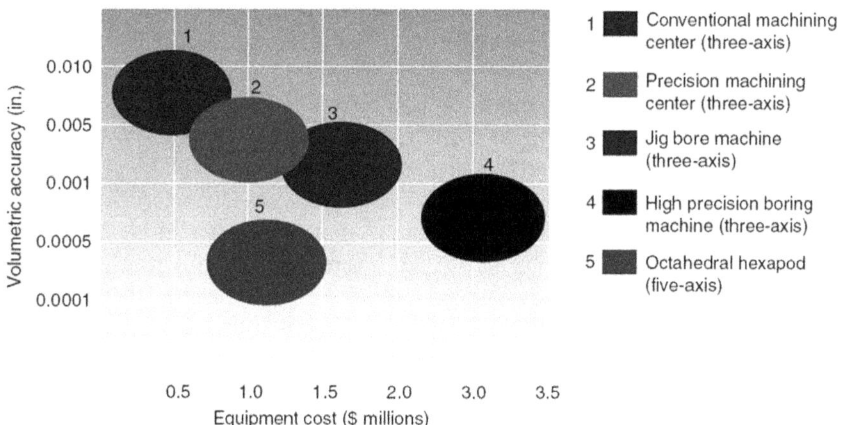

FIGURE 11.28 Cost comparison between hexapods and conventional machine tools (Ingersoll, 1997).

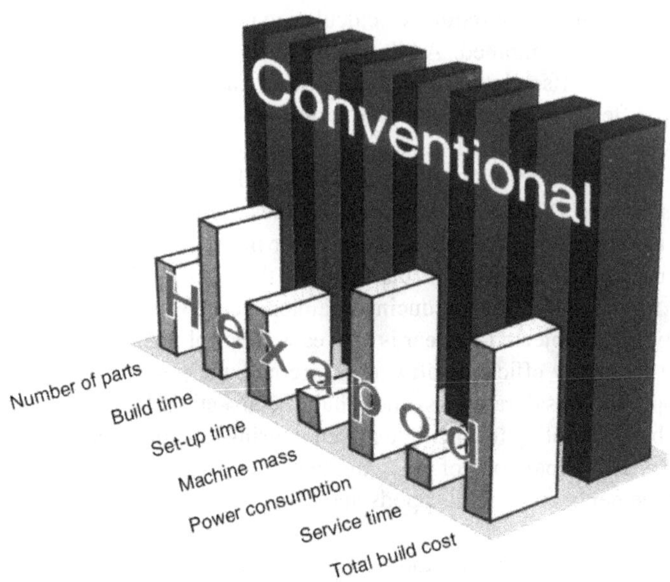

FIGURE 11.29 Total cost, number of parts, power consumption, and time elements of hexapods and conventional machine tools (Geodetic, 1997).

3. *Dynamic thermal growth.* This occurs due to the fast strut movements, as well as high spindle speeds (20,000–30,000 rpm). One way to overcome this problem is by monitoring the struts in a real-time mode, employing FEA that activates an automatic compensation routine into the software as based on the thermal growth induced in hexapod struts (Arafa, 2006). In this regard, Ingersoll has improved the thermal technique by using laser feedback that senses and eliminates deviant length changes of the hexapod struts (Ingersoll Co., 2001).
4. *Calibration.* The hexapod accuracy is not only dependent upon an accurate control of the strut length but also upon knowledge of its geometrical characteristics. According to the fabrication tolerances of hexapod struts, many factors play a role in the final accuracy of the hexapod. Many parameters must be specified to describe the geometrical characteristics of the mechanism. This is done through cumbersome calibration of the hexapod, which is still an open problem. Challenges are faced to develop a calibration system that would guarantee a high level of accuracy, enabling the hexapod to manufacture parts repeatedly within a specified tolerance. A self-calibration system should allow hexapods to check their own performance and correct any detected inaccuracies (Ingersoll Co., 2001).

11.6 MANUFACTURING APPLICATIONS

The hexapods are coming, and they will likely change the manufacturing paradigm. Their applications in industry include:

1. *Machining technology.* Machining is the most suitable and appropriate application making full use of hexapods' attributes. A standard hexapod offers dexterity, stiffness, and precision that is competing with conventional five-axis milling machines. Typical applications in the domain of TM include machining of press tools, mold making, turbine blade cutting, and drilling at inclined angles (Figure 11.30). Stiffness and precision are expensive to achieve in conventional multi-axis grinders. The hexapod grinder offers a cost-saving upgrade and a flexible architecture suited to precise grinding. Typical application of hexapod grinders includes tool grinding and precision grinding of ceramics. Hexapods have a multitude of applications in the domain of NTM. They provide contour machining capability, so they are equipped with lasers for cutting, welding, or hardening. Similarly, they could be set up for high-pressure WJM.

   In other machining domains, hexapods are used for:

   - Machining high-value, low-volume, and high-complexity components such as titanium for use in military aircraft
   - Machining light metal and materials
   - Contouring large surfaces and machining dies for precision sheet metal forming

FIGURE 11.30 A typical hexapod machining application.

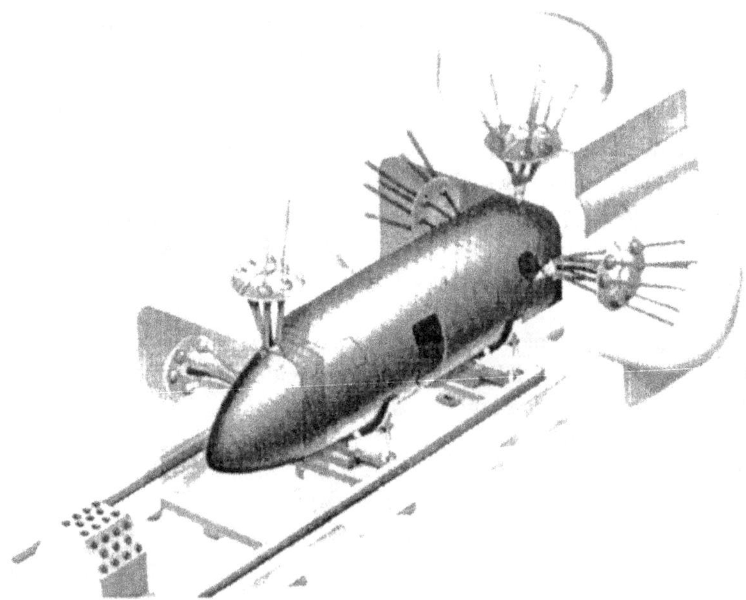

FIGURE 11.31 Hexapod delicate welding in aircraft production lines.

2. *Precision assembly technology.* Hexapods are used for delicate welding in automatic assembly lines and aircraft production (Figure 11.31). Small format, low-cost hexapods, which plug into the back of a PC, are used for such applications. Coupled with the Art-to-Part control software, these smart hexapod centers are easy to operate.
3. *Measuring technology.* Hexapod is an ideal shop-floor CMM. Figure 11.32 indicates an NC-produced part for the National Aeronautics and Space Administration (NASA). A program to determine part location on the hexapod has been developed.
4. *Car-painting station.* A pair of hexapods is mounted on a simple structure (Figure 11.33). This system is expandable for different applications, such as milling, where a milling tool can be used in place of painting nozzle machine.
5. *Electronic industry and fiber handling applications.* Micro-positioning hexapods are used in fields that require very accurate positioning such as the electronics industry, semiconductors, and fiber-handling applications. Figure 11.34 shows a hexapod for fiber alignments. The fiber is attached to the moving platform.
6. *Robotics.* With a move to offline programming, volumetric accuracy is becoming increasingly important. The ability to follow a path to within 25 μm absolute accuracy, while carrying a heavy load, is recognized as exceptional. To combine this with 1 G acceleration is unique.

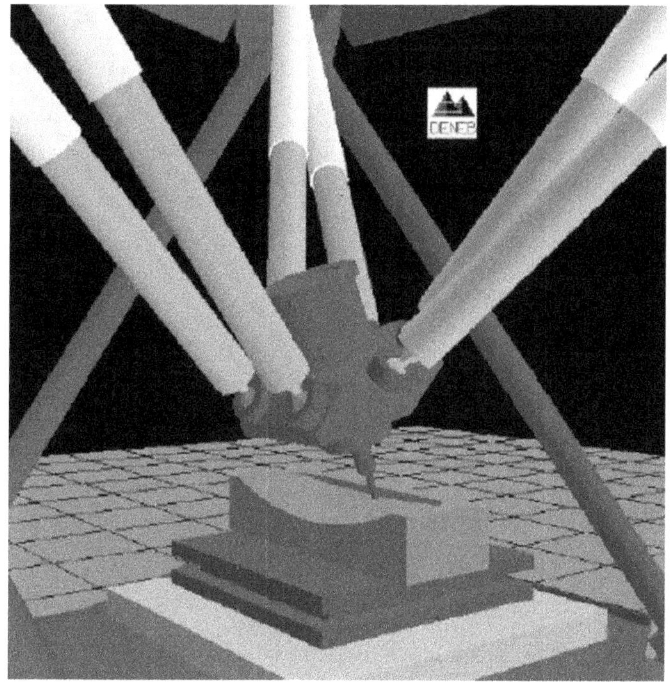

FIGURE 11.32 Verification of NASA test part by hexapod.

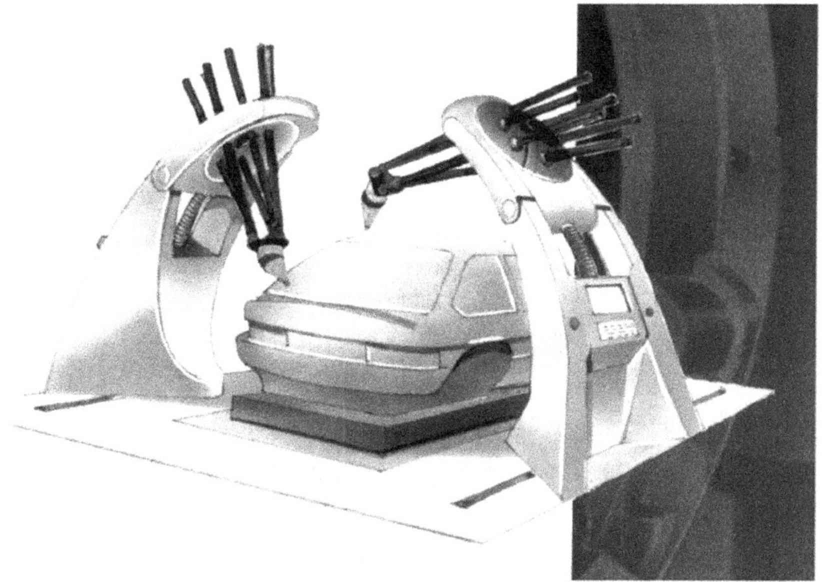

FIGURE 11.33 Hexapod car painting station.

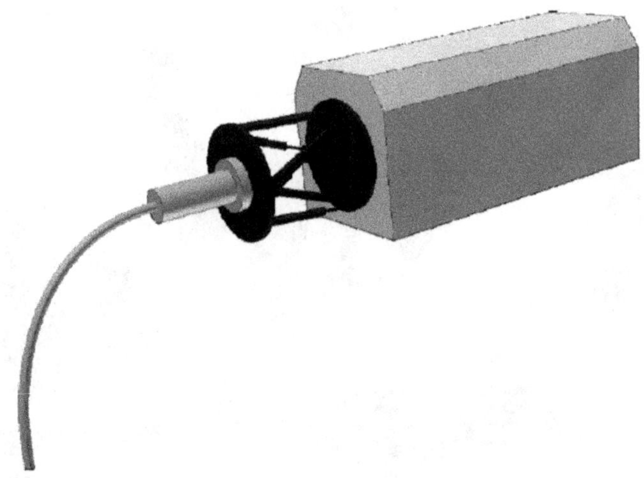

FIGURE 11.34 Fiber adjustments using micro-positioning hexapod.

11.7 REVIEW QUESTIONS

11.7.1 Use neat sketches to differentiate between telescopic and ball screw/strut hexapods.

11.7.2 Draw a neat sketch to illustrate extending, panning, rotating, and twisting of a hexapod mechanism.

11.7.3 Are the bifurcated ball and ball screw elements of a telescopic strut hexapod? To what applications are the magnetic bifurcated ball hexapods best suited and why?

11.7.4 Define the following terms as applied to hexapods: Flexibility, agility, scalability, dexterity, calibration.

11.7.5 What are the basic elements of a ball screw hexapod?

11.7.6 Discuss the main applications of the hexapod in manufacturing technology. List the advantages and limitations of hexapod mechanisms.

11.7.7 What is the main purpose of the Art-to-Part software?

REFERENCES

AKIMA 1997, First European Conference on Advanced Kinematics for Manufacturing Applications, Hannover.

Arafa, HA 2006. 'Six DOF Hexapod'. <www.//E:\Hexapodh\Hexapodl.htm>.

Geodetic 1997, 'Hannover exhibition, hexapod-breakthrough, technical information'.

Ingersoll 1997, 'Hannover exhibition, the next generation in 5-Axis machining technology, octahedral hexapod, technical information, Ingersoll Waldrich Siegen Werkzeugmachinen, GmbH'.

Ingersoll Co. 2001, 'Octahedral hexapod design promises enhanced machine performance, research and data for status report 92–01–0034'.

Kreidler, V et al. 1997, 'A.G. Siemens, hannover exhibition, report, offene objectorientierte CNC- Steuerungsarchitektur am Beispiel der Hexapod-Maschine'.

Stewart, D 1965, 'A platform with 6 DOF', *Proceedings of the Institute of Mechanical Engineers, London*, vol. 180, pp. 371–386.

Index

T

U

V

W

For Product Safety Concerns and Information please contact our EU representative GPSR@taylorandfrancis.com
Taylor & Francis Verlag GmbH, Kaufingerstraße 24, 80331 München, Germany

www.ingramcontent.com/pod-product-compliance
Lightning Source LLC
Chambersburg PA
CBHW060424220326
41598CB00021BA/2279